Lecture Notes in Civil Engineering

Volume 346

Lecture Notes in Civil Engineering (LNCE) publishes the latest developments in Civil Engineering—quickly, informally and in top quality. Though original research reported in proceedings and post-proceedings represents the core of LNCE, edited volumes of exceptionally high quality and interest may also be considered for publication. Volumes published in LNCE embrace all aspects and subfields of, as well as new challenges in, Civil Engineering. Topics in the series include:

- Construction and Structural Mechanics
- Building Materials
- Concrete, Steel and Timber Structures
- Geotechnical Engineering
- Earthquake Engineering
- Coastal Engineering
- Ocean and Offshore Engineering; Ships and Floating Structures
- Hydraulics, Hydrology and Water Resources Engineering
- Environmental Engineering and Sustainability
- Structural Health and Monitoring
- Surveying and Geographical Information Systems
- Indoor Environments
- Transportation and Traffic
- Risk Analysis
- Safety and Security

To submit a proposal or request further information, please contact the appropriate Springer Editor:

- Pierpaolo Riva at pierpaolo.riva@springer.com (Europe and Americas);
- Swati Meherishi at swati.meherishi@springer.com (Asia—except China, Australia, and New Zealand);
- Wayne Hu at wayne.hu@springer.com (China).

All books in the series now indexed by Scopus and EI Compendex database!

Aneetha Vilventhan · Shamsher Bahadur Singh ·
Venkata Santosh Kumar Delhi
Editors

Advances in Construction Materials and Management

Select Proceedings of ACMM 2022

Springer

Editors
Aneetha Vilventhan
Department of Civil Engineering
National Institute of Technology Warangal
Warangal, Telangana, India

Shamsher Bahadur Singh
Department of Civil Engineering
Birla Institute of Technology and Science
Pilani
Pilani, Rajasthan, India

Venkata Santosh Kumar Delhi
Department of Civil Engineering
IIT Bombay
Mumbai, Maharashtra, India

ISSN 2366-2557 ISSN 2366-2565 (electronic)
Lecture Notes in Civil Engineering
ISBN 978-981-99-2551-3 ISBN 978-981-99-2552-0 (eBook)
https://doi.org/10.1007/978-981-99-2552-0

This Springer imprint is published by the registered company Springer Nature Singapore Pte Ltd.
The registered company address is: 152 Beach Road, #21-01/04 Gateway East, Singapore 189721, Singapore

Preface

The papers in the book *Advances in Construction Materials and Management* under Springer book series *Lecture notes in Civil Engineering* are selected presented papers from the national conference on Advances in Construction Materials and Management (ACMM 22). The event was held at National Institute of Technology Warangal on December 16–17, 2022, and organized by the Department of Civil Engineering and supported by team of professors, research scholars and students.

All the presented papers have undergone a double anonymous peer-review process. Each paper was reviewed by two independent expert reviewers from IITs and NITs. The papers were reviewed for its originality and scientific quality. Several criteria such as relationship to literature, use of appropriate research methodology, analysis of results, implications of the research in practice and the quality of communication were assessed during the review process. Based on the reviewers' comments and recommendation, 43 papers were selected for publication in this volume.

The papers in this book are thematically organized into two parts such as Advances in Construction Management and Advances in Construction Materials. There are 14 papers in construction management that cover research topics such as building information modeling, applications of augmented reality and virtual reality, construction human resource management, energy-efficient buildings and contracts and 29 papers in construction materials that cover research topics such as new/alternate/supplementary construction materials, sustainable construction materials, deterioration mechanisms in construction materials, microstructure characteristics of concrete making materials, geo-polymer concrete, fiber-reinforced concrete, recycled aggregate concrete and sustainable materials for soil improvement. The papers in this book will allow students (undergraduate, postgraduate), researchers and faculties to gain insights into the advancements and recent trends of research in construction materials and management. This will also act as a reference for their future research practices.

We would like to express our sincere gratitude to the organizing team comprising of Dr. Aneetha Vilventhan, Dr. B. Kavitha and Dr. S. Anitha Priya Dharshani, research scholars and students. We also express our sincere thanks to Springer Nature for accepting out proposal and agreeing to publish the selected papers in the Springer book series Lecture Notes in Civil Engineering.

Aneetha Vilventhan
Shamsher Bahadur Singh
Venkata Santosh Kumar Delhi

Contents

About the Editors

Dr. Aneetha Vilventhan is working as Assistant Professor in the Department of Civil Engineering at National Institute of Technology Warangal. She obtained her Ph.D. in Building Technology and Construction Management from Indian Institute of Technology Madras. She has experience in academia and construction industry. She has worked in Qatar as Planning Engineer. Her major area of research includes construction project management such as n-dimensional planning, underground utility relocation, building information modeling, construction contracts and lean construction. She has completed an Early Career Research Award project funded by the Science and Engineering Research Board (Department of science and Technology), namely "Development of Road Information Modeling based coordination method for utility relocations." She has published in reputed journals, conferences and books.

Dr. Shamsher Bahadur Singh is P.E. (Civil) License Holder from the state of Michigan, USA, and Postdoctorate (LTU), USA, Fellow of American Society of Civil Engineers (F.ASCE), and currently Senior Professor of Civil Engineering Department at Birla Institute of Technology and Science (BITS), Pilani. He obtained his Ph.D. in Indian Institute of Technology Kanpur and completed his PDF in Lawrence Technological University, Southfield, Michigan. His current areas of research are development of design guidelines for functionally graded composite materials and fiber-reinforced polymer (FRP)-reinforced prestressed concrete structures in particular and composite materials and structures in general including nonlinear finite element modeling. He has published in reputed journals, conferences and books.

Dr. Venkata Santosh Kumar Delhi is Anantrao Jagtap Chair Associate Professor of Construction Management in Civil Engineering Department at IIT Bombay. He has obtained his Ph.D. in Construction Management from Department of Civil Engineering, IIT Madras. His research interests include information management on large engineering projects. Particularly, he works on building information modeling (BIM), AI/ML applications in construction management, construction contract management, AR/VR/MR applications in construction safety and planning. He is also actively involved in a number of projects that aim at providing policy and strategy guidance to various governmental agencies and private companies. He has published in reputed journals, conferences, chapters and technical reports.

Advances in Construction Management

Gaining Competitive Advantage Using Human Resource Management in Indian Construction Industry

Shashank Kumar, Madhumathi Pasupathi, and Rahul Thangeda

Abstract Gaining a competitive advantage in today's world of global competition is a huge issue for all businesses. This is due to the fact that a company's development and performance are determined by whether or not it has a competitive edge over competitors in the same industry Furthermore, strong competitive advantage has a substantial impact on a country's position. This is owing to the success of competitiveness in defining local competition and generating goods, both of which are fundamental factors of a country's economic growth and wealth. A company's competitive edge may be gained through "human resource management methods" that promote employee skill development and motivation. These practices will also boost staff productivity and innovation, which will have an influence on performance improvement. The process of managing an organization's personnel, or human resources, is known as human resource management. It is in charge of attracting, hiring, training, assessing, and rewarding personnel, as well as managing organizational leadership and culture and ensuring compliance with labour and employment laws.

The successful implementation of these practices ensures that all employees are aware of their roles, career paths, and a sense of belonging to the organization, allowing them to manage and reconcile their expectations as well as the organizations and its objectives, such as gaining a competitive advantage. As a result, establishing a knowledge of how views about human resource procedures and their execution impact firm performance in obtaining competitive advantage is critical. A questionnaire has been created based on these characteristics and past research.

S. Kumar
Construction Technology and Management, National Institute of Technology Warangal, Warangal, India

M. Pasupathi
Department of English, Rajiv Gandhi National Institute of Youth Development (RGNIYD), Sriperumbudur, Tamil Nadu, India

R. Thangeda (✉)
School of Management, National Institute of Technology Warangal, Warangal, India
e-mail: rahult@nitw.ac.in

A. Vilventhan et al. (eds.), *Advances in Construction Materials and Management*, Lecture Notes in Civil Engineering 346,
https://doi.org/10.1007/978-981-99-2552-0_1

Data was statistically evaluated, and the article was ended by examining the influence of various human resource management methods on employee well-being and, ultimately, business competitive advantage.

1 Introduction

Gaining a competitive advantage is a fundamental difficulty for all firms in today's world of global competition. This is due to the fact that a company's development and success are decided by whether or not it has a competitive edge over competitors in the same sector.

Furthermore, competitive advantage in firms has a substantial impact on a country's position. This is due to the success of competitiveness in shaping local competition and generating goods, both of which are fundamental factors of a country's economic growth and prosperity. According to the 2021 Global Competitive Index study, India is ranked 43rd, while China is ranked 16th. The impact of a significant expansion in market size in the construction sector as a result of infrastructure development in India and a generally stable macroeconomic environment has enhanced competition. This study intends to provide early data on human resource management techniques for building a work safety culture that increases employee productivity and gives organizations a competitive edge.

1.1 Human Resource Management in Construction

Human resource management is a business function concerned with the recruitment, motivation, and retention of employees. It focuses on the employees of a business. Human resource management in construction is primarily concerned with ensuring that a project has sufficient human resources with the necessary skill sets and expertise to finish it successfully. Human resource management is a business function concerned with the recruitment, motivation, and retention of employees. It focuses on the employees of a business. Human resource management in construction is primarily concerned with ensuring that a project has an enough number of human resources with the necessary skill sets and expertise to successfully execute it. Human resource management is a business function concerned with the recruitment, motivation, and retention of employees. Human resource development, a component of human resource management, is concerned with employee training and development. It includes onboarding an employee, offering opportunity to acquire new skills, providing information relevant to the person's employment, and any other growth activities. A company's capacity to attract, develop, and retain talented personnel is critical to its success People are a company's most precious asset, especially in a labor-intensive industry like construction. Individuals contribute their own viewpoints, attitudes, and qualities to work, and if managed properly, human personality can be immensely valuable to enterprises. However, if not handled properly, they have the ability to hinder organizational advancement.

1.2 Competitive Advantage

Human resource management strategies that promote employee skill development and motivation can help a firm acquire a competitive edge. This leads to improved production and creativity, which Improves performance. Barney (1991) defines a company's resources as valuable when they can be used to carry out plans that result in effectiveness and efficiency. When resources are not held by other firms, particularly competitors, they are called scarce. This implies that competitors cannot replace them with resources of comparable quality to carry out their strategy.

2 Literature Review

People should not be viewed as a company's "cost," but rather as its most valuable resource. HRM involves determining two requirements: what employees expect from their efforts and what a company wants from its employees, and then attempting to reconcile these two sets of requirements.

A priceless "asset." also a fundamental source of competitive advantage. Bambacasa and Kulik's (2013) research objective was to explore how human resource practices embed people in organizations and lessen the likelihood of turnover. In order to explain the relationship between HR procedures and employee turnover intentions, the study examined the function of organizational job embeddedness characteristics (connection, fit, and sacrifice). The results showed that organizational incentive and performance evaluation boosted fit and lowered turnover intentions. A high-quality performance assessment system lowers turnover, but a low-quality one increases employees' chance of leaving the organization, according to research by Brown et al. (2010). One way a company's performance rating system affects turnover intentions is through the embeddedness dimension of linkages. As a result, it was proposed that organizations should concentrate on HR strategies such as performance management and reward systems to reduce employee turnover intentions.

Money, a clean workplace, and a supportive work environment are all things that encourage employees (Pai et al. 2013). By offering competitive salary, prompt payments, upholding a positive workplace culture, and keeping a clean atmosphere, many businesses are able to retain their competent workers. In addition to this, by offering rewards and incentives for achieving goals, employees can be retained. In order to keep its personnel, organisations should concentrate on implementing these factors. The use of performance management systems in significant construction companies was investigated by H.S. Robinson et al. (2004). Performance management systems will be used to enhance organisational procedures and lab management in order to enable continual improvement. The results show that putting performance management systems in place would help construction organisations 1) adopt a consistent strategy for reacting to changes in their organisations and the industry; 2) encourage continuous improvement in the provision of products and services; and 3) recognise the significance of learning and knowledge management in the exploration of innovative solutions in order to maintain a competitive advantage.

Zhai et al. (2014) studied how organisational learning and human resource strategies affected business performance in the Chinese construction sector. Results indicate that organisational learning may operate as a bridge between human resource management activities such as hiring, training, and remuneration of employees as well as employee relations and organisational performance. Therefore, organisational learning must be viewed as a major component of competitive advantage for construction businesses to flourish, and organisational learning should be enhanced by enhancing HR procedures. Lopez et al. (2006) looked examined the relationship between four human resource management practises: hiring, remuneration, decision-making, and organisational learning. The results showed that the adoption of strategic training, selective recruiting, and employee decision-making had a significant influence on organisational learning. (Charles O. Skipper et al. 2006; Ali Nawaz Khan et al. 2020) In order to determine whether there was a correlation between ethical leadership and turnover intention and antisocial conduct, confidence in the leader and procedural fairness were examined in the Hong Kong construction sector. The findings showed that in the construction business, ethical leadership increased antisocial behavior and decreased employee job turnover intentions. These decision-making authorities would learn how ethical leadership may keep workers by boosting confidence and procedural fairness in the construction organizations from these decision-making authorities. Ghodrati et al. (2018) aimed to quantify the effectiveness of set of management strategies which are being implemented. Results indicated that Communication and incentive programs are some of the management practices that have a big beneficial impact on workforce productivity.

Lin et al. (2011) advised using "whole project cost" as the basis for how to allocate human resources in remote building projects. Analysis shows that employing local workers and regular business personnel both require distinct techniques to deal with the issues of globalisation. Regular personnel is, per the study, the best option for lowering management expenses and project risk. Dabiriana et al. (2019) developed a dynamic model for a system dynamics approach to labour allocation. One of the most important and productive activities in human resource management is allocating workers to the project. Project performance can be efficient in terms of time and cost if the right staff is allocated to it. The project staff may be allocated and the necessary planning for on-time supply can be done before and throughout the project, according to the results. Tiwari and Saxena. (2012) discovered that both external and internal factors have an impact on HRM practices, which in turn have an impact on other variables like employees' attitudes, employee-employer relations, employee productivity, etc., and ultimately affect how well the construction organization is performing as a whole. After an extensive literature review, the following research gap has been identified.

Studies have been done in other countries on how human resource management helps in attaining competitive advantage but not in Indian construction industry. From those studies it has been observed that if there is an effective implementation of HR practices, competitive advantage can be achieved. Hence, this study attempts to study the Human Resource Management factors impacting competitive advantage.

3 Objectives

- To identify the important factors of strategic Human resource management for construction industry.
- To determine the human resource management practices which can assist firms to gain competitive advantage by managing the personnel effectively.
- To propose a model to examine the relations between different attributes that is useful to gain Competitive Advantage in construction industry.

3.1 Research Methodology

- In order to have a look at the research goals and gaps, a questionnaire will be developed to conduct an internet-based survey and accumulate observed data. The attributes/factors measuring every assembly will be advanced from the preceding literatures. Some measures are slightly changed in order to match the context of live streaming commerce.
- To collect the primary data for research a five-factor Likert scale closed ended question on Google form, ranging from strongly disagree (1) to strongly agree (5) to measure all of those attributes/factors was used.
- Target population for this research is Gen Z and millennials as the content consumption is really high among Gen Z and millennials. These two groups are relevant to achieve the research objectives.
- Snowball sampling technique is followed for this research which is a chain-referral sampling defined as a non-probability sampling technique. Questionnaire was also shared across the social media platform like WhatsApp, Instagram, Facebook, Twitter and LinkedIn. To get more response various Construction related groups were reached out on Facebook and Instagram.
- Considering the cost and time of the research limited population size with confidence level of 95% and margin of error 10% were considered, so that the sample size is coming 97 but I took 112 for better results.

n = z^2*p*(1-p)

e^2

Where z = Z-score with 95% confidence level p = standard deviation e = margin error

n = 1.96^2*0.5*(1-0.5) = 96.07= 97(approx.)

(0.1) ^2

The responses were recorded and further analysed by an extensive data analysis. The data was tested through hypothesis testing to determine the statistical evidence of the relationship existence. The data was further used to develop a structural equation model to validate the structural relationship between different variable considered.

3.2 *Hypothesis Testing*

It is a hypothesis that may be tested by watching a process that is described by a collection of random variables. A statistical hypothesis testing has been done in the analysis part. There is diverse hypothesis constructed according to context and later on after analysis hypothesis is being approved or rejected accordingly.

3.3 *Descriptive Analysis of Responses*

Descriptive analysis has been done on the responses collected which shows descriptive statistics (mean, median, mode & standard deviation) out of the varied responses. All the results derived from the descriptive analysis are highly opinion based (Table 1).

4 Path Analysis

Here the p-value is much greater than 0.05 means the null hypothesis cannot be rejected i.e., the variable is normally distributed.

Figure 1 shows the structural equation model created in SmartPLS, it comprises of observed variables and latent variables. The square boxes indicate the measured variables (responses in simple language) whereas the circles indicate the latent constructs. Basically, latent constructs are created according to the research objectives and goals. Further, these latent constructs are interlinked by a path according to the purposed theoretical model. Then path analysis is done in SmartPLS software to test its significance and hypothesis testing.

The detailed analysis of paths is explained further (Table 2).

The path analysis findings for the aforementioned model were obtained after utilizing SmartPLS's bootstrapping tool. Through Bootstrapping, the subsample was generated randomly and is then used to estimate the PLS path model. This procedure

Table 1 Descriptive Analysis of responses

	Mean	Median	Min	Max	Standard Deviation
CA1	3.196	3.000	1.000	5.000	1.164
CA2	2.884	3.000	1.000	5.000	0.961
CA3	3.098	3.000	1.000	5.000	1.187
CA4	2.777	3.000	1.000	5.000	1.033
HRM1	2.982	3.000	1.000	5.000	1.232
HRM2	3.009	3.000	1.000	5.000	1.191
HRM3	3.018	3.000	1.000	5.000	1.195
HRM4	3.161	3.000	1.000	5.000	1.207
P1	3.170	3.000	1.000	5.000	1.093
P2	3.143	3.000	1.000	5.000	1.141
P3	3.107	3.000	1.000	5.000	1.190
SC1	3.071	3.000	1.000	5.000	1.223
SC2	3.188	3.000	1.000	5.000	1.065
SC3	3.259	3.000	1.000	5.000	1.148
SC4	3.179	3.000	1.000	5.000	1.167

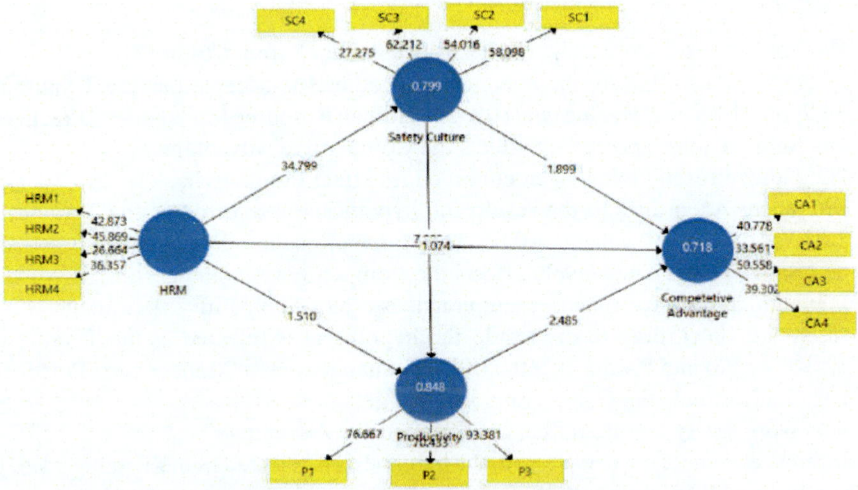

Fig. 1 Results from the Structural Equation Model

is continued until a large number of random subsamples, usually 10,000, have been created.

Table 2 Path Analysis

	Original Sample (O)	Sample Mean (M)	Standard Deviation (STDEV)	T-Statistics (|O/STDEV|)	P Values
HRM_- > Productivity	0.170	0.161	0.116	1.463	0.144
HRM_ - > Safety Culture	0.895	0.895	0.024	36.770	0.000
Productivity- > Competitive Advantage	0.414	0.416	0.153	2.709	0.007
Safety Culture - > Competitive Advantage	0.449	0.449	0.157	2.862	0.004
Safety Culture - > Productivity	0.766	0.775	0.103	7.403	0.000

4.1 Inference of Path Analysis

Through the path analysis the null hypothesis is tested, here the T values and P values showing the significance of path and whether the null hypothesis is rejected or accepted.

H1: Human resource management practices affect Safety Culture.
Inference: According to the result, the hypothesis is accepted as the T value is 34.799 (>1.96) and P value is <0.05. The Human Resource Management Practices and Safety Culture positively effects the Competitive Advantage.
H2: Human resource management practices affect Productivity.
Inference: According to the result, the hypothesis is rejected as the T value is 1.510 (<1.96) and P value is > 0.05. The Human Resource Management Practices and Safety Culture negatively effects the Competitive Advantage.
H3: Human resource management practices affect Competitive Advantage.
Inference: According to the result, the hypothesis is rejected as the T value is 1.510 (<1.96) and P value is > 0.05. The Human Resource Management Practices negatively effects the Competitive Advantage.
H4: Work safety culture influences Competitive Advantage.
Inference: According to the result, the hypothesis is rejected as the T value is 1.510 (<1.96) and P value is > 0.05. Safety culture negatively effects the Competitive Advantage.
H5: Work safety culture influences Productivity.
Inference: According to the result, the hypothesis is accepted as the T value is 7.882 (>1.96) and P value is < 0.05. The Human Resource Management Practices and Safety Culture positively effects the Competitive Advantage.
H6: Productivity effects on Competitive Advantage.

Table 3 Consolidated Hypothesis results were tabulated below

	Original Sample (O)	Sample Mean (M)	Standard Deviation (STDEV)	TStatistics (|O/STDEV|)	P Values
HRM_- > Safety Culture (H1)	0.894	0.893	0.026	34.799	0.000
HRM_- > Productivity (H2)	0.167	0.163	0.111	1.510	0.132
HRM_- > Competitive Advantage (H3)	0.168	0.164	0.156	1.074	0.283
Safety Culture- > Competitive (H4) Advantage	0.333	0.334	0.176	1.899	0.058
Safety Culture- > Productivity (H5)	0.768	0.772	0.097	7.882	0.000
Productivity - > Competitive Advantage (H6)	0.376	0.377	0.151	2.485	0.013

Inference: According to the result, the hypothesis is accepted as the T value is 2.485 (>1.96) and P value is <0.05. The Human Resource Management Practices and Safety Culture positively effects the Competitive Advantage (Table 3).

5 Recommendation

1. The research recommends that frequent implementations of training and development practices like Organizing seminars and trainings for employees can enhance their skills and job-related competencies
2. The research also recommends that organizations should implement performance appraisal systems and performance management systems in order to decrease the turnover intentions of the employees.
3. The research also recommends that organization must pay close attention to strategic practices related to compensation and reward management in order to retain the workforce.
4. The study also suggests that effective implementation of health and safety practices will lead to development of a safety work culture which will enhance organization learning

6 Conclusion

In this research, Structural Equation Modelling (SEM) has been used to explore the factors that determine competitive advantage in the Indian construction sector. The findings demonstrated that safety culture positively and directly impact on productivity and productivity directly impacts competitive advantage. Furthermore, this study found that human resource management approaches have an indirect influence on competitive advantage. The findings of this study have a number of significant implications for India's construction industry's ability to maintain a competitive edge. The importance of culture in the organization is reflected by the safety culture's considerable direct influence. According to the findings of the study, a competitive advantage in the construction industry can be obtained through the effective implementation of human resource management practices that address employee needs, job satisfaction, performance management, career plans, personal development, and practices to improve a safety culture that increases employee productivity. To keep employees, effective performance management techniques must be used.

References

Barney, J. B. (1991). Firm resources and sustained competitive advantage. Journal of Management, *17*(1), 99–120.

Bambacas, M., & Kulik, T. C. (2013). Job embeddedness in China: How HR practices impact turnover intentions. *The International Journal of Human Resource Management, 24*(10), 1933–1952.

Brown, M., Hyatt, D., & Benson, J. (2010). Consequences of the performance appraisal experience. *Personnel review, 39(*3), 375–396.

Pai, S. K., Dheeraj, M. S., & Kristam, M. R. S. (2013). Human resource management techniques dealing with deficit & retention of skilled labor in construction industry. *International Journal of Scientific and Research Publications, 3*(9), 2250–3153.

Robinson, H. S., Carrillo, P. M., Anumba, C. J., & Al-Ghassani, A. M. (2004). Developing a business case for knowledge management: the IMPaKT approach. *Construction Management & Economics, 22*(7), 733–743.

Zhai, X., Liu, A. M., & Fellows, R. (2014). Role of human resource practices in enhancing organizational learning in Chinese construction organizations. *Journal of Management in Engineering, 30*(2), 194–204.

López, S. P., Peón, J. M. M., & Ordás, C. J. V. (2006). Human resource management as a determining factor in organizational learning. *Management Learning, 37*(2), 215–239.

Skipper, C. O., & Bell, L. C. (2006). Assessment with 360 evaluations of leadership behavior in construction project managers. *Journal of Management in Engineering, 22*(2), 75–80.

Nawaz Khan, A., Khan, N. A., & Soomro, M. A. (2020). Influence of ethical leadership in managing human resources in construction companies. *Journal of Construction Engineering and Management, 146*(11), 04020125.

Ghodrati, N., Yiu, T. W., & Wilkinson, S. (2018). Unintended consequences of management strategies for improving labor productivity in construction industry. *Journal of safety research, 67*, 107–116.

Lin, K. L., &Lin, C. L. (2011). Applying utility theory to cost allocation of pavement maintenance and repair. *International Journal of Pavement Research & Technology, 4*(4).

Dabirian, S., Khanzadi, M., & Moussazadeh, M. (2016). Predicting labor costs in construction projects using agent-based modeling and simulation. *Scientia Iranica, 23*(1), 91–101.

Tiwari, P., & Saxena, K. (2012). Human resource management practices: a comprehensive review. *Pakistan business review, 9*(2), 669–705.

Mapping the Project Complexity of Metro Rail Project Using DEMATEL Technique

Sruthilaya Dara, Aneetha Vilventhan, and P R. C Gopal

Abstract Metro rails are the most desired infrastructure for metropolitan cities. These projects are inter- organizational due to its involvement of numerous stakeholders, complex project objectives, and cultural diversity etc., As a result of these specific features, most metro rail projects are facing complexities, which have increased overall project complexity. The analysis of project complexity is a critical step in the project's execution. In this study, a novel assessment technique was developed with the specific goal of understanding complexity factors and their interdependency in metro rail projects. Semi-structured interviews and a questionnaire survey were conducted to identify the complexity and its factors. Decision making trial and evaluation laboratory (DEMATEL) a structured method is used to identify the interdependence among its factors. The analysis revealed that there are dynamic relationships between the complexity factors, with changes in one affecting the other. The complexity map and causal diagram are used in the study to represent the most significant complexity factors in terms of their overall relationship with the other factors. This research has practical implications so that practitioners can focus on complexity factors that have a significant interrelationship for effective management of complexities in metro rail projects. As a result, project owners, contractors, and consultants can use specific findings to create management plans and allocate resources.

Keywords Metro Rail · Complexity · Decision making trial and evaluation laboratory (DEMATEL)

S. Dara · A. Vilventhan (✉) · P. R. C. Gopal
Department of Civil Engineering, National Institute of Technology, Warangal, Telangana, India
e-mail: aneetha@nitw.ac.in

P. R. C. Gopal
School of Management, National Institute of Technology, Warangal, Telangana, India

A. Vilventhan et al. (eds.), *Advances in Construction Materials and Management*, Lecture Notes in Civil Engineering 346,
https://doi.org/10.1007/978-981-99-2552-0_2

1 Introduction

Megaprojects have significantly increased in scope, size, cost, and risk exposure over time. Due to rapid urbanization in this modern age the growth of megaprojects is increasing rapidly. Megaprojects draw more public and political attention because of their larger investments and duration. These both factors exhibit potential impacts on the environment, ecology, economy, nearby communities, and property owners. The impact on these characteristics has increased the complexity of megaprojects.

A complex project exhibits a variety of characteristics with a level of severity, that makes it challenging to manage and predict the project outcomes. Project complexity refers to a project's characteristics that make it challenging to understand, predict, and manage its overall behavior [1]. The management of complexity in the projects has become one of the main concerns [1]. According to [1] project complexity is a natural characteristic because every project contains a certain level of complexity. The occurrence of project complexity has a greater impact on performance and project success [2, 3]. Most of the studies depict that a complex project entails a large number of interdependent activities to achieve the project's overall goal [4]. Mainly, the construction projects are facing numerous challenges due to the rise of complexity [5]. Uncertainty, organizational structure, ambiguity etc., are the causes of the complexity in most of the projects and sometimes lead to project failure [6]. Hence, the study of complexity is required to manage the projects to increase the performance mostly in urban areas [7]. Due to rapid urbanization the construction of metro rail projects is increasing. These projects are considered to be one the most important megaprojects in urban transportation [8] and highly attracts both private and public interests due to their large infrastructure, cost and, time. The metro rail projects also have a contrary impact of environment, property owner, neighboring communities etc., [9]. These projects are commonly associated with cost and time overruns due to uncertainty in the projects [10]. The complexity in metro rail projects is caused due to various interrelated factors, complex nature, and its unique characteristics [8]. Therefore, the management of complexity in metro rail projects is crucial and complex [11]. Numerous studies have considered the construction of metro projects a significant contributor for the social and economic development of the nation [12]. Hence, the study of complexities is required to assist project practitioners in better comprehending and managing complexity, which also helps in the successful implementation of metro projects. Hence, complexity must be prioritized in order for project stakeholders to allocate resources to essential components [13]. It is stated that complexity management requires an understanding of the interdependence among a significant numbers of interrelating components [14]. Due to the difficulty of this task [3], there is very little research on complexity prioritization [14]. However, there is lack of studies in examine the interdependence of complexity factors in the projects.

Therefore, a quantitative study on complexity prioritization and the study of the interdependence among the complexity factors in a systematic manner is thus required. This allows the stakeholders and organizations to comprehend and deal with complexity more effectively and aiding in the successful construction of metro rail projects. This will support organizations for efficient resource allocation during the construction and execution of metro rail projects. This study is concentrated on the Bangalore metro rail project and aims to identify complexity factors, and the interdependence among the identified complexity factors. Academically, the findings of this study will add to what is already known about managing complexity. The proposed framework is a powerful tool that researchers can use to prioritize complexity. By prioritizing complexity criteria in practice and management, project leadership teams can clearly see the challenges and take steps to mitigate their impact. As a result, project owners, contractors, and consultants can use specific findings to develop management plans and decide how resources should be allocated. The organizations can use this study to consider how complicated things appear to different groups when developing future metro rail projects. This study focuses primarily on the Bangalore metro rail project, other transportation projects can use the findings to set their own priorities based on their own characteristics and political situations with only a few minor adjustments.

2 Literature Review

Various definitions of project complexity have been proposed in the literature. However, project complexity is not defined precisely due its uncertainty and ambiguity nature [15]. Various authors have different opinions about complexity from their perspective. According to [7] project complexity consists of many varied interrelated parts and can be quantified based upon its interdependency and differentiation. [16] defined project complexity based upon the number of elements and its interdependence (Structural complexity) and uncertainty in methods and goals. There is a lack of a clear definition of project complexity in the literature according to [13]. The project's components vary in complexity and it is inherently challenging to define complexity and the various perspectives on complexity [17]. [18] in his studies explained the projects as complex adaptive systems or socially constructed entities that are related to structural and dynamic elements of technology or engineering. A review of previous studies revealed that project complexity can be classified into several complexity factors. However, these factors' classifications vary according to the type of project. The project factors are divided into technological and organizational by [19]. Factors contributing to technology complexity are experience with the technology [7] newness in technology [5] technological and quality risks [6] various project management methods [5] numerous tasks [16] the factors causing organizational complexity are size of the project [6] project duration [8, 30], interface among the organizations [6], trust among stakeholders [6, 22] various languages [6, 22] uncertainty in projects [12]. The factors including in environmental complexity are weather condition [6, 8], location remoteness [6] large number of stakeholders [5, 8, 9, 22] stakeholders perspectives [5] risk from environment [6], instability of environment [13].

2.1 Project Complexity Measurement

The study of project complexity is an emerging research area and has attained significant attention in the project management area [3]. The studies reveal that the quantification of project complexity is difficult [3] and has very limited studies on its measurement. Several models have been suggested by earlier research to measure project complexity. A conceptual framework for comprehending the complexity of construction projects was put forth by [21]. Hence, it is necessary to use the multi-criteria decision making (MCDM) method to assess project complexity. Multicriteria Decision Making is another well-liked model for understanding complexity (MCDA). The most frequently employed of these is the analytical hierarchy process (AHP) [19]. The complexity of transportation projects was then evaluated by using a fuzzy AHP. [22] applied the FANP (fuzzy analytical network process) to identify interactions among complex variables (ANP). Although variable interdependence is a key aspect of complexity, little attention has been given to examining interactions between complexity factors, according to a review of the literature. Therefore, it is crucial to give a thorough understanding of how complexity factors interact with one another. The complexity of research and professional practices in the context of metro rail projects has not been adequately understood. Few comprehensive studies have investigated the factors that make it difficult to access complexity in metro rail projects. The main objective of this research is to identify the factors causing project complexity in metro rail projects. The complexity factors are identified through a literature review and the interdependence of project complexity factors are identified using the DEMATEL method. This research helps metro organizations and the stakeholders to focus on the project complexity factors impacting the performance of the project.

3 Research Methodology

Planning, organizing, and managing numerous complex factors and their interdependencies are necessary for a large construction project. Monitoring these complexity factors at once is difficult. According to [23] numerous studies on Complexity factors in construction industry have already been carried out but there is lack of studies in metro rail construction. Identifying and understanding complexity factors, managing, and controlling them has become a tedious task. Therefore, to assess the critical nature of complexities the stakeholders and organizers need to determine the appropriate method to identify and assess the complexity in metro rail projects. The flowchart for the research methodology is shown in Fig. 1.

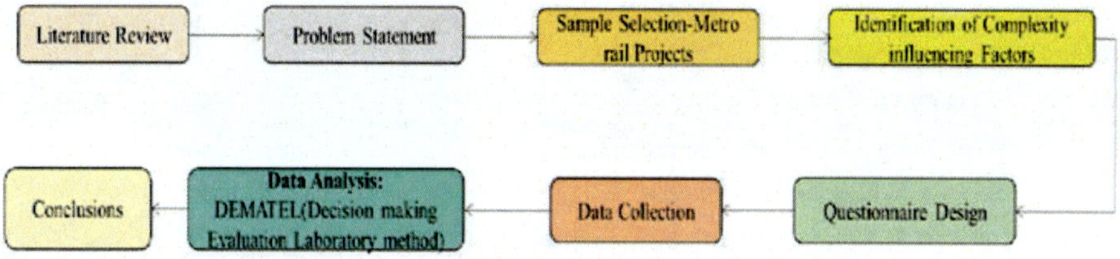

Fig. 1 Research Methodology

3.1 Complexity in the Context of Bangalore Metro Rail Projects

Literature review was conducted to establish a list of 32 project complexity factors represented in Table 1. To fit in the context of metro projects, the factors obtained from literature are reviewed and refined using a group of 5 experienced professionals from metro organization through group discussions and semi structured interviews. Minimum of 8 years' experience professional are chosen for taking the inputs. The factors obtained from the literature are provided to the professionals and are asked to choose the factors applicable for metro rail projects based upon their experience. The questionnaire survey was conducted to identify the relative importance of the factors about project complexity in metro projects. Respondents are asked to rank the complexity factors on likert scale of 1–5 (5 = strongly agree; 1 = strongly disagree). A list of 117 respondents from project engineers, project managers, site engineers, planning engineers, contractors, assistant engineers, executive engineers are collected from the Bangalore metro project. The questionnaire was emailed to the respondents for easy accessibility of answering the questionnaire. As a result, 10 factors chosen by the professionals are considered for the further analysis and highlighted in the Table 1. The reliability test yielded a Cronbach alpha coefficient of internal consistency as 0.83(>0.80) which is considered reliable.

3.2 Application of the DEMATEL Method

From the research it is identified that there is lack of research in providing quantitative analysis of interdependence among complexity factors in construction projects. The research gap can be filled by using the DEMATEL (Decision making Evaluation Laboratory method). This method has become the most common for identifying the interdependence among the complexity factors in a structured pattern [24]. This helps the researchers to represent the interdependence of complexity factors in a structured model consisting of cause and effect group [25]. The participants among are invited to participate in the DEMATEL questionnaire and this helped to increase the value

Table 1 Factors of complexity and their Ranks

Sl. No	Factors of Complexity	Mean	Rank
1	Strenuous Administrative Procedure	3.46	26
2	Use of Multiple Technologies	3.68	24
3	Delay In Project Clearances/Approvals	4.28	4
4	Scope Uncertainties	3.79	19
5	Complex Technologies' Risk	3.82	17
6	Design/Construction Change	4.22	5
7	Interface problems	3.04	27
8	Relationship Between the Various Technological Processes	3.98	12
9	Utilizing New Technologies	4.09	8
10	Project Duration	3.77	20
11	Lack of Experience	3.88	16
12	System Compliance with the Standards	3.96	13
13	Lack of Government Support for the Project	3.75	21
14	Lack of Coordination Between Stakeholders	4.03	10
15	Inadequate Knowledge of Supervision	2.79	32
16	Environment & Working Relationships with Workers	2.85	30
17	Modification of Contractual Clauses	3.74	23
18	Political Influence	4.12	7
19	Breach of Contract	3.92	14
20	Disputes among Project Participants	3.01	28
21	Slow in Making Decisions	3.82	18
22	Lack of Exchange in Information	4.06	9
23	Politics Internal/External	3.62	25
24	Modification of Regulatory Policy	2.90	29
25	Issues in Land Acquisition	4.35	1
26	Delay in the Relocation of Utilities	4.33	2
27	Changes in Exchange Rates	3.75	21
28	Unfavorable Climate Conditions	3.91	15
29	Uncertain Physical and Geotechnical Conditions (UGPC)	4.32	3
30	Location of the Project	2.83	31
31	Design Changes to Suit Non-Divertible Utilities	4.15	6
32	Disruption of Existing Structures	3.99	11

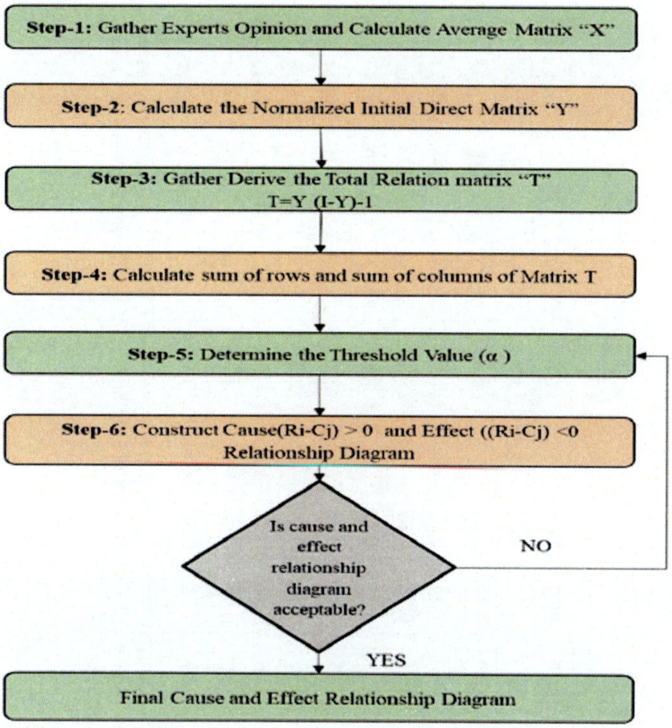

Fig. 2 Flow chart of DEMATEL meth

of the and to identify the interdependence of the complexity factors. To identify the interdependence of the complexity factors, flow chart is represented in Fig. 2.

4 Results and Discussion

In the questionnaire survey, the 5-scale basis i.e., is 0–4 -is considered in the study. 0- "No influence", 1- "low influence", 2- "medium influence", 3- "high influence" and 4- "very high influence" respectively. It is necessary to set a threshold value to sort out insignificant effects [26]. As a result, a threshold of 0.85 as an average mean value was chosen and used in the study. Table 2 highlights values that are equal to or greater than 0.85. R_i and C_i values are obtained by calculating sum of columns and rows of the total relation matrix 'T' and represented in Table 2. Form the obtained values a cause-and-effect diagram is represented in Fig. 3. Figure 4 represents the interdependence of complexity factors as complexity map obtained from total relation matrix.

Table 2 Representation of Total Relation Matrix

Complexity Code	PC1	PC2	PC3	PC4	PC5	PC6	PC7	PC8	PC9	PC10	Ri	Ci	Ri + Ci	Ri-Ci	Impact
PC1	0.775	**0.86**	**0.93**	**0.94**	**0.90**	0.76	**0.87**	0.83	**0.87**	0.78	8.55	8.48	17.03	0.07	Cause
PC2	**0.89**	0.79	**0.92**	**0.95**	**0.92**	0.81	**0.93**	**0.87**	**0.89**	0.81	8.82	8.44	17.27	0.38	Cause
PC3	**0.90**	**0.86**	0.81	**0.93**	**0.89**	0.78	**0.88**	0.83	**0.88**	0.79	8.60	8.90	17.50	−0.31	*Effect*
PC4	**0.94**	**0.91**	**0.98**	**0.89**	0.96	0.82	**0.92**	**0.91**	**0.93**	**0.88**	9.18	9.11	18.29	0.07	Cause
PC5	**0.85**	**0.85**	**0.89**	**0.92**	0.80	0.79	**0.85**	**0.87**	**0.85**	0.82	8.55	8.85	17.41	−0.30	*Effect*
PC6	0.75	0.76	0.80	0.81	0.81	0.62	0.80	0.75	0.75	0.74	7.63	7.72	15.35	−0.08	*Effect*
PC7	**0.88**	**0.90**	**0.92**	**0.93**	**0.89**	0.82	0.81	**0.89**	**0.86**	0.84	8.78	8.68	17.46	0.10	Cause
PC8	0.85	**0.85**	**0.88**	**0.92**	**0.92**	0.77	**0.89**	0.76	0.84	0.81	8.52	8.36	16.88	0.15	Cause
PC9	**0.85**	**0.85**	**0.90**	**0.92**	**0.89**	0.74	0.84	0.82	0.75	0.78	8.39	8.46	16.86	−0.07	*Effect*
PC10	0.77	0.78	0.82	**0.87**	0.84	0.75	0.83	0.79	0.78	0.67	7.95	7.96	15.91	−0.01	*Effect*

PC1-Issues in land acquisition; **PC2**- Delay in the Relocation of Utilities; **PC3**-Uncertain physical and geotechnical conditions; **PC4**-Delay in project clearances/ approvals; **PC5**-Design/construction change; **PC6**- Design changes to suit non-divertible utilities; **PC7**-Political Influence; **PC8**-Utilizing New Technologies; **PC9**-Lack of exchange in information; **PC10**-Lack of coordination between stakeholders.

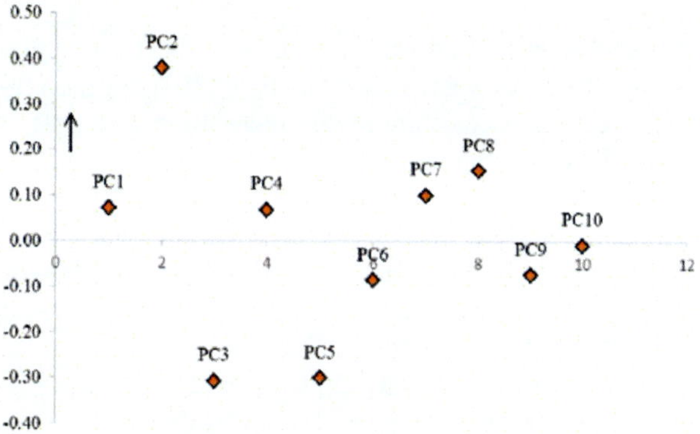

Fig. 3 Cause and Effect Influence Diagram

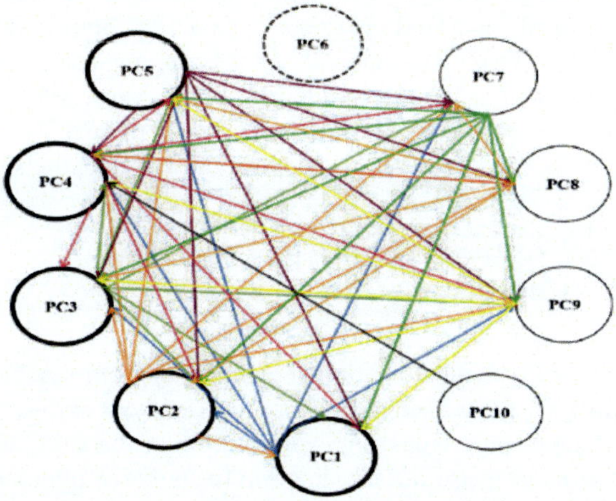

Fig. 4 Complexity Map

From the obtained results the complexity factors having an average mean ≥4.2 are represented with thick circles and a thinner line when mean value ≤4.2. And the factors with zero connection are represented with dotted line in Fig. 4. The study obtained by R_i and C_i values by calculating sum of columns and rows from the total relation matrix T and represented in Table 5. To illustrate the effect of influence cause and effect diagram is represented in Fig. 3. From the ranking index it is obtained that the complexity factors Issues in land acquisition (PC1), Delay in the Relocation of Services (PC2), Uncertain physical and geotechnical conditions (PC3), Delay in project clearances/approvals (PC4), Design/construction change (PC5) have an average mean ≥4.2. From the analysis is observed that the project approvals/clearances have caused the delay in the project and is influenced by other project complexity factors and is net cause group. Issues in land acquisition (PC1), Delay in relocation of utilities (PC2), Delay in project clearance/approvals (PC4), political Influence (PC7), utilizing new technologies (PC8) are in cause group. Uncertain physical and geotechnical (PC3), design/construction changes (PC5), Design changes to suit non-divertible utilities (PC6), lack of exchange in information (PC9) and lack of coordination between stakeholders (PC10) are the effect complexity factors. Delay in project clearances/approvals have the highest ($R_i + C_i$) value in the net cause group. This has the influence of PC1, PC2, PC3, PC5, PC7, PC8 and PC9. Poor site investigations lead to improper drawing and miscalculation in design and have an impact in the diversion of utilities. PC2, PC5 and PC7 have the highest influence on the other complexity factors according to the complexity map. The influence of PC6 is considered nil as it has negligible influence on the other complexity factors. PC10 has been influenced only by PC4.

PC2 has the highest (R_i-C_i) value and is net cause group. In urban areas the metro rail projects mostly face challenges in communication as it is inter-organizational structure because of regulations, time constraint and relationship with the existing structure. In the remote areas there is limited access for the construction hence, approvals and land clearness have a very high impact in causing the delay of the project. The use of machinery needs approvals for the usage in public areas. Coordination among the stakeholders is net effect group contain the least (R_i-C_i) value. Political influence (PC7) is other net cause complexity factor that influences in obtaining the clearance and permissions for utility relocations, transfer of materials etc.

5 Conclusions

The study of identification of complexity factors is the first attempt in the literature of metro projects to conduct a thorough analysis of the complexity factors from the view of developing cities. The contribution of the study is to distinguish the complexity factors of metro rail projects and to study the interdependence among those factors. The study's findings and discussions suggest that DEMATEL is a practical method for understanding how complexity factors interact with one another. It is recognized that different experts may arrive at different conclusions based on their experiences and skill sets, prominent to various outcomes. The nature of metro rail projects is different from other construction projects. Due to its unique characteristics the uncertainty and unpredictable nature made the projects more complex. Hence, to decrease its impact on the performance of the project's identification of complexities and its interdependence is necessary. From the study the metro organizations and the stakeholders can use the research methodology for the development of complexity maps for precise measuring of impact of complexity. The understanding of interdependency of complexity factors for project management is essential because when implementing and managing metro rail projects, the metro organizations and the stakeholders can refer this study. The key stakeholders will be able to develop effective management strategies for project implementation with the help of a distinct understanding of the interrelationships between the complexity factors of metro rail projects. The findings of this research can be applied to future work to choose appropriate project complexity management strategies.

References

1. Bakhshi, J., Ireland, V., & Gorod, A. (2016). Clarifying the project complexity construct: past, present and future. *International Journal of Project Management, 34*, 1199–1213.
2. Kim, S. Y., Nguyen, M. V., & Dao, T. T. N. (2020). Prioritizing complexity using fuzzy DANP: Case study of international development projects. *Engineering Construction and Architectural Management, 28*, 1114–1133.
3. Luo, L., He, Q., Jaselskis, E.J., Xie, J. (2017) Construction project complexity: research trends and implications. Journal of Construction Engineering and Management **143**, 04017019
4. Browning, T. R. (2014). Managing complex project process models with a process architecture framework. *International Journal of Project Management, 32*, 229–241.
5. Vidal, L. A., & Marle, F. (2008). Understanding project complexity: Implications on project management. *Kybernetes, 37*, 1094–1110.
6. Bosch-Rekveldt, M., Jongkind, Y., Mooi, H., Bakker, H., & Verbraeck, A. (2011). Grasping project complexity in large engineering projects: the TOE (Technical, Organizational and Environmental) framework. *International Journal of Project Management, 29*, 728–739.
7. Baccarini, D. (1996). The concept of project complexity - a review. *International Journal of Project Management, 14*, 201–204.
8. Li, W., et al. Optimal Transport in Worldwide Metro Networks. *arXiv preprint* arXiv:1403.7844 (2014)
9. Flyvbjerg, B., Holm, M. K. S., & Buhl, S. L. (2003). How common and how large are cost overruns in transport infrastructure projects? *Transport Reviews, 23*, 71–88.
10. Flyvbjerg, B., Bruzelius, N., Rothengatter, W. (2003). Megaprojects and Risk: An Anatomy of Ambition. Cambridge university press
11. Nguyen, A.T., Nguyen, L.D., Long Le-Hoai, C.N.D. (2015). Quantifying the complexity of transportation projects using the fuzzy analytic hierarchy process. International Journal of Project Management **33**, 1364–1376
12. PMI. Assess and develop talent with a focus on fostering leadership skills. PMI's Pulse of Profession In-Depth Report: Navigating Complexity, pp. 13–15 (2013)
13. Geraldi, J.G., Adlbrecht, G. (2007). On faith, fact and intercation in projects. Project Management Journal **38**, 32–43
14. Luo, J., & Wood, K. L. (2017). The growing complexity in invention process. *Research in Engineering Design, 28*, 421–435.
15. Sinha, S., Thomson, A.I., Kumar, B. (2001). A Complexity Index for the Design Process. WDK Publications, pp. 157–163 (2001)
16. Williams, T. (1999). Modelling Complex Projects. Wiley, Hoboken
17. Parwani, R.R. (2022). Complexity: an introduction. arXiv preprint physics/0201055
18. Cicmil, S., Cooke-Davies, T., Crawford, L., Richardson, K. (2017). Exploring the complexity of projects: Implications of complexity theory for project management practice. in (Project Management Institute)
19. Vidal, L.-A., Marle, F., & Bocquet, J.-C. (2011). Using a delphi process and the analytic hierarchy process (AHP) to evaluate the complexity of projects. *Expert Systems with Applications, 38*, 5388–5405.
20. Xia, W., & Lee, G. (2005). Complexity of information systems development projects: Conceptualization and measurement development. *Journal of Management Information Systems, 22*, 45–83.
21. Cicmil, S., & Marshall, D. (2005). Insights into collaboration at the project level: Complexity, social interaction and procurement mechanisms. *Building Research Information, 33*, 523–535.
22. He, Q., Luo, L., Hu, Y., & Chan, A. P. C. (2015). Measuring the complexity of mega construction projects in China-a fuzzy analytic network process analysis. *International Journal of Project Management, 33*, 549–563.
23. Kiani Mavi, R., Standing, C. (2018). Critical success factors of sustainable project management in construction: A fuzzy DEMATEL-ANP approach. J Cleaner Production **194**, 751–765

24. Ahsan, K., Ph, D., Paul, S. K., & Ph, D. (2018). Procurement issues in donor-funded international development projects. *Journal of Management in Engineering, 34*, 1–13.
25. Si, S., You, X., Liu, H., Zhang, P. (2018). DEMATEL technique : a systematic review of the state-of-the-art literature on methodologies and applications. **2018**, 1–33

Contractor Selection Approaches and Pre-qualification Criteria on Construction Projects: A Review

Mangesh M. Kapote, Avadhut A. Wagh, and Sunil S. Pimplikar

Abstract A prime contracting firm plays a pivotal role in not only executing the construction project but also in coordination and integration of various stakeholders of the entire civil engineering project for its success. Reliable and competent contractor selection is thus a very important need. The best contractor choice is based on a comprehensive multiple criteria selection process that takes into account a contractor's technological, administrative, financial and societal capabilities. The literature review on topics connected with the contractor's selection and prequalification from reliable databases and notable publications is structurally examined in this study over a period of three decades. The aim of this review is to shed light on existing contractor pre-qualification and selection procedure along with various statistical approaches. The findings of this study indicate that though lowest bid consideration is weighted heavily in the construction industry, it is certainly a major contributing factor to many problems like quality issues, time and cost overruns, disputes, etc. The review provides the context for future scientific investigations on contractor pre-qualification and appropriate selection. The study also helps to expand the horizon of professionals and investigators' comprehension of the different attributes used for contractor selection and evaluation.

Keywords Contractor selection · Lowest bid · Multi-criteria decision making · Performance evaluation · Pre-qualification criteria · Tender price

---

M. M. Kapote (✉) · A. A. Wagh · S. S. Pimplikar
School of Civil Engineering, Dr. Vishwanath Karad MIT World Peace University, Pune, India
e-mail: mngshkapote@gmail.com

A. A. Wagh
e-mail: avdhootwagh3@gmail.com

S. S. Pimplikar
e-mail: sunil.pimplikar@mitwpu.edu.in

A. Vilventhan et al. (eds.), *Advances in Construction Materials and Management*, Lecture Notes in Civil Engineering 346,
https://doi.org/10.1007/978-981-99-2552-0_3

1 Introduction

1.1 General

Choosing the finest construction contractor entails a number of difficult decisions for proprietors and consulting firms in the private as well as public sectors.

Pre-qualification of contractors entails the owner evaluating candidate contractors based on a predetermined set of standards to ascertain their suitability for the job should they be granted the construction project. Pre-qualification of contractors and bid evaluation is a decision-making process that happens throughout the whole procurement cycle which are the two most important stages used by many client authorities for contractor selection [1]. Pre-qualification criteria involves the collection of data about the candidate contractor and his firm which mostly relies on theoretical, arbitrary, and vague information. Due to the projects' growing complexity and competition, a contractor should not only be competent in a specific field; he also must be capable to manage projects effectively and efficiently [2]. Since the chosen contractor is in charge of overseeing a project's effective coordination, the client's preference of contractor during the project pre-qualification phase is crucial. Following a careful screening procedure, the client can then assign the work to the contractor for completion [3]. It is essential to devise methods for assessing contractors against various key success metrics during the pre-qualification and bid review phases of the procurement cycle, as well as to specify and create suitable predetermined contractor eligibility requirements in order to call for qualified tenderers [4]. The aim of this study is to focus on various pre-qualification criteria used by clients along with various statistical approaches used for contractor selection and highlights the need of upgradation of existing system of contractor selection. To achieve the set goals literature survey approach was used with collection of significant research papers over a period of three decades. Figure 1 shows the flow chart of traditional bidding process.

1.2 Need of the Study

1. As contractor plays an important role in successful completion of the project, negligence on the topic may leads to failure of the project hence, contractor selection is very important issue and needs to be focused on.
2. There are numerous flaws existing in the current process of contractor selection and pre-qualification which contributes to negative impact on the project.
3. At present many countries including India are not using decision making tools such as MCDM techniques which has many advantages in order to select most suitable contractor.

hence, contractor selection process needs to be investigated.

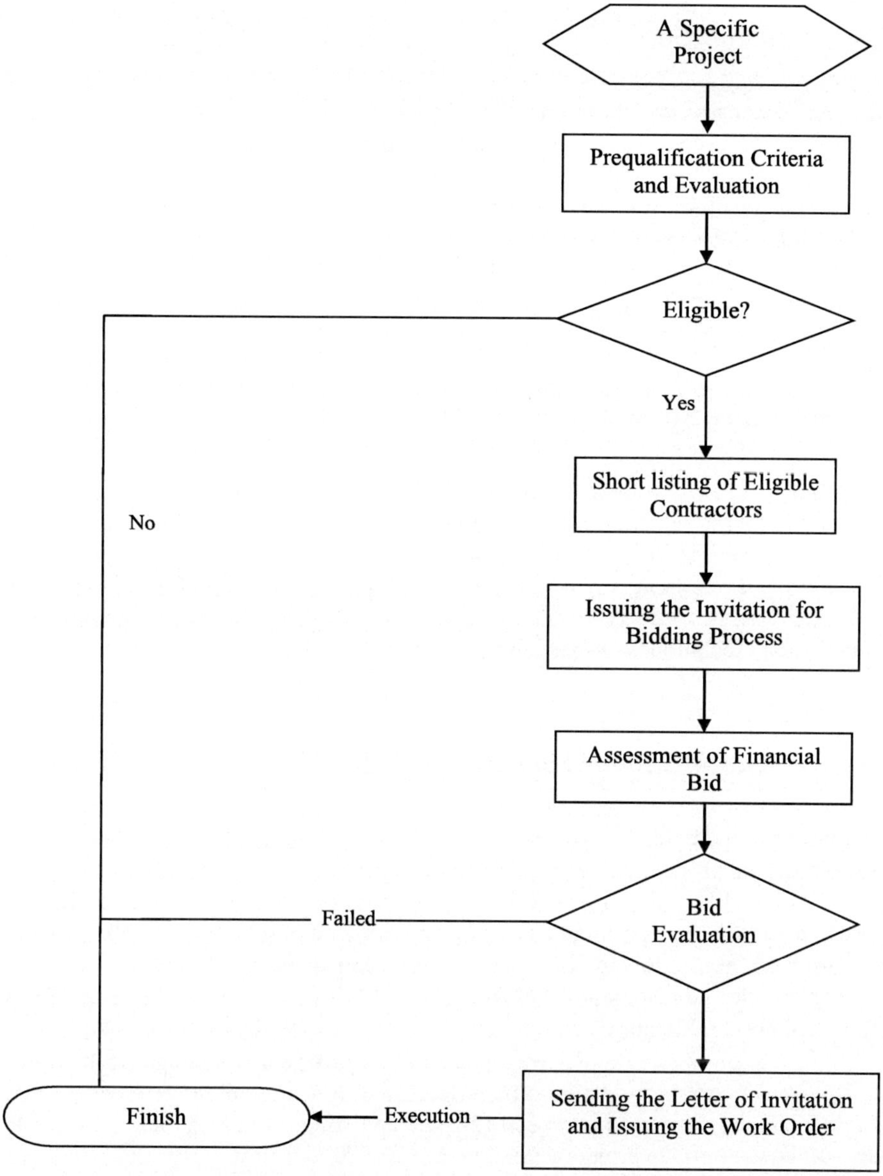

Fig. 1 Traditional bidding process flowchart [18]

1.3 Research Gap

Although extensive investigations being conducted in this area, the concerns listed below have not been explored or resolved in earlier research published as books, journal papers, or reports. For instance, there is currently a dearth of research on the project specific unique pre-qualification criteria and actual on-site implementation of multi-criteria decision making (MCDM) approaches for the selection of competent contractors in the competitive bidding system used in India. Through investigation is therefore required to tackle these concerns in future research.

2 Literature Survey

A thorough literature review has been done in order to acquire a thorough grasp of current contractor selection practices on infrastructure projects in global contexts. Due to the relevance of contractor selection in the construction industry, several specialists and researchers have taken on the responsibility of performing in-depth study and analysis into this issue. It covers things like various pre-qualification standards and techniques, issues with contractor selection etc.

2.1 Pre-qualification Criteria

Pre-qualification standards are a strategy for choosing competent contractors so as to minimize the probability of project failure. The contractor qualifying criteria have numerous benefits, including enabling the client to pick the best contractor for the project, lowering the risk of contractor insolvency, reducing excessive schedule and cost overruns, and reducing poor quality concerns [5]. Pre-qualification is done in order to set minimum standards below which the contractors will not be considered [6]. Every pre-qualification criteria measures a particular contractor's capability to accomplish a job in a particular way hence have a different weightage while evaluating during contractor selection. The procedure of figuring out the relative weights of every selection criterion is crucial [7]. The construction industry's professionals have turned to alternate project delivery methods due to the rapid advancements and needs in all facets of human existence, but the bidding and award procedures have largely remained same [8].

The majority of the seriously offered development contracts in most countries utilize the lowest bid strategy where the contract is awarded to the firm presenting the least bid price [1, 9–14]. Albeit the least bidder framework shields the general population from ill-advised rehearses, it has specific inconveniences. These incorporate nonsensical low offers either inadvertently or intentionally create broad setback cost invade, quality issues and an expanded number of questions. The greatest result might not be obtained by selecting the best contractors merely on the basis of the lowest bid price as it might result in delays and cost overruns [13, 15]. Lo T. Y. et. al [16] in his research highlighted that construction delay factor, "Exceptional low bid", was recognized as the third most huge reason for delay which is practiced even today where L1 bidder is awarded the work amongst all the qualified bidders. The lowest bid criteria are highly weighted in the sector, yet still is key cause of the problem. Moreover, the quality issues, safety standards, experience levels of the contractor are not taken into the consideration [17]. The experience of the contractors and their financial ability are often considered as insufficient factors in selection of contractors. In order to properly select the contractor, it is necessary to give weight to the criteria determined after properly studying it [18]. To ensure that the project is successful overall and that both the customer and the contractor benefit, it is important to choose a contractor based on factors apart from the lowest offer [19]. Therefore, to welcome reasonable bidders, it is important to explain and foster proper pre-determined contractor selection criteria, improve and coordinate the appraisal of data connecting with these, and foster strategies for considering them in contrast to different project success factors at the pre-qualification stage.

Pre-qualification items, fair bidder, interests of the public, and other changes to the L1 (lowest) bidder system over time have made it possible for various evaluation techniques to be used in place of the single criterion system i.e. lowest-cost method. The pre-qualification and selection of just those contractors who are highly conversant due to their safety programs and performance is a crucial step that may be taken to assure project success in terms of safety. A strong safety record lowers construction costs and supports the desired mindset for efficiency and quality. Before selecting, thorough contractor assessments can greatly decrease the dangers a construction project faces [20]. The historical performance of the contractors kept in an online database can be prequalifying factor when judging the contractor [21]. Several researchers and client enterprises have used diverse set of crucial contractor selection criteria to evaluate the suitability of potential contractors [22, 23]. For the purpose of evaluating bids, contractor selection should be based on an objective prequalification mechanism with project-specific micro scale pre-qualification standards [24]. Several pre-qualification procedures use assessments that are unclear or quite subjective [25].

It is frequently more difficult to obtain, assess, and apply evidence to establish expertise in the equally significant pre-qualification domain of managerial ability; in effect [26]. In contrast to Hatush and Skitmore [8], who used financial health, technical skills, management effectiveness, and overall safety and health performance as selection criteria, Holt et. al [27] considered the contractor's existing volume of work, previous experience regarding project size, management assets in terms of a formal training regime, historical performance, and time of year weather. For contractor selection, along with tender price other two most important criteria also should be considered viz time & quality [28]. Despite numerous initiatives undertaken by public authorities to raise awareness and importance of health and safety at all levels of the construction sector, the low degree of significance assigned to the contractor selection criteria pertaining to safety and health indicates a significant lack of awareness of the importance of health and safety on the job.

Contractor selection criteria such as equivalent job experience, qualifications and experience of project leader together with technical and managerial staff, category and magnitude of completed project in the past three years, company liquidity, working capital, and sanctions etc. are relevant and important to settle a wide range of challenges from the perspectives of all field experts from Singapore construction industry [29]. Russell et. al [30] used a variety of selection criteria, including financial security, historical performance, experience, and the availability of key individuals. A centralized contractor assessment process along with the bid price, with the addition of three new selection attributes, viz. time estimated for project completion, the warranty duration estimated, and the contractor's track record of effectiveness helps to make decisions about the contract award [10]. Available bid capacity, total and similar work experience, financial capabilities, equipment and plants available, managerial competency, and quality & safety policies are some of the important criteria that researchers have urged and that are also practically in use.

Every country has its own procedures for construction procurement. In Chinese construction sector there are seven attributes in assessing contractors' competitiveness for awarding contract viz managerial ability, technical know-how, financial ability, organizational design, marketing know-how, social impact, and contribution to project goals which are helpful in identifying strengths & weaknesses of the contractors. The complex characteristics of construction business environment indicates strong governmental influence on construction market in China [31]. Owners have established or implemented standard bidding documents developed by the World Bank, Asian Development Banks, FIDIC, and others all across the world. However, most bidding paperwork and procedures have yet to be standardized in developing countries. Even when the procedure is consistent, the relative importance of different selection factors is a source of debate [18]. Although the civil engineering professionals were familiar with the pre-qualification screening process used in the Ghanaian construction sector, they were unaware with multi-criteria choice strategies [5]. Numerous and frequently contradictory objectives and approaches, such as tender price, completion date, and experience, must be addressed in a multi-criteria framework [32]. Above chapter focuses on major pre-qualification criteria used currently and also various additional criteria considered by various researchers along with their impact on project. Table 1 indicates that out of the 21 criteria which are important in the contractor selection process; few of the criteria such as work progress, human resource management, relation with the sub-contractors, competitiveness, contractor's reputation, political considerations and the like are considered piecemeal only by researchers.

Table 1 Selection factors for the contractor

Criteria considered for selection of contractor	Reference (Past research by various researchers)					
	[4]	[19]	[28]	[46]	[48]	[13]
Financial Status	✓	✓	✓	✓	✓	✓
Technical Competencies	✓	✓		✓	✓	✓
Previous Performance			✓	✓	✓	
Cost Parameter	✓	✓	✓			
Quality Standards	✓	✓	✓			
Health and Safety	✓	✓	✓	✓	✓	✓
Company's Background	✓	✓	✓			
Timely Performance	✓	✓	✓	✓		
Management Capabilities			✓	✓	✓	✓
Work Progress			✓			
Human Resource Management	✓					
Level of Technology			✓	✓		
Relation with Client	✓	✓		✓		
Relation with sub-contractors			✓			
Competitiveness			✓			
Reputation of the Contractor						✓
Bid Price				✓		
Political Considerations				✓		
Similar Project's Experience				✓	✓	
Progress of Existing Projects				✓		
Current Projects in Hand				✓		

2.2 Contractor Selection Approaches

A successful project requires careful planning and the expertise of contractors. The five most crucial requirements for contractor selection are on schedule delivery, compliance with quality standards, safety performance, workplace safety record, and versatility in work plan [33]. There is abundant proof of extensive global attempts and scientific studies aimed at shaping and designing contractor selection [19]. Since the contractor plays a crucial part in the success or failure of projects in this market, choosing the right contractor is a highly crucial issue. In order to identify appropriate suppliers for the projects, it is crucial to define procedures and criteria suitable for this selection challenge [6]. Many of the contractors don't analyze the competitive environment using mathematical or statistical methods. The only factors taken into consideration when choosing a winning bid are past experience, independent opinion, and qualitative evaluation [34]. Primarily there are three main causes of deficient contractor selection. First off, while choosing criteria, the wrong ones are chosen for determining the contractor's suitability for the job. Secondly, some criteria are given an excessive amount of significance (such as bid price) Third, the wrong approach is used for the role of choosing and evaluating contractors [35]. If a suitable and accurate method is not followed to select the most suitable contractor, the success of the project may suffer [32]. The ultimate aim of contractor selection should be to find the "best bidder", not just the "least bidder" [28].

Alternative procurement approaches are increasingly being employed because traditional construction procurement strategies are considered insufficient in order to meet current issues and needs [19]. Closely integrated contractor selection methods are necessary to finish construction projects on schedule. Selecting the "best" contractor is a multi-criteria, multi-group decision [36].

In fact, a multicriteria approach for selecting contractors is heavily influenced by the dynamic nature of construction works and the subjective assessment of the choice makers. A software program called QUALIFIER-1, created in 1990 by Russell and Skibniewski [37], computes statistics and an overall weighted rating for each potential contractor. With a computerized contractor evaluation procedure that enables the user to make uniform and logical prequalification judgments, the system QUALIFIER-2 notifies the participant for relevant contractor details [38]. A decision support system (DSS) is a technique that takes into account the particular circumstances surrounding a proposed project as well as the adaptability of a contractor's present capabilities and qualifications to carry out the proposed project while taking into account the particular circumstances surrounding the project [39]. The process of analytical hierarchy provides a logical structure for determining the benefits of each alternative which define the criteria. The major contractors' prequalification criteria made more realistic and reliable by the use of analytical hierarchy process (AHP) weights. The use of analytical hierarchy model combined with a qualitative evaluation enables owners to minimize their chance of failure and yield the best results [30]. MCDM's AHP approach with six identified criteria's for 'best' contractor selection viz. availability of resources, historical competence, prior experience, present volume of work, and safety compliance will enable project management teams recognize contractors that are most likely to achieve acceptable results in a selection process that is not solely dependent on the lowest bid [40]. In order to complete construction projects on time, highly integrated contractor selection processes are required. Choosing the "best" contractor is a difficult multi-criteria and multi-group decision [36]. Since contractor selection is a multi-attribute group decision with time constraints, many complexities and ambiguities arise; as a result, the fuzzy weighted-average approach may be used to model the grouping of multiple decision-makers. The novelty is that it incorporates subjective judgment into the contractor selection decision using fuzzy logic, as well as multiple decision-makers into the quantitative solution [41]. A fuzzy set may be mathematically defined by giving each person in the universe a value that represents their level of participation in the fuzzy set. The use of fuzzy set theory enables the decision makers to demonstrate their evaluation of contractor's performance on decision criteria in linguistic terms rather than as crisp values [42]. Artificial neural network has ability to map the complex non-linear relationship between contractor performance vector and prequalification decisions. It eliminates the necessity of the construction professionals to give direct correct weights to each of the pre-qualification criteria [43]. If validated using the current method for choosing contractors for previous projects, the Binary Goal Programming Model gives much superior results, supporting the model's credibility [10]. There is room for change in the bidding process in India and other developing countries. Bid documents are not structured, even when they are, they are used indiscriminately for different types of projects without regard for project-specific needs. After conducting thorough analysis, weightage to selection criteria must be allocated [18]. Weighted point score approach with the quantification of several criteria aid in picking the best sub-contractors [44]. Capacity building, policy making, strategies, dialogue, billing, record keeping, contractor performance review, verifications and audits, and settlement of disputes are some of the practices. If these contract tracking elements are all well controlled, there is a huge likelihood of having an effective road infrastructural project [45]. Above chapter highlights different contractor selection approaches and their suitability considered by various researchers so as to choose most competent contractor for a specific project. Table 2 indicates approaches suggested for contractor selection and developments in the same over a period of time.

Table 2 Suggested approaches for selecting contractors

Reference	Modelling mechanism suggested by different researcher
[37]	Dimensional weighted aggregate—QUALIFIER-1
[38]	Knowledge based system—QUALIFIER-2
[27]	Multi-attribute assessment
[8]	Point score method
[4]	Multi-attribute utility theory
[43]	Artificial neural networks (ANN)
[39]	Analytic hierarchy process (AHP)
[28]	Analytical hierarchy process (AHP)
[42]	Fuzzy set theory (FST)
[7]	Hybrid model: using AHP, neural network and genetic algorithm
[21]	Fuzzy-analytic hierarchy process-SMART
[41]	Fuzzy weighted-average
[36]	Fuzzy-AHP & Fuzzy-TOPSIS with BIG DATA
[49]	Pre-qualification model using Quality Function Deployment

3 Discussion

3.1 Prequalification Criteria

The analysis of this study revealed that the bid price continues to be the deciding factor in the selection of bids which may be the key cause of project failure hence, other criteria like health and safety, reputation of contractor, relations with client and subcontractors needs to be considered in order to increase the chances of project's success.

3.2 Contractor Selection Approaches

The literature survey urges client organizations to reconsider adding more prequalification standards and put in greater attempt to examine contractors using weighted standard scores and various MCDM tools such as AHP, ANN, Fuzzy set theory etc.

3.3 Significant Issues in Contractor Selection Process

Many emerging and even industrialized nations, like India, appear to have little attention in the largely understudied topic of contractor selection. Previous studies on the subject have done a great job of identifying the effects of poor contractor selection on infrastructure undertakings and various issues in process of contractor selection like non standardized bidding process, no use of statistical tools for contractor selection, time and cost overruns with quality issues due to incompetent contractor selection etc.

4 Conclusion

Based on above investigation following conclusions are drawn

1. Financial status, past experience, technical competencies, health and safety are the most commonly considered criteria for selecting the contractor.
2. Use of MCDM techniques such as AHP, Fuzzy set theory, ANN etc. provide ease for decision makers to select most appropriate contractor.
3. Contract award to the lowest bidder, non-standardized bidding paperwork and procedures, administrative frauds and bribes etc. are the key flaws in the contractor selection process.

List of Abbreviations

AHP	Analytical hierarchy process
ANN	Artificial neural networks
DSS	Decision support system
FIDIC	Fédération Internationale Des Ingénieurs-Conseils (International Federation of Consulting Engineers)
MCDM	Multi criteria decision making

References

1. Kolekar, P. B., & Kanade, G. N. (2014). Contractor selection in construction industry using fuzzy-logic system. *International Journal of Engineering Research & Technology, 3*(11), 1087–1093. ISSN: 2278-0181
2. Russell, J. S., & Skibniewski, M. J. (1988). Decision criteria in contractor prequalification. *Journal of Management in Engineering, 4*(2), 148–164. https://doi.org/10.1061/(ASCE)9742-597X(1988)4:2(148)

3. Rashvand, P., Abd Majid, M. Z., Baniahmadi, M., & Ghavamirad, F. (2015). Contractor selection at prequalification stage: current evaluation and shortcomings. *Jurnal Teknologi, 77*(16), 81–89. https://doi.org/10.11113/jt.v77.6403
4. Hatush, Z., & Skitmore, M. (1998). Contractor selection using multicriteria utility theory: An additive model. *Building and Environment, 33*(2–3), 105–115. https://doi.org/10.1016/S0360-1323(97)00016-4
5. Ayettey, D. N., & Danso, H. (2018). Contractor selection criteria in Ghanaian construction industry: Benefits and challenges. *Journal of Building Construction and Planning Research, 6*(4), 278–297. https://doi.org/10.4236/jbcpr.2018.64019
6. Araújo, M., Alencar, L., & Mota C. (2015). Contractor selection in construction industry: A multicriteria model. In *IEEE international conference on industrial engineering and engineering management* (pp. 519–523). https://doi.org/10.1109/IEEM.2015.7385701
7. El-Sawalhi, N., Eaton, D., & Rustom, R. (2007). Contractor pre-qualification model: State of-the-art. *International Journal of Project Management, 25*(5), 465–474. https://doi.org/10.1016/j.ijproman.2006.11.011
8. Hatush, Z., & Skitmore, M. (1997). Criteria for contractor selection. *Construction Management & Economics, 15*(1), 19–38. https://doi.org/10.1080/014461997373088
9. Waara, F., Bröchner, J. (2006). Price and nonprice criteria for contractor selection. *Journal of Construction Engineering and Management, ASCE, 132*(8), 797. https://doi.org/10.1061/(ASCE)0733-9364(2006)132:8(797)
10. Padhi, S. S., & Mohapatra, P. K. J. (2010). Centralized bid evaluation for awarding of construction projects - A case of India government. *International Journal of Project Management, 28*, 275–284. https://doi.org/10.1016/j.ijproman.2009.06.001
11. Ioannou, P., & Awwad, R. (2010). Below-average bidding method. *Journal of Construction Engineering and Management, ASCE, 136*(9), 936–946. https://doi.org/10.1061/(ASCE)CO.1943-7862.0000202
12. Ibadov, N. (2015). Contractor selection for construction project, with the use of fuzzy preference relation. *Procedia Engineering, 111*, 317–323. https://doi.org/10.1016/j.proeng.2015.07.095
13. Puri, D., & Tiwari, S. (2015). Efficient contractor selection and bid evaluation methods for construction industry in India. *International Journal of Science and Research (IJSR) ISSN (Online)*, 2319–7064.
14. El-khalek, H. A., Aziz, R. F., & Morgan, E. S. (2019). Identification of construction subcontractor prequalification evaluation criteria and their impact on project success. *Alexandria Engineering Journal, 58*, 217–223. https://doi.org/10.1016/j.aej.2018.11.010
15. El-Abbasy, M. S., Zayed, T., Ahmed, M., Alzraiee, H., & Abouhamad, M. (2013). Contractor selection model for highway projects using integrated simulation and analytic network process. *Journal of Construction Engineering and Management, 139*(7), 755–767. https://doi.org/10.1061/(ASCE)CO.1943-7862.0000647
16. Lo, T. Y., Fung, I. W., & Tung, K. C. (2006). Construction delays in Hong Kong civil engineering projects. *Journal of Construction Engineering and Management, 132*(6), 636–649. https://doi.org/10.1061/(ASCE)0733-9364(2006)132:6(636)
17. Shumank, D., Singh, D., & Ahmad, S. A. (2016). A review of contract awards to lowest bidder in Indian construction projects via case based approach. *Open Journal of Business and Management, 5*(1), 159–168. https://doi.org/10.4236/ojbm.2017.51015
18. Pandit, D., & Yadav, S. M. (2015, October). Contractor selection: A key to project success. *The Masterbuilder*, 84–88. www.masterbuilder.co.in
19. Palaneeswaran, E., & Kumaraswamy, M. M. (2000). Contractor selection for design/build projects. *Journal of Construction Engineering and Management, 126*(5), 331–339. https://doi.org/10.1061/(ASCE)0733-9364(2000)126:5(331)
20. Department of Treasury and Finance (2014). *Guidelines for tender evaluation using weighted criteria for building works and services*, Australia
21. Padhi, S. S., & Mohapatra, P. K. J. (2009). Contractor selection in government procurement auctions: A case study. *European Journal of Industrial Engineering, 3*(2), 170–186.

22. Sönmez, M., Holt, G., Yang, J., & Graham, G. (2002). Applying evidential reasoning to prequalifying construction contractors. *Journal of Management in Engineering, 18*(3), 111–119. https://doi.org/10.1061/(ASCE)0742-597X(2002)18:3(111)
23. Molla, M. M., & Asa, E. (2015). Factors influencing contractor prequalification processes in developing countries. *International Journal of Architecture, Engineering and Construction,* (4), 232–245. https://doi.org/10.7492/IJAEC.2015.024
24. Kapote, M. M., Patil, P. J., & Pimplikar, S. S. (2022). Contract award based on prequalification criteria: A key for project success. *Stochastic Modeling and Applications, 26*(3), 741–748.
25. Hosny, O., Nassar, K., & Esmail, Y. (2013). Prequalification of Egyptian construction contractors using fuzzy-AHP models. *Engineering, Construction and Architectural Management, 20*(4), 381–405. https://doi.org/10.1108/ECAM-09-2011-0088
26. Plebankiewicz, E. (2009). Contractor prequalification model using fuzzy sets. *Journal of Civil Engineering and Management, 15*(4), 377–385. https://doi.org/10.3846/1392-3730.2009.15.377-385
27. Holt, G. D., Olomolaiye, P. O., & Harris, F. C. (1994). Evaluating prequalification criteria in contractor selection. *Building and Environment, 29*(4), 437–448. https://doi.org/10.1016/0360-1323(94)90003-5
28. Topcu, Y. I. (2004). A decision model proposal for construction contractor selection in Turkey. *Building and Environment, 39*(4), 469–481. https://doi.org/10.1016/j.buildenv.2003.09.009
29. Singh, D. A., & Tiong, R. L. (2006). Contractor selection criteria: investigation of opinions of Singapore construction practitioners. *Journal of Construction Engineering and Management, 132*(9), 998–1008. https://doi.org/10.1061/(ASCE)0733-9364(2006)132:9(998)
30. Russell, J. S., & Jaselskis, E. J. (1992). Predicting construction contractor failure prior to contract award. *Journal of Construction Engineering and Management, 118*(4), 791–811. https://doi.org/10.1061/(ASCE)0733-9364(1992)118:4(791)
31. Shen, L. Y., Li, Q. M., Drew, D., & Shen, Q. P. (2004). Awarding construction contracts on multicriteria basis in China. *Journal of Construction Engineering and Management, 130*(3), 385–393. https://doi.org/10.1061/(ASCE)0733-9364(2004)130:3(385)
32. San Cristóbal, J. R. (2012). Contractor selection using multicriteria decision making methods. *Journal of Construction Engineering and Management,* 138(6), 751–758. https://doi.org/10.1061/(ASCE)CO.1943-7862.0000488
33. Hemanta, D. (2009). Analysis of pre-qualification criteria in contractor selection and their impacts on project success. *Construction Management and Economics, 27*(12), 1245–1263. https://doi.org/10.1080/01446190903394541
34. Ahmad, I., & Minkarah, I (1988). Questionnaire survey on bidding in construction. *Journal of Management in Engineering, 4*(3), 229–243. https://doi.org/10.1061/(ASCE)9742-597X(1988)4:3(229)
35. Zala, M. I., & Bhatt, R. B. (2011). An approach of contractor selection by analytical hierarchy process. In *National conference on recent trends in engineering and technology* (pp. 1–6).
36. Taylan, O., Kabli, M. R., Porcel, C., & Viedma, E. H. (2018). Contractor selection for construction projects using consensus tools and big data. *International Journal of Fuzzy Systems, 20*(4), 1267–1281. https://doi.org/10.1007/s40815-017-0312-3
37. Russell, J. S., & Skibniewski, M. J. (1990). QUALIFIER-1: Contractor prequalification model. *Journal of Computing in Civil Engineering, 4*(1), 77–90. https://doi.org/10.1061/(ASCE)0887-3801(1990)4:1(77)
38. Russell, J. S., Skibniewski, M. J., & Cozier, D. R. (1990). QUALIFIER-2; Knowledge-based system for contractor prequalification. *Journal of Construction Engineering and Management, 116*(1), 157–171. https://doi.org/10.1061/(ASCE)0733-9364(1990)116:1(157)
39. Mahdi, I. M., Riley, M. J., Fereig, S. M., & Alex, A. P. (2002). A multi-criteria approach to contractor selection. *Engineering, Construction and Architectural Management, 9*(1), 29–37. https://doi.org/10.1108/eb021204
40. Balubaid, M., & Alamoudi, R. (2015). Application of the analytical hierarchy process (AHP) to multi-criteria analysis for contractor selection. *American Journal of Industrial and Business Management, 5*(09), 581. https://doi.org/10.4236/ajibm.2015.59058

41. Alhumaidi, H. M. (2015). Construction contractors ranking method using multiple decision-makers and multiattribute fuzzy weighted average. *Journal of Construction Engineering and Management, 141*(4), 04014092. https://doi.org/10.1061/(ASCE)CO.1943-7862.0000949
42. Singh, D. A., & Tiong, R. L. (2005). A fuzzy decision framework for contractor selection. *Journal of Construction Engineering and Management, 131*(1), 62–70. https://doi.org/10.1061/(ASCE)0733-9364(2005)131:1(62)
43. Lam, K. C., Ng, S. T., Tiesong, H., Skitmore, M., & Cheung, S. O. (2000). Decision support system for contractor pre-qualification-artificial neural network model. *Engineering Construction and Architectural Management, 7*(3), 251–266. https://doi.org/10.1111/j.1365-232X.2000.00156.x
44. Kapote, M. M., & Pimplikar, S. S. (2014). Suggested mathematical model for specialized subcontractor prequalification scrutiny and ultimately the performance prediction. *IOSR Journal of Mechanical and Civil Engineering, 11*(3), Ver. VI, 43–51. https://doi.org/10.9790/1684-11364351
45. Byaruhanga, A., & Basheka, B. C. (2017). Contractor monitoring and performance of road infrastructure projects in Uganda: A management model. *Journal of Building Construction and Planning Research, 5*, 30–44. https://doi.org/10.4236/jbcpr.2017.51003
46. Idrus, A., Sodangi, M., & Amran, M. A. (2011). Decision criteria for selecting main contractors in Malaysia. *Research Journal of Applied Sciences, Engineering and Technology, 3*(12), 1358–1365.
47. Hassen, S. M., & Al-Tmeemy,. (2017). The impact of incompetent contractor on the project schedule. *Journal of Engineering and Sustainable Development, 21*(03), 87–101.
48. Nieto-Morote, A., & Ruz-Vila, F. (2011). A fuzzy AHP multi-criteria decision-making approach applied to combined cooling, heating, and power production systems. *International Journal of Information Technology & Decision Making, 10*(03), 497–517. https://doi.org/10.1142/S0219622011004427
49. Jafari, A. (2013). A contractor pre-qualification model based on the quality function deployment method. *Construction Management and Economics, 31*(7), 746–760. https://doi.org/10.1080/01446193.2013.825045

Delay Analysis of Residential Construction by Using Augmented Reality and Virtual Reality

Raghavendra S. Sikarwar and Abhaysinha G. Shelake

Abstract According to a survey by Global Data, the size of the Indian construction market was $609.6 billion in 2021 and is projected to increase by more than 6% between 2023 and 2026. However, project delays have a particularly negative impact on the residential sector. The construction industry should minimize activity-related delays to meet these objectives. A project will be deemed successful if it is finished within the allotted time. Researcher indicates that, more than 55% of projects ran over budget and delayed because there was poor coordination and communication across various departments and divisions. Because of the poor coordination and communication project got delayed. Coordination is crucial for reducing delays, and managing activities also guarantees that time is managed. The customer, contractor, and consultant must establish good communication and coordination to maximize project productivity. All employees must be familiar with the project and understand what they are responsible to do and also understands the coordination they are supposed to make. There are some methods like impact-as-planned, collapsed-as-built, time impact analysis, where coordination is not important but these are the methods important for understanding and calculating the delay. While the previous researchers have understood the major problem in the delay on construction site, is the coordination. As the researchers' research that more than 55% projects get delayed. Coordination is really important for the speedy construction among the different departments in the company. Augmented reality (AR) and virtual reality (VR) can be helpful for coordination and these are not used techniques in the previous construction work. AR enables the visualization and on-site modification of architectural designs. Compared to traditional drawings, the VR model offers a better overview and details of the design. AR and VR will aid designers in visualizing and developing their creativity. Author first selected a case study of a residential project and then created a model using AR and VR for further evaluation. The author will be

R. S. Sikarwar (✉) · A. G. Shelake
School of Civil Engineering, Dr. Vishwanath Karad MIT World Peace University, Pune, India
e-mail: raghav.singh322189@gmail.com

A. G. Shelake
e-mail: abhaysinha.shelake@mitwpu.edu.in

A. Vilventhan et al. (eds.), *Advances in Construction Materials and Management*, Lecture Notes in Civil Engineering 346,
https://doi.org/10.1007/978-981-99-2552-0_4

employing BIM-based models for deploying AR and VR technology. G + 20 residential buildings are taken into account in this. The author develops an AR and VR framework to improve collaboration, reduce delays, and simulate the construction process.

Keywords Delay · Construction Industry · Residential Construction · Augmented Reality (AR) · Virtual Reality (VR)

1 Introduction

1.1 General

By 2030, India is anticipated to overtake China as the world's most populous nation, according to the World Economic Forum. India may soon have a housing scarcity as a result of this enormous population growth. The majority of Indians find work in the massive and vibrant construction industry, which also contributes 5% of India's GDP and 78% of gross capital formation. Over the previous ten years, the growth of construction in India has already been well chronicled. This rise has largely been attributed to the effective planning and administration of construction projects. One of the most frequent issues in construction projects is a delay, which has a wide range of detrimental repercussions on both the project and the people involved. Miscommunication between clients, contractors, and consultants causes delays.

In India, delays in activities account for 70% of building projects not being finished in the allotted period [1]. Numerous factors, including approval, funding, worker strikes, contractor and subcontractor schedules, weather, poor communication, etc., cause residential projects in India to be delayed [2]. Among these, coordination between clients and contractors, manager and engineer, and engineer and supervisor, is the main cause of the delay. For the delay analysis, common techniques include time impact analysis, window analysis, as planned vs. as built, as planned vs. as built, and impacted as planned has been used [3]. The bar graph was used for the comparison of these methods. These methods are inefficient and require time to provide a solution.

The world has witnessed significant changes in a wide range of residential building aspects. The industry has made significant progress. New techniques, tools, and tactics are already being used in residential construction by the company to reduce the delay. Building information systems, artificial intelligence, machine learning, 3D printing, modular construction, augmented reality, etc. are some examples of innovative technology [4]. Most professionals rely on augmented reality and virtual reality, two technologies related to three-dimensional space, for the majority of their work [5]. These techniques can be applied to defect, time, and cost management. However, there is a lack of knowledge about how to combine AR and VR technologies for the management of construction activities and personnel coordination. This project will use AR and VR technology together to close this gap. The results of this study will benefit employees' coordination and activity management. This study's overarching goals are to enhance coordination, reduce delays, and reduce construction costs. The specific objectives of this study are:

- To identify factors causing delays in residential construction.
- To implement AR VR technology in reducing time and cost.
- Develop a clear understanding and improve coordination among different personnel.

The next section is all about the literature review. The third section is the methodology used in this research. The fourth section defines AR VR in the construction industry.

2 Literature Survey

The most important causes of construction delays according to the factor analysis were: lack of commitment, bad site management, poor site coordination, incorrect planning, unclear project scope, lack of communication and poor contract [6]. The factors that have a substantial impact on the project's total delay include the owner's slow decision-making, low labor productivity, the architects' resistance to change, and the need for rework as a result of construction errors [6]. Different projects and nations have different methods for identifying and prioritizing delay issues. For road infrastructure projects, limited contractor experience and payment delays were some important delay issues, whereas material shortages, communication, and contractor financial troubles were factors in building projects [7]. Understanding the type and implications of concurrent delays in the Indian industry is crucial. Practices for concurrent conflict settlement vary across India. Some methodologies, including Dominant Cause and Apportionment, are used throughout the world, however, they are not used in Indian industries [8]. Working in areas with high political risk entails particular difficulties, such as learning more about the relative effects of delays and the reasons why projects can create delays. It is impossible to create a methodology that addresses all delay-causing causes. Analysis of quantitative and qualitative data is superior to other approaches [9].

For many construction projects, the weather can pose a serious risk to the timetable, time-related considerations, quality of work considerations, or productivity considerations. The contractor and the owner may be better choices for estimating time extensions and hazards if a weather delay happens [10]. Discrepancies in the analysis of time extensions are caused by seven interrelated factors: normal weather, whether thresholds, type of work, lingering days, criteria for lost days, lost days equivalent due to lost productivity and work days lost versus calendar days lost. These factors might be interpreted in several ways in the delay analysis [11].

For complicated projects with complex systems, a method based on machine learning is preferable. An excellent set of methodologies are provided by machine learning for delay risk analysis. By improving the timeliness of building projects, data analytics solutions can be very beneficial for construction projects [12]. Infrastructure project risks have a significant impact on the timeline. A system dynamic model can be used to simulate different project scheduling hazards. For the investigation of risk implications on a project schedule, a system dynamics (SD) model is built. Six primary hazards were examined in the development of the SD model: client requests delayed project payments, constrained project schedules, faulty information, a lack of ability, and inferior management [13]. Delays are a common issue that affects project schedules and costs. Reason for noncompliance (RNC) indicates scheduling failures while Delay Index (DI) indicates time performance. This information will help project managers to make better decisions about delay causes and also help them in focusing on their work. RNC and DI will prevent the project from impacts of delay [14].

Project management in the construction of public housing and delay factors based on schedule extensions for 184 high-rise apartment blocks are used as examples in this case study. The causes of the 854-month delay include late site handover, variation orders, delays by other contractors, a lack of building materials, bad weather, and others. Additionally, there is a critical need to extend previous efforts to boost precast technology acceptance and automation levels to further raise construction productivity [15]. One of the best and most rapidly developing advanced methods in the construction industry is modular construction. Modern construction techniques must be viewed as an evolution of building employing fresh, cutting-edge approaches rather than a revolution. Traditional construction techniques should not be considered in danger from modern methods [16]. The advancement of new technologies in recent decades has changed the working environment and methods of construction.

AECO is a highly complex industry and the construction industry has already been driven toward new techniques and equipment by safety concerns, energy conservation, time management, and cost control. A survey covering concerns about BIM and its limitations, especially in data transfer, was conducted. The AECO industry lacked AR tools' efficacy and usability [17]. One of the potential technologies in the building sector is augmented reality. Reviewing 546 studies, it was discovered that tangible engagement was more frequently used than hybrid interaction. It also demonstrates how AR may be applied to management, process monitoring, and design evaluation. BIM and AR have been combined for improved visualization [18]. The AEC sector has been hesitant to embrace AR/VR technologies. The elder generations are more comfortable with AR/VR technology, according to responses from various companies. Additionally, they see more advantages in their application. In contrast to other industries, experts predicted that AR/VR technology will rise significantly in home construction [5].

54 experts from 36 industry organizations participated in surveys that were performed. Six AR and VR use cases—stakeholder engagement, design support, design review, construction support, operations, and management assistance, and training—were identified based on the information gathered. The use of AR and VR in the building is categorized and essential information for practitioners is also provided [19]. Applications of AR and VR to various construction issues are

described. Various tools, including AutoCAD, Navisworks, Primavera, etc., were employed for this. In this study, augmented reality is successfully used for scheduling construction projects, following their progress, teaching workers, addressing safety concerns, managing time and costs, and managing quality and defect management. Additionally, the application of virtual reality in visualization, worker training, safety management, and quality and defect monitoring is successful [20].

The application of AR and VR technologies to building project management has attracted a lot of research attention. However, the application of AR and VR combined is not feasible due to a lack of organized analysis. The author has concentrated more on the display technique, interaction device, spatial registration method, and level of immersion for AR and VR systems. Project scheduling and construction progress tracking AR and VR implementations have a significant potential to improve the current approaches [21]. Begins with a review and synthesis of research evidence for several VR/AR prototypes, products, and the related training and evaluation paradigms to better understand the state-of-the-art of VR/AR applications in construction safety (VR/AR-CS) and from which to uncover the related issues and propose possible improvements. While important application fields and trends for the VR/AR-CS research are summarized, certain technical aspects and types that could be used to improve construction safety are deduced and further developed [22].

For Health monitoring AR and VR can be useful along with BIM. Long term-monitoring and analysis of big data is always a major constraint in the field. BIM platforms bring together all the AEC systems that make up a structure in one location, allowing users to properly and effectively incorporate their varied characteristics. Along with AR, BIM has seen widespread adoption in the AEC sector [23].

Underground utility management and maintenance is another challenge in the construction industry. Facilities operation and maintenance (O&M) could advance and change thanks to BIM, which offers the necessary foundation for locating, evaluating, and processing data in a digitalized 3D environment. Before beginning excavation activities, contractors can use AR technology as an inspection tool to detect buried existing utilities in accordance with underground utility construction. This will help the contractors locate and properly identify these facilities [24].

2.1 Research Gap

Although investigations have being conducted in this area, there are some factors acting as constraints for the implementation of AR and VR technology on site. Developed countries seems to have already started adopting the new age technologies; however, developing countries are yet lagging. The limitations that seem to be stopping the implementation is lack of awareness, communication, coordination me and knowledge in India. Thorough study is therefore required to know whether this new-age technology could actually be helpful in future.

3 Methodology

The above framework describes the methodology, author considered for his research work (Fig. 1).

Step1: Main causes of the delay are identified from the previous literature survey.
Step2: Out of which the author specifically identified that first and foremost cause is lack of coordination in between different C's of the project like contractor, consultant and client. In order to support this another step is to take a case study where this reality is validated for this one residential building and commercial building is considered. On which around 20% is caused because of lack of coordination between client, contractor and consultant.
Step3: Here in order to validate the AR and VR work, Panchsil project (G + 22) is considered which is located in Nahur road Mulund(w), Mumbai. Construction was started on 4/11/2020.
Step4: Initially for the reality 2D drawing is formed with the help of AutoCAD 2D software then the model is initially put to 3D model for each percentage of work.
Step5: A 3D model is prepared in Revit software and with the help of Augin Plugin the visualization of model gets better.
Step6: Through enscape plugin, activites monitoring can be observed after wearing VR headset.
Step7: A particular activity is considered on a day for validating whether the coordination is improved in between these spheres or not with respect to utilization of AR and VR.

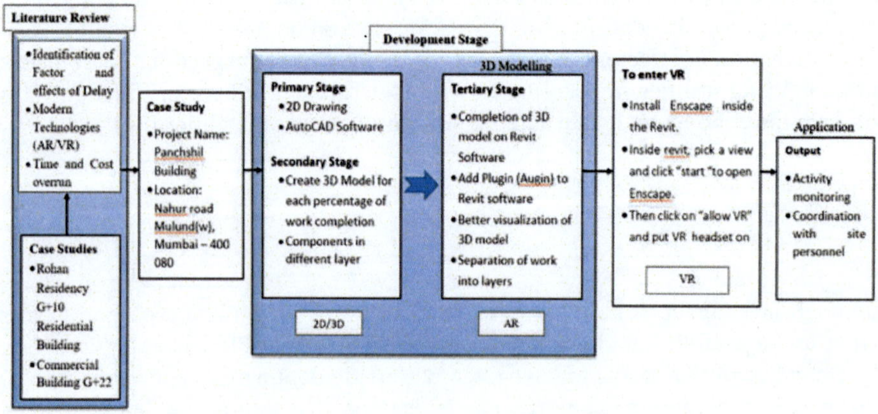

Fig. 1 Research Framework

4 AR and VR in the Construction Industry

4.1 Augmented Reality in Construction

For an engaging experience, augmented reality integrates computer-generated content and the physical world. The potential of augmented reality to deliver real-time information can be used to boost productivity, enhance safety, simplify communication, and control expenses. The major applications of AR are already in project planning, training of employees, and safety management.

- Project Planning: AR has significantly improved the scheduling and planning aspects of building projects. AR may be used to create 3D models on a two-dimensional surface. AR is most effective for keeping an eye on and tracking activity on a PC or mobile device.
- Training of employees: To sustain the pace and standard of residential construction, worker training is crucial. Workers receive instruction using AR technology on how to use and operate equipment like cranes and excavators on the job site. Additionally, it aids in their comprehension of the layouts and designs of the various activities.
- Safety Management: A higher accident risk exists in residential buildings with multiple stories. To manage safety, a framework is created to integrate building information system models with augmented reality. Workers need to recognize the risks that are present on the job site, and AR helps them do so by educating them about the risks [20].

4.2 Virtual Reality in Construction

VR is a visualization tool used in the construction sector for a diverse range of purposes, including design, management, training, etc. The virtual world can be experienced thoroughly by VR users. Users can see a project before it is built thanks to VR technology. Most residential projects frequently fail due to the field personnel's incapacity and lack of expertise, as well as that of the engineers and workers.

Virtual reality (VR) is made up of an interactive computer simulation that lets users engage with a created world physically by tracking their movements and positions. Architecture and design, building, training and education, landscape and urban planning, engineering, facility management, and lifecycle integration are some of the application areas of VR in the built environment that can be categorized. In contrast to AR, which overlays additional visual information on the real world, VR completely removes the real aspects so that the user is immersed in a virtual environment with virtual objects. The ability of VR and AR to enhance the delivery of construction projects is a key driver in the industry's adoption of these technologies.

5 Case Study

All data generated and analyzed during this study are available from a construction project named Panchshil which is Construction of Bldg. on Land bearing CTS No. 555 (pt.) of village Mulund situated at Nahur road Mulund(w), Mumbai – 400 080). The deadline of the project is December 2023.

Total area: 6333.38 sq.mt.
Total height: 69.85 M
Flat number: 100 Nos.
Parking floors: Separate Parking tower will be constructed for 30 Parking.
Refuge floor: Partial area reserved at 7th, 14th & 21st floor.
Project starting and ending date: 4/11/2020 to 12/2022.

A detailed 2D plan is prepared in AutoCAD of Panchsil reality floor plan which is mentioned in Fig. 2.

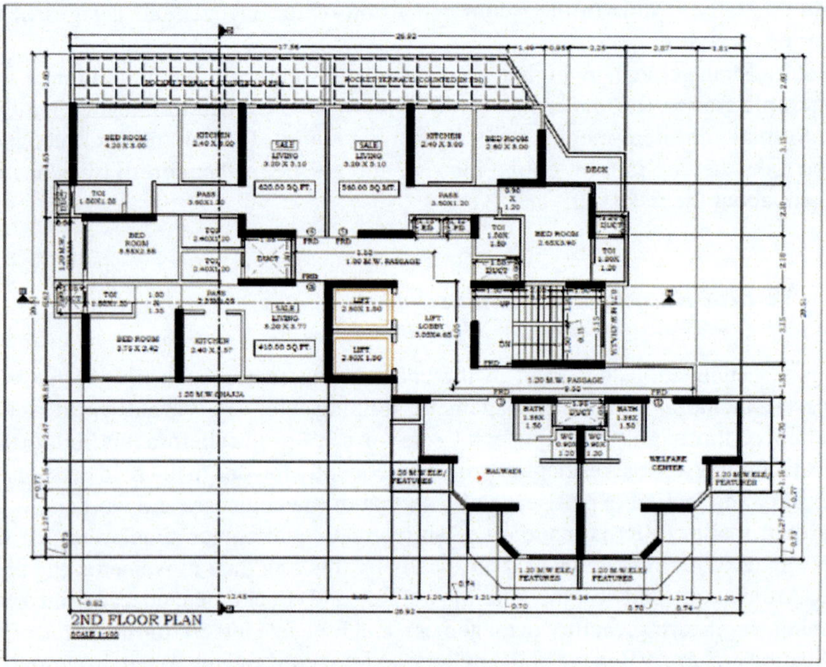

Fig. 2 2D Floor Plan of the project Panchshil

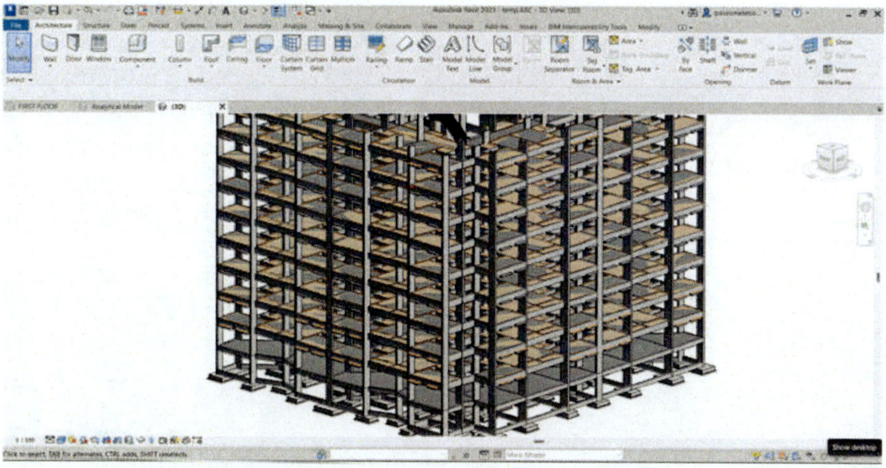

Fig. 3 View through augin plugin

5.1 Observations and Results

5.1.1 AR/VR Models

The author created a structural model of the building using BIM-based modelling. The 3D modelling process uses the Revit program. In that Augin and Enscape plugins have been installed. The Enscape plugin represents virtual reality, whereas the Augin plugin represents augmented reality. Site personnel can see the external work of the building through Augmented reality (augin plugin) and the internal walk-through by virtual reality headset.

Figure 3 depicts the project's structural work, which comprises both substructure and superstructure. Additionally, it contains structural drawings and architectural drawings such detail elevations and plans, among others. We may also keep an eye on the progress of the project and compare it to the schedule.

Virtual reality systems create a realistic perspective of building that a user can experience digitally using virtual reality headsets. One can travel about the site and engage in the activities taking place there while wearing a VR headset. Every single activity can be observed in front of the VR headset's display. Figure 4 illustrates the project's internal detail work.

5.1.2 Delay Analysis

The author created a detailed percentage-wise model and divided it into 100 small models representing each 1% of the work. The project manager asked to take a 7 -day review concerning this model. The models will help the manager in monitoring

Fig. 4 Inside view of the building after wearing a VR headset

the activities. On 4th October around 27% of the work was delayed on the 700th day. This delay was before the use of AR and VR framework implementation. As per planning, only the 510th day of work was finished on the 700th day from the start of the site (Table 1).

The researcher considered the next 30 days for validation of the VR framework on the given residential construction site. On the 700th day of the site i.e. 4th October, the AR VR framework is given to the contracting firm as they were interested to follow it for one month to see the changes. For these 30 days, activities mentioned in table number 1 were planned as per the original planning. However, as already site was delayed the project manager planned to pace up the work and finish work on to 552nd day. Delays at the end of the 700th and 730th days are calculated below,

$$\frac{Actual\,Days - Planned\,Days}{Actual\,Days} \times 100$$

- o Delay on the 700th day of the construction

$$\frac{700 - 510}{700} \times 100 = 27\%$$

- o Delay on the 730th day of the construction

$$\frac{730 - 552}{730} \times 100 = 24\%$$

Here, 1% of work represents 7 days of work. As per planning, 63% to 67% of work is to be done between the 510th day to 540th day. As per framework guidelines, they are

Table 1 Difference of work planned vs work performed

Sr. No	Activity Work	Work complete till 510th day	Planned work 510 to 540	Actual work completed 510 onwards to date	AR/VR framework additional work
1	R.C.C Superstructure	17th floor	18th floor	19th floor	On the 7th, 14th, 21st, and 28th, a day less work was detected by PM by using VR Framework, and on the 8th, 15th, 22nd, and 29th, day respectively additional work is assigned to the workforce
2	Blockwork and Door frames	11th floor	14th floor	16th floor	
3	Electrical Work	9th floor	12th floor	14th floor	
4	Plumbing Work	8th floor	10th floor	12th floor	
5	Internal Plaster	6th floor	7th floor	8th floor	
6	Gypsum Work	5th floor	6th floor	7th floor	
7	External Plaster	5th floor	6th floor	7th floor	
8	External Painting	3rd floor	4th floor	5th floor	
9	Internal Work	1st floor	3rd floor	4th floor	
10	Site Handover	-	-	-	
	510th day		540th day	552nd days	

supposed to monitor the site after 7 days i.e. 4 times a month by which the PM can easily understand what activities are remaining at the end of 64%, and accordingly, he can increase the resources and speed up the pace of the work to appropriate duration. As at the end of the 730th day from the start of the site project manager succeed in achieving the planned work of 30 days, the validation of the VR and AR framework is proved for minimizing delay on the construction site.

6 Conclusion

- The most common reason behind the increase in the project cost is delay; which itself has numerous reasons behind. With the current advancement in technology the construction industry can utilise it for its benefit and try to solve this issue.
- AR/VR could be one of the best approaches to solve this issue.
- Keeping in mind the future development & rapid growth, it is extremely necessary to make utmost use of such modern tools and technologies in the construction sector.

- AR and VR are technologies that can be used for improving coordination among employees and help the manager in reducing the delay. Preparation of one major model and its sub models will give realistic experience of exterior and interior activities or work of the construction.
- AR and VR technology has various applications but it was absent in the construction sector. This paper has focused on the implementation of AR and VR in the construction field.
- Through the above study it is found that by the application of AR/VR in construction industry has some benefits to offer and hence it could be a game changing thing in the future for the construction industry.
- AR and VR is the best technology for monitoring the activities on site. If applied before the start of construction, it can be helpful in minimizing the delay. It is also best suited for training and management of the activities.
- User just required a VR headset to experience the reality of the construction.

References

1. Assaf, S. A., & Al-Hejji, S. (2006). Causes of delay in large construction projects. *International Journal of Project Management, 24*(4), 349–357.
2. Durdyev, S., Hosseini, M.R. (2019). Causes of delays on construction projects: a comprehensive list. International Journal of Managing Projects in Business **13**, 20–46.
3. Al-Gahtani, K. S., & Mohan, S. B. (2011). Delay analysis techniques comparison. *Journal Civil Engineering Architecture, 5*(8), 740–747.
4. Abioye, S. O., Oyedele, L. O., Akanbi, L., Ajayi, A., Delgado, J. M. D., Bilal, M., & Ahmed, A. (2021). Artificial intelligence in the construction industry: A review of present status, opportunities and future challenges. *Journal Building Engineering, 44*, 103299.
5. Noghabaei, M., Heydarian, A., Balali, V., & Han, K. (2020). Trend analysis on adoption of virtual and augmented reality in the architecture, engineering, and construction industry. *Data, 5*(1), 26.
6. Doloi, H., Sawhney, A., Iyer, K. C., & Rentala, S. (2012). Analysing factors affecting delays in Indian construction projects. *International Journal of Project Management, 30*(4), 479–489.
7. Sánchez, O., Castañeda, K., Mejía, G., Pellicer, E. (2020, March). Delay factors: a comparative analysis between road infrastructure and building projects. In Construction Research Congress 2020: Project Management and Controls, Materials, and Contracts, pp 223–231. American Society of Civil Engineers.
8. Munvar, C., Mengistu, D. G., & Mahesh, G. (2020). Concurrent delay analysis: methods, case law, and expert perception. *Journal of Legal Affairs Dispute Resolution Engineering. Construction, 12*(1), 04519035.
9. Kadry, M., Osman, H., & Georgy, M. (2017). Causes of construction delays in countries with high geopolitical risks. *Journal Construction Engineering Management, 143*(2), 04016095.
10. Ibbs, W., & Kang, J. M. (2018). Weather-related delay provisions in public transportation construction contracts. *Journal of Legal Affairs Dispute Resolution Engineering Construction, 10*(3), 04518009.
11. Nguyen, L. D., Kneppers, J., García de Soto, B., & Ibbs, W. (2010). Analysis of adverse weather for excusable delays. *Journal of Construction Engineering and Management, 136*(12), 1258–1267.

12. Gondia, A., Siam, A., El-Dakhakhni, W., & Nassar, A. H. (2020). Machine learning algorithms for construction projects delay risk prediction. *Journal of Construction Engineering Management, 146*(1), 04019085.
13. Wang, J., & Yuan, H. (2017). System dynamics approach for investigating the risk effects on schedule delay in infrastructure projects. *Journal of Management in Engineering, 33*(1), 04016029.
14. Gonzalez, P., González, V., Molenaar, K., & Orozco, F. (2014). Analysis of causes of delay and time performance in construction projects. *Journal of Construction Engineering Management, 140*(1), 04013027.
15. Kog, Y. C. (2018). Project management and delay factors of public housing construction. *Practice Periodical on Structural Design and Construction, 23*(1), 04017028.
16. Kyjaková, L., Mandičák, T., & Mesároš, P. (2014). Modern methods of constructions and their components. *Journal of Engineering and Architecture, 2*(1), 27–35.
17. Sidani, A., Dinis, F. M., Duarte, J., Sanhudo, L., Calvetti, D., Baptista, J. S., & Soeiro, A. (2021). Recent tools and techniques of BIM-based augmented reality: a systematic review. *Journal of Building Engineering, 42*, 102500.
18. Chen, K., Xue, F. (2020). The renaissance of augmented reality in construction: history, present status and future directions. Smart and Sustainable Built Environment **11**(3), 575–592
19. Delgado, J. M. D., Oyedele, L., Demian, P., & Beach, T. (2020). A research agenda for augmented and virtual reality in architecture, engineering and construction. *Advanced Engineering Informatics, 45*, 101122.
20. Ahmed, S., Hossain, M. M., & Hoque, M. I. (2017). A brief discussion on augmented reality and virtual reality in construction industry. *Journal of System and Management Sciences, 7*(3), 1–33.
21. Albahbah, M., Kıvrak, S., & Arslan, G. (2021). Application areas of augmented reality and virtual reality in construction project management: A scoping review. *Journal Construction Engineering, 14*, 151–172.
22. Li, X., Yi, W., Chi, H. L., Wang, X., & Chan, A. P. (2018). A critical review of virtual and augmented reality (VR/AR) applications in construction safety. *Automation in Construction, 86*, 150–162.
23. Sadhu, A., Peplinski, J. E., Mohammadkhorasani, A., & Moreu, F. (2023). A Review of Data Management and Visualization Techniques for Structural Health Monitoring Using BIM and Virtual or Augmented Reality. *Journal of Structural Engineering, 149*(1), 03122006.
24. Shekargoftar, A., Taghaddos, H., Azodi, A., Nekouvaght Tak, A., & Ghorab, K. (2022). An integrated framework for operation and maintenance of gas utility pipeline using BIM, GIS, and AR. *Journal of Performance Constructed Facilities, 36*(3), 04022023.

Identification of Challenges Influencing the Adoption of Building Information Modelling (BIM) and Facility Management for Metro Rail Projects in India

Ahmad Alothman, Saiteja Kudikala, and Aneetha Vilventhan

Abstract The architecture, engineering, and construction industry have recently bccome more aware of building information modeling (BIM). Facilities management (FM) is one particular application of BIM. However, existing literature has shown that the applications of BIM in FM practices are yet to be widely adopted specifically for large-scale projects. This study aimed to identify challenges and barriers influencing the adoption of building information modeling (BIM) and facility management for metro rail Projects in India. Thus, the authors suggested different sequence research methods to gather the data, and both qualitative and quantitative approaches were used. Qualitative data were collected and analysed based on a literature review of similar studies, and semi-structured interviews followed by an online survey with a total of 81 participants. As a result of this research, factors that influence the adoption of BIM to enable FM for metro rail projects have been investigated. These findings may assist BIM implementation and FM professionals to establish a baseline and find solutions to enable implement BIM successfully in the area of FM activities for metro rail projects in India.

Keywords BIM · Facility management · Metro rail projects · BIM implementation · Construction industry

A. Alothman (✉) · S. Kudikala · A. Vilventhan
Department of Civil Engineering, National Institute of Technology, Warangal 506004, India
e-mail: aa712190@student.nitw.ac.in

A. Vilventhan
e-mail: aneetha@nitw.ac.in

A. Vilventhan et al. (eds.), *Advances in Construction Materials and Management*, Lecture Notes in Civil Engineering 346,
https://doi.org/10.1007/978-981-99-2552-0_5

1 Introduction

The AEC industry has undergone a paradigm shift as a result of the advent of BIM. Building information modeling (BIM) is a process for digitally representing a building's physical and functional attributes. The breadth of BIM extends beyond this; it is a comprehensive strategy for planning, executing, managing, and collaboratively maintaining the construction project while utilizing a single, cohesive digital modeling framework. BIM is currently recognized as helpful during the building's design and construction phases [1].

To manage the building's life cycle and make any necessary decisions, the facilities manager must have the right information on the building's design and other components. BIM makes it possible for professionals to access a centralized database that can serve as the basis for prolonging the life of the construction. BIM provides information on the structure, plumbing system, floor plan, space requirements, furniture, equipment, and electrical systems of the building, as well as details on the service life, manuals, and replacement schedules for each component. With this level of specificity and at the lowest possible cost, managers may achieve operational efficiency [2].

Facility maintenance management (FMM) during the operational stage of a facility's life cycle has grown in significance as a subject of inquiry and academic study. Managing facility maintenance data is essential to effective facility management (FM). Due to the varied types of equipment and facilities, managing FMM work efficiently during the operation phase can be very challenging. By integrating people, place, process, and technology, the multidisciplinary field of facilities management ensures the functionality of the built environment. Facility management is subject to growth and continuity of business, development and innovation, customer relation, and cost reduction. A variety of property and user-related responsibilities are grouped under the term "facilities management" (FM) for the benefit of the business as a whole and its workforce [3].

Currently, there is little use of digital technologies in Indian metro rail projects. Except for the Nagpur metro, India's other metro rail projects have only just begun to explore digital technology. One of the most significant aspects of a typical metro rail project is operations and maintenance, which has been made more apparent by the rapidly expanding metro market in India and the rest of the world. Maintenance is the most fundamental component of a successful and well-maintained metro rail system. Everyone is aware that a metro train has a large number of moving parts in addition to tasks and operations to coordinate. The equipment must always be in good working order if the Metro Station administration wishes to be dependable and secure for its clients. This goal can be achieved by planning proper facilities management [4].

This paper investigates the potential of BIM to enable FM for metro rail projects and explores the barriers that have prevented their adoption. The central objective of this study was to determine the challenges and barriers influencing the adoption of building information modelling (BIM) and facility management for Metro Rail

Projects. To obtain this goal, the authors conducted an in-depth literature review to identify the critical factors that influence the adoption of BIM and FM, followed by an expert interview for evaluation and exploratory process. Eventually, the online questionnaire survey was developed to analyze and rank the factors. Finally, this study discusses the results obtain from SPSS software and highlight the significant factors that affect the adoption of BIM and FM for metro rail projects.

2 Literature Review

A fundamental understanding of facility management (FM) and how it might be combined with building information modeling (BIM) is provided by this literature review. It also describes the limitations and peculiarities of the traditional approaches used for BIM-FM integration. It shows the need for research into BIM-enabled FM for metro projects in India and how leveraging BIM-FM platforms can enhance the integration of BIM and FM.

2.1 Review of the Indian Metro Rail Projects

The use of building information modeling (BIM) in metro rail projects is in the initial stage in India. By using BIM, examined the embodied and operational energy consumption of an elevated metro rail station in Ahmadabad, India. Listed the four most important and influential KPIs. They simplified access to real-time data, ensured data interoperability and compatibility, reduced claims and disputes, and increased information accessibility and accuracy through BIM [5].

[6] stated that one of the main difficulties in building metros was making design information accessible and reliable, especially when it came from many sources. The only metro rail project in India at the moment that has made a considerable effort to use BIM in its planning and design phase is Nagpur Metro. Even though the client, Nagpur metro rail Corporation Limited (NMRCL), did not require the inclusion of digitalization, private technology providers and the general contractor took the initiative in delivering a BIM-integrated project [7]. Digital information management systems are used throughout the Nagpur metro Project's lifetime. The project's BIM plan was modeled after the BIM strategy created for the Cross rail project and was based on the PAS1192 standards. Except for the Nagpur metro, India's other metro rail projects have only just begun to explore digital technology. Metro projects confront several difficulties that can be resolved by digitization. Digital land use maps can aid with planning, automated or generative design can quickly provide alternatives, 3D and 4D models can ensure that work is completed on the sites promptly, and so on [5].

2.2 Overview of BIM in FM Applications

The architectural, engineering, construction and operation (AECO) phases of a project are where building information modeling (BIM) has been employed most frequently. Building information modeling (BIM) aims to integrate activities across a building's whole lifecycle as a lifecycle evaluation concept [2]. Geometry, geographical linkages, geographic information systems, and the number of different building components are all included in building information modeling (BIM). As construction projects require sustainability, efficiency, and cost reductions throughout the project life cycle, BIM is incorporated inevitably into the AEC industry. Three phases the pre-construction phase, the construction phase, and the post-construction phase can be used to categorize the implementation of BIM. Early cooperation, a productive platform for information sharing, and accurate cost and material take-offs based on the 3D model are all used in the pre-construction phase of BIM [8].

Since the development of building information modeling (BIM) procedures and the claim that BIM data collected throughout the facilities lifecycle can enhance the effectiveness of facility management. This argument has taken on new significance (FM), [9] highlighted that BIM's benefit in FM comes primarily from: (1) Improvements to the manual information transfer techniques now in use and to the correctness of FM data, (2) An improvement in the speed of data access and intervention location during the execution of work orders. The usage of BIM for building maintenance operations will be greatly influenced by how the model is populated, the information that is present in the model about electrical, M&E, and equipment issues, and how this data in the model is managed over time. Looked into the difficulties in transferring information between FM systems and building information modeling (BIM). Building SMART created IFC, an open-source data format, to address the interoperability issues in building projects [10]. COBie (Construction Operations Building Information Exchange) is one the most frequently employed standards for information exchange from the planning, and construction stages to the Operations and Maintenance (O&M) phase. There are two ways to produce COBie data when utilizing BIM software: 1) Transfer an XML file from BIM software to an Excel file and back again by exporting an IFC or IFCXML file. 2) Produce a COBie file directly from the BIM software, in which case, based on the BIM software, a file conversion module may be added as a plug-in for use [11]. [12] proposed that outlines the procedures used to determine the needs, create, and put into use a central handover and operation-specific model (BHIM). The framework based on project characteristics contextualizes the Construction Operation Information Exchange (COBie) and the Level of Development requirements, two well-known open BIM schemas. The data is more accurate and reliable, and the higher the LOD. [13] highlighted that of the large data sizes present in large-scale facilities and the limits of the BIM authoring tools, it was determined that using BIM models as data banks or connecting CMMS systems with BIM models for displaying data options was not technically viable. [8] highlighted that Maintenance personnel onsite could not use high-end desktop computers for operating BIM models successfully during the maintenance and inspection procedure. BIM models must be converted into smaller files and sent via mobile devices, which are more frequently utilized on construction sites. [14] according to the author, BIM technology has a lot of potentials to improve FMM work. BIM and web technology integration enables FM managers and maintenance employees to efficiently monitor and manage the entire FMM process. For the BIM model to fully deliver value to facility owners after handover, [15] identified four qualities or requirements: In addition to being data-centric, the model also has accurate and full graphics, is systems-centric, and enables direct integration with the owner's facilities management system.

3 Research Methodology

3.1 Identification of the Challenges of BIM-FM for Metro Rail Projects

Since previous studies have shown that the implementation of BIM for FM in metro rail projects is still in the early stage, and there are limited empirical data available in this research area. Thus, the authors suggested different sequence research methods to gather the data, and both qualitative and quantitative approaches were used. Qualitative data were collected and analysed based on a literature review of similar studies, and semi-structured interviews followed by an online survey with 81 participants to identify the current state of BIM implementation in FM, challenges, and barriers to of utilization BIM and FM for metro rail projects. Detailed descriptions of each method are provided in the following subsections. The extensive literature review of similar studies has been discussed in the previous part.

3.1.1 Expert Interviews

The face-to-face interviews and online meetings with (10) industry professionals in various BIM and FM applications areas, conducted from September to October 2022, provided significant information on the metro rail project operation and maintenance stage. The experts who would be involved in the interviews would have to satisfy the following selection criteria:

- The interviewee should have a minimum of 5 years of experience in FM for Metro rail projects
- The interviewee should implement BIM in the metro rail project.

The main aim of this interview is to (1) identify the current facility management practices in Metro Rail Projects, (2)to explore the level of awareness of using BIM and FM for metro rail projects, and (3) challenges and barriers to the adoption of BIM as a platform for transferring design, operational, and maintenance information

Table 1 List of the experts for the interview survey

Expert	Occupation	Employer	Years of work experience	Number
A	Manager	Facility company	8–10	2
B	Manager	BIM company	12–16	3
C	Engineer	BIM company	10–15	2
D	Engineer	Construction company	15–20	3

Table 2 Summary of factors influencing the adoption of BIM and FM for Metro Projects

Code	Factors	References
A1	Incomplete national standard	
A2	The high initial cost of implementing BIM for FM	
A3	Lack of awareness of the benefits of BIM in the FM	
A4	Using BIM only for Design and construction, not for FM	
A5	Current contractual methods might not be appropriate for BIM in the FM industry	[5–15]
A6	High cost of training and education	
A7	Inadequate software interoperability	
A8	lack of FM consideration at the early design stage	
A9	Lack of FM data required in BIM format	
A10	FM stakeholders trustworthy the traditional strategies of FM. So, there's little motivation to use BIM	

of metro project components. Table 1 shows the list of experts and their affiliations. All the selection experts had previous experience with BIM and FM and had closely worked with the BIM implementation in operations and maintenance. Furthermore, most of them had more than 10–20 years of experience. After analysis of the crucial factors Table 2 shows the summary of factors influencing the adoption of BIM and FM for Metro Projects.

3.1.2 Online Survey

A new questionnaire was created based on the results of the expert survey. This survey was distributed through a variety of channels to ensure maximum participation by a wide range of BIM and FM professionals with rich experience in metro projects to increase its representativeness, starting through (1) lists of BIM and FM associations, (2) online networks of industrial associations, including India BIM Association, facility management association of India. During the two monthly survey periods, respondents were able to participate between September 2022 and November 2022. A total of (89) responses were collected and (81) responses were eventually extradited and included in the analysis with the total response (91%) rate. To ensure the survey was credible, omitted respondents who had a non-background in BIM, which is important to all participants also the responses that had a poor experience in FM, or metro rail projects were excluded from the survey. Moreover, it is significant to note that more than (70%) of the responses were from a professional background from academic and recognized organizations and most of them have close connections to the government, industry, and private agencies for consultancy services regarding metro rail projects. The online survey was intended to compile the standpoint of BIM, and FM professionals from industry practitioners to understand the challenges of integrating BIM and FM for Metro rail projects, the type of problems that occur during the operation and maintenance stage, and the possibility to solve it in the early stage of the design. The focus of this survey was to answer the following question:

1. What are the current practices for facility management in the Metro rail project?
2. What are the current practices in the Handover process of the Metro rail project?
3. What challenges/limitations/barriers exist to hinder the application of BIM technology to enable FM?

To achieve the objective of the study, a Relative Importance Index (RII) was adopted as an appropriate analytical method to explain the ranking of responses, which is relatively familiar in the construction management literature. As a result, the questionnaire ratings analyzed the influence factors that affect of adoption of BIM and FM for metro rail projects, the calculation was carried out using the RII formula in Equation

$$\text{Relative Importance Index, } \mathbf{RII} = \frac{\sum \boldsymbol{W}}{\boldsymbol{A} \times \boldsymbol{N}}$$

W represents the ratings given by respondents to each factor, (from 1 to 5), A is the highest score (i.e., always 5) and N is the total number of responses [16].

4 Research Results

4.1 Questionnaire Results

Demographic Survey Distribution

According to the study's results, Table 3 shows the respondents' professions, the type of company where they work, and the building categories. Around 24% are Architect and Engineers (Civil or MEP), 21% are BIM Professionals, 12% are Facilities Managers, 11% are Facilities Engineer, 10% and 7% are Consultant and Contractor while 9% are Academic and 6% for other domains. Where about 44% of the respondents are from BIM and FM rich experience backgrounds. In addition, around 41% of the respondents are from BIM companies and FM private organizations. Furthermore,

Table 3. Shows the respondent's backgrounds.

Position	Percentages (%)	Type of company	Percentages (%)	Building categories	Percentages (%)
Architect	9	Architecture	6	Infrastructure	15
Engineer	15	Government	7	Metro Projects	20
Consultant	10	Private sector	16	Airports	6
Contractor	7	Private FM	19	Hospitals	9
BIM Professional	21	BIM company	22	Industrial	5
Facilities Manager	12	Contractor	7	Multi-Residential	16
Facilities Engineer	11	Construction	9	Universities	11
Academic	9	Education	7	Residential	14
others	6	others	6	Other	5

35% of the respondents had their project in the Infrastructure metro rail projects, while 20% of respondents had their project in the Airports, Hospitals, and Industrial, and 41% of the respondents had their project in Multi-Residential Universities Residential. In the survey, respondents were asked about their organization's size, around 26% were small, 35% from medium, and 39% form large.

Reliability Analysis

For this analysis, a coefficient alpha (Cronbach's alpha) test was applied to examine the reliability of the factors extracted from the rating questions. The SPSS software was used to examine the 10 variables and their associated Likert scales to see if they consistently represented the construct studied. It was found in the analysis that the results of the five-point Likert-type test were reasonably reliable based on Cronbach's coefficient alpha of 0.923, which represents excellent internal consistency. An acceptable Cronbach's alpha coefficient is 0.7 [17].

Mean Ranking

According to the respondents, the mean rank reveals the top four most important factors that influence the adoption of BIM and FM which is related to the strategy of using BIM for design and construction only, the initial cost of adoption of BIM, the traditional methods of contracting does not appropriate BIM for FM and finally the Incomplete national standard.

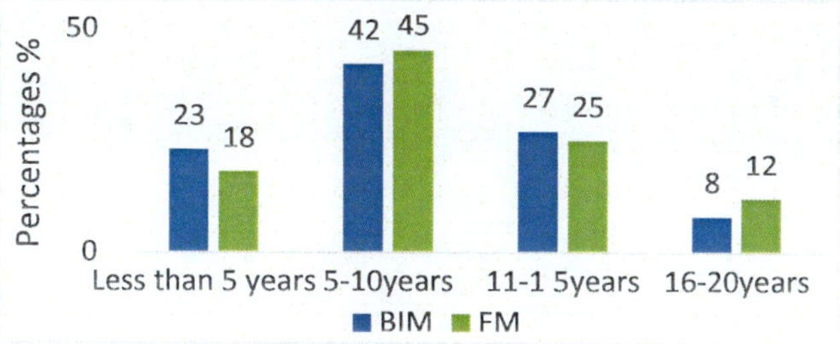

Fig. 1 The respondent's years of experience in BIM and FM

Fig. 2 Shows the awareness of using BIM and FM for metro projects

4.2 *Status of BIM Enables FM for Metro Projects Practices*

BIM and FM Experiences

In the survey, respondents were asked about their years of experience and knowledge in BIM, which is shown in Fig. 1. That around 27% of the participants have 11 to 15 years of rich experience in implementing and using BIM in their projects, and over 65% of them have less than 10 years, while there are around 8% of them have more than 16–20 years of experience. Furthermore, the respondents were asked about their years of experience and knowledge in FM, and the results indicate that around 25% of the respondents have 11 to 15 years of experience in FM, while 63% of the respondents have less than 10 years and around 12% of them have less than 20 years.

Awareness of using BIM

Respondents were asked to scale their awareness of using BIM tools to support the facility management practices and software during the operation stage. Figure 2 shows, that around 64% of the respondents were not satisfied, about 17% were Somewhat Satisfied with the current practices of delivering BIM-FM projects, and around 19% of the respondents were satisfied to extremely satisfied. This indicates that approximately 81% of respondents hope to raise the project's stakeholder awareness to enable BIM for FM in metro projects.

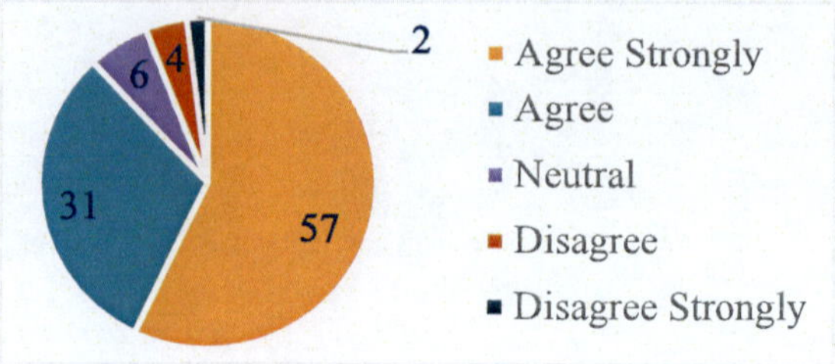

Fig. 3 Shows the ability of BIM and FM to improve the handover process of metro project

BIM-FM Improves the Handover

In addition, respondents were asked to scale their opinion of improving the FM handover process of metro rail projects using BIM. Figure 3 shows that around 57% of the respondents are agreeing strongly and about 31% Agree, 6% are Neutral and 6% Disagree. The results mention that approximately 88% of the respondents are supporting the use of BIM as a rich database, a more effective coordination method, and permit a more efficient handover of information at the end of the project that can be referred them as the basis for maintaining the building's life.

5 Discussion and Analysis

As this study focuses on determining the factors that influence adopting BIM and FM in metro rail projects, the respondents were requested to scale the significance of each factor using the Likert scale with a five-point scale [18].

Relative Importance Index (RII)

As this study focuses on determining the factors that influence adopting BIM and FM in metro rail projects, 10 factors were identified (from the extant literature reviewed, and expert interview). SPSS software was used to enter the data obtained and analyse it using the RII method. RII ranges from (0 to 1), with 0 not included shown in Table 4. A higher RII value indicates a more important consideration of the criteria, and vice versa argues that the transformation matrix represents an evaluation of RII, including appropriate significance levels and importance levels as a function of RII [16].

Table 4 Classification of RII

Scale	Level of contribution	RII
1	Very low	$0.0 \leq RII \leq 0.2$
2	low	$0.2 < RII \leq 0.4$
3	Average	$0.4 < RII \leq 0.6$
4	High	$0.6 < RII \leq 0.8$
5	Very high	$0.8 < RII \leq 1.0$

The respondents were requested to scale the significance of each factor using the Likert scale with a point scale. According to the respondents, the four influence factors are, first using BIM only for design and construction, not for FM, second, the high initial cost of implementing BIM for FM, third current contractual methods might not be appropriate for BIM in the FM industry, and finally incomplete national standard with an RII of (0.886%), (0.863%), (0.834%), and (0.806%) respectively. Table 5 shows the mean, standard deviation, and RII for each factor.

Ranking of factors.

According to the RII value, the influence factors had been ranked from (1 to 10) a higher RII value gets the number 1 while the lower gets 10. In addition, the level of contribution had been mentioned with the reference to the RII classification Table 6. With a very high level of contribution. This is followed by the influence factors with high levels of contribution.

The above Table 6 shows that "Using BIM only for design and construction, not for FM" is ranked in the first position with RII 88.6%. The level of contribution of this factor is hence considered very high. This result shows the current application of BIM in the construction industry in India. Owners and facility managers lack sufficient BIM expertise, These results are in line with the conclusion of [4].

"The high initial cost of implementing BIM for FM" is ranked in the second position with RII 86.30%. Thus, the level of contribution of this factor is also very high. The cost of implementing BIM is largely determined by a number of cost factors, including the purchase of software, technical support, hardware, services, training, and implementation contingencies. On average, software expenses made up around 55% of all implementation costs. These results are in line with the conclusion of [19].

"Current contractual methods might not be appropriate for BIM in the FM industry" is another factor that is ranked in the third position with an RII 83.40%. Level of

Table 5 Shows the mean, standard deviation, and RII for each factor

S/N	Factors influence adopting BIM for FM in metro projects	Mean	Std. dev	RII (%)
1	Incomplete national standard	4.03	0.95	0.806
2	The high initial cost of implementing BIM for FM	4.31	0.89	0.863
3	Lack of awareness of the benefits of BIM in the FM	3.94	1.02	0.789
4	Using BIM only for Design and construction, not for FM	4.43	0.84	0.886
5	Current contractual methods might not be appropriate for BIM in the FM industry	4.17	0.95	0.834
6	High cost of training and education	3.46	0.91	0.691
7	Inadequate software interoperability	3.80	1.12	0.76
8	lack of FM consideration at the early design stage	3.69	0.98	0.737
9	Lack of FM data required in BIM format	3.63	0.90	0.726
10	FM stakeholders trustworthy the traditional strategies of FM. So, there's little motivation to use BIM	3.54	1.08	0.709

Table 6 Shows the ranking and level of contribution for all factors

S/N	Factors influence adopting BIM for FM in metro projects	RII (%)	RII-ranking	Level of contribution
1	Using BIM only for Design and construction, not for FM	0.886	1	Very H
2	The high initial cost of implementing BIM for FM	0.863	2	Very H
3	Current contractual methods might not be appropriate for BIM in the FM industry	0.834	3	Very H
4	Incomplete national standard	0.806	4	Very H
5	Lack of awareness of the benefits of BIM in the FM	0.789	5	H
6	Inadequate software interoperability	0.760	6	H
7	lack of FM consideration at the early design stage	0.737	7	H
8	Lack of FM data required in BIM format.	0.726	8	H
9	FM stakeholders trustworthy the traditional strategies of FM. So, there's little motivation to use BIM	0.709	9	H
10	High cost of training and education	0.691	10	H

contribution of this factor is also very high. Implementation of BIM has the problems of contractual framework and this issue is more sensible for BIM-FM integration. The BIM-FM traditional contract deliverable must be updated [4].

"Incomplete national standard" is ranked in the fourth position with RII 80.60% with the level of contribution as very high. There are not any specific national standards for the implementation of BIM for facilities management in India. Because the research is still at a the beginning stage. There are currently no substantial legal restrictions on the widespread use of BIM-based design. This could result in confusion throughout the investor-designer-executive chain's findings that support these statements [14].

"Lack of awareness of the benefits of BIM in the FM" is ranked in the fifth position with an RII 78.90% and level of contribution as High. Hard data and best practice case studies that demonstrate the advantages of BIM for FM are completely lacking. Concerning opportunities and rewards, the current case studies are untrustworthy and lacking in science; they fail to mention the difficulties. Facility managers need concrete, reliable data to convince owners to use BIM [6].

"Inadequate software interoperability" is ranked sixth with an RII of 76%. Because BIM and FM technologies have such different life cycles, there is only a limited amount of interoperability between them, which is problematic. There is also a lack of shared interests among software vendors. There are limitations for standards used [9].

"lack of FM consideration at the early design stage" is the factor that is ranked seventh with an RII of 73.70%. The level of contribution is hence high. In the construction industry, few efforts have been made to involve facility managers in the design process. The early consideration of FM could reduce the efforts for maintenance during the operation stage.

"Lack of FM data required in BIM format" is the next factor that is ranked eighth with an RII of 72.6%. Although COBie is presented as the major method of data exchange, it is rarely utilized in actuality. The 3D BIM data is converted into a spreadsheet format via COBie. The challenges of data acquisition, data exchange formats, data management for FM systems, etc., are the subject of heated discussions. For instance, it is unclear whether COBie can effectively share data for FM systems.

"FM stakeholders trustworthy the traditional strategies of FM. So, there is little motivation to use BIM" is ranked second from last with an RII of 70.90%. Facility owners confront two significant challenges: first, bringing together various segments of a fragmented sector; and second, investing in BIM technology to avoid wasting time, effort, and money.

"High cost of training and education" is ranked last with an RII of 69.10%. Any level of BIM adoption requires training, and the context and content must be in line with corporate objectives. Evidently, various staff levels would call for various training programs [19]. Therefore, there is a need for a high initial investment on training and education of the personnel.

6 Conclusions and Recommendations

This paper suggests BIM be a technological innovation that mutates the FM of metro rail projects and analysis the challenges and limitations of integration of BIM in the operation and maintenance stages. A total of 10 factors influencing the adoption of BIM and FM for metro rail projects in India were identified through a literature review and a semi-structured interview, furthermore quantitatively analysed based on a questionnaire survey. The results highlighted the critical factors to be considered by professionals while adopting BIM at the FM phase in metro rail projects. Furthermore, further research should be conducted to identify the data requirements for BIM-enabled facilities management and the challenges faced by the FM stakeholders at the stage of the project handover process. Moreover, academics and practitioners should take the feature of combing BIM and FM in innovative ways to boost the solutions that improve the large-scale project delivery process, and operation and maintenance performance.

References

1. Aldowayan, A., & Dweiri, F. T. (2020). A review on current status of facility management practices in building industry and prospective BIM intervention to manage the facilities effectively during its service life (pp. 831–846).
2. Arayici, Y., Onyenobi, T., & Egbu, C. (2012). Building Information Modelling (BIM) for facilities management (FM). *International Journal of 3-D Information Modeling (IJ3DIM)*, 55–73.
3. Naghshbandi S. N. (2016). BIM for facility management: challenges and research gaps. *Civil Engineering Journal*, 679–684.
4. Senthilvel, M., et al. (2018). Towards digital delivery of metro-rail projects in India.
5. Liu, Y., et al. (2020). Research on application of BIM 5D in communication of project muti-participants - a case study of Nagpur metro project. *IOP Conference Series: Earth and Environmental Science, 568.*
6. Pakhale, P. D., & Pal, A. (2020). Digital project management in infrastructure project: a case study of Nagpur metro rail project. *Asian Journal of Civil Engineering*, 639–647.
7. Liu, B., & Sun, X. (2018). Application analysis of BIM technology in metro rail transit. *IOP Conference Series: Earth and Environmental Science, 128.*
8. Shin, S., Moon, H., & Shin, J. (2022). BIM-based maintenance data processing mechanism through COBie standard development for port facility. *Applied Sciences.*
9. Kassem, M., Kelly, G., Serginson, M., Lockley, S., & Dawood, N. (2013, January). BIM for facility management: a review and a case study investigating the value and challenges. In *Proceedings of the 13th international conference on construction applications of virtual reality.*
10. Matarneh, S., Danso-Amoako, M., Al-Bizri, S., Gaterell, M., & Matarneh R. (2019). BIM-based facilities information: streamlining the information exchange process. *Journal of Engineering, Design and Technology*, 1304–1322.
11. Alvanchi, A., TohidiFar, A., Mousavi, M., Azad, R., & Rokooei S. (2021). A critical study of the existing issues in manufacturing maintenance systems: can BIM fill the gap? *Computers in Industry,* 103484.
12. Sadeghi, M., Elliott, J. W., Porro, N., & Strong, K. (2019). Developing building information models (BIM) for building handover, operation and maintenance. *Journal of Facilities Management*, 301–316.
13. Kula, B., & Ergen, E. (2021). Implementation of a BIM-FM platform at an international airport project: case study. *Journal of Construction Engineering and Management,* 1–14.
14. Lin, Y. C., & Su, Y. C. (2013). Developing mobile- and BIM-based integrated visual facility maintenance management system. *The Scientific World Journal.*
15. Ensafi, M., Harode, A., & Thabet, W. (2021). Developing systems-centric as-built BIMs to support facility emergency management: A case study approach. *Automation in Construction, 133.*
16. Kassem, M.A., Khoiry, M. A., & Hamzah, N. (2020). Using relative importance index method for developing risk map in oil and gas construction projects. *Jurnal Kejuruteraan*, 441–453.
17. Taber, K. S. (2018). The use of Cronbach's Alpha when developing and reporting research instruments in science education. *Research in Science Education,* 1273–1296.
18. Joshi, A., Kale, S., Chandel, S., & Pal, D. (2015). Likert scale: explored and explained. *British Journal of Applied Science & Technology,* 396–403.
19. Wood, G., Davis, P., & Olatunji, O. A. (2011). Modelling the costs of corporate implementation of building information modelling. *Journal of Financial Management of Property and Construction*, 211–233.

Evaluation of Operational Energy for Institutional Building – A Case Study

Nishath Aliya and Suchith Reddy Arukala

Abstract The urbanization is unavoidable and the construction industry infrastructure development is facing major concern of energy utilization. Due to heat island and climate change phenomena, ambient temperatures have increased drastically and energy resources have been strained. This discomfort has resulted in quadruple peak power usage and increased the building cooling load. Therefore, a diverse and scalable approach is required to mitigate the environmental consequences of energy utilization during the life cycle of the building projects. The developing countries like India's reliance on energy imports is predicted to surpass 53% of total energy consumption by 2030. Keeping this in view, the present study is carried to find the effect of different parameters on energy efficiency of an existing building (an institutional G + 2 building) considering thermal coefficients of materials and calculated the peak cooling demand on the hottest summer day using design builder tool. The study varied the materials usage in terms of lighting, glazing and roofing and observe the energy performance at various spaces in the considered building. From the findings, it is observed that with the change in thermal properties for different materials, the energy required during operation of building has decreased. The study also considered various cases of combinations of material usage and recommended suitable materials for improving efficient design capacity to make the building more operation energy efficient.

Keywords Energy Efficiency · thermal comfort · sustainability · embodied energy · life cycle energy

N. Aliya · S. R. Arukala (✉)
Kakatiya Institute of Technology and Science Warangal, Warangal, Telangana, India
e-mail: asr.ce@kitsw.ac.in

N. Aliya
e-mail: m20sc021@kitsw.ac.in

A. Vilventhan et al. (eds.), *Advances in Construction Materials and Management*, Lecture Notes in Civil Engineering 346,
https://doi.org/10.1007/978-981-99-2552-0_6

1 Introduction

The majority of our time is spent indoors either at work or at home. A country's overall energy consumption is largely attributed to the use of building energy (both residential and commercial) [1]. In many countries, operation of buildings consumes more energy in transportation. It is predicted that the building sector utilizes 42% more electricity than any other sector globally [1]. As countries get wealthy, building efficiency becomes more important as a result of development. The thermal transmission (U-Value), which measures heat loss through a substance, is also used to calculate a building's energy performance. (R-Value) is a thermal attribute and a measurement of a temperature differential by which an item or substance opposes the flow of heat. The concept of energy efficiency in buildings is related to the supply of energy necessary to achieve desirable environmental conditions that minimize energy consumption [2]. Indecent light increases a building's energy consumption, which increases its cooling design capacity. To reduce this energy consumed in lighting, LED is used, which reduces the overall energy demand and cooling design capacity. The glazing design of a building has a large impact on the heating and cooling design capacity. To reduce this energy consumed in lighting, LED is used, which reduces the overall energy consumption of building.

The glazing design of a building has a large impact on the heating and cooling design capacity. It is stated that glass can lose up to 40% of a building's heat energy. The cooling design capacity of a building is also affected by the roofing materials. At high temperatures, sunlight passes through the roof, increasing the heat/temperature inside the building. As a result, the increase in energy consumption and cooling design capacity is influenced. Different simulation tools are used to observe the energy consumption and performance of the building like Integrated Environmental Solution-Virtual Environment (IESve), Climate Consultant Software, ECOTECT Weather Tool, ArchiCAD, REVIT, Allplan, Aecosim, Edificius, EE-ShaDe [3–6], and Design Builder. Design Builder is an Energy Plus-based software application for measuring and controlling energy, carbon, lighting, and comfort. Design Builder combines rapid three- dimensional architectural modelling with dynamic energy calculations, in addition to its distinctive ease of use. It is regarded as a revolutionary software tool for creating and evaluating architectural designs. Energy Plus is a modern building performance simulation program that combines the best features from BLAST and DOE-2 along with some new capabilities [7]. The present study focus on measuring the design capacity and energy consumption of an existing building and suggest suitable changes to reduce the overall temperature inside the building with the changes in lighting, glazing and roofing.

2 Literature Review

Simulation is used to optimize cooling design calculations for the warmest day of the year and to optimize the size of the cooling design plant of educational building based on climate figures, tenancy, and construction factors such as heat transferred by various facades due to inside and solar heat gains [1]. The adoption of appropriate building orientation, shape, envelope system, passive heating and cooling, shading, and glazing factors forms the basis for these requirements to reduce heating and cooling loads [2]. E. Mushtaha et al. [3] Using Climate Consultant and IESve as assessment software, three passive design parameters—shading devices, natural ventilation, and thermal insulation—are investigated, along with the effects of the building's indoor temperature and cooling load. Different simulation tools are used to observe the energy consumption and performance of the building like ECOTECT Weather Tool, Design Builder, ArchiCAD, REVIT, Allplan, Aecosim, Edificius, EE-ShaDe [4–6]. Energy Plus is a modern building performance simulation program that combines the best features from BLAST and DOE-2 along with some new capabilities [7]. According to Bhagyesh [8], the Energy Conservation Building Code (ECBC) determines the energy parameters for a commercial building such as total electric load, carbon footprint, total gain in solar heat, infrared greatest gain in solar heat numbers, and optimizes the best results among the seven cases using the design builder. The simulation is also used to estimate the impact of various parameters on interior air temperature, such as window-to-wall ratio, building orientation, total height, and building products. & Greatest temperature drop was discovered to be 2. 76 °C [9, 10], identify alternative passive techniques for producing the required interior microclimate circumstances. The effect of mechanical ventilation on discomfort was explored, along with overnight ventilation, further summer overheating, the justification for using expensive heat pump systems with a ground heat exchanger are investigated. Safa et al. [11] evaluated the energy consumption & life cycle expenses of structural lightweight concretes (SLWC) having lesser thermal conductivity over typical weight concrete (NWC) to evaluate the strength of concrete, mass weights, and thermal performance coefficients of SLWCs and NWCs and the heat and cooling energy usage.

Variable air volume (VAV) [12] systems are pitted against constant air volume (CAV), photoelectric dimming control systems are pitted against general illumination, and double-glazed minimal emittance windows are pitted against single glazed windows. S. Semahi, et al. [13] utilized the current meteorological statistics, examine the bioclimatic potential of the eight representative Algerian climate zones (2003–2017). M.-V. Belmonte et al. [14] Analyzes the impact of different parameters set in the beginning stage of a single-family home's design on its energy efficiency and investigates the potential role of a computer tool in this stage to increase energy efficiency. Bernardo H et al. [15] A calibration building energy simulation model of a school building is created in order to investigate the influence of changing the ventilation system on efficiency. Rallapalli H S [16] opines that the capabilities of programs such as eQUEST and EnergyPlus for doing whole-building energy analyses and comparing the findings to entire building energy performance.

3 Methodology

The present study investigates the effect of lighting, glazing and roofing parameters on energy efficiency of an existing building (Block VI of Kakatiya Institute of Technology and Science) using design builder. The building is G + 2 in design with a ground floor, first floor, and second level. The dimensions of the structure are observed to be 284 ft × 56 ft. The study considered the meteorological data (Begumpet airport) including dry bulb temperature, solar radiation, moisture and humidity which are useful in evaluating energy consumption within the building. The model required weather data files based on location of building were incorporated into Design. A 3-Dimensional model geometry of the building is created using AutoCAD and design builder software (Figs. 1 and 2). The study investigated the material properties of the building like external wall, partition wall, floor, roof, doors type, density, dimensions etc., to generate the building envelop model for simulation (Table 1).

The actual thermal design capacity for peak hot summer season day is observed, which is assumed to be the warmest day of the year for understanding the energy

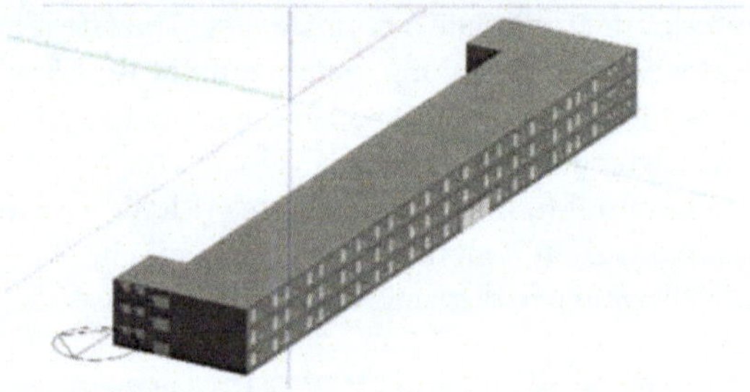

Fig. 1 A geometrical representation of the existing structure

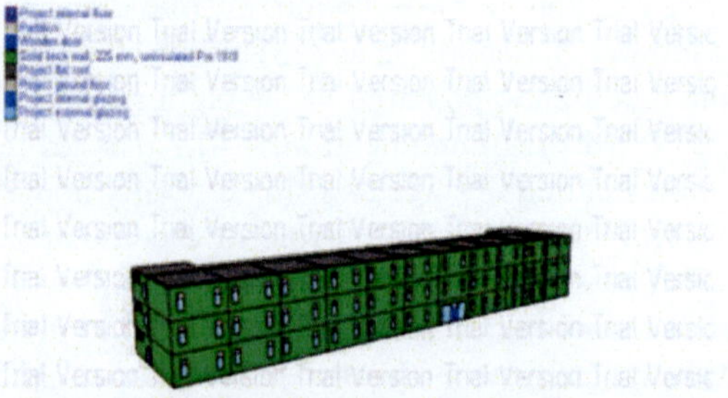

Fig. 2 The visualize view of the building incorporated with the building materials, components dimensions (like flooring, partitions, doors, roofing, glazing, etc.)

Table 1 Building data of existing educational building

Description	Materials	All values(w/m2k)	Dimensions
External wall	Lightweight mud brick	U value-0.299 R value-0.421	9 inch
Partition wall	Lightweight masonary brick	U value-0.398 R value-2.512	4 inch
Floor	Concrete (M20 grade)	U value-0.516 R value-1.940	4 ½ inch
Roof	Concrete (M20 grade)	U value-0.516 R value-1.940	6 inch
Doors	Painted oak	U value-0.299 R value-3.342	6ftx7ft, 3ftx6ft 6inch

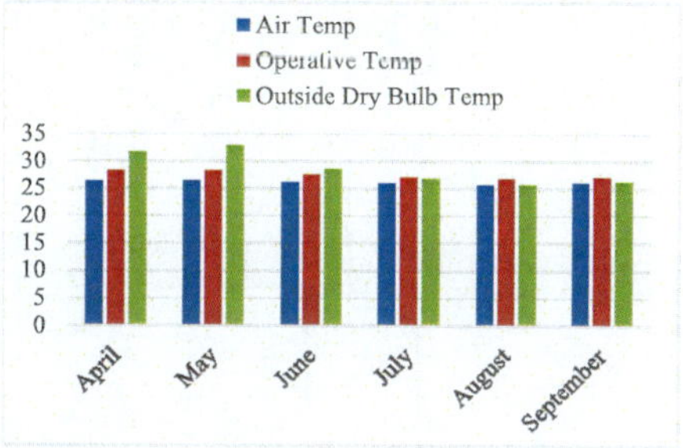

Fig. 3 All summer graph

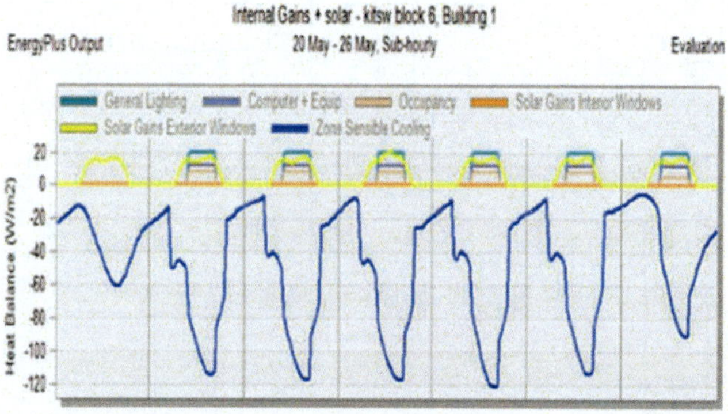

Fig. 4 Simulation results for summer design week

requirement and thus the findings instantly were mapped in for several other days of the year Figs. 3 and 4.

3.1 *Methodology Flow Chart*

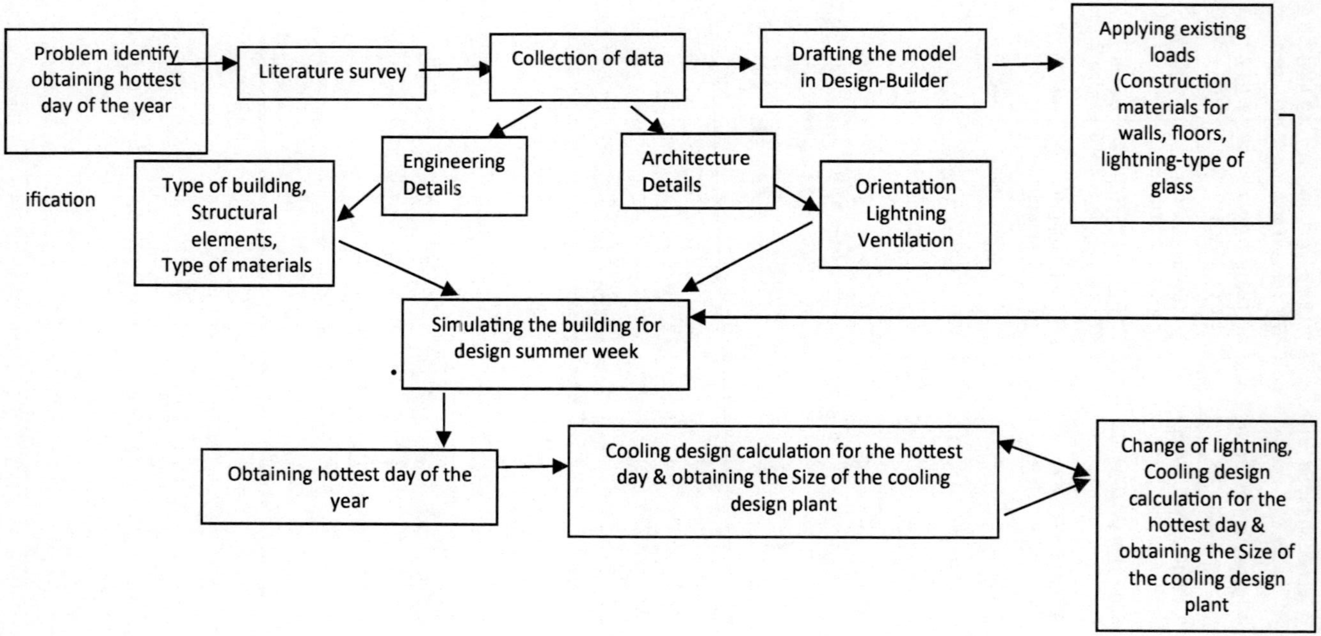

Table 2 All summer data

Date	Air Temp (°C)	Operative Temp (°C)	Outside Dry Bulb Temp (°C)
01-04-2021	26.33	28.18	31.70
01-05-2021	26.35	28.23	32.91
01-06-2021	26.09	27.46	28.58
01-07-2021	25.90	27.03	26.77
01-08-2021	25.69	26.74	25.69
01-09-2021	26.00	27.14	26.19

Note: Air temperature is the temperature of the air nearby an individual and is characteristically measured in degrees Celsius (°C) or degrees Fahrenheit (°F); Operative temperature is the acceptable temperature range of the local ambient situation at which an electrical/mechanical device function; The Dry Bulb Temperature refers basically to the ambient air temperature. It is called "Dry Bulb" because the air temperature is indicated by a thermometer not affected by the moisture of the air

4 Results

4.1 All Summer Simulation

Based on the input data provided to the developed building envelop, the building is simulated using design builder software given the site's location and specific climatic data and existing building conditions. Keeping this view, the study adopted closest airport weather data (i.e., Begumpet airport) and simulated between April to September which is a peak summer duration for the location. The findings revealed that May month has a high operative and dry bulb temperature (Table 2).

4.2 Simulation for Hottest Week

The study focused on identifying hottest week peak demand energy performance based on general lighting, occupancy, solar gains exterior windows. From Fig. 4, it is observed that the general lighting, occupancy, computer + equip values are constant, fluctuation is observed on solar gains exterior windows and heat gains caused by solar radiations passing through the glass of window. The results demonstrate how the solar gains from exterior windows are typically consistent for five days of the week (20th to 26th May 2021) and being highest on 23th May, as a result, it is regarded as the warmest day of the year.

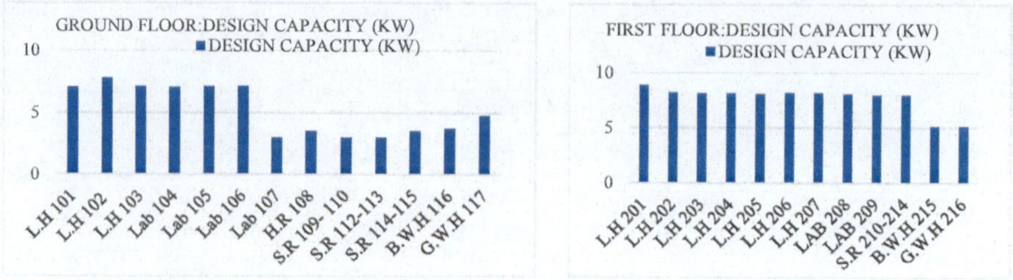

Fig. 5 Base case design capacity graphs for ground floor & first floor

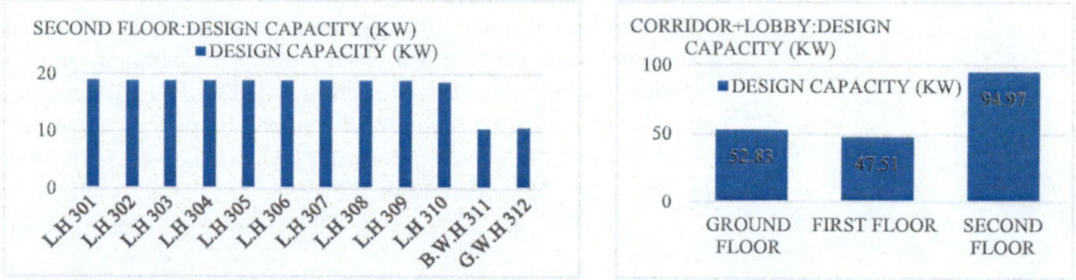

Fig. 6 Design capacity for second floor and corridor + lobby for all floors

4.3 *Base Case*

The cooling design calculations are performed for the building and identified the peak demand on the hottest day of the year and calculated the cooling design capacity required as 569.29 kW (or) 161.87 ton (Figs. 5 and 6).

Based on the observed data of the case study on the existing institutional building the study sensitizes the passive design strategies and adopted three different strategies on the base case.

4.4 *CASE 1: Change of Lighting (LED)*

In case 1, the general lighting (incandescent light bulb) is replaced with LED lighting. The Table 3 shows the specifications of the LED light replaced with incandescent light.

Table 3 Specifications of the LED light

Description	Type of lighting	Watts	All values
Lighting	LED	20	Normal power density-2.5 (w/m^2-100 lx)

The Figs. 7 and 8 represent the cooling design requirement with change in LED lighting for all the spaces (rooms) on the hottest day of the year. The calculation of cooling design for the design summer day i.e., 23th May the capacity of the cooling plant is estimated to be 540.34 kW or 153.64 ton.

The Figs. 9 and 10) represent the percentage change in design capacities (D.C) and change in maximum operative temperature (max op temp). The percentage change of maximum operative temperature at ground level is observed to be high (0.17%) and percentage change in design capacity is observed to be 21.93% (i.e., reduced from 3.51 kW to 2.74 kW). Similarly at first level, change in design capacity is 23.68% i.e., reduced from 5.15 kW to 3.93 kW.. From Figs. 11 and 12), the change in design capacity is observed to be 3.45%. For corridor + lobby, the percentage change of max op temp at ground level is high and percentage change in design capacity is 6.56% (reduced from 52.83 kW to 49.36 kW).

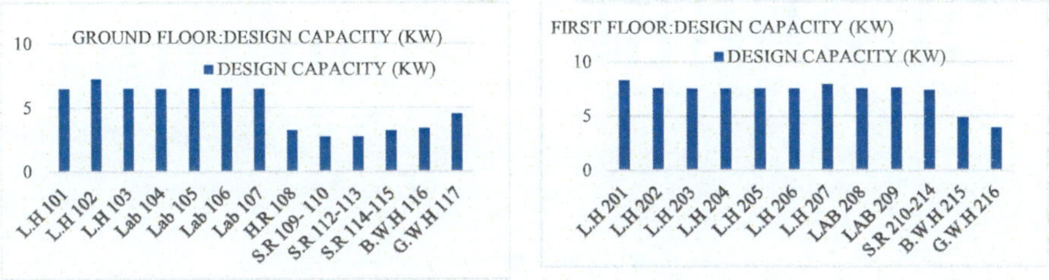

Fig. 7 Case1 Design capacity for ground floor & first floor

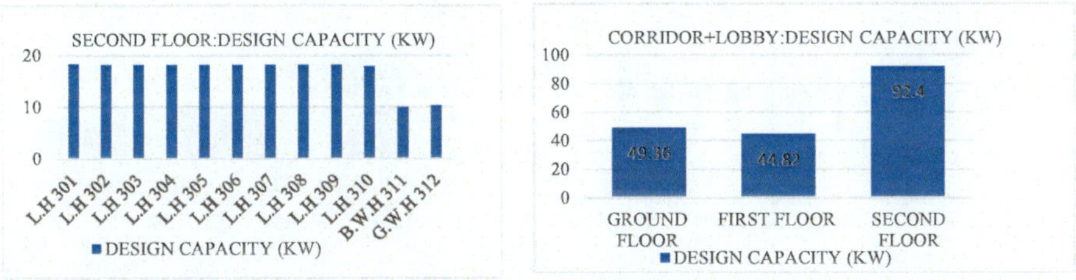

Fig. 8 Case 1 Design capacity for the second floor & corridors + lobby for all floors

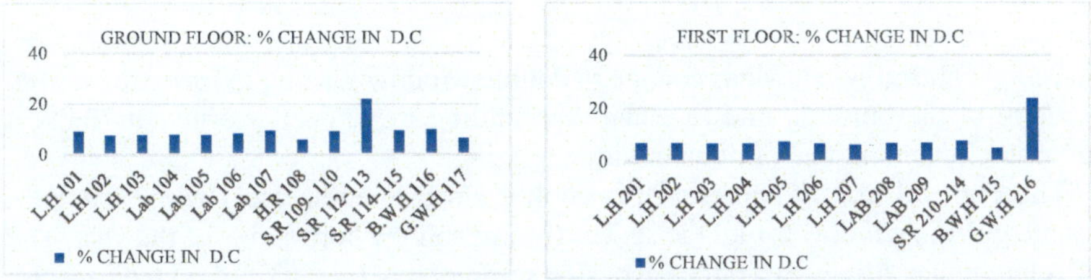

Fig. 9 Case 1 Percentage change in Design Capacity of Ground and First floors

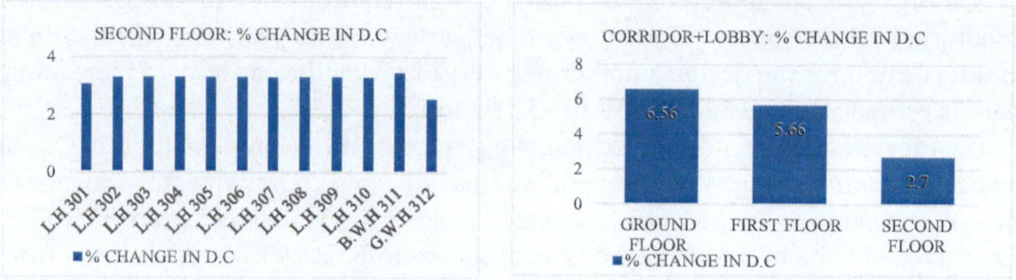

Fig. 10 Case 1 Percentage change in Design Capacity of Second floor, corridor + lobby of all floors

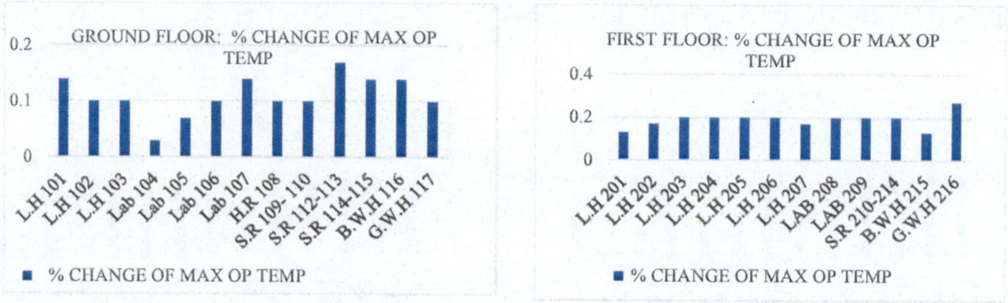

Fig. 11 Case 1: % change of Max op temp % of ground floor and first floor

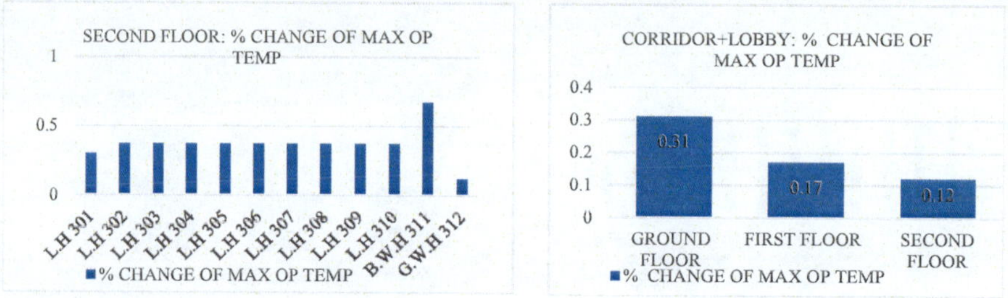

Fig. 12 Case 1: Percentage Change of Max op temp % of second floor, corridor + lobby of floors

4.5 *Case 2 Change of Window Glazing*

The single glazed window has change to double window glazing to lower the design capacity of the building and to make the building more energy efficient. Table 4 shows the specifications of double widow glazing.

The Figs. 13 and 14 represent the cooling design requirement with change in double glazing window for all the spaces (rooms) on the hottest day of the year. The calculation of cooling design for the design summer day (i.e., 23th May) the capacity of the cooling plant is estimated to 562.42 kW or 159.92ton.

Table 4 Specifications of double window glazing

Description	Materials	All values (w/m^2 k)
Glazing (Double) Dblclr 6 mm/ 13 mm Air	Sage glass clima plus grey no tint	U value = 1.267

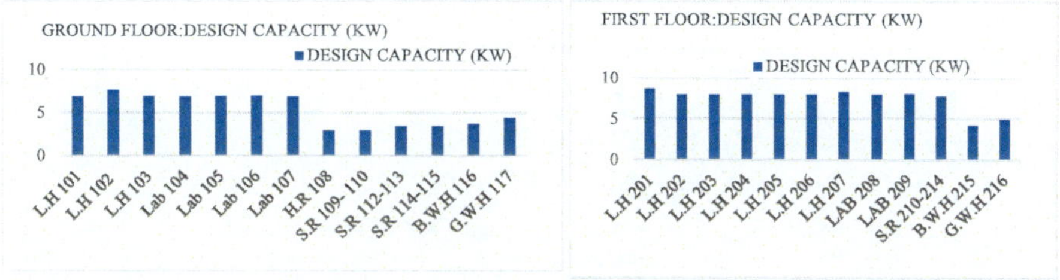

Fig. 13 Case 2 Design capacity for Ground floor & First floor

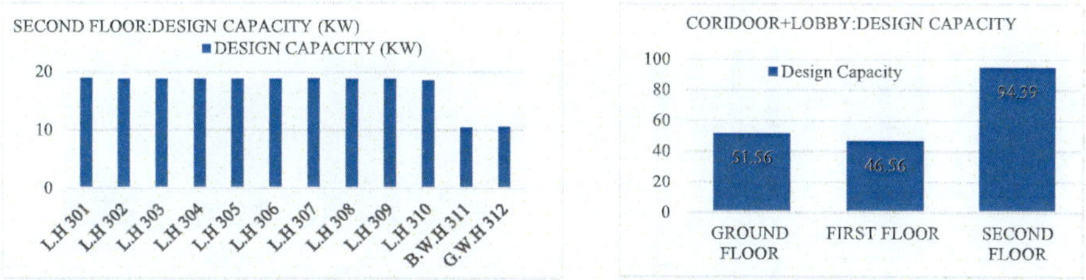

Fig. 14 Case 2 Design capacity for the second floor & corridors + lobby

From the Figs. 15 and 16 it is observed that with the percentage change in design capacities and change in maximum operative temperature is high at ground floor (i.e., 0.24%) and changes in design capacity is 13.41% (i.e., reduced from 3.43 kW to 2.97 kW. Similarly at first level, the percentage change in design capacity is observed to be 15.88% (i.e., 5.1 kW to 4.29 kW).. From the Figs. 17 and 18 the percentage change of max op temp for the second level is high in second floor percentage change in design capacity is 0.53%. For corridor + lobby, the percentage change of max op temp at ground level is high and percentage change in design capacity is 2.4% (i.e., reduced from 52.83 kW to 51.56 kW). Above the values are at peak because the sun radiations enters in the spaces during afternoon thought convention and radiation heat transfer phenomenon.

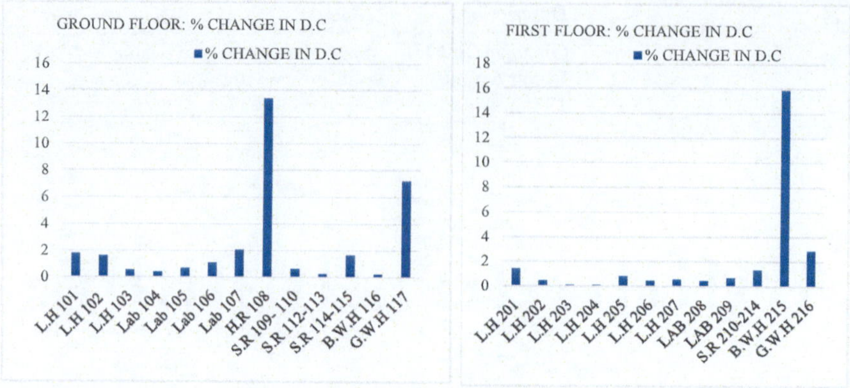

Fig. 15 Case 2: % change in D.C of ground floor and first floor

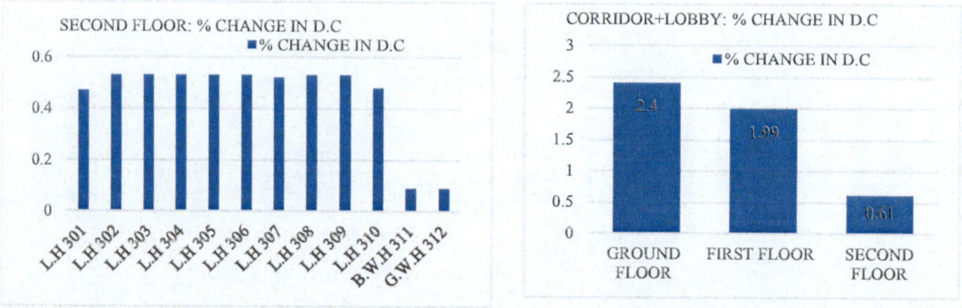

Fig. 16 Case 2: % change in D.C of second floor, corridor + lobby of all floors

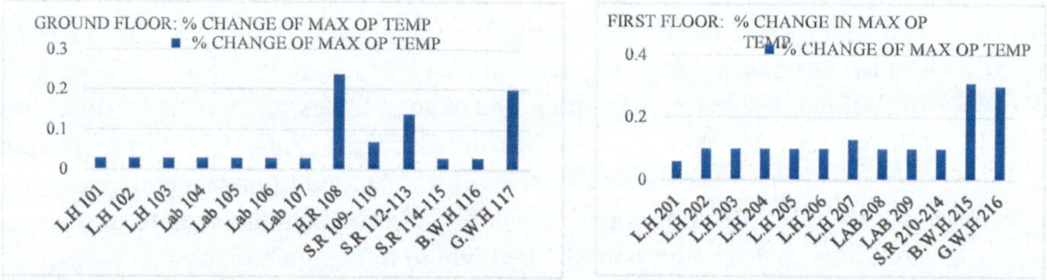

Fig. 17 Case 2 Percentage change in Maximum Operative temperature in ground floor and first floor

4.6 Case 3 Change of Roof Materials

To observe the reduction in the capacity of the structure and make it more energy efficient (Table 5).

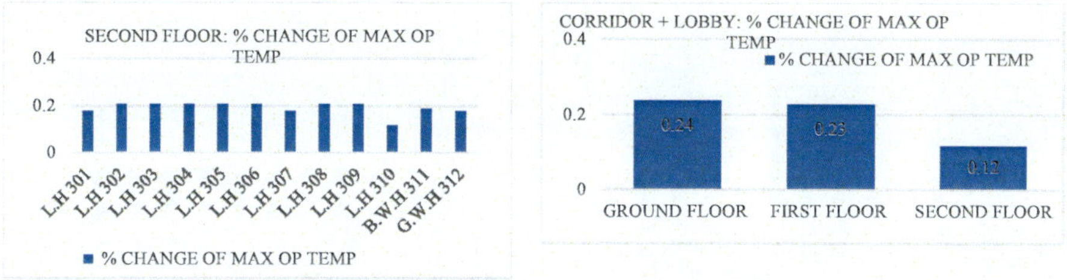

Fig. 18 Case 2 Percentage change in Maximum Operative temperature in second floor and corridor + lobby

Table 5 Specifications of the Roof part L 2013 reference building

Description	Materials	All values (w/m^2 k)
Roof part L 2013 reference building	Min wool quilt 100 mm, NCM Plaster board (wall board)	U value = 0.250 R value = 4.001

The Figs. 19 and 20 represents the cooling design capacity with Roof part L 2013 reference building. For the case 3 the cooling design calculation for the design summer day i.e., 23th May is carried out and the size of the cooling plant was estimated to be 386.69 kW or 109.95 ton.

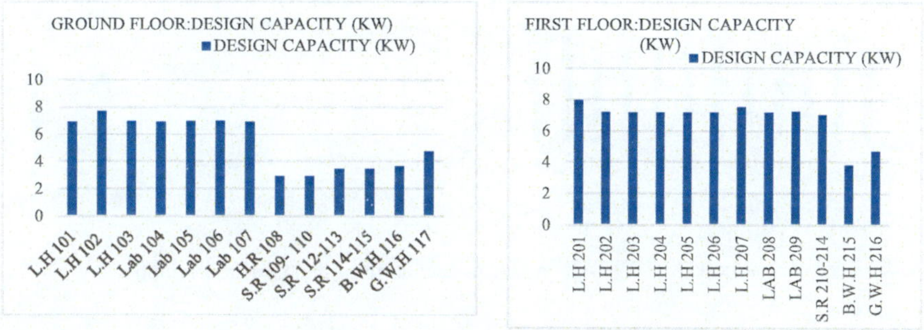

Fig. 19 Case 3 Design capacity for Ground floor & first floor

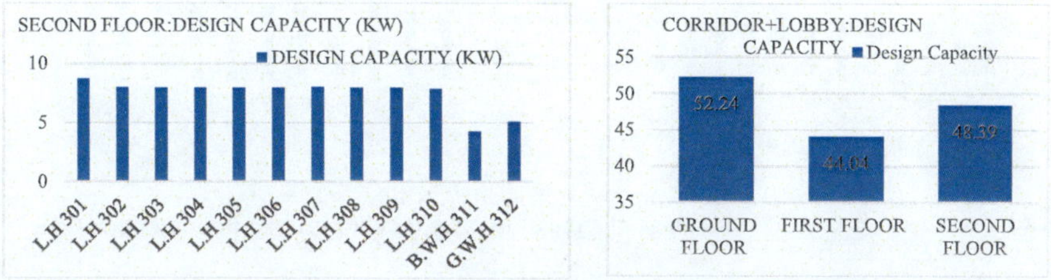

Fig. 20 Design capacity for the Second floor & Corridors + lobby

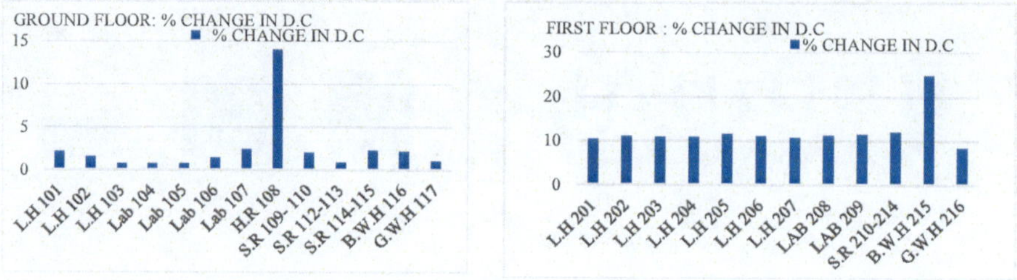

Fig. 21 Base case & case 3: % Change in Design capacity of ground floor and first floor

From the Figs. 21, 22, 23, and 24), it is observed that percentage change in design capacities and change in maximum operative temperature at ground level is high and change in design capacity is 13.99% (i.e., reduced from 3.43 kW to 2.95 kW). Similarly at first level, percentage change in design capacity is 24.70%.From the Figs. 20 and 22, percentage change of max op temp for the second level is high (i.e., 12.54%) and percentage change in design capacity is 59.30% (i.e., reduced from 10.42 kW to 4.24 kW). Similarly, for corridor + lobby percentage change of max op temp at second level is high (i.e., 11.35%) and percentage change in design capacity is observed to be 49.04% (i.e., reduced from 94.97 kW to 48.39 kW).

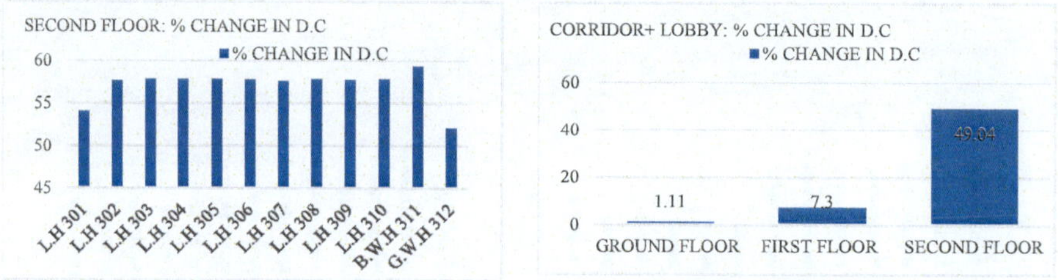

Fig. 22 Base case & case 3: % Change in D.C of second floor and corridor + lobby

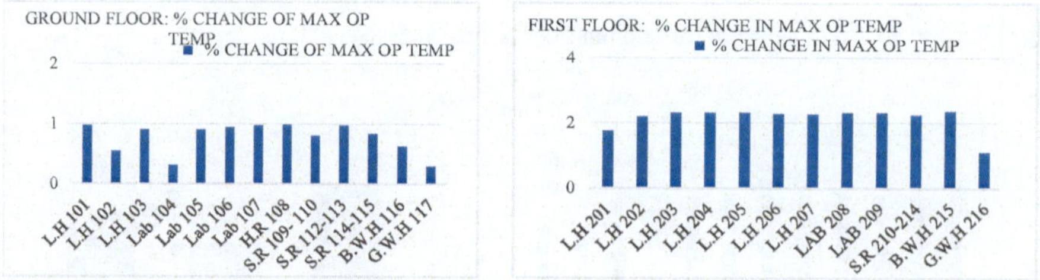

Fig. 23 Base case & case 3: Percentage change of max op temp of ground floor and first floor

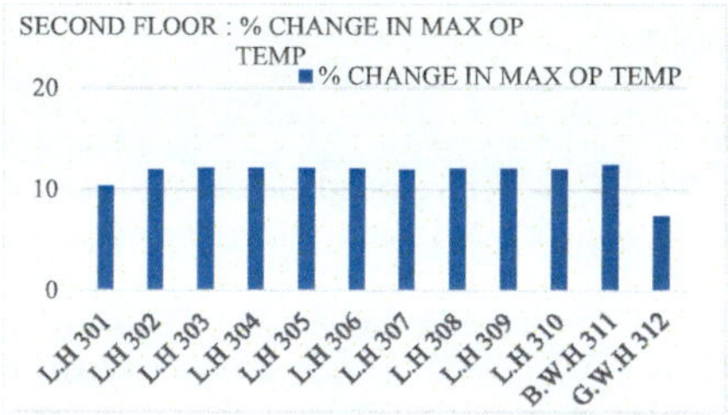

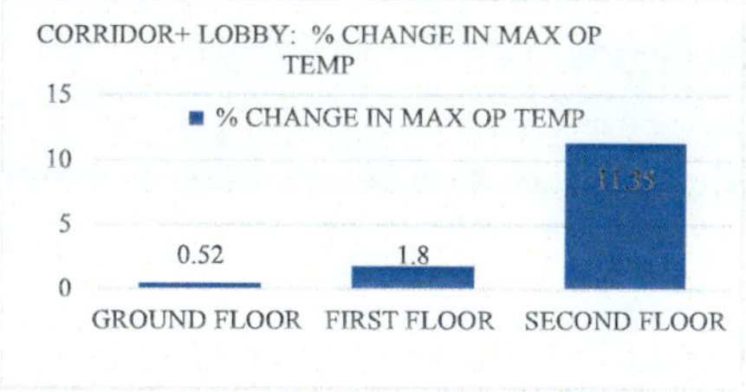

Fig. 24 Base case & case 3: Percentage change of max op temp of second floor and corridor + lobby

5 Conclusions

The study invested the effect of lighting, glazing and roofing parameters on energy efficiency of an existing building (Block VI of Kakatiya Institute of Technology and Science) using design builder. The study considered the meteorological data (Begumpet airport) including dry bulb temperature, solar radiation, moisture and humidity which are useful in evaluating energy consumption within the building. The study investigated the material properties of the building like external wall, partition wall, floor, roof, doors type, density, dimensions etc., to generate the building envelop model for simulation. The simulation is performed to analyze the solar heat gains of the building and cooling design for the hottest day of the year is observed to fall on 23th May and the design capacity value considered as base case is observed to be 569.29 kW or 161.87 ton. The materials like LED, window glazing, roof materials are chosen based on their U, R values (thermal transmission and thermal resistance). In Case 1, The LED lighting is selected by the lowering the normal power density i.e., 2.5 (w/m2-100 lx) with respect to incandescent light bulb as 5.0 (w/m2-100 lx). The design capacity has reduced to 530.34 kW or 153.64 ton (reduced by 38.95 kw). In Case 2, single glazed window (clear glass) is replaced by double glazing window, for single glazed window U values are 5.9 w/m^{2}k and for double glazed window the values are u value = 1.267(w/m^{2}k). In Case 3, standard roof materials are changed to Roof part L 2013 reference building i.e., standard roof U, R values are 0.516,1.940 with respect to Roof part L 2013 reference building U, R values are 0.25 and 4.0 respectively. The design capacity value with modification of materials is noticed as 562.42 kW or 159.92 ton (reduced by 6.87 kW). In Case 3, the materials for roofing are altered with Roof part L 2013 reference building. The design capacity value of the design cooling is noticed as 386.69 kW or 109.95 ton. The design capacity for the cooling design plant are reduced by 182.6 kw (or 51.92 ton). From simulation, it is observed that the combination of all 3 cases the size of the cooling design plant is 340.98 kw (or) 96.95 ton. The design capacity for the cooling design plant are reduced by 228.31 kw (or 64.91 ton) when compared with base case. The findings facilitate the building to replace the existing energy loadings with energy efficient and other design strategies.

References

1. Bharath, B., Sami, A., & Rastogi, K. (2020). Analysis of a structure for efficient energy utilization using design builder. In IOP Conference Series: Materials Science and Engineering, volol 1017, p. 012016 (2020)
2. Pacheco, R., Ordo´nez, J., & Martínez, ˜G. (2012). Energy efficient design of building: a review. *Renewable and Sustainable Energy Reviews, 16*, 3559–3573. https://doi.org/10.1016/j.rser.2012.03.045
3. Mushtaha, E. et al. (2021). The impact of passive design strategies on cooling loads of buildings in temperate climate. *Case Studies in Thermal Engineering, 28*, 101588. https://doi.org/10.1016/j.csite.2021.101588
4. Autodesk, Ecotect Weather Tool. Version 8, (2019)
5. Acampa, G., García, J. O., Grasso, M., & Díaz-Lopez, C. (2019). ´ Project sustainability: criteria to be introduced in BIM. *Valori e Valutazioni, 119–128*, 2019.
6. Mandow, L., P´erez-de-la-Cruz, J.L., Rodríguez-Gavil´ an, A.B., & Ruiz-Montiel, M. (2020). Architectural planning with shape grammars and reinforcement learning: habitability and energy efficiency. *Engineering Applications of Artificial Intelligence, 96*, 103909. https://doi.org/10.1016/j.engappai.2020.103909
7. Lawrie, L.K., Fredrick, C. Dr., Winkelmann, Pedersen, C.O. Dr. (2000, April). Energy Plus: energy simulation program. *ASHRAE Journal, 42*, 49–56
8. Pawar, B.S., Kanade, G.N. (2018). Energy optimization of building using design builder software. *(IJNTR), 4*(1), 69–73 (2018). ISSN: 2454-4116
9. Thaha, S.M., Bennet Kuriakose, R. (2022). Thermal efficiency analysis of buildings with phase change. *Sustainability, Agri, Food and Environmental Research, 10*(X), 2022. ISSN: 0719-3726
10. Dudzi´nska, A., Kisilewicz, T. (2021). Alternative ways of cooling a passive school building in order to maintain thermal comfort in summer. *Energies 14*, 70. https://doi.org/10.3390/en14010070
11. Nayir, S., et al. (2021). The effects of structural lightweight concrete on energy performance and life cycle cost in residential buildings. *Periodica Polytechnica Civil Engineering, 65*(2), 500–509
12. Rahman, M.M., et al. (2008). Energy conservation measures in an institutional building by dynamic simulation using design builder. In 3rd IASME/WSEAS Int. Conf. on Energy & Environment, University of Cambridge, UK, February 23–25, ISSN: 1790-5095
13. Semahi, S., et al. (2019, March). Comparative bioclimatic approach for comfort and passive heating and cooling strategies in Algeria. *Building and Environment, 161*, 106271 (2019). https://doi.org/10.1016/j.buildenv.2019.106271
14. Belmonte, M.-V., et al. (2021). Introducing passive strategies in the initial stage of the design to reduce the energy demand in single-family dwellings. *Building and Environment, 197*, 107832. https://doi.org/10.1016/j.buildenv.2021.107832
15. Bernardo, H., Quintal, E., & Oliveira, F. (2017). Using a calibrated building energy simulation model to study the effects of improving the ventilation in a school. *Energy Procedia, 113*, 151 – 57
16. Rallapalli, H.S. (2010). A comparison of Energy Plus and eQuest whole building energy simulation results for a medium sized office building 2010, (U.S.A.: Arizona State University).

Interactions of Lean and BIM Integrated Augmented Reality in Underground Utility Relocation Projects

R. Rajadurai and Aneetha Vilventhan

Abstract Building Information Modelling (BIM) and lean construction were being extensively used in construction projects to improve project performance and eliminate lean waste associated in the process. Though being different in concepts and applications, their integrated application was found to provide better advantages than implementing individually. With the advancement of technologies, the current BIM applications are evolving to consider technologies such as Augmented Reality. Several studies have addressed the interactions or synergies that exist between BIM functionalities and lean principles. However, such interactions were limited to standalone BIM approaches. Other approaches of BIM, such as BIM with AR and lean were seldom targeted. Hence, this paper aims to establish interactions between AR integrated BIM and lean, and identify how AR integration can improve BIM-lean interactions. To establish such interaction, a case study of an underground utility relocation project is considered, and BIM models are developed and applied in practice. The identified BIM and AR functionalities are used then to establish interaction with the lean principles. 46 AR-BIM-lean interactions and 36 BIM-lean interactions have been identified. From analysis, it is evident that AR integration with BIM not only reinforces BIM-lean interactions but also enables newer interactions with lean and enables achieving lean in construction.

Keywords BIM · Lean construction · Augmented Reality (AR) · underground utility · utility relocation

R. Rajadurai · A. Vilventhan (✉)
Department of Civil Engineering, National Institute of Technology Warangal, Warangal, Telangana, India
e-mail: aneetha@nitw.ac.in

A. Vilventhan et al. (eds.), *Advances in Construction Materials and Management*, Lecture Notes in Civil Engineering 346,
https://doi.org/10.1007/978-981-99-2552-0_7

1 Introduction

The construction industry is infamous for its lower productivity, poor safety, unfavourable working conditions, poor management of labour, materials, and severe schedule and cost overruns [1]. Over the years, several techniques, technologies, and concepts were evolved to improve construction project management. Despite several advancements, two concepts created a paradigm shift in the way construction project/process is being handled. They are Building Information Modelling and Lean Construction.

Building Information Modelling (BIM) is a process of creating or developing, and managing coordinated digital information of a construction project throughout its lifecycle with the use of appropriate tools and technology. The use of BIM eliminated the traditional silos of workflow in design and construction and shifted the construction process into a collaborative workflow, thereby improving the overall performance of the project. Lean construction, on the other hand, is a production principle adopted from the Toyota Production System and focuses on eliminating waste and improving the value of the process. The implementation of lean in construction imparts significant improvements in the delivery of the project and improves the overall performance of the process [2] and improves coordination and communication among project teams in construction.

Unlike BIM assisting in the production/execution of construction projects, lean construction designs the production system of construction projects focusing on the individual process involved in the delivery of the project. Though BIM and lean are different paradigms in concepts and application, they both have common synergies in terms of functionality and principles. Studies have identified the integrated applications of BIM and lean were found to provide better advantages than they are implemented individually [3]. Since BIM and lean have common synergies, studies have also applied BIM as a lean tool in achieving lean in construction projects [4].

With the advancements in technology, the construction industry is witnessing the adoption of newer technologies such as Augmented Reality (AR) for managing construction projects. Studies are adopting BIM integrated with AR for managing construction projects for improved visualisation [5, 6]. Despite research focusses on using BIM-lean and BIM-AR, limited research has focussed on the perspective of using lean with BIM-AR. Nevertheless, the focus on establishing synergies between lean and BIM-AR remains unexplored. The current interactions between lean and BIM reinforce the consensuses that, BIM enables lean application. However, if the application of BIM-AR can enable such consensuses or the integration of AR with BIM enable achieving a higher degree of interaction with the lean principles is not explored yet. Hence, this paper aims to identify the interactions that exist between lean and BIM integrated AR in construction projects. To identify interactions, a case study of an underground utility relocation project is considered and the existing BIM-lean interaction matrix (developed by Sacks [7]) is extended to incorporate AR functionalities.

2 Literature Review

The concept of BIM lean interaction was first envisaged by Sacks [7]. The study identifies lean construction principles and existing BIM functionalities and establishes interactions between them. This study depicted how BIM functionalities can enable achieving lean and how the interactions can be used for planning BIM and lean adoption strategies in construction [7]. Later studies started exploring BIM and lean interactions for different project applications and extended the existing BIM-lean interaction framework.

In a study, the interactions between the BIM practice by the general contractors and the lean principles were established using BIM capability maturity model [8]. The study initially considers the existing interactions [7] and utilised BIM maturity model for measuring and highlighting areas of improvement for Lean. Apart from BIM-lean interaction for design and construction projects, studies have also established BIM-lean interaction for the construction of MEP works and demolition projects. These studies adopted a similar interaction framework and expanded it to consider BIM functionalities during the demolition process and established interactions representing how it can achieve lean [9]. In a similar approach, BIM functionalities were expanded to support MEP works [10].

The current studies on BIM-lean interactions were examined for the designing, construction, and demolition of building projects. Limited efforts were made to consider BIM-lean interaction for infrastructure projects. Also, the current studies limit the interaction only with BIM and does not consider advancing BIM integrated technologies such as BIM-AR. This paper fills this gap by establishing interactions between AR integrated BIM and lean in underground utility projects.

3 Methodology

To establish interactions between lean and integrated BIM-AR the methodology shown in Fig. 1 is adopted in this paper. A case study of utility relocation projects was considered. 3D BIM models were developed and applied for practice. The identified BIM functionalities from the case study were then used in identifying interactions with the lean principles. To establish interaction between AR BIM-lean, the extant literature on AR was refereed and BIM AR functionalities were identified. These identified functionalities were then extended to the BIM lean interaction framework and possible interactions between lean and BIM-AR were established.

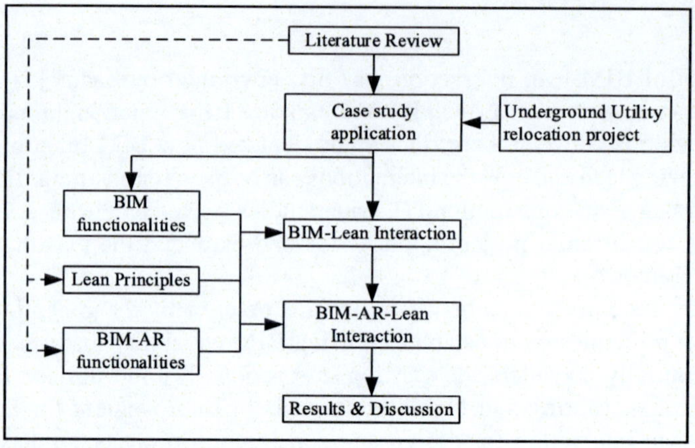

Fig. 1 Research methodology

4 BIM Model Development and Application

An underground utility relocation project was considered as a case study in this paper. The considered project VAN (name modified due to confidentiality reasons) is located in Chennai, India. The scope of the project is to construct a flyover to overcome traffic congestion and widen the existing urban roads. The required project information such as 2D CAD drawings, utility maps, and utility-specific information such as utility type, material, depth, age, conditional status, and the alignment of the right of way was collected. The collected information was then used in developing a 3D BIM model of underground utilities. Software such as Autodesk Civil 3D was used for modelling 3D elements of utility pipelines and roads and Autodesk Navisworks was used in visualizing and coordinating with the project stakeholders. The developed 3D models were used during the relocation process to facilitate 3D visualisation (shown in Fig. 2), locating and identifying underground utilities, and clash detection (shown in Fig. 3).

5 BIM Lean Interaction

The study considers the framework developed by Sacks et al. (2010), to propose identified BIM and lean interaction, specific to underground utility. The similarities posed by BIM practice with respect to the lean construction principles were marked in the interaction matrix and an overall 32 of such interactions were mapped as shown in Table 1.

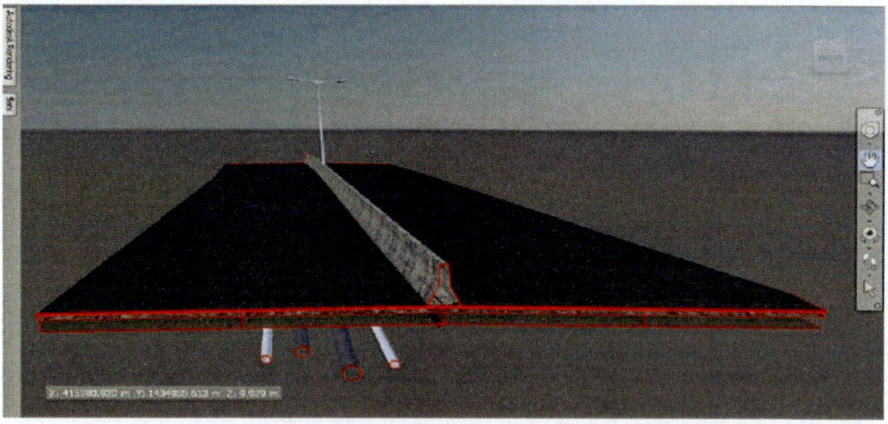

Fig. 2 Developed 3D model of underground utilities

Fig. 3 Clash detection during relocation process

6 BIM-AR Functionalities

Augmented Reality (AR) is an advanced visualizing technique that enables superimposing virtual 3D models into physical reality [5]. AR differs from other visualisation techniques, it enables the connection between the physical and the virtual environment [11]. For analysis, the applications of BIM-AR in the literature were referred to identify BIM-AR functionalities in construction projects. The applications of BIM-AR in both building and utility infrastructure projects were reviewed and their functionalities were obtained as shown in Table 2.

It can be noted from Table 2 that both BIM and AR have some similar functionalities such as visualisation, and information management. However, AR differs from BIM by providing an immersive experience of visualisation by superimposing virtual objects over the real word entities, thereby creating a sense of physical touch for the users.

Table 1 BIM-lean interaction matrix

Interaction matrix		BIM functionality*							
		A	B	C	D	E	F	G	H
Lean Principles	Reduce variability				✓	✓			✓
	Reduce cycle times		✓	✓		✓	✓	✓	✓
	Increase flexibility		✓			✓			
	Select appropriate control approach	✓	✓	✓			✓	✓	
	Standardize				✓		✓		✓
	Continuous improvements	✓					✓		
	Visual management	✓							
	Design the production system for flow and value	✓	✓			✓	✓		
	Verify and validate	✓					✓		✓
	Go and see for yourself								
	Decide by consensus						✓		
	Cultivate an extended network of partners						✓		

Table 2 BIM-AR functionalities

Literature source	Application	Functionality
[12–14]	Visualisation of construction MEP works, AR mock-ups	AR Visualisation
[12]	AR walkthrough, augmented walkthrough, safety orientation or training to workers,	AR Walkthrough
[15, 16]	Defect/damage identification, custom information retrieval, custom software application	AR Visualisation, information management, context awareness
[17]	Multi-user collaboration	Collaboration
[18]	Remote Visualisation of on-site inspection process	AR visualisation
[19]	Geo-visualisation of underground utilities	AR visualisation
[12, 20]	On-site quality management, AR walkthrough	AR visualisation, AR walkthrough
[1]	Site visualisation, site planning, context based visualisation and information retrieval, AR walkthrough, schedule progress management	AR visualisation, AR walkthrough, context awareness
[21]	Site layout planning	AR visualisation, context awareness

Table 3 AR integrated BIM lean interaction matrix

Interaction matrix		BIM-AR functionality*										
		A	B	C	D	E	F	G	H	I	J	K
Lean Principles	Reduce variability				✓	✓			✓	✓		✓
	Reduce cycle times		✓	✓		✓	✓	✓	✓	✓		
	Increase flexibility		✓		✓	✓						
	Select appropriate control approach	✓	✓	✓			✓	✓			✓	✓
	Standardize				✓		✓		✓			
	Continuous improvements	✓					✓					
	Visual management	✓								✓	✓	✓
	Design the production system for flow and value	✓	✓			✓	✓				✓	
	Verify and validate	✓					✓		✓		✓	✓
	Go and see for yourself										✓	
	Decide by consensus						✓					✓
	Cultivate an extended network of partners						✓			✓		

7 AR-BIM Lean Interactions

The obtained BIM-AR functionalities were then added to the identified BIM-lean interaction matrix (Table 1) and newer interactions between the lean principles and BIM-AR functionalities were established as shown in Table 3.

8 Discussion

On reviewing Table 1 and Table 3, the number of interactions between BIM-lean and BIM-AR lean can be identified. It can be identified (from Table 1), there are 32 interactions between BIM and lean for utility relocation projects. A higher number of interactions were identified between BIM functionality and the lean principle was collaboration and coordination in design and construction (8 interactions), and visualisation of the form (5 interactions). Similarly, lean principles such as reduce cycle time (6 interactions) and selection of appropriate control approach (5 interactions) were identified to have higher interactions with BIM functionalities.

On reviewing the interactions between lean and BIM-AR functionalities, it can be identified, there are 46 such interactions. A higher number of interactions were identified for collaboration and coordination in design and construction (8 interactions), visualisation of form (5 interactions), and AR walkthrough (5 interactions). Similarly, lean principles such as reduce cycle time (7 interactions), selection of appropriate control approach (7 interactions) have a higher level of interactions with BIM-AR functionalities.

On comparison of the two interaction matrices (Table 1 and Table 3), it is evident that the total number of interactions between BIM and lean increased due to the integration of AR. It can be noted that the integration of AR has increased the level of interactions with lean principles such as reduce cycle times and select appropriate control approach. One of the reasons for such an increase in interaction is that the application of AR visualisation and context awareness enables the user to overlay digital information into the physical environment and allow interaction with the objects and obtain the required information at the right time. For example, in site monitoring activity the AR-BIM enabled obtaining site progress information in real-time [1, 17, 21], thereby allowing the project team to obtain the right information at the right time (Reduce variability – lean principle). Similarly, in geo-visualisation of underground utilities, the AR-BIM enabled real-time visualisation of utilities on site with the required information for individual utilities [19].

It is also evident that AR integration enables achieving newer interactions that were not possible with BIM, such as the interaction between "Go and see yourself" and "AR walkthrough". For example, the integration of AR enables visualising the existing underground utilities with the onsite surrounding context, thereby enabling the project team in making effective decisions during utility maintenance works [19].

At the outset, the integration of AR not only reinforces existing BIM-lean interactions such as reduce variability, and information management but also enabled achieving newer interactions between BIM and lean. The integration of AR increases the existing interactions making the process leaner.

9 Conclusion

Being different paradigms, both BIM and lean construction are used in conformity in the current practice due to their profound benefits. In some cases, BIM is used as a lean tool in achieving lean construction. Though several studies exist on identifying how BIM functionalities conform or interact with lean principles, studies exploring if BIM integrated with AR conforms or interacts with lean principles and if AR integrated with BIM can improve current BIM-lean interactions is not explored yet. This paper attempted to identify such interactions between BIM-AR and lean in construction projects. To establish such interactions, a case study of underground utility relocation project was considered, and BIM model was developed and applied in practice. Further, interaction matrices between BIM-lean and AR integrated BIM-lean was established. It was observed that AR integrated BIM-lean poses higher interactions (46) when compared with the non AR BIM-lean interactions (36). A higher number of interactions of BIM-AR has been found mapping lean principles such as reduce cycle times and select appropriate control approach. The integration of AR with BIM enabled newer interactions with lean principles such as go and see yourself and reinforces existing BIM-lean interactions such as reduce variability, visual management and decide by consensus. The identified interactions in this paper can be used for enhancing BIM and lean adoption with AR in the construction industry.

References

1. Ratajczak, J., Riedl, M., & Matt, D.T. (2019). BIM-based and AR application combined with location-based management system for the improvement of the construction performance. *Buildings, 9*(118).
2. Castillo, G., Alarc, L.F., & González, V.A. (2007). Implementing lean production in copper mining development projects : case study. *Journal of Construction Engineering and Management, 14,* 05014013.
3. Mahalingam, A., Yadav, A. K., & Varaprasad, J. (2015). Investigating the role of lean practices in enabling BIM adoption: Evidence from two Indian cases. *Journal of Construction Engineering and Management, 141*(7), 1–11.
4. Vilventhan, A., Rajadurai, R. (2018). Application of 4D bridge information model as a lean tool for bridge infrastructure projects: A case study. In: *26th Annual Conference of the International Group for Lean Construction: Evolving Lean Construction Towards Mature Production Management Across Cultural Frontier* (pp. 1229–1239).
5. Shekargoftar, A., Taghaddos, H., Azodi, A., Tak, A., & Ghorab, K. (2022). An integrated framework for operation and maintenance of gas utility pipeline using BIM, GIS, and AR. *Journal of Performance of Constructed Facilities, 36*(3), 1–177.
6. Tarek, H., & Marzouk, M. (2022). Integrated augmented reality and cloud computing approach for infrastructure utilities maintenance. *J Pipeline Syst Eng Pract, 13*(1), 1–11.
7. Sacks, R., Koskela, L., Dave, B. A., & Owen, R. (2010). Interaction of lean and building information modeling in construction. *Journal of Construction Engineering and Management, 136*(9), 68–980.
8. Hamdi, O., Leite, F. (2012). BIM and Lean interactions from the bim capability maturity model perspective: A case study. *International Green Lean Construction 512.*
9. Elmaraghy, A., Voordijk, H., Marzouk, M. (2018). An exploration of BIM and lean interaction in optimizing demolition projects. In *Annual Conference of the International Group for Lean Construction: Evolving Lean Construction Towards Mature Production Management Across Cultural Frontier* (pp. 112–122)
10. Clemente, J., Cachadinha, N. (2013). BIM-lean synergies in the management on MEP works in public facilities of intensive use - a case study. In *21st Annual Conference of the International Group for Lean Construction* (pp. 751–759).
11. Fenais, A. S., Ariaratnam, S. T., Ayer, S. K., & Smilovsky, N. (2020). A review of augmented reality applied to underground construction. *J Inf Technol Constr, 25,* 308–324.
12. Nassereddine, H., Hanna, A. S., Veeramani, D., & Lotfallah, W. (2022). Augmented reality in the construction industry: use-cases, benefits, obstacles, and future trends. *Front Built Environ, 8,* 1–17.
13. Dudhee, V., Vukovic, V. (2021). Building information model visualisation in augmented reality. *Smart Sustain Built Environ, 12*(4), 919–934.
14. Diao, P.H., Shih, J. (2019). BIM-based AR maintenance system (BARMS) as an intelligent instruction platform for complex plumbing facilities. *Applied Science, 9*(8)
15. May, K.W., Chandani, K.C., Ochoa, J.J., Gu, N., Walsh, J., Smith, R.T. (2022). The Identification, development, and evaluation of BIM-ARDM: A BIM-based AR defect management system for construction inspections. *Buildings, 12*(140)
16. Nguyen, D.C., Nguyen, T.Q., Jin, R., Jeon, C.H., Shim, C.S. (2021). BIM-based mixed-reality application for bridge inspection and maintenance. *Construction Innovation.* (ahead-of-print

17. Garbett, J., Hartley, T., & Heesom, D. (2020). A multi-user collaborative BIM-AR system to support design and construction. *Automation in Construction, 122*, 103487.
18. Liu, D., Xia, X., Chen, J., & Li, S. (2021). Integrating building information model and augmented reality for drone-based building inspection. *Journal of Computing in Civil Engineering, 35*(2), 04020073.
19. Stylianidis, E., et al. (2020). Augmented reality geovisualisation for underground utilities. *Journal of Photogramm Remote Sens Geoinformation Science, 88*(2), 173–185.
20. Mirshokraei, M., De Gaetani, C.I., Migliaccio, F. (2019). A web-based BIM-AR quality management system for structural elements. *Applied Science , 9*(3984)
21. Singh, A. R., & Kumar, V. S. (2018). User behaviour in AR-BIM-based site layout planning. *International Journal of Production Lifecycle Management, 11*(3), 221–244.

Inventory Management of Construction Project Through ABC Analysis: A Case Study

P. Murthi, K. Poongodi, and M. Geetha

Abstract An effective inventory management is the prime parameter for success of any construction projects. The issues related to inventory management in construction particularly medium budget projects are not monitored properly and lead to increase in construction cost. However every project have a detailed project schedule and reviewed periodically with technical persons. The insufficient review process affecting the inventory management and also cost effectiveness of construction projects. The improper material management extends the project completion duration which directly affects the finance management. A volume of such kind of issues are prevailing in the developing areas particularly the proximity of Metros. In order to evaluate the problem, the proposed analysis was identified private construction projects and examined the status at completion stage. The questionnaire based survey was identified five completed projects with the project cost of more than 100 Lakhs. The net profit after completion of the project was considered as the benchmark factor and identified a project which was produced 5.14% loss when compared to the estimated cost. The impact of inventory management for the poor performance in the selected project was predicted through ABC analysis and the suggestions were given for the sustainability of construction.

Keywords Inventory management · Finance management · ABC analysis · Construction projects

P. Murthi (✉)
Centre for Construction Methods and Materials, SR University, Warangal, India
e-mail: p.murthi@sru.edu.in

K. Poongodi
Department of Civil Engineering, SR University, Warangal, India

M. Geetha
School of Business, SR University, Warangal, India

A. Vilventhan et al. (eds.), *Advances in Construction Materials and Management*, Lecture Notes in Civil Engineering 346,
https://doi.org/10.1007/978-981-99-2552-0_8

1 Introduction

The managerial task in a typical construction project can be clustered into two distinct categories such as material management and labour management. The success of completing such an unorganized construction project has mainly formulated with finance management though all sorts of resources available for construction. Based on the experiences and continuous monitoring of cash flow in a construction project, material cost including the equipment hire charges constitutes 70% of the total project cost and the remaining 30% has to be allocated to the labour cost [1]. The effective material management can be reducing the risk in the major part of finance management and able to complete the project within the stipulated budget.

The literatures mentioned that the material management is nothing but the inventory allocation according to a preplanned schedule. The effectiveness of material management is further controlling waste generation during construction which improves the outturn of works during execution. The proper implementation of inventory made simple the completion of project in the scheduled time. The prime responsibility of material management is to diminish the project inventory cost.

The major factors which increasing the project cost are varied from the type of project executed. The relationship between the contractor and the respective administration of the government projects was observed as one among the dominant factors affect the project duration which leads to additional investment beyond the estimated cost. It was observed that nearly 42% of government own projects are finished with overtime and cost [1]. Santu and Neeraj (2010) conducted questionnaire based survey for the effective implementation of material management techniques and concluded that the improper scheduling of inventory and improper participation due to various factors such as improper procurement, over stock and improper coordination among the workers [1–3]. The construction progress has to be monitored and rescheduled periodically to minimize the extension of project duration through proper scheduling techniques for predicting the undesirable fluctuations in resource utilization and eliminate the conflicts for procurements [4]. It was observed from the previous research findings that the scheduling delays occurred in 40%, 50% and 70% of government construction projects in India, UAE and UK respectively [5].

The project delay can create more issues towards implementing the financial management policies which affects in both direct and indirect cost. This can be witnessed in mega projects implanted in Indian construction industry such as metro rail projects, airport extension, marine and power projects [6]. The prolonged project duration causes increasing the construction cost which is the result of improper material management. Though the various management tools effectively utilized for monitoring the project execution, the inventory management needs an alternate tool to monitor the resource allocation. The network analysis is the popular inventory control tools which can be adopted to improve the efficient use of resources in material procurement.

1.1 ABC Analysis

ABC analysis, also called as Pareto analysis, permits the inventory managers to segregate the materials requirements according to the mandatory item of materials for execution of all kind of activities without any delay and it can be used to assist the evaluation according to the expenditure spent for procurement. Through the ABC analysis, the essential materials required for a construction project grouped based on cost per unit quantity and inventory. To improve the material management, the construction materials are segregated in to three categories, according to ABC analysis, as most important (A), less important (B) and least important (C). The details of these three categories of inventory management are mentioned in Table 1.

2 Objectives of the Study

The drive of this survey is to evaluate the current management practice for material handling in the sub-urban area of Erode Municipal Corporation and Namakkal Municipality in Tamil Nadu. The specific proposal of this study to examine the inventory monitoring approaches implemented in the current practice and suggests the suitable mechanism to optimize the construction process in both construction period and cost of the project through the effective utilization of ABC analysis.

3 Research Methodology

A structured questionnaire was prepared to collect the various construction procedure practiced in order to minimize the wastages and effective utilization of all resources and concluded with the case study. The study was carried out in two stages; in the first stage, the data were collected through interview and identified the suitable materials management practices followed during execution. In order to ascertain the effectiveness during the project execution, the final profit earning was considered as

Table 1 Categories of inventory management

Category	Inventory (% of total inventory)	Value of Inventory (% of total value)	Remarks
A	Less than 20%	80%	Keen attention required
B	30%–40%	15%	Formal attention sufficient
C	50%	5%	Relaxed inventory sufficient

benchmark. The client, builder and project engineer were interviewed for getting the detailed information about the method of execution, material management practices and finance management. The documents related to the inventory management was primarily considered to validate the data collected during the survey process. The budget utilization details were collected from the pay roll register and stock register which can also be considered to investigate the material management procedure implemented.

4 Results and Discussion

4.1 Earning Level of Selected Projects

The macro level information gathered during data collection were highlighted in Table 2. The project III and V were constructed by the method of build by own and the remaining projected were executed by the contractors. The construction scheduled duration and completed duration of all the selected projects are shown in Fig. 1 and noticed that the project V exceeds the completion duration by 5 months. All other projects were completed before the specified time schedule. The detail about benchmark parameter, called profit or earning, are shown in Fig. 2. Though the estimated cost of the projects II was found nearly double the amount of project I, the final profit was observed as 6% each. It was also observed during investigation that the projects executed by contractors were monitored by the technical supervisor and the duty of the supervisor was maintaining the suitable register and reported everyday regarding the outturn of work and material requirements for the further progress to the contract forum.

The monitoring system in comparison with the program schedule was followed as office job of the forum to avoid the delay and conflict during the execution. Further, it was observed from the project III was constructed as a time bound project and completed in the stipulated time without any delay. The final earning was found

Table 2 Earning level of selected projects

Project Identity	Nature of Project	Location	Construction Procedure	Estimated Cost (₹. Lakhs)	Actual Cost (₹. Lakhs)
I	Villa	Namakkal	Contract	125	117
II	G + 3 Apartment	Erode	Contract	215	202
III	Duplex House	Erode	Own	107	105
IV	House	Namakkal	Contract	110	98
V	G + 3 Apartment	Erode	Own	175	184

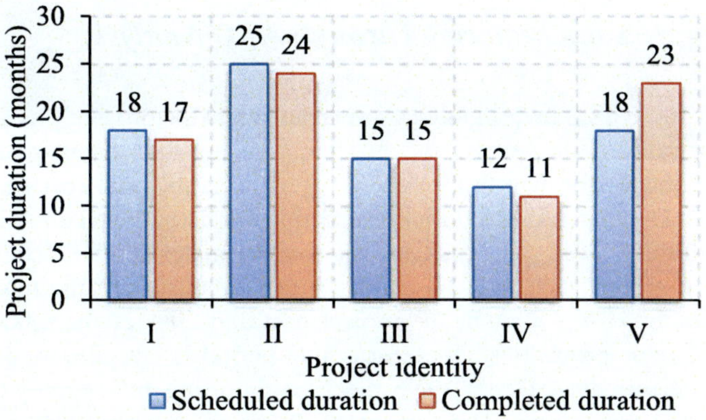

Fig. 1 Project duration

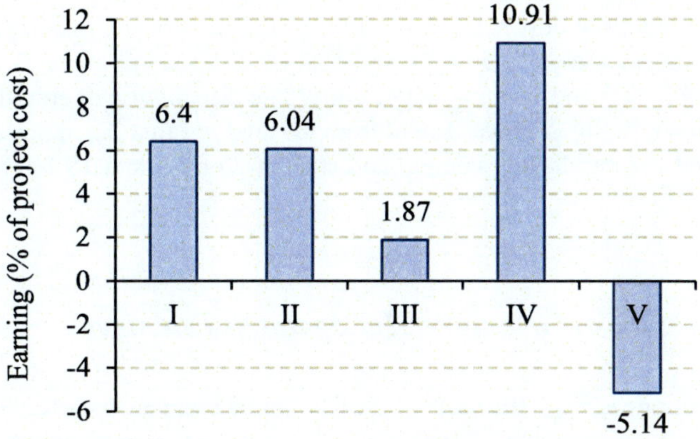

Fig. 2 Earning level of selected projects

as 1.87%, the lowest among the other profit earned projects and it was due to the improper waste management procedure. Since, it was also executed build by own concept, the generated construction recyclable wastes were not able consumed and considered as waste and affects the profit earned at the end. From Fig. 2, it was clearly understand that the project V was not monitored during the execution or it was under estimated before construction. Hence, the project V was considered for the stage II micro level analysis.

4.2 Categories the Material Through ABC Analysis

The detailed micro level investigation was carried out to the project V as a case study by collecting detailed answers for the questionnaire and subsequently verifying the available documents for determining the category wise material cost and total cost of the budget. The materials used for construction are listed based on ABC analysis as shown in Table 3. The ABC analysis is a tool for material administration which involves three groups such as A, B and C. A group consists of the most valuable items accounting 70% - 80% of the project cost and 20% of the total inventory. As per the IS code provisions, stock for a long period is not permitted due to the reduction of performance. Hence, the cement require stringent inventory control and more protected storage area. The cement stock has to be monitored carefully and re-ordered frequently and hence selected in group A.

The group C consists of least valuable items and able to maintain required inventory. The group B contains the materials ranging between A and C. The group wise distribution of material cost is shown in Fig. 3. The cost of the materials identified in group A was calculated as 53.53% and cost of materials in group B and C were calculated as 31.52% and 14.95% respectively. The distribution of cumulative material cost is shown in Fig. 4. It was found from the analysis that the major expenditure was consumed for purchasing cement and steel at 27.69% and 18.1% of total cost respectively.

4.3 Evaluation of Inventory Management Practices

The answers obtained from the part two questionnaire were analysed considering the various inventory management and shown in Table 4. The answers which are directly applicable to the selected Project V, were considered for discussion. Accordingly the answer acquired from owner of the project, who build the building by own and the selected lead labours involved in this project were categorized and comparable with the earlier publications [7–11]. The impact of improper material management which affect the outturn of the project execution and increasing the completion duration are listed as follows:

1. Improper storage space in the proximity of the construction site particularly in the category A materials
2. Improper planning of material procurement through purchasing inadequate quantity and maintaining poor inventory in category A materials.
3. The delay of material delivery by the supplier due to the poor stock and purchasing only the limited and selected supplier.

Table 3 Classification of Materials based on ABC analysis

Material	Category/ Group	Total material cost	% of usage value	Category wise total cost	% of total cost
Cement	A	5,095,000	27.69	9,850,000	53.53
Steel		3,330,000	18.10		
Timber		1,425,000	7.74		
River sand	B	2,050,000	11.14	5,800,000	31.52
Brick		2,300,000	12.50		
Aggregate (20 & 40 mm)		1,250,000	6.79		
Size stones		200,000	1.09		
Readymade Doors and Windows	C	335,000	1.82	2,750,000	14.95
Plywood		230,000	1.25		
Plumbing items		325,000	1.77		
Electrical items		450,000	2.45		
Flooring tiles and Wall tiles		305,000	1.66		
Cub board and Kitchen granite slabs		175,000	0.95		
Doors and window fittings		95,000	0.52		
Window and Ventilator grills		132,000	0.72		
Staircase grill		83,000	0.45		
Water supply and Sanitary fittings		145,000	0.79		
Paintings		215,000	1.17		
Elevation tiles		30,000	0.16		
Paver blocks		45,000	0.24		
Hollow blocks for compound wall		50,000	0.27		
M-sand		25,000	0.14		
Filling soil		25,000	0.14		
Cover block		5000	0.03		
Miscellaneous		80,000	0.43		

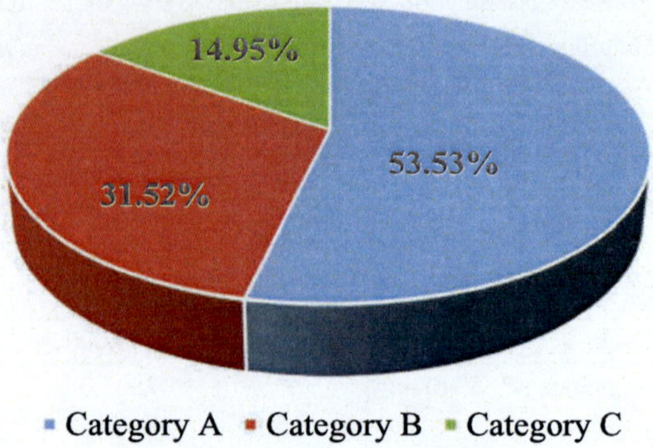

Fig. 3 Material cost distribution

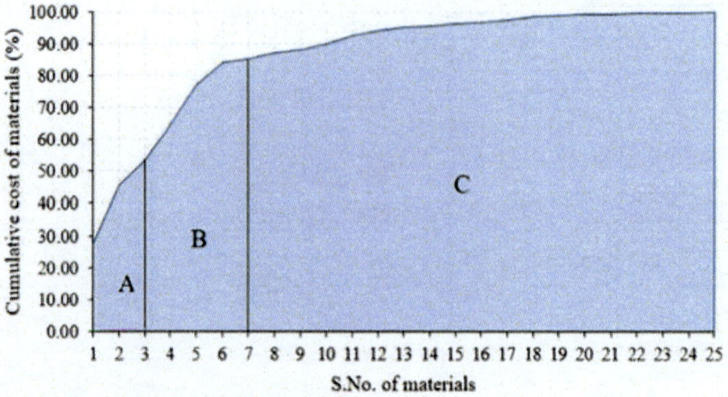

Fig. 4 Distribution of cumulative material cost

4. The mechanism towards avoiding the wastage and damage of the category A materials both stock and transportation was not implemented during construction due to the lack of knowledge.
5. The knowledge of technical supervisor appointed by the owner about the material performance and inventory management were observed as not up to the expected level.
6. The preparedness of execution of some specific works were not monitored properly.
7. Improper coordination among the workers were noticed and developed wastage at the construction site due to improper site management particularly in category A materials.
8. The purchasing price for some major materials were found in higher order than the market rate due to the delay in payment of previously supplied materials and further observed it subsequent delay in next supply of materials.
9. The improper finance management through the submission of inadequate document to the bank which caused delay in money sanction and found delayed in execution of construction and delay in purchasing of materials in some occasions.

Table 4 Evaluation of inventory management

Material management factors	Description of factors covered in the questionnaire	Level of impact in material cost		
		Mostly	Moderate	Unlike
Insufficeint planning of materials	Delay in taking decision of material selection	Y		
	Shortage of materials in market		Y	
	Changes in material prices during construction	Y		
	Inadequate order quantity of materials	Y		
	Lack of knowledge with respect to lead time of materials from specific supplier		Y	
	Purchasing from only selected suppliers	Y		
	Not receiving materials at time of requirements	Y		
Inappropriate distribution of materials	Receiving incorrect type of materials			Y
	Lack of knowledge about the materials		Y	
	Improper handling during transport and site			Y
	Receiving incorrect type of material			Y
	Supply of poor quality of materials		Y	
	Damages and Loss of materials during delivery		Y	
Financial issues in purchase	Delay in payment to supplier	Y		
	Shortage of fund to purchase material	Y		
	Delay in sanction of bank loan	Y		
Changes in scope of materials	Changes in plan and design during construction		Y	
	Changes in materials during construction		Y	
	Delay in exchange of improper supply	Y		

5 Conclusions

A rapid growth in construction industry had been observed when investigating the constructed and under-construction building in the sub-urban areas. However the construction schemes often affected due to the postponement and overall cost involved. Improper inventory management has been reported as a critical cause of this failure in execution as per the planned schedule. It is necessary to identify the critical inventory based on the ABC analysis and evaluated by critical analysis for adopting suitable methods to minimize the impact of poor inventory management on the total cost of the project. However, the following are selective conclusions and these parameters have to be focused in further projects:

1. Once the project plan and design were finalized, it should not be revised and if revision is needed, it should be within the budget allocation.
2. A suitable experienced technical person has to be appointed to monitor the execution and he has to be reported frequently and conducted regular meeting to brief the growth and deviations if any in the progress. If any deviations found in this review, suitable modification should be made in the schedule and complete the task within the time limit.
3. The procurement policy should be improved and hence the suitable quality and quantity only purchased as per the preplanned schedule.
4. The monitoring in the labour system is a complex one in construction sector since different category of workers are involved simultaneously. Making coordination among the various crews are mandatory in order to avoid the delay in project execution and this can be achieved proper inventory management.
5. The payment policy for purchasing the materials are to be standardized unless the unit rate may be increased which leads to improve the considerable increase in the inventory cost.

References

1. Kar, S., & Neeraj Jha, K. (2020). Examining the effect of material management issues on the schedule and cost performance of construction projects based on a structural equation model: Survey of Indian experiences. *Journal of Construction Engineering and Management*, *146*(9). https://doi.org/10.1061/(ASCE)CO.1943-7862.0001906.
2. Said, H., & El Rayes, K. (2011). Optimizing material procurement and storage on construction sites. *Journal of Construction Engineering and Management*, *137*(6). https://doi.org/10.1061/(ASCE)CO.1943-7862.0000307.
3. Jun, D.H., & El Rayes, K. (2011). Fast and Accurate risk evaluation for scheduling large-scale construction projects. *Journal of Construction Engineering and Management*, *25*(5). https://doi.org/10.1061/(ASCE)CP.1943-5487.0000106.
4. Hegazy, T., & Menesi, W. (2008). Delay Analysis under multiple baseline updates. *Journal of Construction Engineering and Management*, *134*(8). https://doi.org/10.1061/(ASCE)0733-9364(2008)134:8(575).
5. Ibbs, W., Nguyen, L.D., & Lee, S. (2007). Quantified impacts of project change. J. Profess. Iss. Eng. Educ. Pract. *133*(1). https://doi.org/10.1061/(ASCE)1052-3928(2007)133:1(45).
6. Randolph Thomas, H., Riley, D.R., & Messner, J.I. (2005). Fundamental principles of site material management", *Journal of Construction Engineering and Management*, 131(7). https://doi.org/10.1061/(ASCE)0733-9364(2005)131:7(808).
7. Gurmu, A.T. (2019). Tools for measuring construction materials management practices and predicting labor productivity in multistory building projects", *Journal of Construction Engineering and Management*, *145* (2). https://doi.org/10.1061/(ASCE)CO.1943-7862.0001611.
8. Venkat Reddy, P., Siva Krishna, A., & Ravi Kumar, T. (2017). Study on concept of smart city and its structural components. *International Journal of Civil Engineering and Technology*, *8*(8), 101–112.

9. Baba, S. K. V., & Haripriya, S. (2017). Performance analysis of black cotton soil treated with granite dust and lime. *International Journal of Civil Engineering and Technology, 8*(10), 1341–1350.
10. Singh, S.P., Satheesh Raju, G., & Shravan, M.: Waste management in Construction - A study with reference to India. *International Journal of Civil Engineering and Technology, 9*(9), 533–538.
11. Shravan, M., Satheesh Raju, G., Singh, S.P., Yamsuani, N., & Mahesh, D. (2018). Construction materials management on construction sites. *International Journal of Civil Engineering and Technology, 9*(3), 809–818.

Crisis Management Due to Covid-19 in Indian Construction Industry - An Overview

P. Murthi, K. Poongodi, and V. Mahesh

Abstract The crisis developed due to the Covid-19 was ever seen un-expected situation in construction industry. The construction industry is the leading un-organised sector next to agriculture in India and leader in working with migratory labours. The sudden lockdown and various restrictions implemented day by day by the government took the construction industry in to pitiable situation. In this contextual, this paper was intended to understand the crisis in construction industry and its impact due to Covid-19 pandemic condition. The threat situation in construction sector like loss of life, economy setback etc., was highlighted along with possible remedial measures to be carried out during the future scenario of construction industry after resuming the execution in a phased manner.

Keywords Covid-19 · Crisis management · Construction industry · Loss of life · Economic setback

1 Introduction

The construction industry has considered as the second largest industry after agriculture in India [1]. Lot of crises has affected the Indian economy during the last two decades. The construction industry is the most affected industry [2]. The crisis developed in construction industry had not focused properly compared to other industry. The latest crisis is Covid-19 which affects all the industries including construction sector. The important resources required in construction industry are men, materials,

P. Murthi (✉)
Centre for Construction Methods and Materials, SR University, Warangal, India
e-mail: p.murthi@sru.edu.in

K. Poongodi
Department of Civil Engineering, SR University, Warangal, India

V. Mahesh
Department of Mechanical Engineering, SR University, Warangal, India

A. Vilventhan et al. (eds.), *Advances in Construction Materials and Management*, Lecture Notes in Civil Engineering 346,
https://doi.org/10.1007/978-981-99-2552-0_9

machineries and money. The availability of all above said resources are affected tremendously due to Covid-19 pandemic situation. Beyond that, the supply chain among the resources is also fully destroyed in an unbelievable way. Due to the effects of Covid-19, construction business was delayed and some of them were cancelled and most of the new projects are dropped / stopped. Further, possible supply chain blockages of all sorts of resources to small screw from various part of the country can root the project interruptions in various projects. As per the literatures, a crisis is a sudden change in the normal procedures.

According to Heath [3], the crisis is a sudden variation in regular activities and because of that a totally new system is come to rule. Though the vaccination will be invented in the earliest future, the impact of the crisis due to Covid-19 is extended in a cumulative manner. The construction industry was already affected by demonetization to some extent since the construction sector is leader in un-organized industrial sector. Though new strategies are being formulated to manage the crisis, the impact will be prolonged.

The construction industries are not having crisis management strategies in the fundamental execution procedure except the unexpected accidents [4]. However, no one think about this extended lockdown situation after Covid-19 was found various part of the world. The most immediate impacts were found in the construction process like struggling of subcontractor to complete the time bound task due to the construction site shutdowns. The subcontractors may be mostly affected even the site shutdown extended for one week. With this background, the objectives of this paper, the factors such as the meaning of crisis, the impact of Covid-19 on construction industry and the different approaches to manage the situation were considered and the results are discussed.

2 Crisis Management

The crisis is nothing but as a major risk to functional operations if it is not handled properly [4]. Only Covid-19 itself is not at all crisis, accidents, natural calamities like flood, earthquakes are also considered as crisis [5]. The crisis management is an important one for all type of industries. It should be dynamic and continuous that includes the identification of the problems, planning to answer the clients, to face the challenges against the crisis and solve it. The crisis management strategies are varying from industry to industry and also the nature of crisis. It should take the challenges of the past, implementation for the current situations and match for the future also. Unfortunately, no construction company has the crisis management policy even for the frequent accidents.

3 Impact of Covid-19 on the Construction Industry

Covid-19 was a disaster, spreading all over the world at high speed. All most all industries faced and affected this crisis. Engineering and construction industry are also affected without any exception. The expected growth of construction industry in 2020 has been revised from 3.1% down to 0.5% [6]. Where there is a big problem there will be a big opportunity, its counterparts for this Covid-19 situation also. The extraordinary opportunities are existing in the mere future which necessitated to solve some of the industrial historical challenges and to prepare for a more digital future. In most cases, crises can create four related threats:

- Loss of life
- Economic setback
- Reputation loss
- Labour loss

3.1 *Loss of Life*

Covid-19 lockdown situation is most likely creating all kind of threats at a time for all industries can ever face. The positive corona creates negative impact in all industries reputation get downward graph and un-believable financial loss. The covid-19 is disturbing the workplace environment. Employees have to be adjusted in to new ways of working environment and the employers are finding the ways to ensure the safety of their employees [7]. All these sudden changes are impacting the employees comfort and mental health. Everyone from top to bottom level of management is felt stressed, insecure and worried. The employees are worried about their jobs, the viability of their company and their ability to take care of their families and studies of the children. The organization set up of all industries is always triangle. The peak is for owners and CEOs, the middle of the triangle is always allotted for technocrats or white collars and the bottom most is for workers or the blue collars. The Covid-19 pandemic affects all the levels of an industry like an Earth quake or Tsunami. The impact of earth quake is sudden but able to predict using various systems and evacuation is possible but the Covid-19 impact being observed as continuous it terms of month. One of the latest statistics had exposed the loss of life up to August in India shown in Fig. 1. This rate is now growing cumulatively [8].

Though no loss in infrastructure due to Covid-19, the impacts are more like upsetting the survival and economic development similar to other disaster. In any one of the other crisis situations, one can affected and the other can extend their help. But in this Covid-19 crisis, such strategy is not applicable since everybody in pandemic situation. This situation had created panic in the entire human beings' mind since it leads to their loss of life.

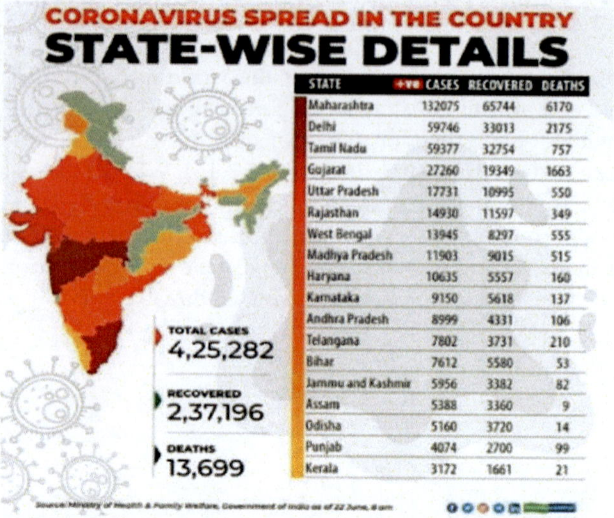

Fig. 1 State-wise situation of Covid-19 in India up to August 2020. (*Source*: https://www.deccanherald.com/national/coronavirus-news-highlights)

3.2 Economic Setback

Corona virus was accepted as pandemic by the World Health Organization on 11th March 2020. The pandemic has steered to multiple challenges almost all the industries including the construction and infrastructures. The latest statistics had shown a clear picture on the worst situation in the GDP in India as shown in Fig. 2. In June-quarter 2020, India affected worst among the other nations in the world and found 23.9% negative gross Domestic Product (GDP) growth rate. Among the various industries in India, the construction sector was extremely affected by 50.3% negative growth [9]. The construction industry is one in which more than 50% migratory labours are involved. In India the unskilled workers depends mostly on the construction industry for employment [10]. All kind of labours involved in construction industry are struggled to survive since mostly all are below poverty level. The loss of current employment and more uncertainty in future employment in construction sector is directing such group of people in to alternate job.

3.3 Reputation Loss

Because of the restrictions enforced by the State and Central Governments, construction industry and its associated profession across the country have stopped. Time is the essence of any contract agreement in addition to the utilization of all other resources. Time management is the prime importance in construction and clients

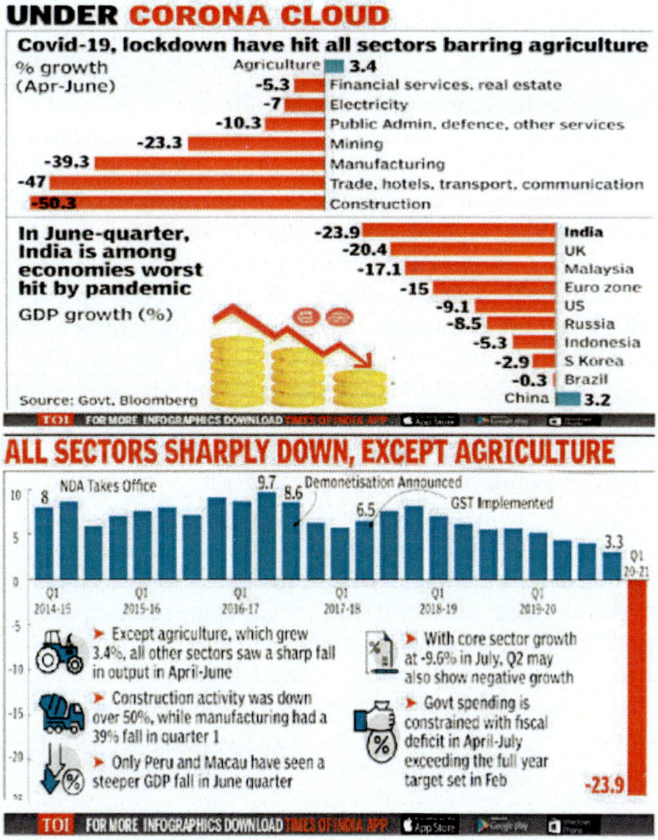

Fig. 2 The impact of Covid-19 on GDP in India. (*Source*: https://timesofindia.indiatimes.com/business/india-business/india-faces-dwindling-policy)

are very specific in the time for completion. Section 5 of Indian Contract Act, 1872 provides for the effect of failure to meet the specified time limit [11]. However, the law permits extension of time in certain situation to the contractor. In the meantime, a contractor fails to perform even in the extended period as per the Sect. 5 of contract act 1872, would be summoned against the defaulting contractor [12]. This build a negative status on the reputation of the contractor such scenario was developed due to the Covid-19 lockdown situation in most of the projects especially industrial structures.

3.4 Labour Loss

Covid-19 is setting unbelievable pressures of different magnitude in all parts of construction. The manpower is the most important for any construction industry. The workforce management is the major part of the construction industry and the supervisory level personnel are concentrating an equal importance in both execution and labour management. The entire construction industry are struggled to resume operations without the workforce though all other resources are manageable during this Covid-19 pandemic situation. In order to operate all sorts of special equipments, sufficient skilled manpower is required. The workforce in construction industry can be grouped in to two categories namely local labours and migratory labours. Nearly 70% of workers in the selected area are employed in migratory category. It was observed that, the labours migrating from Odisha, Jharkhand and Bihar are mostly employed in Telangana state [10]. The majority of this kind of workforce is employed in the big construction projects and migrated to their native places during this issue. A group of migratory labours are assigned to complete a specified job within the stipulated time and they are called as bonded labours. The construction industry at present facing this labour problem leads to the labour loss in construction which affect the growth of construction industry though the execution are initiated in a phased manner. There was a survey conducted among the migratory labours about the issues during Covid-19 pandemic situation and the results are shown in Fig. 3. More than 27% of construction workers worried about their job security and more than 20% of labours expressed their health aspects due to this Covid-19 pandemic situation.

4 Coronavirus Crisis Management Opportunities

The authorities of construction companies are planning to prevent the effects caused due to the Covid-19 lockdown situation. In order to understanding the present scenario and future strategies to be implemented for managing the present condition, a questionnaire based survey was made and the way in which the contractors and company management highlighted their possibilities depends up on the nature of projects [13]. The following points are the few opportunities highlighted by the construction industry authorities:

1. Establish a designated technical team to ascertain the impact of lockdown situation.
2. Develop a strategic plan about the progress of work and the quantum of works to be completed within the project duration.
3. Establish a revised or an alternate network diagram incorporating all sorts of possibilities.
4. Appoint additional managerial and administrative team to monitor the progress.

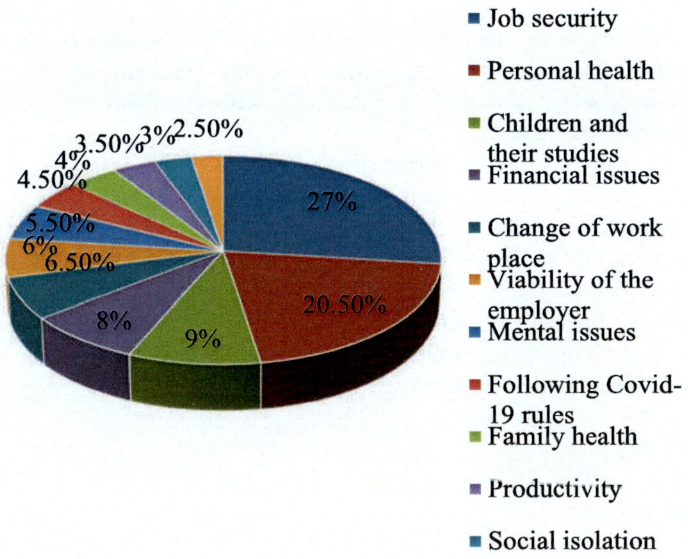

Fig. 3 Labour issues during Covid-19

5. Adopt a suitable methodology to operate the construction activities in overtimes and provide the additional incentives to the workers for the overtime works.
6. Ensure the availability of labour force for different works and a special supervisor may be appointed to monitor the labour issues.
7. Establish a concrete confidence among the workers against the Covid-19.
8. Extent the facilities provided to the construction workers like personal protective devices, accommodation at site, health care, insurance, avoiding the occupational hazards etc.
9. Update employees contact numbers and emergency contact details.
10. Ensure the availability of required inventory at site and take measure to avoid the down time [14].
11. Providing the proper private transport facilities for the workers to avoid the delay in construction progress.

5 Conclusions

The following conclusions were arrived from this survey-based investigation on the impact of Covid-19 crisis in construction industry:

1. Covid-19 crisis was ever experienced un-expected situation. The construction industry was affected extremely higher hit rate than any other industry.

2. Loss of life, economy and opportunities for construction labours were acknowledged particularly the migratory labour crew was worst affected.
3. The loss of economy in construction and real estate sector was observed as highest among any other industry.
4. The reputation of the construction industry was observed in poor status particularly in small scale construction industry.
5. Majority of construction labours were switch over to alternate possible employment opportunities and rate of return to the construction industry was very less particularly the migrated labours, which is observed from the survey report.
6. The future perspective of such crisis has to be included in the contact management and the listed suggestions were observed from various organization in order to overcome the Covid-19 crisis.

References

1. Maneesh, P., & Jasna, P. T. (2017). Socio- economic condition of women workers in Kannur District Kerala. *Indian Journal of Economics and Development, 5*(8), 1–11.
2. Poongodi, R., & Revathi, K. (2010). Economics of women construction workers with special reference to Thuraiyur Taluk Trichirappalli District in Tamil Nadu. *Asian Journal of Management Research, 2*(1), 684–687.
3. Heath, R. (1998). Working under pressure: Crisis management, pressure groups and the media. *Safety Science, 30*, 209–221.
4. Ocal, E., Laptali, E., & Oral, E.E. (2006). Crisis management in Turkish construction industry, *Building and Environment*, ***41,*** 1498–1508
5. Poongodi, K., Archana Reddy, R., Murthi, P., Udaya Sree, Ch., & Praneeth Paul, B. (2020). Studies on the status of the women construction workers before and present Covid-19 situation in Warangal Districts. *IOP: Materials Science and Engineering Proceedings*, unpublished accepted article.
6. https://www.accenture.com/no-en/insights/industrial/coronavirus-engineering-construction-impact-esponse
7. Vaidya, V.G., Mamulwar, M.S., Ray, M.S., Beena, R., Bhathlawannde, P.V., & Ubale Lokmanya, S. (2015). Occupational health hazards of women working in brick kiln and construction industry. *International Journal of Applied Management and Business Utility*, ***4***(1), 45–54.
8. https://www.deccanherald.com/national/coronavirus-news-highlights-andhra-pradesh-records-new-single-day-high-of-1933-covid-19-cases-taking-the-aggregate-tally-to-29168-856206.
9. https://timesofindia.indiatimes.com/business/india-business/india-faces-dwindling-policy-options-after-record-gdp-slump/articleshow/77871746.cms
10. Poongodi, K., Murthi, P., & Gobinath, R. (2022). Impact of Covid-19 pandemic lockdown on the socio-economic condition of brick manufacturing unit – A case study, *IOP: Materials Science and Engineering Proceedings*, unpublished accepted article.
11. The Indian Contract Act, 1872, https://www.indiacode.nic.in/handle/123456789/2187
12. Rengarajan, L. (2017). Living and working conditions of women construction workers. *Shanlax International Journal of commerce, 5*(2), 10–18.

13. Singh, S.P., Satheesh Raju, G., & Shravan, M. (2018). Waste management in construction - A study with reference to India. *International Journal of Civil Engineering and Technology, 9*(9). 533–538.
14. Shravan, M., Satheesh Raju, G., Parminder Singh, S., Yamsuani, N., & Mahesh, D. (2018). Construction materials management on construction sites. *International Journal of Civil Engineering and Technology, 9*(3), 809–818.

Studies on the Factors Influencing Occupational Accidents on Health Hazards of Labours in Thermal Power Plant Construction

K. Poongodi, P. Shivakrishna, and P. Murthi

Abstract One of the most accident-prone industries in the Polish economy is construction. Accidents have numerous root causes, many of which are events that take place at work. In favorable conditions, these elements can considerably raise the likelihood that a danger will become operational and result in an occupational accident. The article lists the key elements that the surveys revealed were responsible for developing most accidents in the construction sector. Due to their placement within the construction work environment and its surroundings, these elements are split into seven groups. Epidemiology and retrospective methods were used to identify essential factors of occupational accidents and their correlation between occupational accidents and the socio-demographic profile of a laborer.

Findings - The study identifies the essential influencing factors of occupational accidents. The factor loading of these factor is as follows FRFP (80.6%), followed by SAOC (77.4%), CMM (77.2%), CFO (71.7%), FFS (70.8%), HOL (70.2%), and BC (67.0%). Here study also identifies the socio-demographic factors such as age, gender, and work experience affecting the occupational accidents at the thermal power construction site in Telangana. As per Telanga workers, the most common age group of injured workers is 15–25 years. Occupational accidents are most likely to occur in male workers with less experience. The likelihood of an accident occurring is highest in the first year of employment, which is considered an active workplace compared to subsequent years of employment. However, male workers with less experience are the ones who get affected easily.

Practical Implications – This paper gives insight to construction managers in two ways. Firstly, how to reduce the number of occupational accidents at the thermal power site, and secondly, it will help the managers segregate the accidental issues under one cluster to work out a more detailed training and safety program for workers.

K. Poongodi (✉) · P. Shivakrishna
Department of Civil Engineering, SR University, Warangal, India
e-mail: k.poongodi@sru.edu.in

P. Murthi
Centre for Construction Methods and Materials, SR University, Warangal, India

A. Vilventhan et al. (eds.), *Advances in Construction Materials and Management*, Lecture Notes in Civil Engineering 346,
https://doi.org/10.1007/978-981-99-2552-0_10

Originality – This research enhances the discussions on workers' safety and training needs in the thermal power construction site and will also support the type of workers who need the training most. The paper will also help increase labor efficiency and boost the plant's economy, adding value to the facility and helping the utility establish a solid reputation in the industry.

Keywords Thermal power plant construction · Occupational accidents · Factors influencing the occupational accidents

1 Introduction

Workplace accidents have a serious negative impact on laborers' health. Industrial accidents are not only dangerous for organizations and the nation, but they also have social repercussions, psychological and emotional connections [15]. According to BSS (Brazilian Social Security) workplace accidents are defined as accidents that occur during task performance at work-site. These accidents result in physical harm for a long or short-term functional impairment that may cause death, loss, or reduced ability to work. As per the statistics, in every 15 s, laborers die from an accident or have got occupational sickness. International Labour Organization (ILO) estimates about 2.3 million laborers worldwide die from illnesses or accidents caused at work-construction sites yearly, or more than 6000 people per day. Over 160 million work-related disease cases have been filed annually, and 340 million work-related accidents worldwide. These numbers are periodically updated by the ILO, and the updates show a rise in accidents and illness.

Compared with the 5,850 documented cases, the projected number of fatal occupational accidents in the Commonwealth of Independent States (CIS) countries is over 11,000. The severity of the issue is being misrepresented by the egregious underreporting of occupational diseases and accidents, including fatal accidents. According to ILO's, the most recent statistical data on occupational accidents, illnesses, and deaths globally are as follows:

- The majority of worker fatalities are caused by occupational diseases. According to estimates, hazardous substances alone result in 651,279 deaths annually.
- Accident rates in the construction sector are abnormally high.
- Workers in their younger and older years are particularly at risk because of the aging population in developed nations.

Further, according to [21] lower the probability of workplace accidents, it is vital to design efficient safety management systems. Organizations are using a variety of instruments, such as management systems that can be assessed and accredited, to assess their products, services, and operational procedures to enhance their risk management strategies. Furthermore, an occupational accident is a multidimensional problem with several predicted issues, such as falling from the roof-floor platform, falling from a scaffold, get contact with a falling object, getting in touch with affecting

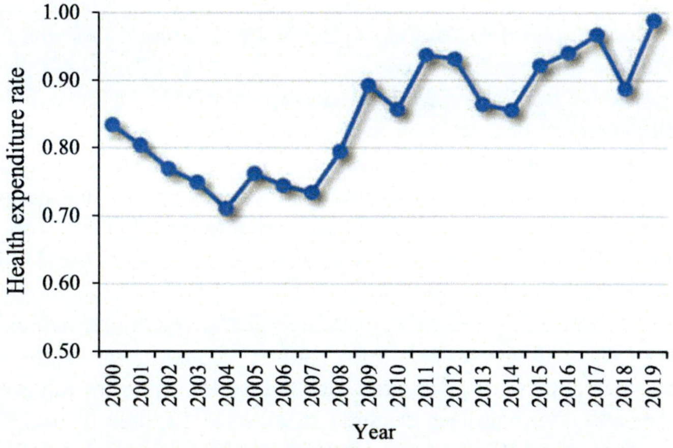

Fig. 1 Growth Chart of Domestic General Government Health Expenditure

parts of a machine, and getting hit by an object during lifting, etc. Humans have a fundamental need for safety on both a physical and psychological level. Over 720,000 workers suffer injuries, and 950 people die due to industrial accidents each day. Further, as per the researcher, almost 37 million work-related accidents result in at least 4 days off from work each year in India, where more than 48,000 people die as a result of these mishaps. Economically speaking, the ILO has estimated that the expenses of work-related illnesses and accidents is equal to 4% of the country's GDP. According to the World Bank (2019) database, Indian domestic general government health expenditure has increased to 0.99 in FY2019 from 0.83 in FY2000 as shown in Fig. 1.

After agriculture, the construction sector employs the second-highest number of people in the country [7]. In addition, the writer said, according to the national president of the Builders' Association of India, over 1-lakh construction businesses, of which 250 are prominent, and approximately 4-lakh sub-contractors and contractors work in this trade. That is why the accident figures for the Indian construction industry are not accurately and consistently disclosed. However, due to the characteristics of Indian construction and its dynamic nature, the involvement of numerous stakeholders in a project, including migrant laborers, and a less controlled environment, it is projected that many deadly and non-fatal accidents will occur.

In the article published by [17], 80% of Indian construction sites are unsafe, which is why the death rate is 20 times higher in India compared with Britain. The author further said, despite efforts done by the Indian government to position the country's construction industry is the 2nd largest employer in the country post-agricultural industry, employing more than 44-million people and contributing close to the 9% GDP, it is ironic that the industry's workforce is less protected than that of any other sector in the nation. According to data, there is a five-times more significant chance of death in the construction sector than in the manufacturing sector and a 2.5-times greater chance of a catastrophic injury.

A survey done by Bandhkam Majoor Sangathan (BMS) has revealed that India is not the only country lacking legal protection; there are other countries as well, like Britain and many more. 25% of deaths are from workers falling from a height, and nearly 80% of the deaths are preventable. Unfortunately, the International Labor Organization and the government of India lack reliable data that could shed light on the predicament of a sector with such a sizable labor population and the deaths that happen due to occupational accidents.

In 2018, there were 137 fatal accident-related deaths in Gujarat, which makes up approximately 13% of all investments in the country's construction sector, the highest in the last ten years. According to data collected by Pandya with the assistance of BMS volunteer and civil engineer Darshan Patel. In 2017, there were 67 fatalities, compared to 55 in 2016, 62 in 2015, and 69 in 2014, as shown in Fig. 2. This information was obtained by submitting RTI (Right to Information) requests to the Gujarat Police and media clippings. The report also reveals that 21% of deaths in Gujarat were caused by being buried in rubble; 49% were caused by falling from a height etc. 74% of accidents resulted in no FIR being filed. FIRs are only filed in fatal accident instances (Noora et al., 2014). According to Patel, no FIR is filed, even if the harm renders a person permanently disabled.

According to other data, 84% of victims worked at private sector sites, while the remaining victims worked at state-owned sites. While Ahmedabad accounted for the majority of victims, 38% of construction site accident victims were in the age range of 19 to 28, followed by 16% of victims in the age range of 29–38 years; 37% of the victims were locals; 21% were from another village or town, and the remaining 17% were not from Gujarat.

There are several causes for so many deadly incidents, such as worker ignorance, contractor refusal to require workers at construction sites to wear safety equipment,

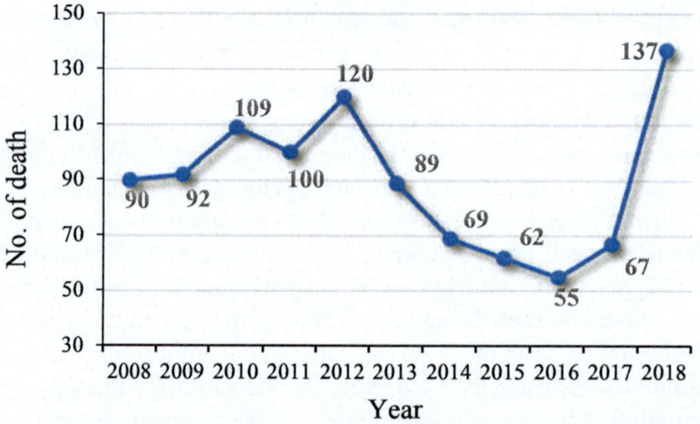

Fig. 2 Number of Construction Workers' Deaths in Gujarat (FY08-18)

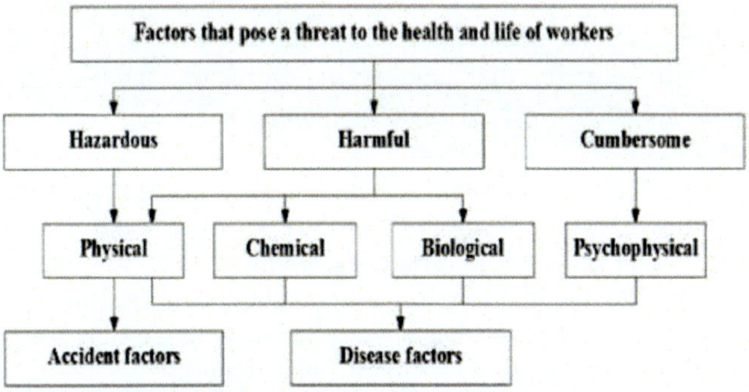

Fig. 3 Accidental Factors that Pose a Threat to the Workers' Life

worker ignorance, and lack of government inspection (as per [17], there is only one inspector for every 506 registered sites in India). This all happens due to the lack of safety issues. According to structural engineer Rajendra Desai, "civil engineers, who are the crucial individuals to monitor the building construction, are not trained in safety measures". In addition to the risk of fatal accidents, those employed in the construction sector also risk contracting several occupational diseases, such as dermatitis, silicosis, respiratory illnesses, muscular-skeletal disorders, asbestosis, etc., all of which can cause slow death and disability. If there were enough physicians who were knowledgeable about occupational sickness, these could be avoided.

2 Study Objectives

The study's main objective is to assess the most critical factors influencing the development of occupational accidents in the construction industry. The identified factors will be based on a survey done in the thermal power construction industry, especially in Telangana (Fig. 3).

3 Research Methodology

The method employed to assess the factors affecting occupational accidents in the thermal power construction industry is exploratory factor analysis (EFA). This is a quantitative research approach where data has been taken from laborers from different construction sites and time zones. The survey method has been used to collect data, wherein secondary data from the previous literature have been used to design the structured questionnaire. Major accidental incidents have been taken from the journal written by and [13]. Overall there are 25 incidences have discussed, and responses on the same have been collected in the survey form. Apart from 25 statements, eight statements are on the demographic profile of a laborer. There were 225 laborers contacted, out of which 214 had responded to all the 33 statements. Questions on accidental incidents are based on a 5-points Likert scale wherein laborers are expected to respond from strongly disagree to strongly agree. Further impact of demographic profiles such as age, gender, and experience, have been evaluated on occupational accidents via a non-parametric Chi-square test. SPSS-IBM software is used to evaluate and interpret data. Before assessing the data, the internal consistency of the responses of 214 data points has been calculated via Cronbach alpha. For data collection, the purposive sampling technique has been used.

3.1 Labor Demographic Profile

Laborer data profiling has been worked out to understand workers profiling at the thermal power plant construction site in Telangana. Here data infers that 18.7% of workers' age group is 18–20 years, followed by 21–25 years (33.6%), 26–30 years 14%, 31–35 years (24.3%), and more than 36 years 9.3%. Among all, 62.1% of workers are male and the remaining 37.9% are female. Regarding years of working experience, 56.5% of workers have less than 5 years of experience, and 43.5% have more than 5 years of working experience at a construction site. Further, 69.6% of workers have children, and the breakup is as follows 1-child (25.7%), 2 children (29.9%), and more than or equal to 3 children 14%. Apart from that, 81.3% of workers work with construction sites where the number of laborers is more than 1000 daily. While discussing, workers informed that 71.5% of them had met an accident while working at a construction site and would send their children to work (66.8%) (Fig. 4).

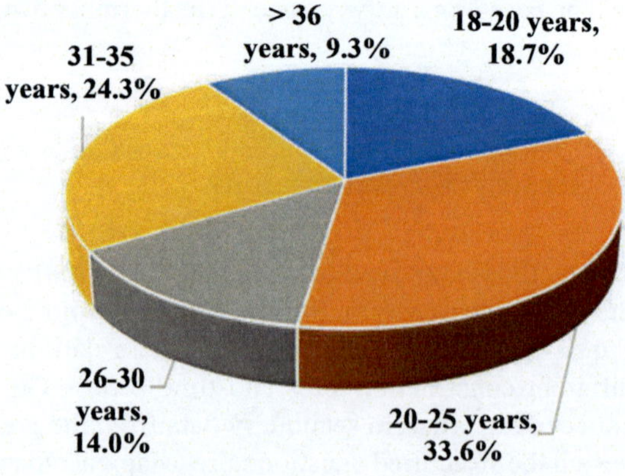

Fig. 4 Laborers' Age

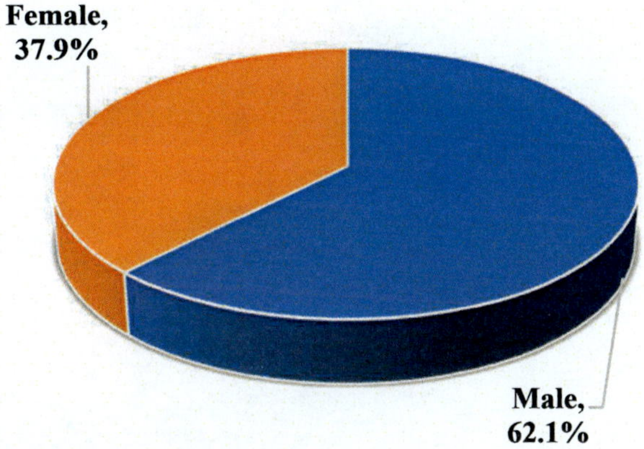

Fig. 5 Laborers' Gender *.Source: Collected Data from Thermal Power Construction Site*

3.2 Data Internal Consistency

The internal consistency of the data is calculated via Cronbach alpha, where how closely each item is related to each other has been evaluated. In the data, 214 workers have shared their experiences on occupational accidents. According to the results, the Cronbach alpha value is close to 0.782, greater than the tolerance limit of 0.70. Data also revealed that none of the statements have been excluded from the study. Further ANOVA with Cochran test shows a significance level at 95% confidence level. As per the data, the Cochran Q-value is close to 1394.979 and has a significance value of 0.000. Hence it can be concluded that there is an internal consistency in the data of 214 workers. Therefore data can be used for further analysis at a 5% significance level. Overall gathered data is found to be reliable and consistent for further analysis (Figs. 5, 6, 7, 8, 9, 10 and 11) (Tables 1, 2 and 3).

3.3 Socio-Demographic Wise Occupational Accidents

The retrospective review assessing the socio-demographic effect on occupational accidents will give a unique insight on the targetted workers to focus on reducing accidents at the construction site, especially in Telangana. A non-parametric chi-square test has been used to find the significant impact of the workers' demographic profile on workplace accidents at 95% confidence level. In the socio-demographic profile of a worker, age, gender, and working experience have been taken, wherein at the construction site, the minimum age of a worker is 18 years. The age breakup of these workers is as follows 18–20 years 18.6%, 21–25 years (33.6%), 26–30 years (14%), 31–35 years (24.2%), and > 36 years (9.3%).

Hence it can be concluded that most of the workers (52.3%) at the construction site fall in the age group of 15–25 years. Regarding gender, 62.1% of workers are male, and the female contribution is 37.8%. In addition, the working experience of

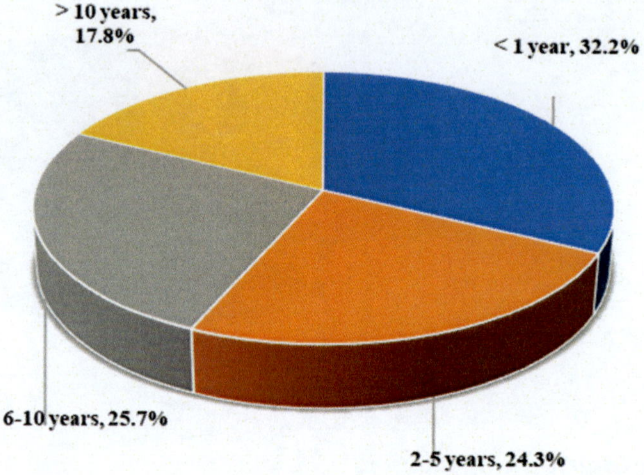

Fig. 6 Laborers' Years of Experience

Fig. 7 Laborer ever met any accident in Thermal Power Construction Site. *Source: Collected Data from Thermal Power Construction Site*

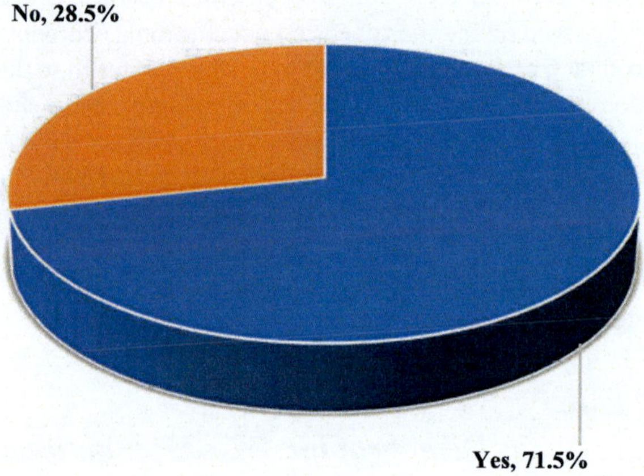

most of these workers (32.2%) is less than a year, followed by 2–5 years (24.2%), 6–10 years (25.7%), and more than 10 years (17.7%). Further, the chi-square and significant values of the demographic profile of the workers are as follows age ($\chi 2 = 8.214$, $\alpha = 0.041$), gender ($\chi 2 = 9.580$, $\alpha = 0.008$), and ($\chi 2 = 9.949$, $\alpha = 0.012$). All the significance values are significant. Hence it can be concluded that age, gender, and work experience significantly positively impact workers' occupational accidents at the thermal power construction site in Telangana.

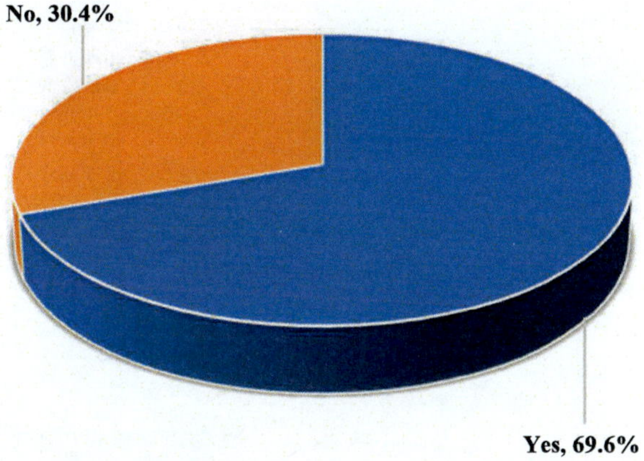

Fig. 8 Laborers have Children

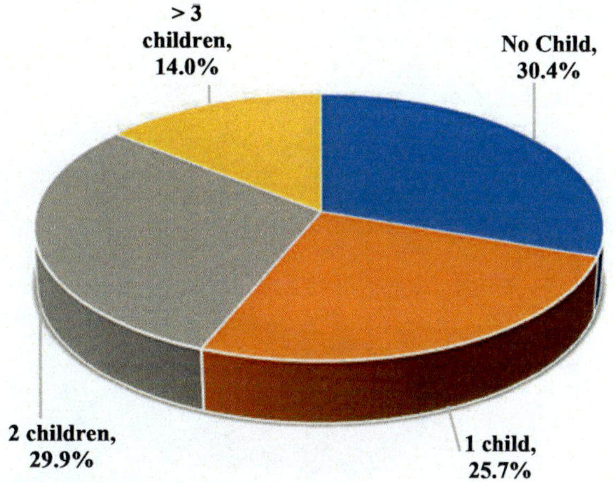

Fig. 9 Laborers' No. of Children. *Source: Collected Data from Thermal Power Construction Site*

4 Results

Factor analysis has been used to comprehend how factors influence occupational accidents in the thermal power construction industry. The objection of the factor analysis is to assess the factors associated with the 25 statements. Generally, factors are methodically calculated using SPSS software. The main component matrix and varimax technique, with a cutoff of 0.30, have been used to calculate the factors.

The Scree plot is also presented with the main factors associated with it. The respective hypothesis has been formed to find factors aligned with 25 statements, wherein the null and alternate hypotheses are presented below. In the null hypothesis, there are no factors associated where there is no association found between the statements and vise-versa.

H_{null}**:** No factors influence occupational accidents in the thermal power construction industry. Vs.

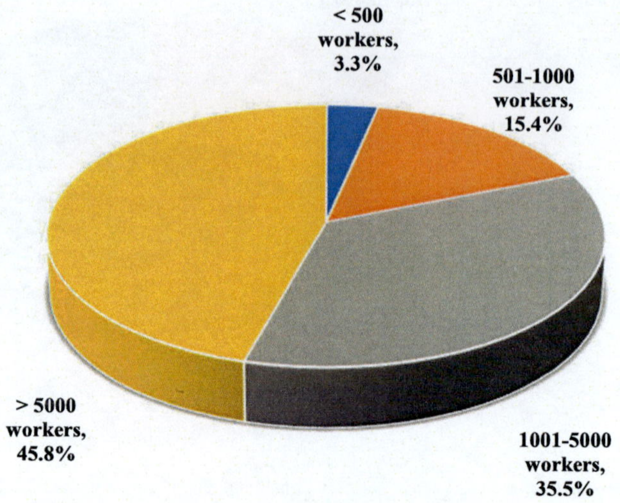

Fig. 10 Laborers' Size of an Enterprise

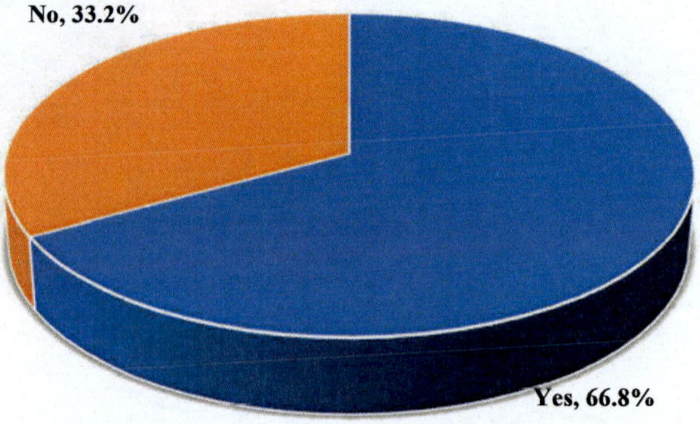

Fig. 11 Children working while accident. *Source: Collected Data from Thermal Power Construction Site*

Table 1 Reliability of 214 Data Points

Case Processing Summary				Reliability	
		N	%	Cronbach's Alpha	N of Items
Cases	Valid	214	100.0	0.782	25
	Excluded[a]	0	0.0		
	Total	214	100.0		
a. Listwise deletion based on all variables in the procedure.					

Table 2 Significance Level of the Overall Data

ANOVA with Cochran's Test						
		Sum of Squares	df	Mean Square	Cochran's Q	Sig
Between People		513.901	213	2.413		
Within People	Between Items	1003.950	24	41.831	1394.979	0.000
	Residual	2692.370	5112	0.527		
	Total	3696.320	5136	0.720		
Total		4210.221	5349	0.787		
Grand Mean = 3.84						

Table 3 Occupational Accidents as per Labor Profile

Category	Sub-Category	Occupational Accidents			Total	Chi-Square	Sig
		Disagree	Neutral	Agree			
Age	18–20 years	4 (1.8%)	35 (16.3%)	1 (0.4%)	40 (18.6%)	8.214	0.041
	21–25 years	8 (3.7%)	62 (28.9%)	2 (0.9%)	72 (33.6%)		
	26–30 years	7 (3.2%)	22 (10.2%)	1 (0.4%)	30 (14%)		
	31–35 years	12 (5.6%)	39 (18.2%)	1 (0.4%)	52 (24.2%)		
	> 36 years	6 (2.8%)	14 (6.5%)	0 (0%)	20 (9.3%)		
	Total	37 (17.2%)	172 (80.3%)	5 (2.3%)	214 (100%)		
Gender	Male	31 (14.4%)	100 (46.7%)	2 (0.9%)	133 (62.1%)	9.58	0.008
	Female	6 (2.8%)	72 (33.6%)	3 (1.4%)	81 (37.8%)		
	Total	37 (17.2%)	172 (80.3%)	5 (2.3%)	214 (100%)		
Years of Experience	<1 year	7 (3.2%)	59 (27.5%)	3 (1.4%)	69 (32.2%)	9.949	0.012
	2–5 years	13 (6%)	39 (18.2%)	0 (0%)	52 (24.2%)		
	6–10 years	9 (4.2%)	46 (21.4%)	0 (0%)	55 (25.7%)		
	>10 years	8 (3.7%)	28 (13%)	2 (0.9%)	38 (17.7%)		
	Total	37 (17.2%)	172 (80.3%)	5 (2.3%)	214 (100%)		

Source: Collected Data from Thermal Power Construction Site

$H_{alternate}$: Factors do influence occupational accidents in the Thermal power construction industry.

According to the Fig. 12, the model is appropriate for inclusion in exploratory factor analysis since the KMO value is 0.735, higher than the tolerance limit of 0.50, and the significance value of Bartlett's test is 0.000 at a 95% confidence level. The KMO and Bartlett's values are more than the tolerance limit. Hence it can be inferred that data is significant for the identified factors with a degree of freedom close to 300. Further, as per the writer (Kakung Prasetyo et. al., 2018), a KMO value greater than 0.50, eigenvalue $\geq$1, and loading factor $\geq$ 0.3, is good to continue with the factor analysis. In the study, all the values are above the tolerance limit to continue with factor analysis (Table 4and 5).

Further, the rotated component matrix has been formed, wherein factor loading is presented at each statement level. Here data reveals that 60.7% variance can be explained by the gathered factors, and many more factors affect occupational accidents. Further, the factor loading and data reliability of the identified components

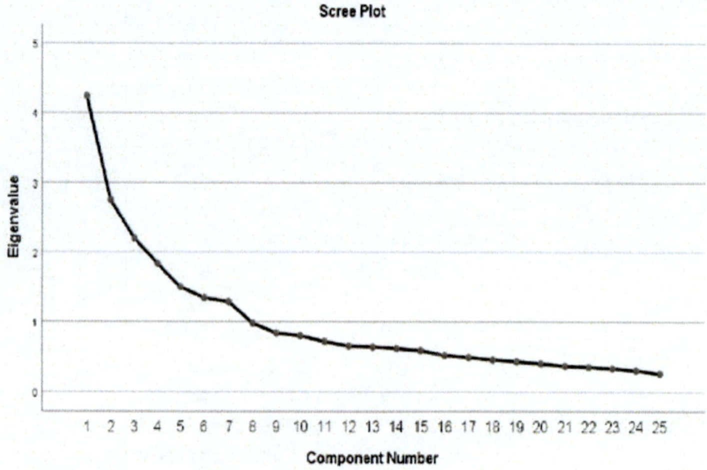

Fig. 12 Scree Plot of the identified factors

Table 4 KMO & Bartlett-Test for Factor Identification

KMO and Bartlett's Test		
Kaiser-Meyer-Olkin Measure of Sampling		0.735
Bartlett's Test of Sphericity	Approx. Chi-Square	1477.238
	df	300
	Sig.	0.000

Source: Collected Data from Thermal Power Construction Site

Table 5 Total Variance Explained

Factors	Initial Eigenvalues			Rotation Sums of Squared		
	Total	% of Varian	Cumulative %	Total	% of Variance	Cumulative %
1	4.246	16.985	16.985	2.288	9.152	9.152
2	2.755	11.018	28.004	2.245	8.981	18.134
3	2.200	8.801	36.805	2.239	8.958	27.091
4	1.839	7.358	44.163	2.172	8.689	35.781
5	1.503	6.012	50.175	2.137	8.546	44.327
6	1.345	5.382	55.556	2.066	8.265	52.592
7	1.289	5.156	60.713	2.030	8.121	60.713

Source: Collected Data from Thermal Power Construction Site

is as follows fall from roof floor, platform (FRFP) (FL = 80.6%, α = 0.782), scaffolding accident organizational causes (SAOC) (FL = 77.4%, α = 0.756), contact with moving parts of a machine (CMM) (FL = 77.2%, α = 0.766), contact with falling objects (CFO) (FL = 71.7%, α = 0.715), fall from the scaffold (FFS) (FL = 70.8%, α = 0.722), hit by an object during lifting (HOL) (70.2%, α = 0.767) and building collapse (BC) (FL = 67.0%, α = 0.768). None of the statement loadings was found non-significant. All the loading is higher than 0.30, and the reliability values are greater than 0.70. Hence alternate hypothesis is accepted and concludes that factors influence occupational accidents in the Thermal power construction industry.

5 Discussion

On a national, regional, or business level, descriptive epidemiology is a strategy that is frequently employed in the examination of construction accidents (Jonathan et al., 2019). This epidemiology examines the distribution of accident outcomes to summarise circumstances based on person, place, and time. This study employs an approach known as "incident concentration-analysis", which is done by EFA. Finding factors of episodes with similar features is the goal of incident concentration analysis. The concentrations help choose the factors affecting the incidences or accidents. In the study, 25 incidences were there, out of which 7 factors were identified. In industrial contexts, incident-concentration analysis is conducted on various "dimensions". The fundamental principle is that each industrial system has its unique clusters of accidents, primarily determined by the kinds of accidents at construction sites. Listed below are the seven identified clusters or factors from top to bottom as per their factor loading (Table 6).

On a national, regional, or business level, descriptive epidemiology is a strategy that is frequently employed in the examination of construction accidents (Jonathan et al., 2019). This epidemiology examines the distribution of accident outcomes

Table 6 Identified factors with their Mean ± Std., Factor Loading & Reliability

Factors	Item	Mean ± Std		Factor Loading		Reliability
		Item	Factor	Item	Factor	
Fall from scaffold (FFS)	Floor is lacking	3.33 ± 0.84	4.21 ± 0.63	0.759	0.708	0.722
	Working outside the scaffold or on or underneath the railing	3.15 ± 0.93		0.730		
	Edge defense/fall prevention	3.31 ± 0.92		0.691		
	Moving, toppling, and collapsing scaffolds	4.17 ± 0.55		0.651		
Contact with falling objects (CFO)	loss of load or equipment control	4.18 ± 0.63	4.23 ± 0.63	0.808	0.717	0.715
	Inadequately connected objects	4.14 ± 0.62		0.706		
	labors in the danger area	4.31 ± 0.67		0.696		
	lifting apparatus malfunctioned	4.19 ± 0.56		0.660		
Fall from roof floor, platform (FRFP)	roof or floor collapse	4.20 ± 0.64	3.27 ± 0.901	0.831	0.806	0.782
	a crack or hole in the building	4.17 ± 0.62		0.808		
	Edge defense/fall prevention	4.35 ± 0.67		0.778		
Contact with moving parts of a machine (CMM)	Unfocused worker "bumped into" the blade	3.35 ± 0.81	3.29 ± 0.88	0.799	0.772	0.766
	loss of control over the saw or object	3.16 ± 0.92		0.783		
	Use hands rather than a push stick	3.33 ± 0.90		0.734		
Hit by object during lifting (HOL)	Edge defence	3.99 ± 0.82	4.10 ± 0.74	0.798	0.702	0.767
	Employees in the danger area	4.11 ± 0.68		0.723		
	a loss of load control	4.00 ± 0.75		0.719		
	Crane lost control	4.27 ± 0.68		0.566		

(continued)

Table 6 (continued)

Factors	Item	Mean ± Std		Factor Loading		Reliability
		Item	Factor	Item	Factor	
Scaffolding accident organizational causes (SAOC)	Each step is not kept carefully by workers leads to accident	4.05 ± 0.78	3.28 ± 0.89	0.801	0.774	0.756
	Employees used their personal protective equipment during their working hours	4.08 ± 0.80		0.762		
	Many unexpected happenings happened at the site	4.01 ± 0.80		0.759		
Building Collapse (BC)	unsuitable soil of the construction building	4.24 ± 0.78	4.10 ± 0.79	0.730	0.670	0.768
	Earthquake	3.35 ± 0.83		0.703		
	Collapse may be caused because of the heavy wind	3.17 ± 0.92		0.668		
	The height of the building is higher than the normal site	3.31 ± 0.92		0.580		

Source: Collected Data from Thermal Power Construction Site

to summarise circumstances based on person, place, and time. This study employs an approach known as "incident concentration-analysis", which is done by EFA. Finding factors of episodes with similar features is the goal of incident concentration analysis. The concentrations help choose the factors affecting the incidences or accidents. In the study, 25 incidences were there, out of which 7 factors were identified. In industrial contexts, incident-concentration analysis is conducted on various "dimensions". The fundamental principle is that each industrial system has its unique clusters of accidents, primarily determined by the kinds of accidents at construction sites. Listed below are the seven identified clusters or factors from top to bottom as per their factor loading.

Fall From Roof Floor Or Platform (FRFP) (FL = 80.6%): The most frequent barrier failure in the construction site falls from the roof, floor, or platforms. Most of the apertures are for ventilation systems, staircases, and levels. The majority of these mishaps occurred when there were temporary apertures between floors or openings for staircases during building construction or renovation. Some of the cracks had plates or other coverings on them that could not support the weight, and some of the plates were not firmly fastened. The affected workers in most of these occurrences are unaware of the openings. In nine accidents, "edge protection" is determined as the primary barrier failure. Some employees slipped and fell, while others were distracted by something else and moved to the side. Edge protection or fall arrest are not used in these accidents. In three instances, "floor/roof collapse" is the primary barrier failure with a factor loading of 0.831, followed by a crack or hole in the building (0.808), and edge defense/fall prevention (0.778).

Scaffolding Accident Organizational Causes (SAOC) (FL = 77.4%): The second critical barrier failure is SAOC, wherein workers don't put the step properly or use their personal protective equipment during working hours. In many cases, it has been seen that there are several unexpected happenings happened at the site because of floor insufficiency. In one accident, the workers found it challenging to complete the task while standing on the scaffolding and instead opted to work outside or beneath the barrier. The author further said, these accidents are very different from one another and involve barrier failures, deviations, moving scaffolds with employees on them, attachment issues, uneven terrain, and too much weight on the platform that caused it to fall.

Contact with the Moving Parts of a Machine (CMM) (FL = 77.2%): The third important barrier is CMM. This occurs when the worker comes in contact with the moving machinery. Most of the injuries in this event could be severe. Despite the severity of the wounds, this accident is not considered a fatal accident [4]. As per workers, the major incidences are unfocused workers "bumped into" the blade (0.799), loss of control over the saw or object (0.783), and using hands rather than a push stick (0.734). To avoid this type of accident, the author has suggested a few strategies, such as.

- use of pre-cut materials can reduce saw-related risks,
- saw manufacturers should make modified bladed which can stop when they come in contact of human skin
- avoid wearing items that could come into touch with the saw, such as gloves, jewelry, and long sleeves, to lessen the risk
- by ensuring that the saw configuration and materials are stable etc.

Contact with Falling Objects (CFO) (FL = 71.7%): The fourth factor is CFO. This involves workplace accidents where the falling objects is big, massive things toppling during construction or disassembly. Supporting beam, power pole, main rafter, mesh reinforcement, and concrete wall element (fatal accident) are examples of heavy things. Variations in the objects' weight, size, and height of descent all impacted the likelihood of fatalities. One of these accidents is deadly, while most are considered likely or potentially fatal. Large objects in most construction projects make it difficult to eliminate, modify, or lessen this hazard. According to (Chi et al., 2014), many mishaps occur when constructing structures and materials is assembled and disassembled. As per the workers, accidents sometimes happen when there is severe wind. In the study, four types of accidents have been discussed whose factor loading are as follows loss of load or equipment control (0.808), inadequately connected objects (0.706), labor in the danger area (0.696), and lifting apparatus malfunctioned (0.660). According to (Jonathan et al., 2019), in most accidents, the workers fail to recognize the danger zone, and the accident happens.

Fall from the Scaffold (FFS) (FL = 70.8%): The fifth barrier failure is FFS. This happens where the floor is lacking (0.759), working outside the scaffold or on or underneath the railing (0.730), edge defense or fall prevention (0.691), and moving, toppling, and collapsing scaffolds (0.651). This all could happen due to the scaffold floors that weren't securely fastened or shifted when the worker stood on them. In four scaffold accidents, "working outside or underneath scaffold" and "fall prevention" is determined to be the primary barriers that failed. Some fell, while others "forgot" that there was no safety and moved to the side. In some cases, workers find it difficult to complete the task while standing on the scaffold. A scaffold swaying, toppling, or falling is the primary barrier of accidents.

Hit by an Object During Lifting (HOL) (70.2%): The sixth important factor of the occupational accident is HOL. It is clear that incidents involving workers being struck by objects while lifting occurred. The main and significant reasons for HOL are edge defense (working in height) (0.798), employees in the danger area (0.723), a loss of load control (0.719), and crane lost control (0.566). In all the given incidents, the lifting apparatus (hook, strap) broke, causing the cargo or apparatus to strike the worker.

Building Collapse (BC) (67.0%): The seventh important factor of the occupational accident is BC. It occurs when there is unsuitable soil of the construction building (0.730), an earthquake (0.703), collapse may be caused because of the heavy wind (0.668), and the height of the building is higher than the normal site (0.580). Building collapses frequently result in driven management of construction sites, which forces employees to complete structures rapidly at the expense of safety and endangers the health and lives of the person who must continue to work in these unsafe, incomplete structures. Injury caused by a building collapse is severe and frequently fatal. When a building collapses, some workers become trapped in the rubble for a while as rescue efforts begin. This kind of construction mishap in Fort Worth frequently results in severe injuries to the arms, legs, spine or neck.

6 Conclusion

A wide range of criteria for implementation exists in the building business, which is unique. Throughout the year, construction projects are carried out in various weather conditions, both during the day, at night, and in the evening. Additionally, construction employees frequently put in extra time beyond the standard eight-hour workday. Working people's health and lives are particularly at risk when such conditions occur. The worker, who in the accident process has a triple function as the decision-maker, the accident's perpetrator, and the victim, is the most crucial factor in workplace safety matters. This article aims to determine factors influencing occupational accidents in the thermal power construction industry. A survey form is used to collect data from 214 laborers. The study identified seven factors: FRFP, FFS, CFO, CMM, HOL, BC, and SAOC. These are those factors that influence occupational accidents at the construction site. According to the workers, the first important factor is FRFP (80.6%), followed by SAOC (77.4%), CMM (77.2%), CFO (71.7%), FFS (70.8%), HOL (70.2%), and BC (67.0%). Several studies have been done before to confirm the authenticity of the identified factors.

Further, as per the secondary data, there is a slight upward trend in the number of accidents in Gujrat (Fig. 2). In 2018, there were 137 fatal accident-related deaths in Gujarat, which makes up approximately 13% of all investments in the country's construction sector, the highest in the last ten-year high. According to data collected by Pandya with the assistance of BMS volunteer and civil engineer Darshan Patel. In 2017, there were 67 fatalities, compared to 55 in 2016, 62 in 2015, and 69 in 2014.

In the study, the most common age group of injured workers is 18-25 years. Occupational accidents are most likely to occur for male workers with less experience. The likelihood of an accident occurring is highest in the first year of employment at an active workplace compared to subsequent years of employment. Here study reveals that, there are factors influencing occupational accidents. However, male workers with less experience are the ones who get affected easily. To reduce the number of the accidents, the first step is to understand the accidental severity of each component. The second step would include finding the work safety plan and training initiatives. The third step would be monitoring of the construction site. A group of male laborers with little work experience needs special consideration (up to 1 year).

References

1. Ho, D. C. P., Ahmed, S. M., Kwan, J. C., & Ming, F. Y. W. (2000). Site safety management in Hong Kong. *Journal of Management in Engineering, 16*(6), 34–42.
2. Elsler, D. (2011). "Innovative solutions to safety and health risks in the construction health care and HORECA sectors". *European Agency for Safety and Health at Work (EU-OSHA).*
3. Bożena, H., & Mariusz, S. (2017). An Occupational profile of people injured in accidents at work in the polish construction industry. *Procedia Engineering, 208*, 43–51. https://doi.org/10.1016/j.proeng.2017.11.019
4. Bożena, H., Tomasz, N., Iwina, S., Jacek, S. (2017) "Identification of factors affecting the accident rate in the construction industry". https://doi.org/10.1016/j.proeng.2017.11.018
5. Chi, C.F. (2014)."Graphical fault tree analysis for fatal falls in the construction industry", *Accident Analysis & Prevention, 72*, 359–369. https://doi.org/10.1016/j.aap.2014.07.019.
6. Robles, C., & Mirosevic, V. (2011). "Social protection systems in Latin America and the Caribbean: Brazil". *Social Protection Scetion.* https://www.cepal.org/sites/default/files/publication/files/4062/S2013126_en.pdf
7. Deccan, (2012). "Construction industry nation's second largest employer", *Deccan Herald,* 2012. https://www.deccanherald.com/content/222581/construction-industry-nations-second-largest.html

8. Patel, D.A., & Jha, K.N. (2016). "An estimate of fatal accidents in indian construction". *Association of Researchers in Construction Management, 1*, 577–586, https://www.researchgate.net/figure/State-wise-estimate-of-average-annual-fatal-accidents-in-Indian-construction-sector-from_fig1_308155592/download
9. Gurcanli, G.E., & Munge, U. (2013) Analysis of construction accidents in Turkey and responsible parties". *Industrial Health, 51*(6), 581–595.
10. Huang, X., & Hinze, J. (2003)."Analysis of construction worker fall accidents". *Journal of Construction Engineering and Management, 129*(3), 262–271.
11. Jonathan, F., Christopher, B.M., Koehoorn, M. (2019). "Descriptive epidemiology of serious work-related Injuries in British Columbia, Canada". *National Library of Medicine, 7*(6). https://doi.org/10.1371/journal.pone.0038750
12. Prasetyo, K. (2018). "Development of Mathematical Literation Instruments Based on Class", *Journal of Educational Research and Evaluation.* http://journal.unnes.ac.id/sju/index.php/jere
13. Kavya, K., Pradeep, T. (2019). "Causes and effects of construction accidents". *International Journal of Innovative Technology and Exploring Engineering, 9*(10). https://doi.org/10.35940/ijitee.L3917129219
14. Koehn, P. E., & Datta, N. K. (2003). Quality, environmental, and health and safety management systems for construction engineering. *Journal of Construction Engineering and Management, 129*(5), 562–569.
15. mponsah, K. (2013). "Examining psychosocial and physical hazards in the Ghanaian mining industry and their implications for employees' safety experience". *Journal of Safety Research, **45***, 75–84.
16. Nawaz, T., Ishaq, A., Ikram, A.A. (2013). "Trends of safety performance in construction and civil engineering projects in Pakistan". *Civil and Environmental Research, **3***(5), 23–40.
17. Noora (2014). "*Global estimates of occupational accidents and work-related illnesses 2014, made for the ILO Report at XX World Congress, Frankfurt, 2014*". https://doi.org/10.13140/2.1.2864.0647
18. Shah, R. (2019). "India's 80% construction sites "unsafe", deaths 20 times higher than those in Britain". https://www.counterview.net/2019/05/indias-80-construction-sites-unsafe.html
19. Wings, S., Albrechtson, E. (2018). "Accident types and barrier failures in the construction industry". *Safety Science, 105*,158–166. https://linkinghub.elsevier.com/retrieve/pii/S0925753517317721
20. Szóstak, M. (2019). Analysis of occupational accidents in the construction industry with regards to selected time parameters. *De Gruyter.* https://doi.org/10.1515/eng-2019-0027
21. Fatih, Y. (2018)."Evaluation of safety trends in construction, mining and transportation sectors in Turkey". *60*(1), 13–23. https://doi.org/10.31306/s.60.1.2

Studies on the Status of the Women Construction Workers Before and During Covid-19 Situation in Warangal Districts

K. Poongodi, R. Archana Reddy, P. Murthi, Ch. Udaya Sree, and B. Praneeth Paul

Abstract The service of women workers in the construction industry is inevitable. The contribution of unorganized crew in the development of the construction sector is highly appreciable and to be encouraged. The total number of workers involved in construction industry in India is more than a population of seven crores in 2019. Among the total construction workers, more than 30% of the population is women workers, and they are all manual laborers and also unskilled workers. The role of construction women workers is including the family and personal commitment in addition bridging the family economic with overall expectations of family members. The occupational hazards and their nature of habitation affect their regular life naturally. The present Covid-19 pandemic situation creates a pathetic situation in their employability and also in living style. This investigation comprises the detailed survey about the before and present Covid-19 condition for their socioeconomic situation and health condition. The survey was carried out in Warangal Rural and Urban district, Telangana State, India and the samples are collected from 200 women workers. This investigation is concluded with the recommendations for improving the socioeconomic and the health conditions of women construction workers.

Keywords Covid-19 · Women Construction Workers · Socioeconomic situation · Health Condition

K. Poongodi (✉) · Ch. Udaya Sree · B. P. Paul
Department of Civil Engineering, SR University, Warangal, India
e-mail: k.poongodi@sru.edu.in

R. Archana Reddy
Department of Mathematics, SR University, Warangal, India

P. Murthi
Centre for Construction Methods and Materials, SR University, Warangal, India

A. Vilventhan et al. (eds.), *Advances in Construction Materials and Management*, Lecture Notes in Civil Engineering 346,
https://doi.org/10.1007/978-981-99-2552-0_11

1 Introduction

The population of Great Warangal Municipal Corporation (GWMC) is more than 8.3 Lakhs as per the 2011 census. GWMC includes three municipal cities namely Warangal, Hanamkonda and Kazipet. GWMC is now comes under the smart city project undertaken by the Government of India which promote the rate of growth of the city [1]. The status of any country or any part of the country is benchmarked by the prestige of building and infrastructural development. Considering the economic growth of any developing country, majority of the workers are employed in unorganized sectors like agriculture and construction.

The number of women workers in unorganized sector is more than the population of organized sector due to the literacy and driven force of poverty. The construction industry contribution is more than any other sector for the development of any part of the country and solely mentioned that the construction is the pillar for development of any country. The usage of cement and mineral admixture and other chemical products used in construction industry are inevitable and also not to be stored for a long period. In India, GDP from construction industry is increased to ₹2670 billion in the first quarter of 2020 from ₹2593.57 billion in the fourth quarter of 2019 [2]. However, the construction industry is not yet properly organized by judicially [3]. Construction industry comprises of managerial and skilled workers naturally. Various skilled and unskilled labours are predominantly working in this unorganized sector such as mason, bar bender, carpenter, plumber, electrician and welder and mazdoor respectively. In the above said category, women workers fall under mazdoor - II and men workers are considered as mazdoor- I category.

But construction workers especially the women workers situation is still stagnated [4]. The total number of workers in the construction sector in India is more than 6 crores in 2019. Only 3.5 crores construction workers are registered with welfare boards. Among them, women contribution towards the construction workforce is less than 25%. Out of the total construction worker, 1.25 crores population are belongs to women workers [5]. Most of the women workers are in manual unskilled jobs in this industry. In India large group of female unskilled workers works in the rural areas as agriculture laborer as soon as the season ends, they shift to the construction industry with their husband for income generation and to face the family needs [6]. Women workers are mostly affected by the occupational hazards and reported that the women worker have a different pattern of fatal injuries and some difference in patterns of non-fatal injuries than man workers in construction industry [7–9]. This construction sector had been affected more horrible than any other industry due to the covid-19 pandemic situation. All of the sudden the construction workers are losing their job and wages. However, such unorganized category labours are established their life depends only on their daily wages and this situation affects the social and economic conditions of the construction workers [10, 11].

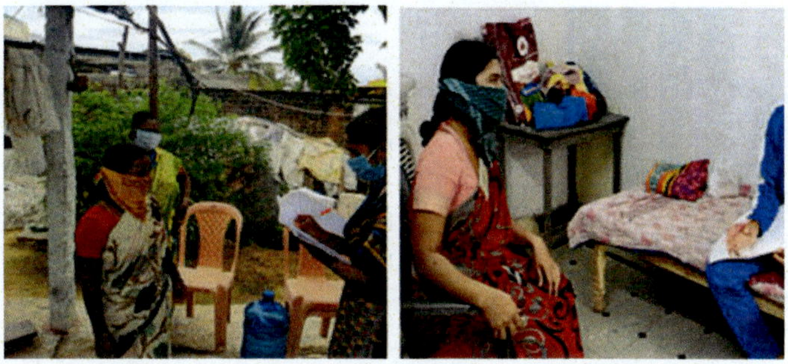

Fig. 1 Survey being carried out with construction women workers and supervisor

2 Methodology of the Study

A study was initiated among the women construction workers to evaluate the present situation when compared with before covid-19 situation. A personal survey was carried out with a detailed set of questionnaire and micro level study about their present position. Chi-square test has been applied for establishing the relationship between wage and family expenditure. The study was concentrated on the socio-economic conditions of women workers engaged in construction sector as shown in Fig. 1 and presented in this article.

3 Results and Discussion

3.1 Age Composition

The women workers from the age of above 20 years old were considered in this study and assembled in to nine groups. The higher percentage with 26% of workers is belongs to the age group of 35–39 years old and it is followed by the age group of 30–34 in the account of 18.5%. There are 15.5% of women workers belonging to the age group of 40–44 participated for this survey and observed more than 65% of women workers are in the range of 30–45 years old [12]. From the total samples of 200 women surveyed, 2.5% of women workers are more than of 60 years old. The classification based on the age composition is depicted in Fig. 2. It was observed that the age old women workers (more than 60 years) are working in the construction industry and new generation young women belong to below 30 years old are also working by 23%.

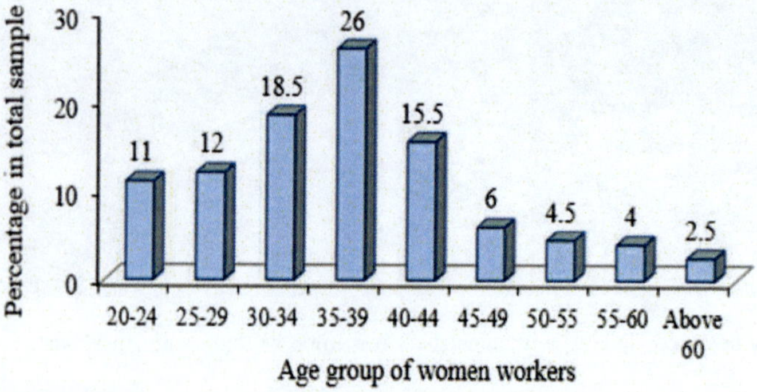

Fig. 2 Age composition

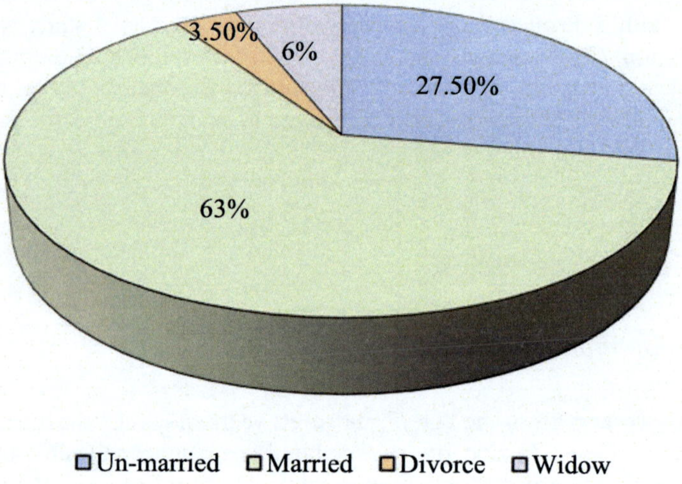

Fig. 3 Marital status

3.2 *Marital Status*

As far as the marital status is concerned, it is revealed that 63% of women workers are married and mostly they are working along with their spouse at the construction site. It was noted that 27.5% un-married women workers are working in construction industry along with their parents. They are mostly in the age group of below 29 years old and assigned to involve in tough nature of activities. On the other hand, 6% women workers are comes under the category of widow and 3.5% of women are divorce category and most of the women workers above 50 years old are falls in this category. These group of people expressed during the survey that the poverty is the prime force to involve in this nature of work. The distribution of marital status of construction women workers contacted in this survey is shown in Fig. 3.

3.3 Educational Status

Table 1 shows the education level of the crew participated in this study combined with the age group. The majority of women with less than 44 years old are minimum secondary educated. However, the higher percentage of illiterate category women is noted in more than 40 years old group. The age group below 24 years is scoring higher percentage of secondary educated which clearly stated that the young women are having minimum educational back ground and are able to calculate the construction quantities and wage details. Further, it is noted that four women graduate are working in the construction field as site supervisors and they are below 34 years old. Fig. 4 had shown the overall literacy status of all women consider in this study. It clearly shows that more than 50% of the women workers are completed their secondary education and most of them are monitoring the economic aspects of their home expenditure. The working women population of about 20% with more than 45 years old are not known to read and write and they are fully depended their spouse.

3.4 Present Scenario of Women Workers Due to Covid-19 Lockdown

Sudden lock down executed by the respective authority due to the covid-19 pandemic situation, the construction sector is one of the worst affected industries since it is the notable unorganized labour driven industry. The results obtained from the construction women works responded this study is shown in Fig. 5. It was clearly seen that the construction women workers lies in the age group 30 to 45 manage themselves with alternate petty jobs with lower wages to meet the domestic expenditure. The same time, a group of women below 40 years are staying in home without job. The majority of young women labours below the age of 30 are having no strain though they are staying in home due to quarantine. In comparison, the rate of young women workers affecting the health quarantine is lower than the aged women. The data are clearly shown that the aged women workers are affected their health in higher rate than young women. The family managed women workers who are in quarantine during this observation period were severely affected both socially and financially. They are panic about their future job than Covid-19 and income to run the family in addition to their health condition. The recoup of construction is only the solution

Table 1 Education level of responders

Education Level	Age group in years (No. of responders)										
	20–24 (22)	25–29 (24)	30–34 (37)	35–39 (52)	40–44 (31)	45–49 (12)	50–55 (9)	55–60 (8)	Above 60 (5)	Total (200)	(%)
Illiterate	0	1	4	9	4	5	4	6	4	37	18.5
Primary	5	9	10	17	6	4	4	2	1	58	29
Secondary	17	12	21	26	21	3	1	0	0	101	50.5
Graduate	0	2	2	0	0	0	0	0	0	4	2

for such women workers with proper precaution. From this survey, it was observed though most of them are uneducated they have lot of dreams about their children education and their future. This situation pulls their legs in the reverse direction.

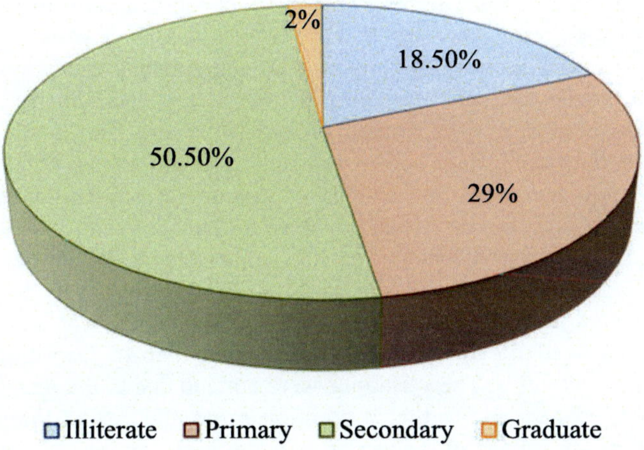

Fig. 4 Educational status

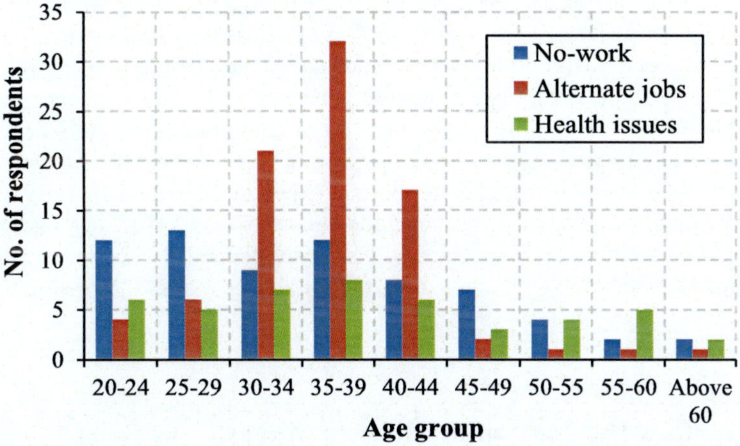

Fig. 5 Present scenario of women workers

3.5 Present Income Scenario of Women Workers

The construction crews are categorized as un-organized group of labours. There is no assurance of work for all days and in some occasions, the construction works being carried out without suspension. Further it was observed that the wages for construction workers are not uniform and varies depends upon nature of task. The women workers are getting 50–60% of wage compared to male workers and it varies among the women workers according to their assignment. It was noted that the dispersion of wages normally happened once in a week particularly on Saturday considering the number of days worked during that period. The variation of women worker wages with age group before and during Covid-19 are shown in Fig. 6. The average wage details before Covid-19 was inquired from the survey among the responders and found that the women worker in the age group 35–39 years have earned more than ₹380 as an average wage in a day followed by an average wage of more than ₹327 received by the age group 40–44. The average wage for the women workers below 34 years old had drawn less than ₹310. The lower wages have also been given to the aged category women workers (more than 45 years old) involved in the construction activities. The variations were noticed due to the managerial skills of middle aged women and further it was noted that majority of the middle aged women workers are spouse of the senior male workers employed in the particular construction. The wage during this pandemic situation abruptly reduced to below 50% of the regular average wage. This data was collected during May and June 2020 period and majority of the women workers are jobless and though they had realize the impact of corona virus, few of them are involved some petty works due to the poverty. In order to survive, this minimum wage itself is very important; a statement given by the women workers who are comes under the age group of more than 50 years old. The scenario of wage reduction particularly women workers shakes the regular life style of their entire family since these kind of family is fully depends on the income from the women workers. The situation was observed due to diverting the wage from male workers to some unimportant expenditure and not directly supports the regular family expenses. It was concluded that the situation of such unorganized women working labour sector, particularly who are depended only the construction related works, was affected pathetically and not able to express their situation in a statement.

3.6 Families Depends on the Women Wages

One of the important observation of this study was the number of construction workers family met the domestic expenditures with the women construction labours wage. It was very surprise to note that more than 70% of the family associated with women workers in this survey and it was depends on their wage in addition to the male workers of the same family. The domestic expenses, children education, medical

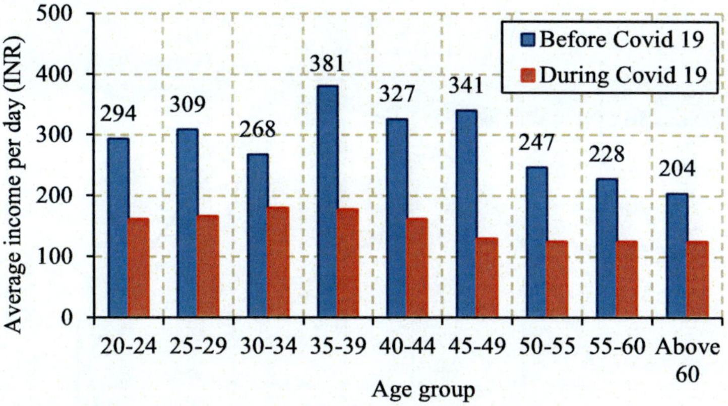

Fig. 6 Present average daily income scenario of women workers

expense, fuel expense and savings are the various factors which seeks the women contribution for the family income. Nearly 35% family has depends on the 50–70% of women wage and 27% of family has associated with 70–90% wage of the women workers. The higher percentage of women wage to sustain their life was unanimously noted. It was concluded that majority of the construction workers family is blended with both male and female workers wage.

In order to find out the role of women worker's wage to manage the domestic expenditure, the chi-square test has been applied to predict the significant relationship between need of women wage for their family. The contribution level of women wage in percentage to their family was consider as the variable (A) and observed frequency in terms of number of responders are the other variable (B) in this chi-square test. There are five range of contributions were selected as shown in Table 2. The degree of freedom was calculated as 4 in this test (Fig. 6).

The hypothesis was developed by considering the variables as per the following:

H_A = Both the variables A and B are independent.

H_B = Variable A and B are not independent.

The expected variables are calculated as mean value of total responders to the number of ranges and equal to 40 as shown in Table 2. The significance level (α) was selected as 0.05 (ie.5%). After calculation, the chi-square statistic value (X^2) was found as 30. From the chart, the critical value of chi-square distribution (${X_c}^2$) with 4 degree of freedom was found as 11.07. Since, critical chi-square statistic value is less

Table 2 Variables of Chi-square test

Categories of Variables	Contribution level of women wage (%)				
	Above 90%	70–90%	50–70%	30–50%	Below 30%
Observed	23	54	61	37	25
Expected	40	40	40	40	40

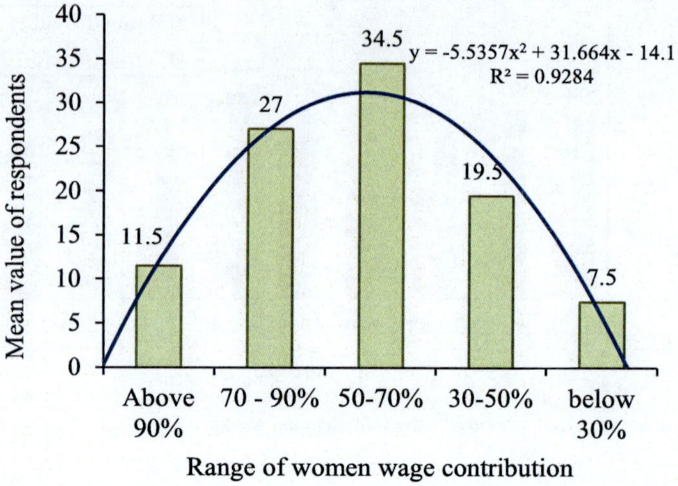

Fig. 7 Families depends on the women wages

than the chi-square statistic test value $(X_c^2 < X^2)$, the X^2 lies in the rejection region. Hence, the null hypothesis (H_A) is rejected and concluded that there is a significant relationship among the wage contribution of women workers in their family [13, 14]. A polynomial regression trend line was established using the observed responded values shown in Fig. 7. It shows that the variables are properly fit in the normal distribution curve with higher co-efficient of correlation ($R^2 = 0.9284$).

4 Suggestions

The following are the various suggestions that can be derived from the survey of the responders:

1. A system has to be developed for encouraging the saving practice of women construction workers by developing awareness about such lockdown or unwaged situation and adequate insurance facilities through employers.
2. The women workers are enrolled in welfare board of construction workers and extend support to any such critical condition and related health issues.
3. Construction women grievance redresses body has to be created in order to handle the various issues and problems developed in construction site against women workers.
4. Providing technical training and skill development programs can help the construction women workers to improve productivity and minimizing occupational hazards.
5. Establishing safety culture and providing safety measures will help the crew for better involvement in the construction industry.

6. Ensuring the transparent and uniform wage system in this unorganized sector for betterment of both employer and employee.

5 Conclusions

The construction women workers are severely affected due to the impact of covid-19 and its follow up action of lockdown. The daily wage category of such workers are struggled to manage their domestic expenses including the medical expenses. The women workers wage plays a vital role to meet all kind of expenditure and irrespective of the age of the workers, all are searching alternate works to survive during this critical situation. The covid-19 situation make majority of the women workers are jobless and some of the aged workers are panic about their forthcoming opportunities in the construction industry than the covid-19 impact. The chi-square test result had shown that there is a perfect relationship between women construction workers wages and its contribution to the development of their family. It should be recognized from this study during this lockdown situation that the construction women workers demands the minimum wage and social security and assured job opportunities in the post covid-19 lockdown.

References

1. https://gwmc.gov.in
2. https://tradingeconomics.com/india/gdp-from-construction
3. Rasheedha Banu, S., & Sampath Kumar, S. (2018). "Working conditions and issues of women workers in an unorganized sector - Special reference to construction sector of Thuraiyur Taluk, Tirchirapalli". *International Journal of Trend in Scientific Research and Development, 2*(3).
4. Maneesh, P., & Jasna, P. T. (2017). Socio- economic condition of women workers in Kannur District, Kerala. *Indian Journal of Economics and Development, 5*(8), 1–11.
5. Poongodi, R., & Revathi, K. (2010). Economics of women construction workers with special reference to Thuraiyur Taluk, Trichirappalli District in Tamil Nadu. *Asian Journal of Management Research, 2*(1), 684–687.
6. Kalpana Devi, U.V., & Kiran, V. (2013). "Status of female workers in construction industry in India: A review". *IOSR Journal of Humanities and social Science (IOSR-JHSS), **14***(4), 27–30.
7. Laura, S. (2000). Women in construction: "Occupational health and working condition." *Jamwa, 55*(2), 89–92.
8. Adane, M.M., Gelaya, K.A., Beyera, G.K., Sharma, H.R., Yalew, W.W. (2013) "Occupational Injuries among building construction workers in Gondar City, Ethiopia". *Occupational Medicine and Health Affairs*, Working conditions and issues of women workers in an unorganized sector - Special reference to construction sector of Thuraiyur Taluk, *Tirchirapalli, 1*(5), 2329–2339.
9. Sultana, N., Ferdousi, J., Shahidullah, Md. (2014). "Health problems among women building construction workers". *Journal of Bangladesh Society Physiologies, **9***(1), 31–36.
10. https://www.financialexpress.com/industry/coronavirus-shakes-foundation-of-construction-sectorexecution-to-beslow-as-migrant-workers-flee/1928188
11. https://timesofindia.indiatimes.com/blogs/bunt-frank/coronavirus-economy-a-viewpoint-part-2.

12. Vaidya, V.G., Mamulwar, M.S., Ray, S.B., Beena, R., Bhathlawannde, P.V., Ubale Lokmanya, S. (2022). "Occupational Health hazards of Women working in brick kiln and construction industry". *International journal of applied management and Business Utility*, ***4***(1), 45–54.
13. Rengarajan, L. (2017). Living and working conditions of women construction workers. *Shanlax International, Journal of Commerce, 5*(2), 10–18.
14. Ocal, E., Laptali Oral, E., Erdis, E. (2006). "Crisis management in Turkish construction industry". *Building and Environment*, ***41***, 1498–1508.

Comparative Study on the Cost Analysis of Embodied Energy of Construction Materials: Cellular Lightweight Concrete (CLC) Versus Conventional Brick Systems

Rajiv Nehru and Purva Mujumdar

Abstract Usually, carbon emission during execution of construction activities and processes are not considered, such as the embodied energy (energy required in manufacture and transportation of building materials). Hence, this study focuses on conceptualizing carbon neutrality by comparative cost analysis of traditional construction material with upcoming green material in terms of their embodied energy. Here, the construction activity chosen to estimate the costs of the embodied energy is walling. Conventional burnt bricks walling systems are compared with Cellular Lightweight Concrete (CLC) walling in building projects. In addition to embodied energy value, the economical aspect of the two walling systems is also highlighted. To achieve this objective, literature pertaining to carbon emissions during construction activities is studied and advantages of using CLC as a construction material are discussed. For cost comparison of the two walling systems, data collection is done from an ongoing residential project being constructed with 1200 apartments. The results obtained reveal that energy consumption for construction activity i.e. walling with CLC as a building material reduces by almost 50% by obtaining cost savings on embodied energy required in material manufacturing and transportation thereby ensuring more sustainable construction.

1 Introduction

Sustainable construction integrates utilization of renewable and recyclable materials along with varied processes and methods especially during the planning, execution and operation stages of construction projects [1]. The selection and usage of

R. Nehru · P. Mujumdar (✉)
Sushant University, Gurugram, India
e-mail: purvamujumdar@sushantuniversity.edu.in

R. Nehru
e-mail: rajivnehru.phd21@sushantuniversity.edu.in

A. Vilventhan et al. (eds.), *Advances in Construction Materials and Management*, Lecture Notes in Civil Engineering 346,
https://doi.org/10.1007/978-981-99-2552-0_12

green building materials, products and processes during construction phase plays the most crucial role in ensuring the sustainability of any building [2, 3]. There have been many benefits of green building processes and materials identified such as reduced maintenance/replacement costs over the lifecycle of the building; reduced energy consumption; improved worker / occupant health and productivity, lower costs associated with altering space configurations, greater design flexibility and so on [3, 4].

Globally, construction projects consume 3 billion tons of natural resources and materials each year that is almost 40 percent of total global energy usage [5]. Adopting sustainable building processes and methods and green building materials may result in efficient utilization of finite natural and non-renewable resources. Also, integrating green building practices can help reduce the environmental impacts associated with the extraction, transport, processing, fabrication, installation, reuse, recycling, and disposal during construction [6]. The main objective of this paper is to assess and estimate two walling processes during construction, one using Cellular Lightweight Concrete (CLC) block and the other using Burnt Brick by cost analysis of the two systems [7, 8]. To achieve this, data is collected from an ongoing residential project in Pune, India which is being constructed on a 10.5-acre of land with 1200 apartments. The purpose of comparing these two walling systems shows the benefits earned in terms of money as well as reveals that the carbon emissions can be reduced by less energy consumption and thereby improving construction productivity and sustainability.

2 Green Processes

In order to attain carbon neutrality, fifteen major construction processes are identified in which modifications/changes in materials, processes and methods may lead to less carbon footprint. The processes are highlighted in the Table 1.

Among all the construction processes and activities, walling work in a structure is selected for this paper as it is large in volume (with respect to building components) and has more duration of execution during construction. The primary focus is on change of material, its manufacturing processes and execution techniques. A large number of materials can be categorized as green materials, however, this study assesses replacement of Cellular Lightweight Concrete (CLC) blocks in place of burnt brick which is the traditional material for walling process.

Cellular Lightweight Concrete (CLC), as shown in Fig. 1., is a new age sustainable material that has been utilized for construction of numerous residential, commercial, industrial, healthcare buildings etc. in almost 45 countries globally for past 40 years [9]. In India, CLC has been used primarily with a modified form constituting 25% pond ash that is more environment friendly and economical [9]. Another reason for using CLC block is the inherent shortcomings of burnt bricks during their manufacture, use and maintenance. The next section discusses the major inadequacies of burnt brick walling systems.

Table 1 Construction processes identified for modifications

S. No	Processes	Activities
1	Sub Surface	Excavation
		Back Filling
		Dewatering
2	Structure	PCC
		RCC
		Walling Work
		Curing
		Hacking
		Reinforcement
		Plastering
3	Finishing	Painting
		Doors/Windows
		Waterproofing
		Grills and Railing
		Tiling

Fig. 1 Cellular Lightweight Concrete (CLC) Blocks

3 Limitations in Burnt Bricks Walling System

Bricks are the most popular building material all over the world. In construction sector, brick represents the most basic unit of construction i.e. one unit. However, even though brick has been in usage since long, there are many limitations of bricks systems [10] as listed.

3.1 *Erosion of Top Soil*

To manufacture bricks, a lot of cultivable land is used to set up brick kilns. In addition, bricks are manufactured using agricultural top soil that results in soil erosion. Hence, utilization of CLC can prevent ruining of agricultural land.

3.2 *Power and Fuel Consumption*

Almost 25.77 metric ton of coal is employed to produce quantity of one lakh bricks [10], while the energy consumed for manufacturing CLC is almost negligible. Another advantage of using CLC is cost and fuel saving in its transportation for being a lightweight material as well as it can be manufactured at the project site also.

3.3 *Small Size*

Small sized bricks require more quantity of mortar and labour for masonry and plastering work. A CLC block amounts to almost 14 bricks for an external wall and 7 bricks for a partition wall respectively.

3.4 *Inadequate Supply*

Usually, requirement of bricks for construction throughout the year is not met with the supply as manufacturing of bricks is done only six months in a year. On the other hand, CLC can be produced continuously to fulfil construction sector requirements.

3.5 *Different Quality*

Many times, bricks manufactured are not of uniform good quality. As the CLC blocks are manufactured on the project site itself, more quality control can be exercised during its production.

4 Advantages of CLC Walling Process

The advantages of CLC Walling process are as listed below [10].

4.1 *Reduced Dead Weight of Building Components*

Regular dense concrete is a heavyweight material to be used especially in case of additional floor requirements on existing floors or in unstable ground conditions. This results in limited application of the concrete as well as underutilization of available floor area ratio (FAR). Lightweight concrete such as CLC is an excellent alternative due to reduced overall dead weight of building components particularly in severe earthquake zones as well as it is an economically viable option [11].

4.2 *Cost and Material Savings*

The main constituents of CLC are cement, pond ash, sand, water and foam. The absence of gravel in Cellular Lightweight Concrete makes it a lightweight material. On the other hand, it leads to material and cost savings particularly in areas where gravel is difficult to procure and/or is very expensive.

4.3 *Reduced Costs of Transportation*

CLC, being a lightweight material i.e. produced at the project site, incurs almost nil transportation expenses.

4.4 Ease of Handling

Building components of large dimensions constructed using CLC can be manually and easily handled as compared to those of normal concrete because of reduced deadweight.

4.5 Hilly Construction Sites

Transportation of bricks and other materials from plain regions to construction sites in hilly areas is difficult. Using CLC is advantageous in such cases due to CLC being produced at the site itself.

4.6 Environment Friendly

Cellular Lightweight Concrete is an environmental friendly material. It saves depletion of the top soil, while at the same time it can use pond ash as one of its major component which is a waste material from industrial sector. The manufacturing process of CLC does not release any harmful effluents to ground, water or air i.e. dissimilar from smoke of brick kilns and erosion of top soil during production of bricks. CLC, due to its lightweight characteristics, is ideal for making partitions.

4.7 Thermal Insulation

Air is known to be the best available insulation material. Air voids, if smaller than 2 mm each, consequently, increase thermal insulation considerably. Normal aggregate concrete has a specific thermal conductivity of 2.1 W/mK, compared to only 0.405 W/mK for 1200 kg/m^3 of Cellular Lightweight Concrete. To provide identical thermal insulation in a 100 mm thick CLC wall, the equivalent thickness of dense concrete wall needs to be more than 5 times thicker (i.e., 500 mm) and ten times heavier [10].

4.8 Fire Protection

Fire rating of a 100 mm thick wall constructed with CLC that has a density of 1200 kg/m^3 offers 3 h of fire endurance which is much more than that of brickwork or dense concrete [10].

4.9 Fast-track Construction

The presence of foam in place of gravel provide denser and consistent Cellular Lightweight Concrete. No compaction is required because of its even distribution that results in uniform density all over the material.

4.10 Saving of Steel

With the use of CLC blocks for internal walling, almost 15 to 20% steel can be saved that results in reduced costs as well as environmentally sustainable construction in comparison to a high embodied energy material such as steel. This saving is possible due to the density of CLC i.e. 460–600 kg/m^3 as compared to that of burnt brick with a density of 1800–2000 kg/m^3 [9]. Hence, in case of CLC blocks the dead weight of the Concrete building frame reduces by almost 5.4 kg/m^2 of constructed area 10].

4.11 Embodied Energy in CLC

As depicted in Fig. 2. the embodied energy in CLC blocks is half that of burnt brick [5] thereby making it more sustainable alternative. The other energy saving parameters like saving in fossil fuel due to onsite production, use of fly ash and saving of aggregates can also reduce the embodied energy of CLC.

5 Methodology and Case Application

This section discusses the methodology and case example selected for evaluating carbon offset of CLC blocks and burnt brick for Walling process. An ongoing residential project in the city of Pune, India was chosen for this study. In order to weigh the advantage of replacing burnt brick by CLC as a walling system, a step-by-step selection, evaluation and verification process is followed as depicted in Fig. 3.

Post selection of CLC as a walling material, the material was produced on site. The effect of changing the burnt brick to CLC was majorly in the large size of the CLC blocks, low density and more work quantum per day. The CLC were produced and used in a project site whose parameters are given here.

The project site is a 10.5-acre land with over 1200 residential apartments being constructed (1 BHK, 2 BHK, 3 BHK and town houses) where BHK is defined as Bedroom-Hall-Kitchen. A CLC plant of capacity 4000 blocks per day was installed on the site with air and steam curing facility. The plant was set up in an area of 75 × 40 m with two curing tanks having a capacity of 5000 L each. The blocks produced were of size 600 × 200 mm × 225 mm and 600 mm × 200 mm × 150 mm. The most crucial part of the block manufacture was the curing process which was done either by steam curing or by air curing (7 days). The CLC block was one of the components for process improvement.

Fig. 3Methodology for CLC Walling selection, execution and verification

The other components of this process improvement were reduction in dead weight of the structure due to low density of CLC blocks (almost half of burnt bricks)

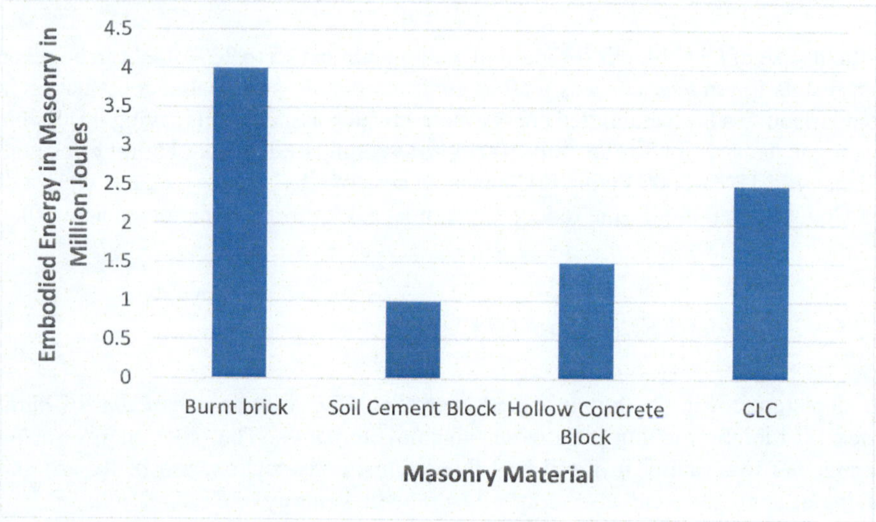

Fig. 2 Embodied Energy in Masonry Materials in Million Joules (MJ)

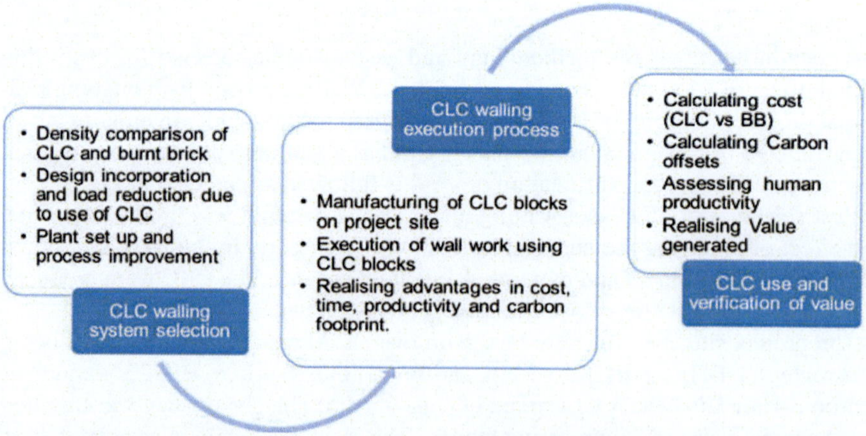

Fig. 3 Methodology for CLC Walling selection, execution and verification

hence saving almost 5.4 kg of reinforcement per square metre of construction (if incorporated at design stage); improved productivity (almost 3 times) as the size of CLC blocks was large and weight was less; reduction in quantity of water required for curing; homogeneity with concrete building frame thus reducing cracks; saving fossil fuels because of onsite manufacturing.

6 Results and Discussions

The cost of manufacturing CLC blocks on site was considerably less. As depicted in Table 2, almost 625 rupees per m^3 can be saved when steam curing is not used. The unfavourable aspect of not using steam curing is that the block has to be air cured and used after 7 days only as against steam cured CLC block which can be used in 24 hours. Apart from the cost, there were savings in transport miles, energy cost (when steam curing is avoided) and equipment capital cost.

6.1 Direct Cost Savings Due to Onsite CLC Block Production and Walling System

As evident in Table 3, a saving of 15–25% of reinforcement steel can be obtained as the weight of CLC blocks is about 40% as that of burnt brick thus reducing the dead weight of the building.

As highlighted in Table 2, process improvement was recorded due to savings in;

- Almost **₹500 per m^3** fossil fuel for un-steamed CLC blocks.
- Quantity of water for curing by about 15–20 L per m^3

Table 2 Cost of manufacturing of CLC Blocks

Components	With Steam Curing ₹/m^3	Without Steam Curing ₹/m^3
Material Cost	2000	2000
Consumables	100	100
Fixed charges for mould, DG, Mixer	100	60
Labour charges	500	500
Steam Curing Equipment cost	50	0
Steam Curing operation	500	0
Wastage (manufacturing, handling and transportation 1%)	32.5	27
Overheads 5%	162.5	133
Total Cost: ₹/m^3	3445	2820

Table 3 Direct cost saving using CLC block replacing burnt brick

Parameter	Burnt Brick per m^3	CLC Blocks per m^3	Savings
Basic Cost/^{3}	₹000	₹2800 – 3500	₹500 - 1200
Density	2000 kg/cum	450-800 kg/ cum	
Steel saving per m^3 of concrete	0	15–25% 5.3 kg /sqm	₹290–300 per sqm
Curing Water	40–60 L	20–30 L	₹600 – 900 per m^3
Energy Cost saving	₹1200 per m^3	₹500 per m^3	₹700 per m^3

- 60% less number of joints due to large size of CLC blocks saving cement, sand and water.
- Increased construction progress recorded per day due to large size and light weight. Construction up to 3 m height as against 1.2 m for a burnt brick walling can be done.
- Onsite production reducing transport miles at an average of 3–5 L of diesel per cum of CLC blocks (saving of ₹400).

6.2 *Indirect Cost Savings*

Indirect cost saving was obtained by replacing burnt brick wall with CLC blocks [12] in terms of

- Cost of transportation
- Cost of disposal of fly ash and its environmental hazards
- Cost of work stopped due to unavailability of burnt brick
- Cost of additional mining for the reinforcement steel saved
- Cost of work delays due to unavailability of burnt bricks.
- Cost of rework due to poor quality of burnt bricks.
- Carbon neutrality of CLC blocks

The embodied energy in CLC blocks is half of that of burnt brick while manufacturing, hence if the entire structure is constructed with CLC blocks, a -50 percent carbon neutrality in the walling quantity in terms of energy use is attained. Also, cost savings in quantity of water needed for curing, reinforcement steel, transportation etc. can be obtained.

7 Conclusions

Utilizing of CLC blocks in construction projects may result in aforesaid process improvement along with a saving of about ₹2000–2500 per m^3 of walling and can also result in reduction of steel from 53.82 kg/m^2 to about 43 kg/m^2. For example, in a building of 10,000 m^2, 100 Metric ton of reinforcement steel (**₹55–60** lakhs) can be saved. In addition, CLC if used for other components in building projects may lead to reduced carbon emissions and an eco-friendly, sustainable and available material as and when required. Hence, Cellular Lightweight Concrete is expected to be a future construction material in terms of carbon neutrality thereby achieving sustainable construction.

References

1. Luo, Q., & Zhan,Q. (2012). Best practice in Low-carbon community planning. *Advanced Materials Research, **450-451***, 1082–1085
2. Elena, P., Maria, P., Gorshkov, A., & Rakova, X. (2014). Concept project of zero energy building, 25th DAAAM international symposium on intelligent manufacturing and automation, DAAAM 2014. *Procedia Engineering, 100*, 1505–1514.
3. Gupta, M.S. (2017). A path towards net zero energy buildings. *International Journal of Research in Engineering & Technology, **5***(2), 5-14.
4. Sumateja Reddy, V. (2016). Net zero energy building movement in India - an overview, international journal of scientific research in science. *Engineering and Technology, 2*(5), 360–363.
5. Venkatarama Reddy, B.V., & Jagadish, K.S. (2003). Embodied energy of common and alternative building materials and technologies, *Energy and Buildings, **35***, 129–137
6. Venkatarama Reddy, B. V. (2009). Sustainable materials for low carbon buildings. *International Journal of Low Carbon Technologies, 4*(3), 175–181.
7. Nangare, P., & Warudkar, A. (2015). Cost analysis of green building. *International Journal of Scientific Engineering and Research, **3***(6), 2347-3878
8. AlSadi, A., Cabrera, N., Faggin, M, He, Y., Patel, M., Trevino, F., Boyajian, D., et al. (2019). Comparative study on the cost analysis of a green versus conventional building. *Advanced Civil Engineering Technology, **3***(5).
9. Mustapure, N. (2016). A study on cellular lightweight concrete blocks. *International Journal of Research in Engineering and Technology, 5*(5), 188–191.
10. Bhavani Sriram, K. (2012). Cellular light-weight concrete blocks as a replacement of burnt clay bricks. *International Journal of Engineering and Advanced Technology, 2*(2), 149–151
11. Kumar, M.P. (2007). Cements and concrete mixtures for sustainability. *Proceedings of Structural Engineering World Congress*, Bangalore, India.
12. Deshmukh, R., & More, A. (2014). Low energy green materials by embodied energy analysis. *International Journal of Civil and Structural Engineering Research **2***(1), 58–65

An Exploratory Study on the Integration of Digital BIM and IOT in Structural Health Monitoring Practices

Karthik Dasari, Aaditya Dogra, and Huzefa Adeel

Abstract Utilization of data recording sensors for analyzing the distress patterns of structural members is gaining significance in the global construction industry. Decision-making on structural integrity requires real-time huge data visualization of the structural members. A framework to integrate the Building Information Modelling (BIM) and Internet of Things (IoT) to improve the process of real-time data monitoring and virtual visualization of the structure is developed. The present study makes special emphasis on a digital BIM model of a structure to visualize the structural integrity and develop a workflow modelling which is least focused in the structural engineering domain literature. Digital BIM model connected the virtual sensors created in BIM to visualize the real-time data monitoring from the data received through IoT sensors. A laboratory scale model is tested using IoT-based sensors. The outcome of the study showed that the virtual model created in digital BIM can respond to the physical model performance in real time. The study concludes that the decision-making using the integrated framework of BIM and IoT could introduce a digital BIM model that helps in leading professional SHM practices.

K. Dasari (✉)
Civil Engineering Department, National Institute of Technology Srinagar, Srinagar 190006, J&K, India
e-mail: dasari.karthik@nitsri.ac.in

A. Dogra · H. Adeel
Structural Engineering, Civil Engineering Department, National Institute of Technology Srinagar, Srinagar 190006, J&K, India
e-mail: 2021mcivst016@nitsri.ac.in

H. Adeel
e-mail: huzefa_2021mcivst012@nitsri.ac.in

A. Vilventhan et al. (eds.), *Advances in Construction Materials and Management*, Lecture Notes in Civil Engineering 346,
https://doi.org/10.1007/978-981-99-2552-0_13

1 Introduction

Continuous monitoring of civil engineering structures is required during all stages of a construction project. Efficient engineering practices such as Structural Health Monitoring (SHM) provide support for many infrastructure projects to monitor their structural performance [1]. Besides the successful implementation of SHM practices, there are several unsatisfactory factors that still need to be focussed towards live monitoring and inspection of civil engineering structures. SHM is a fast-growing domain and is proved to be the most reliable concept for the study of structural integrity and assessment for several construction and facility management projects [2]. Further to achieve effective monitoring data, smart sensors are being used [3]. When installed these smart sensors collect the performance data of a structure. These sensors could easily collect tons of real-time data containing various issues like cracks, deflections, temperature, moisture content etc., when placed at desired locations within the member or in whole structure. Sensors like ultrasonic, infrared, LVDT, optical fiber, piezometric sensor are widely used because of their feasibility to procure the data in any kind of environment [4]. Also, with the sensors shall act as a better medium for the exchange of data throughout the service life of the structure.

A large volume of data is generated through the sensors to analyse and monitor the structural members. Systematic data management strategies would aid the live data analysis of structures. The Internet of Things (IoT) is a widely utilized strategy for monitoring live data and it has been recently introduced into SHM practices [5]. The combination of SHM, cloud computing, and the IoT-enabled pervasive services and strong processing of sensing data provides a capable solution for rapid, accurate, and low-cost SHM processes. IoT used for data extraction from the sensors helps to detect live information such as the size and location of damage in the structures [6]. The SHM process contains huge amounts of monitoring data that helps to detect the behaviour of the structure during several task operations. Further with the focus on Artificial intelligence (AI) in engineering, the recorded information thus can be processed by intelligent algorithms which then become capable of extracting relevant information to the initial and future state of health of the infrastructure [7].

Building Information Modelling (BIM) which is one of the emerging digital technologies in the construction field transformed conventional civil engineering infrastructure into digital infrastructure enabling real-time virtual visualization by improving the reliability and sustainability of civil engineering structures [8]. Further digital BIM technology provides a virtual visualization of a field practice (and its related processes), that requires to be continuously updated from the exchange of data between the real model and virtual model [9].

BIM and IoT together with SHM practices can create a virtual interaction between the physical entity and a digital model of a structure resulting in a real-time representation of their performances [10]. The present study made an attempt to explore the possibilities of integrating BIM and IoT technologies to monitor the structural parameters of a laboratory model and developed a framework for live data monitoring of structural elements.

1.1 Objectives

1. To obtain the deflection data of a laboratory model by using sensor technology and integrate the data transfer by connecting to IoT devices
2. To develop a digital BIM model that shall link to the physical sensors for virtual visualization of structural data analysis.
3. To develop an integrated framework to aid the automation of the SHM process by linking physical sensor data to the digital BIM model.

2 Literature Review

The visualization output of SHM so far is done in the form of intuitive charts or images by processing data with procured from the original monitoring process. Computer software processes the information in the back end to display the output on the interactive interface [11]. The internet-based computer software is the main tool for SHM visualization and is dependently established for various projects. Data visualization output with 3D models [12] is lacking in some of the monitoring systems as they are directly focused on live data processing and visual presentation for the intended purpose of a construction project [13]. For most of the computer tools in India at this stage, live 3D model analysis can be done using customized plugins due to the lack of independent commercial BIM software. For example, visual monitoring of ta bridge by using a plugin on BIM software [14]. For high-level structural health management, there is an Open GL-based visualization system is available [15]. The GIS-based integrated software was developed to assess historical buildings [17] and obtain automatic monitoring of bridges [16]. A web-based SHM system is proposed with the help of web-available applications which also displays 3D model [19]. Most of the SHM monitoring systems with visualization output like deep foundation pits and safety systems in buildings are working with the tools and technology that use graphical displays [20]. Many of them gather data with the help of distributed sensors and the output will be displayed in two-dimensional planes [18]. A conceptual framework proposal to visualize the live output based on SHM is made but not fully developed [21].

In the field of SHM, there are many types of sensor technologies available which can gather the data and successfully link to BIM [6] as physical sensors are arranged at suitable locations to understand the structural fatigue and these are linked to BIM software tool to realize construction management [22]. In the construction field, laser sensors are linked to BIM along with videos to provide useful navigation for tower crane operators [23]. The integration of BIM with sensor technology is commonly implemented in the facility management phase of a construction project. To inspect the structural fatigue of a bridge deck under varying climatic conditions, thermal sensors were connected in series on the deicing system [24]. With the same concept, combining BIM and fire emergency management system lead to the development of a visual aid fire rescue warning system [25]. By integrating BIM and wireless sensing technology, a prototype framework was developed for obtaining real-time thermal data monitoring in close work environments [26]. In the meantime, IoT technology being widely used in many fields can be a good opportunity to include with BIM and monitor the live data of SHM of structures. Application of this IoT technology can provide a better and efficient method of gathering the required data in the field of SHM. They shall be utilised to get reference templates in gathering the data [27], transforming the data [28], security encryption [29], data storing and related applications [30].

3 Methodology

To quickly find the prime source of damage and for continuously monitoring the structure from the received data using smart sensors, efficient data transfer through wireless system, visualizing real-time state of the monitoring area corresponding to the data and a visual based decision-making monitoring system for structural integrity and health assessment is developed. The overall integrated framework of the system is shown in Fig. 1.

The study methodology is divided into 4 parts. First stage includes identifying the SHM parameters based on the actual needs for the monitoring system, several

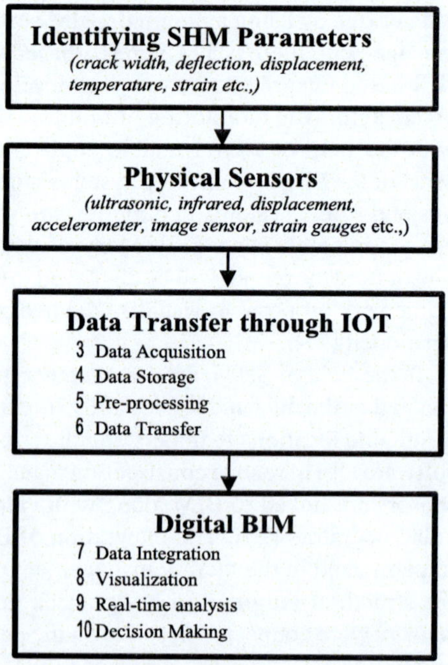

Fig. 1 Proposed Methodology for Smart Structural Health Monitoring System

parameters (like crack width, deflection, displacement, temperature, strains, etc.) to be required for the effective monitoring of the structure are being selected.

Second stage is selection of sensors. Various types of sensors are adopted to get the real time information of the member or whole structure. These sensors include ultrasonic, infrared, displacement, accelerometer, image sensor, strain gauges, and many more. These sensors were selected according to their feasibility for the case study. These sensors can either be attached or embedded within the member for monitoring purpose.

Third stage is linking IoT transfer with sensor data. This phase contributes to the acquisition of the data and transfer it through the wireless system to the cloud storage. The information recorded via sensor technology is preprocessed from raw data to understandable format. The IoT technology facilitates extraction of the live data at any location that aided the real time monitoring of the study.

Last stage is to integrate the sensor data and develop a digital BIM model. The BIM technology is used to create a virtual illustration of a physical system (and its associated environment and processes), which is continuously updated through the exchange of information between the physical and virtual system by means of IoT data transfer. The virtual 3D models are created and these models are linked to virtual sensors which processes data initially collected by physical sensors. The whole digital BIM system integrates the physical system and virtual model towards developing a real-time visualization of the structure which is being analysed by SHM process.

3.1 Integration of IoT and BIM

The present study links the BIM with IoT data transfer and created a smart visualization monitoring system. The field information of the structure performance will be detected through physical sensors and these performance parameters are used to visualize the behaviour of the structure at any phase of its life. The physical sensors' data are then sent to a cloud database which in turn can be extracted to BIM tool. Virtual sensors are developed using BIM tools such as Revit, SAP2000, etc. Thus, BIM tools model virtual sensors by processing the data gathered from the physical sensors received through IoT data transfer. The physical data which was sent will be typically embedded into software applications that combine the input with physical data from various sources and execute analytical algorithms on the combined set of data. When the BIM model shows abnormal behaviour, the IoT platform controls the monitoring system, and the model will get updated to visualize the affected location and the predictions for any possible damage or failure are made. Usually, the digital BIM models of the structure are done by general BIM modelling software and can be converted to the target formats according to industry standards to smooth completion of analysis on other software platforms. Most commonly used BIM application extensions are IFC, DGN, RVT, FBX, etc.

4 Case Study

A laboratory model has been setup to validate the proposed framework (Fig. 2). The procedure developed in the study enables real-time monitoring of the central deflection of a structure. For this purpose, a test is performed on a portal frame to find the central deflection data of the beam for SHM process through the physical sensor and is linked to IoT device for further process. The whole arrangement consists of an ultrasonic sensor HC-SR04 to measure the distance between the beam intrados and ground surface, Arduino Uno (microcontroller board to process the data), Wi-Fi communication module connected to Arduino for transmission, a cloud storage to accumulate the sensor data, digital model with virtual sensor created to represent the structural fatigue and for the real-time monitoring through digital BIM processing.

4.1 Data Capturing and Analysis

HC-SR04 is a sensor generally used to measure the distance and consists of two ultrasonic transducers. The first transducer acts as a transmitter and generate the ultrasonic pulses and the second one acts as a receiver and hears for the transmitted ultrasonic pulses. The transmitter produces ultrasonic sound pulses at 40,000 Hz which travel through the air and if any obstacle comes in its path, ultrasonic pulses bounce back to the module. When these pulses are received by the receiver, it produces an output pulse whose width is proportional to the distance of the obstacle from the sensor. HC-SR04 work with an accuracy of 3 mm and can measure the distance between 2 to 400 cm. Further using high precision sensors in the study could help to get higher accuracy.

In this study, ultrasonic sensor HC-SR04 has been fixed at the bottom center of the beam to measure the distance from the ground and it is connected to a microcontroller

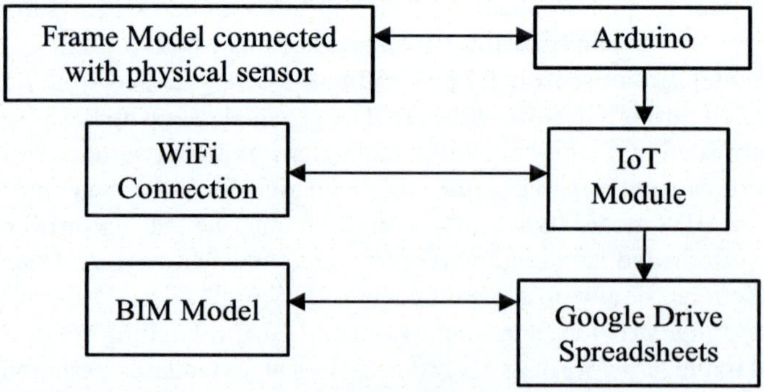

Fig. 2 Process showing the connection between the physical model and BIM model

Fig. 3 Laboratory model showing the sensors connected to a portal frame

board (Arduino UNO) which provides USB-serial conversion. A Wi-Fi module ESP-8266 has also been connected with Arduino UNO to transmit the sensor data to the cloud storage so that the communication of sensor data to the cloud storage can be done. A point load (F) is applied to the center of the beam and deflection data is measured through sensor (Fig. 3).

4.2 Data Transformation

For the transformation of data Arduino UNO and a Wi-Fi module ESP-8266 are used as shown in Fig. 4. The data acquired from the sensor are sent to a web server through an internet connection. The requirement of internet connectivity or Wi-Fi connection is fulfilled by ESP-8266 module. The deflection data is extracted using ultrasonic sensor. This sensor uses transmitter (T) which excites ultrasonic pulses toward the ground surface, and the receiver (R) picks up the reflected waves back from the surface. The distance between the hit surface and the physical sensor is the measurement done by estimating the latency time between the pulse emission and its return to the receiver. These measurements can be recorded and read simultaneously through the serial monitor of Arduino work environment. In order to determine the deflection of the frame, a mathematical model is used as shown.

First, the distance travelled by the wave is measured as

$$\boldsymbol{Distance} = \boldsymbol{speed} \times \boldsymbol{time}$$

As the sensor uses sound waves thus the speed of wave is 343 m/s (0.343 mm/μs), ts be the total time (in μm) taken by the pulse to travel from T to R. therefore

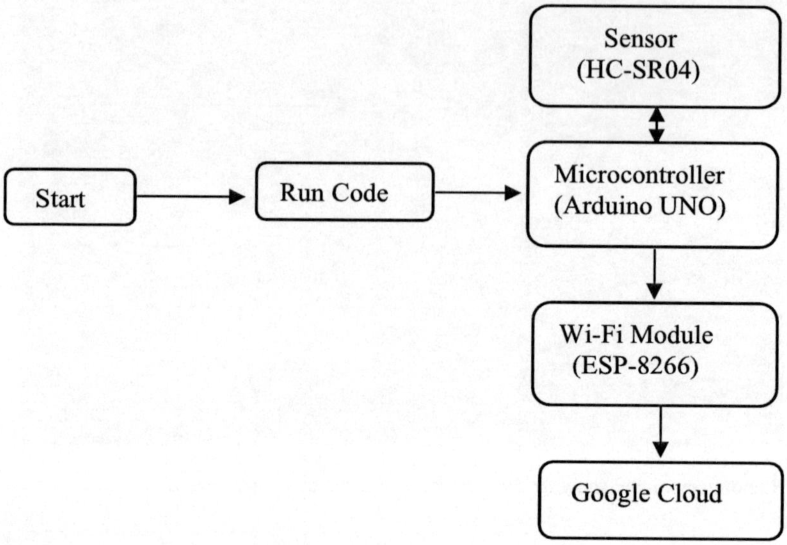

Fig. 4 Workflow for Data Transformation

the measured distance is calculated as:

$$d = \frac{t_s}{2} \times 0.343\, mm$$

The value is multiplied by 1/2 because it is the time for go-and-return distance.

For measuring the deflection values using the recorded distance values, following expression is adopted:

$$\delta(mm) = d_i - d_{i+1}$$

Further to test for dynamic loading with different loads, the value of distance varies with load and hence the deflection of beam varies accordingly. Once the sensor HC-SR04 acquires distance measurement, the data are sent to a remote server through an internet connection. However, Arduino Uno has no Wi-Fi connection, so it required to connect it to a wireless module for appropriate functionality. In this case, the ESP-8266 module is used for this process which is a Wi-Fi module capable of using the TCP/IP protocol and connected with the microcontroller. With the native firmware provided by Espressif Systems, it responds well to attention commands (the command set originally developed for modems). Attention commands were considered in this experiment due to their ease of application and integration with the programming code saved in the Arduino. The programming for the process of data collection through the ultrasonic sensor is for arduino will convert the values into millimetres. A TCP connection along with the server is established by the ESP-8266 module, any physical data received from the Arduino are then immediately transferred to the network.

4.3 Monitoring, Communication and Hardware Setup

In order to evaluate the proposed technique. A test frame is monitored to get the deflection at the centre of the model frame using ultrasonic sensor positioned at the downside of the frame at centre. The device is connected to Arduino UNO and is supplied with 5 V DC supply/500 mA power. Data from the device are integrated with digital BIM technology connected with IoT board ESP8266, which setups a communication between the integration platform with the physical setup. Sensor setup is codified with a code as mentioned above in Arduino IDE 2.0.2 application. The input is given to the physical setup through the code to extract the deflection values. The data collected is then sent to the cloud database over internet connection, which stores this information for the monitoring purpose and can be extracted anytime and anywhere (Fig. 5).

4.4 Developing a Digital BIM Model with Dynamo

As there is a variety of BIM software available in the field i.e. AutoCAD, Revit, SketchUp, Archicad, Navisworks, etc. in this experimental study we have used Autodesk Revit for simulation purpose. For creating a digital BIM model of the structural member, a generic model has been chosen from the software families and modified in its properties and dimensions according to the physical model.

Once the 3D modeling has been done in Revit software, virtual sensor is created by using Dynamo and then linked with the physical sensor data. For creating the virtual sensor in Dynamo physical data response were added in the object categories. Dynamo is the graphical programming tool for design in Revit. Sensor data from google sheets linked to the dynamo using BIMOne add-on. Dynamo tool works on basis of connecting required nodes. Each node upon linking to other node according to the function parameters perform operations through input and output nodes. The dynamo programming is made in two stages. First stage programming links the google sheet to dynamo and the later stage links the data input to the BIM model (Fig. 6).

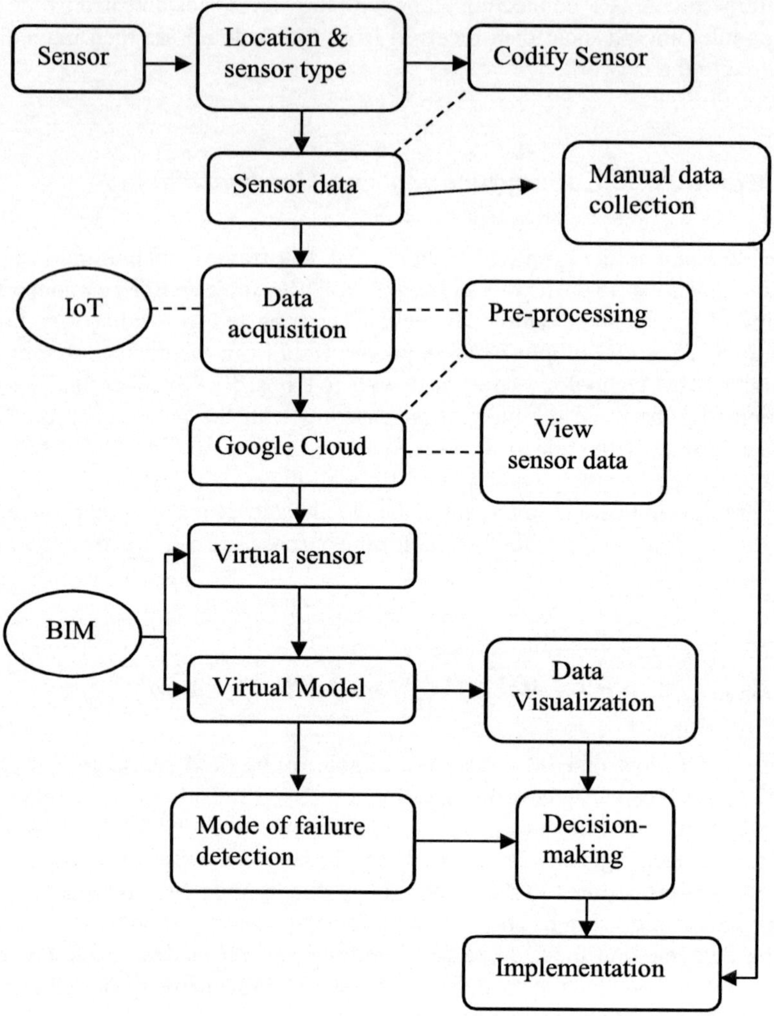

Fig. 5 Data Workflow for the Study

4.5 Results and Summary

This part describes the visual monitoring of SHM process based on BIM-IoT integration. This integration can be utilized on various structural projects and can be further expanded with richer functions to meet the requirement of engineering analysis of a structure. As shown in the Fig. 7, the study links the sensor data to a digital BIM model creating a novel framework for live data monitoring in SHM process in

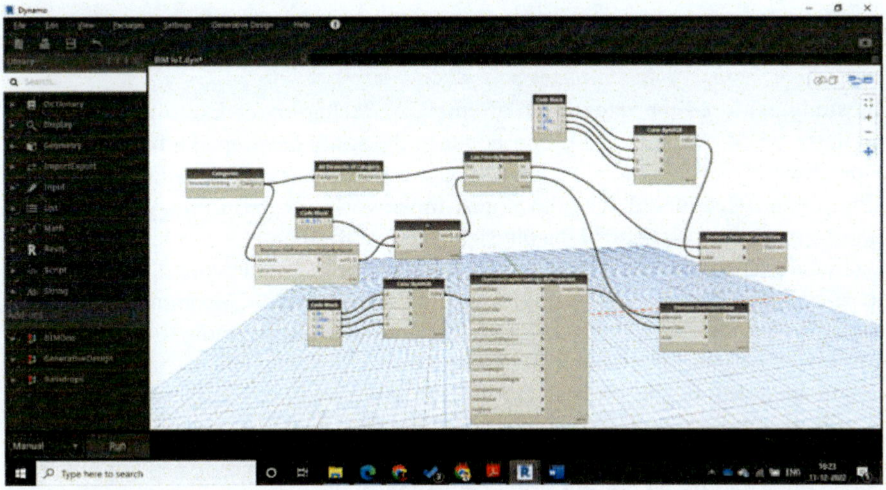

Fig. 6 Dynamo programming in Revit software

structural engineering field. The study further developed IoT based platform for efficient data management and decision process but it still requires to model a complete IoT-Digital BIM framework.

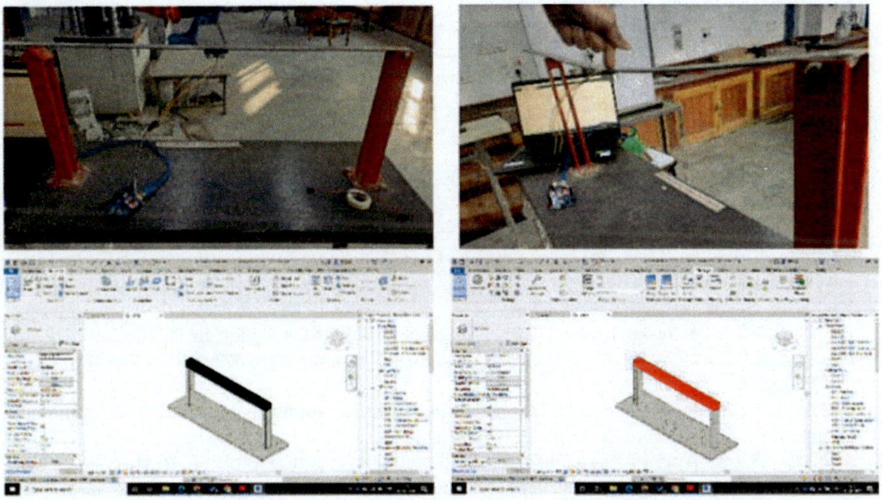

(a) (b)

Fig. 7 BIM model showing deflection simulation **a** deflection < threshold, **b** deflection > threshold

5 Conclusion

The study explored the integration of emerging technologies like digital BIM and IoT in the SHM practices. As a consequence the study developed a framework that helped towards virtual visualization of structural members and live data monitoring using digital BIM model. The developed framework directed to monitor the structural integrity that can control the physical and environmental variables affecting the functional performance of a structure. The context of the framework is to interact the virtual and physical sensors through BIM-IoT integration. A laboratory model was tested to visualize the digital BIM based output as well as to explore the possobilities of linking the data for efficient SHM monitoring process. IoT technology is in a continuous development process and therefore the proposed framework should be periodically updated and modelled to cater the technological advancements. Thus, a conceptual framework for model development and technological integration like the one proposed in the present study could contribute to add further support to the future digital BIM development even after the technological advancements.

References

1. Ngeljaratan, L., & Moustafa, M. A. (2020). Structural health monitoring and seismic response assessment of bridge structures using target-tracking digital image correlation. *Engineering Structures, 213*, 110551.
2. Hou, G., Li, L., Xu, Z., Chen, Q., Liu, Y., & Mu, X. (2022). A visual management system for structural health monitoring based on web-BIM and dynamic multi-source monitoring data-driven. *Arabian Journal for Science and Engineering, 47*(4), 4731–4748.
3. Martin, D., Kühl, N., & Satzger, G. (2021). Virtual sensors. *Business & Information Systems Engineering, 63*(3), 315–323.
4. Ciciriello, P., Mottola, L., & Picco, G. P. (2006, November). Building virtual sensors and actuators over logical neighborhoods. In *Proceedings of the International Workshop on Middleware for Sensor Networks* (pp. 19–24).
5. Scianna, A., Gaglio, G. F., & La Guardia, M. (2022). Structure monitoring with BIM and IoT: the case study of a bridge beam model. *ISPRS International Journal of Geo-Information, 11*(3), 173.
6. Abdelgawad, A., & Yelamarthi, K. (2017). Internet of Things (IoT) platform for structure health monitoring. *Wireless Communications and Mobile Computing, 2017*. https://doi.org/10.1155/2017/6560797
7. Salehi, H., & Burgueño, R. (2018). Emerging artificial intelligence methods in structural engineering. *Engineering Structures, 171*, 170–189.
8. Mihindu, S., & Arayici, Y. (2008, July). Digital construction through BIM systems will drive the re-engineering of construction business practices. In *2008 international conference visualisation* (pp. 29–34). IEEE.
9. Chiachío, M., Megía, M., Chiachío, J., Fernandez, J., & Jalón, M. L. (2022). Structural digital twin framework: formulation and technology integration. *Automation in Construction, 140*, 104333.
10. Cao, Y., Miraba, S., Rafiei, S., Ghabussi, A., Bokaei, F., Baharom, S., Haramipour, P., & Assilzadeh, H. (2020). Economic application of structural health monitoring and Internet of Things in efficiency of building information modeling. *Smart Structures and Systems, An International Journal, 26*(5), 559–573.

11. Guo, H., & Wang, R. (2017). Study on BIM-based health monitoring information and its visualization realization. *Construction Technology, 46*(S1), 510–513.
12. Li, H., Ou, J., Zhao, X., Zhou, W., Li, H., Zhou, Z., & Yang, Y. (2006). Structural health monitoring system for the Shandong Binzhou Yellow River highway bridge. *Computer-Aided Civil and Infrastructure Engineering, 21*(4), 306–317. https://doi.org/10.1111/j.1467-8667.2006.00437.x
13. Desjardins, S. L., Londono, N. A., Lau, D. T., & Khoo, H. (2006). Realtime data processing, analysis and visualization for structural monitoring of the confederation bridge. *Advances in Structural Engineering, 9*(1), 141–157. https://doi.org/10.1260/136943306776232864
14. Delgado, J. M. D., Butler, L. J., Gibbons, N., Brilakis, I., Elshafie, M. Z. E. B., & Middleton, C. (2017). Management of structural monitoring data of bridges using BIM. *Proceedings of the Institution of Civil Engineers-Bridge Engineering, 170*(3), 204–218. https://doi.org/10.1680/jbren.16.00013
15. Ni, Y., Lin, K., Wu, L., & Wang, Y. (2017). Visualized spatiotemporal data management system for lifecycle health monitoring of large-scale structures. *Journal of Aerospace Engineering, 30*(2), B4016007. https://doi.org/10.1061/(ASCE)AS.1943-5525.0000622
16. Ciampoli, L. B., Gagliardi, V., Calvi, A., D'Amico, F., & Tosti, F. (2019). Automatic network level bridge monitoring by integration of InSAR and GIS catalogues. In *Multimodal sensing: technologies and applications 2019* (p. 110590I). International Society for Optics and Photonics
17. Tsilimantou, E., Delegou, E. T., Nikitakos, I. A., Ioannidis, C., & Moropoulou, A. (2020). GIS and BIM as integrated digital environments for modeling and monitoring of historic buildings. *Applied Sciences, 10*(3), 1078. https://doi.org/10.3390/app10031078
18. Chen, B., & Liu, W. (2012). A web-based structural health monitoring sensor network. *International Journal of Computer Applications in Technology, 44*(3), 188–197. https://doi.org/10.1504/ijcat.2012.049082
19. Providakis, C., & Liarakos, E. (2014). Web-based concrete strengthening monitoring using an innovative electromechanical impedance telemetric system and extreme values statistics. *Structural Control and Health Monitoring, 21*(9), 1252–1268. https://doi.org/10.1002/stc.1645
20. Zhu, C., Yan, Z., Lin, Y., Xiong, F., & Tao, Z. (2019). Design and application of a monitoring system for a deep railway foundation pit project. *IEEE Access, 7*, 107591–107601. https://doi.org/10.1109/ACCESS.2019.2932113
21. Wang, S., Du, J., & Song, J. (2016). A framework of BIM-based bridge health monitoring system. In *Proceedings of the 2016 International Conference on Civil, Transportation and Environment 2016*. Atlantis Press
22. Riaz, Z., Arslan, M., Kiani, A. K., & Azhar, S. (2014). CoSMoS: a BIM and wireless sensor based integrated solution for worker safety in confined spaces. *Automation in Construction, 45*, 96–106. https://doi.org/10.1016/j.autcon.2014.05.010
23. Lee, G., Cho, J., Ham, S., Lee, T., Lee, G., Yun, S.-H., & Yang, H.-J. (2012). A BIM- and sensor-based tower crane navigation system for blind lifts. *Automation in Construction, 26*, 1–10. https://doi.org/10.1016/j.autcon.2012.05.002
24. Kensek, K. (2014). Integration of Environmental sensors with BIM: Case studies using arduino, dynamo, and the revit API. *Informes de la Construcción, 66*(536), 31–39. https://doi.org/10.3989/ic.13.151
25. Chen, X.-S., Liu, C.-C., & Wu, I. C. (2018). A BIM-based visualization and warning system for fire rescue. *Advanced Engineering Informatics, 37*, 42–53. https://doi.org/10.1016/j.aei.2018.04.015
26. Smarsly, K., & Tauscher, E. (2016). Monitoring information modeling for semantic mapping of structural health monitoring systems. In *Proceedings of the 16th International Conference on Computing in Civil and Building Engineering*
27. Iwendi, C., Moqurrab, S. A., Anjum, A., Khan, S., Mohan, S., & Srivastava, G. (2020). N-sanitization: a semantic privacy-preserving framework for unstructured medical datasets. *Computer Communications, 161*, 160–171. https://doi.org/10.1016/j.comcom.2020.07.032
28. Faheem, M., Fizza, G., Ashraf, M. W., Butt, R. A., Ngadi, M. A., & Gungor, V. C. (2021). Big data acquired by internet of things-enabled industrial multichannel wireless sensors networks

for active monitoring and control in the smart grid industry 4.0. *Data Brief 35*, 106854–106854. https://doi.org/10.1016/j.dib.2021.106854

29. Shabbir, M., Shabbir, A., Iwendi, C., Javed, A. R., Rizwan, M., Herencsar, N., & Lin, J.C.-W. (2021). Enhancing security of health information using modular encryption standard in mobile cloud computing. *IEEE Access, 9*, 8820–8834. https://doi.org/10.1109/access.2021.3049564
30. Pan, Y., & Zhang, L. (2021). A BIM-data mining integrated digital twin framework for advanced project management. *Automation in Construction, 124*, 103564. https://doi.org/10.1016/j.autcon.2021.103564

Quantitative and Qualitative Benefits of BIM Implementation in Hospital Management: A Case Study Analysis

Apoorv Mishra and Aneetha Vilventhan

Abstract The current studies of BIM implementation have extensively proclaimed the benefits of BIM. However, they lack the explicit justification of direct cost savings due to BIM implementation. Hence, this paper aims to identify the economic benefits of BIM in the post-construction stage of hospital buildings. This paper presents a comparison of two case studies, with one utilizing BIM approach for hospital management and the other using non-BIM approach for hospital management. Cost-benefit analysis for both the projects was performed using a time-effort distribution curve and the results were analysed quantitatively and qualitatively. The results indicate that the aggregated value for BIM-based projects is lower than that of non-BIM-based projects. It was found that the BIM-based operation and maintenance saves 343 INR per square meter when compared with non-BIM-based operation and maintenance practice. Studies that have used time-effort distribution methodology to identify BIM benefits, are mostly focused on the construction stage. The findings provided in this publication are unique because it focuses on quantifying BIM benefits in the post construction stage of hospitals.

Keywords Building Information Modelling (BIM) · Cost-Benefit analysis · Time-effort distribution curve · Building maintenance and management

1 Introduction

Hospital building projects are crucial, need a substantial investment, require extensive planning, and have very high maintenance costs. Owing to the complex need for sanitation, safety, professional tools, special equipment and a huge amount of data to be handled, hospital building projects are quite complex. The construction process is quite dynamic and includes iterative stages and interim adjustments. Serious issues,

A. Mishra · A. Vilventhan (✉)
Department of Civil Engineering, National Institute of Technology Warangal, Warangal, Telangana, India
e-mail: aneetha@nitw.ac.in

A. Vilventhan et al. (eds.), *Advances in Construction Materials and Management*, Lecture Notes in Civil Engineering 346,
https://doi.org/10.1007/978-981-99-2552-0_14

like work order delays, inadequate documentation, and poor communication management still hinder healthcare building projects in the operating phase. The operational costs of hospitals are the greatest among public service organizations due to the complicated interaction of individuals with different organizational functions, a diversity of materials and facilities, and a variety of delicate medical operations [12].

The fundamental goal of the operation and maintenance system is to offer a systematic process for managing the condition of existing buildings and facilities, maximizing their potential for optimum efficiency, and lowering their breakdown and disability rate.

The operation and maintenance of healthcare facilities are more complicated and expensive and require continuous support to manage emergency and life-saving situations. It was identified that the patients' comfort and happiness were linked with the degree of maintenance practices performed in healthcare institutions [20].

In the AEC sector, Building Information Modelling (BIM) is used as "an information technology-enabled method that uses and maintains a data repository as an essential digital representation of all building information for various stages of the project lifecycle". The use of BIM can save capital expenditures roughly by 20% throughout the operating phases of a project, which can have a positive impact on costs overall [10]. Though several studies have developed a framework to estimate the advantages of BIM, its use has not yet been empirically established. As a result, owners/authorities must choose BIM implementation only based on its potential advantage rather than its economic benefit. This reluctance to implement BIM is also addressed in prior studies [5]. Hence this paper aims to identify the cost-benefits of BIM implementation in the post-construction stage of healthcare facilities.

The rest of the paper is organized as follows: the next section discusses on literature review of BIM applications in operation and maintenance phase of buildings later followed by cost-benefit analysis. Further, the paper discusses on research methodology, quantitative and qualitative assessment and discussion on results and finally conclusion is provided.

2 Literature Review

BIM adoption in the Architectural, Engineering, and Construction (AEC) sector has received attention in recent years. [9] and [4] described BIM as a group of technologies and related techniques used to represent and manage information that is generated during the building design, construction, and operation processes. BIM has been used, to enhance management, decrease construction costs and delays, increase design quality, and enable AEC teaching at universities [21]. BIM, when combined with other technologies, can be used to analyze energy use, estimate carbon emissions, and assess sustainability [1, 11, 23].

During the operation and maintenance phase of buildings BIM was employed to perform space planning, security management, maintenance of services, remodeling of existing, and documentation management as shown in Fig. 1. BIM provides detailed, accurate, and up-to-date information about facilities and allows owners, operators, and contractors to access and share information at any time during facility management [3]. BIM-based energy modeling allows for tracking energy usage of the constructed facility and enhances the energy performance of the building [5]. BIM also enhances building management's capacity for emergency response. It can quickly deliver vital information and identify the exact location of the issue. BIM tools allow developing and managing document systems coupled with online IFC models and enable storing all projects in a single data repository, thereby collaborating project stakeholders and workflows from design to construction and demolition [13]. By performing real-time monitoring of the operational parameter of the equipment using the BIM model, equipment functions can be assessed and predicted for failure and future maintenance [2, 13].

Similarly, BIM was used in supporting operation and maintenance of healthcare facilities [16]. BIM is used for improving technological and environmental performances of operating rooms in hospitals [18]. BIM allowed better visualization of hospital condition and improved decision-making process during maintenance tasks planning and records management. BIM integrated with lean techniques and Building Energy Performance Systems and Big data were used to facilitate data management and increase the efficiency and eliminate waste in FM practices [7]. Studies have

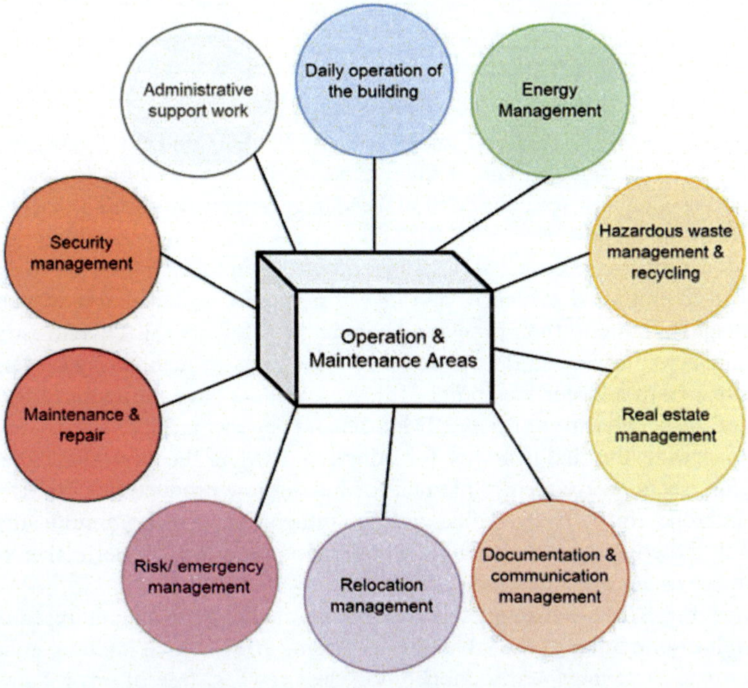

Fig. 1 Different areas of operation & maintenance buildings

also examined using BIM with Internet of Things to facilitate object (wheel chairs, beds) based tracking and locating in hospitals during facility management [8].

The current approaches of using BIM for healthcare management attempts to explore possible applications to improve the current facility management practices [7, 8]. Studies on examining the effectiveness or economic benefit of BIM implementation for healthcare facility is rarely explored. The main challenge in identifying BIM effectiveness is the difficulty in separating the costs from the benefits of BIM. Many AEC projects implement BIM for fostering better collaboration and communication, however, it is difficult to identify the benefits of coordination in terms of cost. This requires performing cost-benefit analysis to measure the benefits of BIM implementation. In extant literature, studies have used cost-benefit analysis to identify economic benefits of BIM implementation in construction projects [15]. They include cost-benefit analysis of BIM implementation for clash detection (Chahrour et al. 2020) and for planning and construction of railway sites [22]. However, the current approaches are limited to designing and construction phases of the project and the use of cost-benefit analysis of BIM implementation in post construction phases is not explored yet. Hence, to bridge the gap, this paper aims to use cost-benefit analysis to identify economic benefit of BIM implementation in the operation and maintenance phase of healthcare facilities.

2.1 Cost-Benefit Analysis

The implementation of BIM and analysis of its benefits and disadvantages have always been closely related. This is to be expected because a technological project needs a strong economic foundation to succeed in a competitive commercial environment. One of the biggest barriers to BIM adoption is convincing stakeholders of the benefits of the increased costs [6]. Users who want to embrace BIM require encouragement from empirical evidence, and investors require unmistakable evidence of its advantages to support their time and financial investments [3]. This investigation can be connected to past studies examining the effects of technology on business performance from a larger viewpoint [15]. Researchers have particular difficulties because of the peculiarities of how BIM costs and benefits are measured.

Understanding the justifications for adopting BIM is the first challenge. The construction sector is frequently blamed for having low productivity. Better information interoperability-based collaboration, integration, and communication are required in the construction sector [6, 21]. BIM is seen to be an effective way to address these issues and, increase productivity.

Recognizing BIM's advantages is the second challenge. Previous research has an in-depth examination of the advantages of using BIM, which include, enhanced productivity, cost savings, early cooperation, less rework, free of error, and better predictability [6, 13, 14, 21]. However, existing measurement techniques are difficult to separate the part of costs-benefits that the implementation of BIM contributes. As an example, clash detection is widely used as an illustration for mainstreaming BIM adoption in the AEC sector. The potential cost of a clash happening without being detected by BIM is typically quantified and assigned to BIM as one of its advantages. However, more and more on-the-job engineers think that this attribution overstates the advantages of BIM because they may also utilize their experience to spot a clash. Recognizing the advantages of BIM in management AEC processes, where it is utilized to enhance communication, promote cooperation, and facilitate knowledge exchange, is more difficult.

The third and final challenge is that it is frequently difficult to obtain the data required to assess the costs and advantages of BIM deployment in the AEC sector. While few empirical studies have been recorded, studies frequently cite anecdotes to present their statements about the advantages of BIM. Studies frequently use case studies, which are primarily controlled experimental contexts. The lack of authentic data also makes it challenging to apply rigorous mathematical techniques (such as time-series econometric models) that can lessen the impact of random events. This is due to the limits of current measurement methods. To account for total BIM costs and advantages, the model must be "extensive" enough, while balancing out the uncontrollable variables that affect the deployment of BIM in practice. In this regard, the time-effort distribution curve given by MacLeamy was found to be suitable for analyzing the cost and benefits of BIM implementation.

2.2 *Time-Effort Distribution Method*

MacLeamy's [17] time-effort distribution curve has four components (as shown in Fig. 2), (1) showing how a project's progress can affect the cost and functional capacity; (2) depicting the price of design change; (3) showing the distribution of design effort in conventional AEC methods; and (4) illustrating how design effort is distributed in BIM-enabled AEC processes. In the conventional AEC practices, designers and contractors put out independent efforts in construction documentation and administration (Curve 3), while BIM-enabled methods encourage the entire project team to put in the additional effort during the phases of schematic design and design development (e.g., early collaboration and open information sharing), (Curve 4).

Studies have adopted MacLeamy's set of curves as an example of how BIM might improve construction projects [3]. The determination of BIM's total costs and advantages deployment in building projects is also impacted methodologically by time-effort distribution curves. It is possible to think of the region bounded by Curve 3, the X-axis, and the Y-axis as the overall work required to complete a project using conventional AEC techniques. The area bounded by Curve 4, the X-axis, and the Y-axis also depicts the overall effort put in to complete the project via BIM. Subtracting both areas equals the work that BIM implementation has saved. This can be expressed in mathematical terms as Eq. (1).

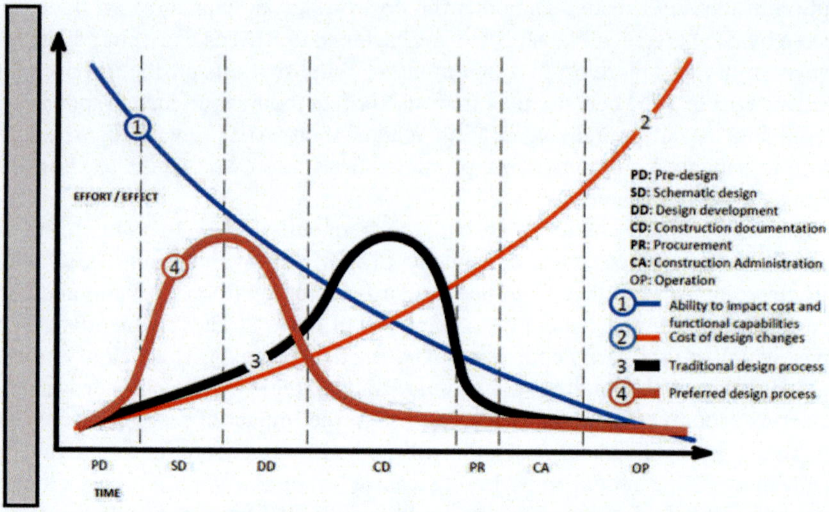

Fig. 2 Time-effort distribution between BIM-enabled and traditional construction processes. (Adapted from MacLeamy [17])

$$\int f(x4)dx - \int f(x3)dx = \text{BIMs' Benefits} \tag{1}$$

where the indefinite integrals represent the regions enclosed by curves 3 and 4.

3 Research Methodology

To perform cost-benefit analysis and identify the benefits of BIM implementation a framework was adopted as shown in Fig. 3.

3.1 Collecting Data to Develop the Time-Effort Distribution Curves

To examine the BIM implementation during the operation and management phase, real-time projects with and without BIM implementation were identified. Required time and effort information were collected from the projects to develop time-effort distribution curves. The term "time" refers to the duration of a project's process (like on a daily, weekly, or monthly basis), and "effort" refers to the total amount of billable time that each project participant has contributed [4]. The term "Priced Efforts (PE)" includes the range of efforts that have been valued using market procedures such as competitive bidding and tendering. With this approach, input effort data

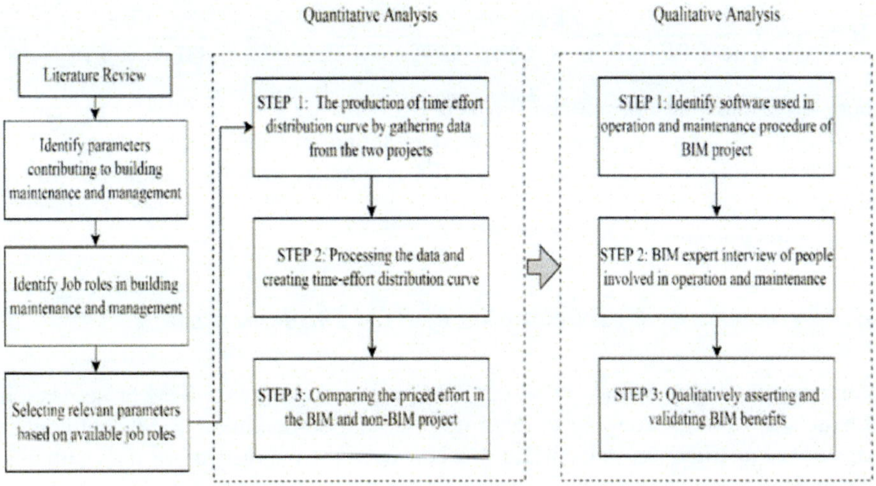

Fig. 3 Research methodology

were gathered through referring payment history analysis. Time-effort data for BIM and non BIM-based projects were gathered and expressed in mathematical terms as shown in Eq. 2.

$$E = \{PE_1,\ PE_2,\ PE_3, \ldots\ldots\ldots\ldots.,\ PE_T\} \quad (2)$$

where PE_T stands for all of the input-priced efforts in period T, and E is the set of priced effort data. Further, a time-effort distribution curve is developed by vertically aggregating the priced efforts for both BIM and non-BIM projects.

3.2 Data Processing and Developing Time-Effort Distribution Curves

To create time-effort distribution curves, more processing must be applied to the priced efforts as illustrated in Eq. (2). The construction projects are distinct from one another by differences in site location, condition, gross floor area, contract amounts, procurement processes, starting dates, and other factors. To compare any two projects it is required to minimize the effect of these differences and make them similar through normalization. To normalize different projects, two factors such as Gross Floor Area (GFA) and Consumer Price Index (CPI) were considered. The collected datasets (BIM and non-BIM) were then divided by the respective project's GFA and CPI to obtain normalized priced attempts as shown in Eq. (3). As the gross area and start date and duration of the two projects may vary, GFA and the influence of inflation (CPI) are considered.

$$E = \frac{PE_1}{GFA_1 \times CPI_1}, \frac{PE_2}{GFA_1 \times CPI_1}, \frac{PE_3}{GFA_1 \times CPI_1} \cdots\cdots\cdots \frac{PE_T}{GFA_1 \times CPI_1} \quad (3)$$

where E denotes the normalized proceed effort of the project, GFA is the corresponding gross floor area and CPI is the corresponding consumer price index. The total effort required to build each unit of GFA is considered as the normalized priced efforts.

3.3 Determining Costs/Benefits of BIM Implementation

Comparing the input effort used in the BIM project and the non-BIM project is the fundamental strategy adopted to determine the costs/benefits of BIM deployment based on the justification stated in Eq. (1). However, the majority of the data gathered from real-life projects are discontinuous, which prevents the creation of an accurate and continuous curve. In this regard, instead of employing the indefinite integral, this approach analyses costs and benefits by accumulating discrete data. Hence, the resulting equation is shown in Eq. (4).

$$K = \frac{\sum EN't - \sum ENt}{\sum EN't} \times 100\% \quad (4)$$

where K denotes the proportion of normalized priced efforts saved over the comparison period as a result of the installation of BIM when comparing non-BIM projects, $\sum ENT$ is the accumulated normalized priced efforts of BIM project and $\sum EN'T$ represents normalized priced efforts in the non-BIM project.

3.4 Data Collection

To evaluate the benefits of BIM implementation during the operation and maintenance phases of healthcare facilities, case scenarios with a BIM-based hospital management approach and a non-BIM-based hospital management approach were considered as shown in Table 1.

To collect data for time-effort calculation, identification of individuals involved in hospital management is necessary. Hence, activities involved in the operation and maintenance of a hospital were grouped as shown in Fig. 4, and job roles involved under each parameter (such as security, and maintenance) were identified for both projects. Three sets of data such as salaries of people involved in operation and maintenance activities of different departments, cost of different safety inspections, security awareness/training programs, and cost incurred for any repair/rework/breakdown were gathered. All these expenses that are made from October 2020 to 2021 were tallied and shown in Table 2.

Table 1 Demographic information hospital buildings

Particulars	Non-BIM Hospital	BIM Hospital
Name & Location	Jaya Maxwell Hospital Haridwar, Uttarakhand	Tertiary Cancer Care Hospital Ambala, Haryana
Type of building	Institutional Building	Institutional Building
Gross site area	970 sq. m	2830 sq. m
Gross floor area	875 sq. m	2525 sq. m
Start date	July, 2017	Jan, 2017
Completion date	Feb, 2019	May, 2020
Total construction cost	8.53 Crore	24.91 Crore
Tools used for hospital management		Revit, E-work orders, MediXcel EMS

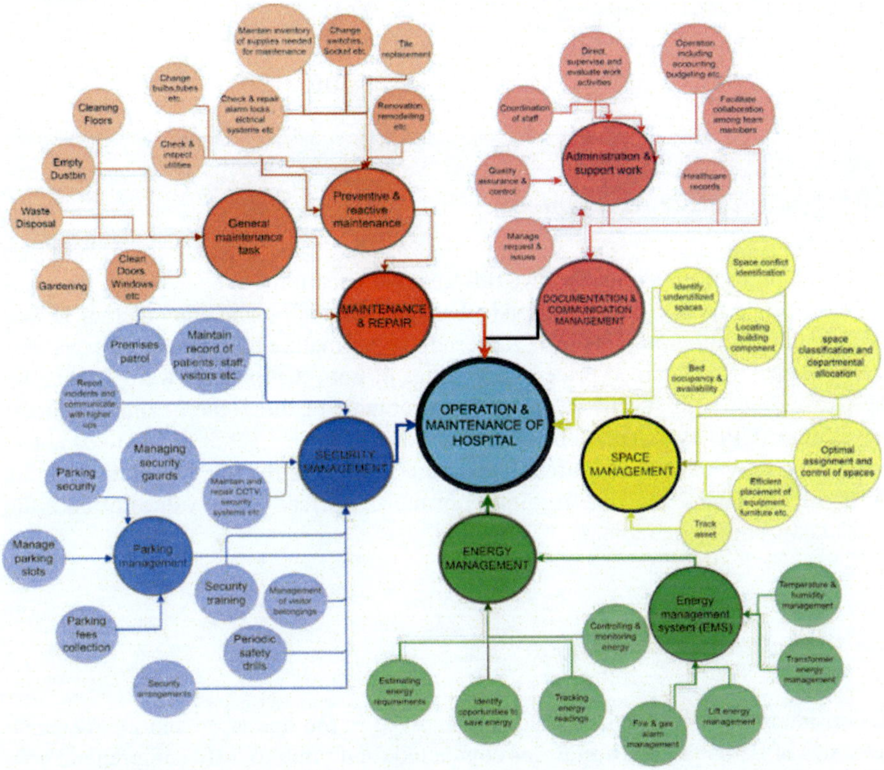

Fig. 4 Bubble diagram representing various O & M parameters

Table 2 Priced effort values of the projects

Months	Non-BIM Hospital maintenance cost in Rupees	BIM Hospital maintenance cost in Rupees
Oct-20	447,140	1,174,660
Nov-20	454,450	1,174,720
Dec-20	441,700	1,174,720
Jan-21	469,950	1,246,420
Feb-21	464,000	1,244,720
Mar-21	446,400	1,256,220
Apr-21	492,040	1,322,210
May-21	531,940	1,352,210
Jun-21	472,540	1,330,010
Jul-21	502,540	1,322,210
Aug-21	472,540	1,480,210
Sep-21	472,540	1,322,210
Oct-21	480,040	1,382,210

3.5 *Quantitative Assessment*

Construction projects typically have different site locations, site environment, gross floor area, construction cost, inflation in that year, start dates, and other factors. Hence to compare both the projects, it is important to normalize their price effort values. To normalize, the obtained price effort values of both the projects were divided by the Gross Floor Area (GFA) of the building (obtained from Table 2) and Consumer Price Index (CPI) by using the Eq. 4 as mentioned in the research methodology. The values of CPI were obtained through referring website of the Reserve Bank of India (RBI) for the year 2021. The resulting normalized priced effort values of both the projects are shown in Table 3.

3.6 *Qualitative Assessment*

Expert interview with the professionals involved in the operation and maintenance activities of BIM-based hospital was conducted for 45 min to fully comprehend how BIM is being implemented in the project. During the interview, it was made sure that the interviewee's interpretation of cooperation in BIM projects was compatible with the definition of collaboration in the current study. The professionals in the BIM-based hospital management project use software for the operation and maintenance activities as shown in Fig. 5.

Table 3 Normalized priced effort values of the projects

Months	Non-BIM Hospital maintenance cost in Rupees	BIM Hospital maintenance cost in Rupees
Oct-20	86.61	57.43
Nov-20	88.03	57.44
Dec-20	85.56	57.44
Jan-21	91.03	60.94
Feb-21	89.88	60.86
Mar-21	86.47	61.42
Apr-21	95.31	64.65
May-21	103.04	66.11
Jun-21	91.53	65.03
Jul-21	97.34	64.65
Aug-21	91.53	72.37
Sep-21	91.53	64.65
Oct-21	92.99	67.58

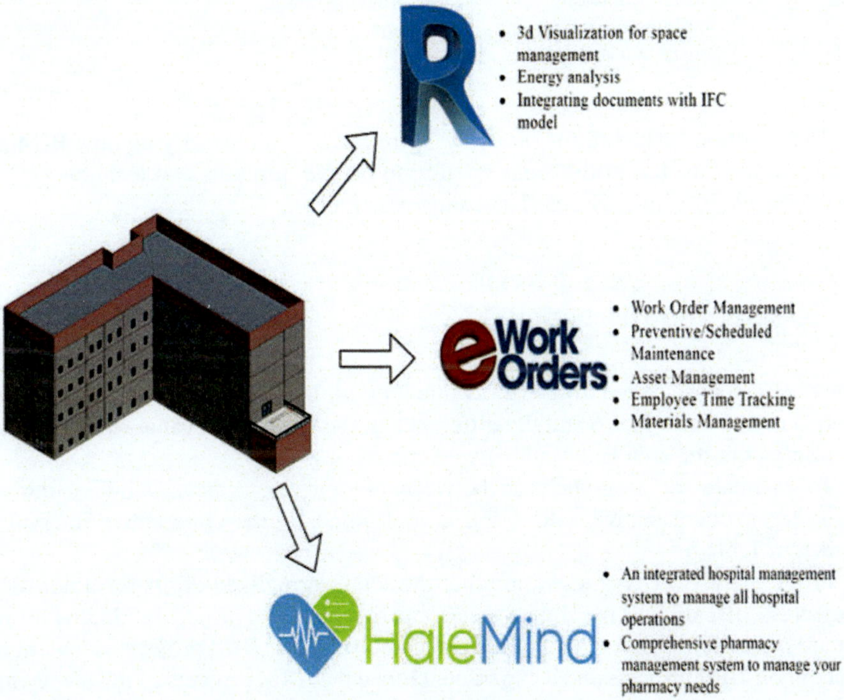

Fig. 5 Software used in BIM based hospital management

Table 4 Benefits of BIM observed by employees

BIM Focal Area	Facility Manager (Interviewee 1)	System solution Engineer (Interviewee 2)	Administrative Manager (Interviewee 3)
People	• Greater emphasis on teamwork when tackling projects and fixing issues	• Enhanced skill for multi-tasking	• More efficiency at work and increased staff productivity
Process	• Clearer building maintenance requirement and requirements being written into contract documents • Less effort is spent making adjustments and resolving mistakes	• Process has become more simplified. Data from systems can be measure easily and more accurately • It maintains openness in the process and promotes collaboration and trust	• Reduced data re-entry requirements • Quicker response times during operations • Faster data retrieval, more proactive maintenance, and improved emergency readiness
Technology	• Better management of activities leading to reduction in waste • Aid in simulating building operations and energy consumption	• The control of information flow between monitoring and real-time operation systems	• Encourage the use of energy-efficient clean technology

The questions asked in the interview were focused on identifying how BIM has benefited the hospital in the areas relating to people, process, and technology. The responses obtained were briefed and shown in Table 4.

4 Results and Discussion

The obtained normalized time-effort value for both BIM and non-BIM hospitals was plotted as shown in Fig. 6 and the difference in the time-effort value was calculated by finding out the area enclosed between the curves.

To calculate the area enclosed between two curves, a mean value is founded between two consecutive values and is multiplied by the consecutive interval as shown in Table 5.

From analysis, it was identified that the total normalized effort for a non-BIM project is 1,101.06 INR per square meter, and that of a BIM project is 758.06 INR per square meter. It can be seen that the aggregated value for BIM projects is lower than that of non-BIM projects and it can be concluded that BIM projects spend fewer costs per square meter for operation and maintenance operations. It is also identified that a cost of 343 INR per square meter can be saved while using BIM-based management for healthcare facilities.

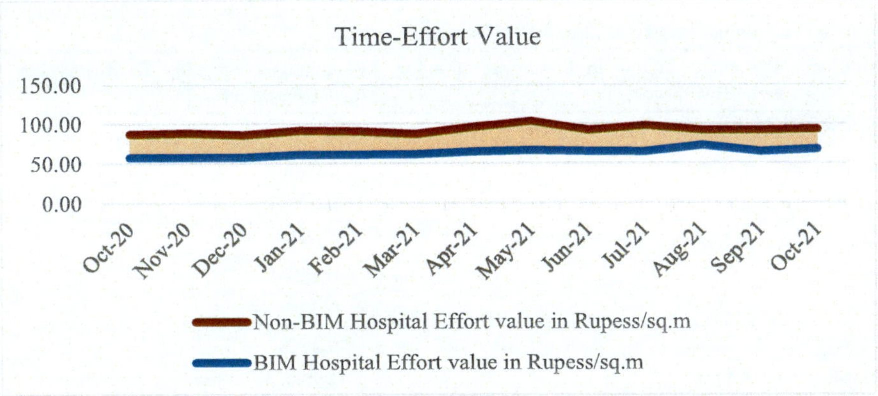

Fig. 6 Time-effort curve of non-BIM & BIM Project

Table 5 Area enclosed between two curves

X	Y1 (Non-BIM)	Y2 (BIM)	Mean Value (Non-BIM)	Mean Value (BIM)
1	86.61	57.43	87.32	57.44
2	88.03	57.44	86.79	57.44
3	85.56	57.44	88.30	59.19
4	91.03	60.94	90.46	60.90
5	89.88	60.86	88.17	61.14
6	86.47	61.42	90.89	63.03
7	95.31	64.65	99.17	65.38
8	103.04	66.11	97.29	65.57
9	91.53	65.03	94.44	64.84
10	97.34	64.65	94.44	68.51
11	91.53	72.37	91.53	68.51
12	91.53	64.65	92.26	66.11
13	92.99	67.58		
SUM			1101.06	758.06
Area between curves = Absolute difference between sum			343 Rupees/sq. m	

On referring to the time–effort distribution curves of the two projects (Fig. 6), it was identified that the BIM-based project curve is more uniform than that of the non-BIM project throughout the considered time. On calculating percentages in the time-effort distribution curves (shown in Table 6), it was identified that non-BIM projects experienced several peaks, for example, a rise of 10.22% in April 2021 and 8.11% in May 2021. Whereas, the BIM project experienced a maximum rise of 6.94% in August 2021 due to various inspection costs that are scheduled in that month.

Table 6 Percentage change in Time-Effort Values

Months	Percentage change in time effort value for non-BIM project	Percentage change in time effort value for BIM project
20-Oct	–	–
20-Nov	1.64%	0.02%
20-Dec	−2.81%	0.00%
21-Jan	**6.39%**	**6.09%**
21-Feb	−1.26%	−0.13%
21-Mar	−3.79%	0.92%
21-Apr	**10.22%**	**5.26%**
21-May	**8.11%**	2.26%
21-Jun	−11.17%	−1.63%
21-Jul	**6.35%**	−0.58%
21-Aug	−5.97%	**6.94%**
21-Sep	0.00%	−10.67%
21-Oct	1.60%	4.53%

Compared to the non-BIM project, the BIM project experiences fewer changes in normalized effort. Every variation in a BIM project is followed by a downgrade, showing that the project resolves the uncertainties in the operation and maintenance activities in the following month. This may be attributed to BIM's role in providing a better platform for management and communication for the staff members involved in the process.

4.1 Result Significance

The current studies of performing cost-benefit analysis of BIM implementation were focused on BIM for the construction stages of the project [15, 19]. Similar efforts on BIM implementation for the operation and maintenance phase are not explored yet. This paper performs cost-benefit analysis of BIM implementation for hospital buildings during their operation and maintenance phase. It was identified that BIM implementation enabled saving 343 INR per square meter during facility management.

5 Conclusion

This paper develops time–effort distribution curves for BIM and non-BIM-based projects of healthcare facilities. Both projects were compared and quantitative and qualitative benefits of BIM implementation were identified. It is identified that the total normalized effort for the non-BIM project is higher (1,101.06 INR per square meter) when compared with the BIM project (758.06 INR per square meter). Also, the use of BIM approaches contributed to savings of 343 INR per square meter. This demonstrates that the use of BIM reduces the cost of operation and maintenance of healthcare facilities. It is also identified that the use of BIM, enabled better management of information and enabled a reduction in wastage of time for acquiring the required information. BIM-based projects are more transparent and require fewer efforts for editing and modifications. Thus the time–effort distribution curves are a useful graphical tool for assessing cost-benefit trends and the advantages and disadvantages of a BIM implementation approach.

The results obtained from this paper could be utilized as a benchmark for comparison, however, it must be noted that they were drawn from a limited number of samples. Although the two example projects' comparability has received a lot of attention, some elements may not have been taken into account. However, the cost-benefit analysis in this study is based on up-to-date accurate information.

In the future, research can explore the time–effort distribution for characterizing effort using hourly data by type of staff and initial cost, normalizing depending on complexity or timetable, and gathering complete data for analysis. Further research can be done to determine parameters that result in producing the most effort while implementing BIM as well as the process and organizational changes brought about by BIM implementation based on the time–effort distribution curve.

References

1. Ahn, K. U., Kim, Y. J., Park, C. S., Kim, I., & Lee, K. (2014). BIM interface for full vs. semi-automated building energy simulation. *Energy and Buildings, 68*, 671–678. https://doi.org/10.1016/j.enbuild.2013.08.063
2. Atkin, B., & Bildsten, L. (2017). Editorial: a future for facility management. *Construction Innovation, 17*(2), 116–124. https://doi.org/10.1108/CI-11-2016-0059
3. Azhar, S., Hein, M., & Sketo, B. (2008). Building information modeling (BIM): benefits, risks and challenges. In *Proceedings of the 44th ASC Annual Conference.*
4. Baldauf, J. P., Formoso, C. T., & Tzortzopoulos, P. (2021). Method for managing requirements in healthcare projects using building information modelling. *Engineering Construction and Architectural Management, 28*(8), 2090–2118. https://doi.org/10.1108/ECAM-12-2020-1040
5. Barlish, K., & Sullivan, K. (2012). How to measure the benefits of BIM - a case study approach. *Automation in Construction, 24*, 149–159. https://doi.org/10.1016/j.autcon.2012.02.008
6. Chahrour, R., Hafeez, M. A., Ahmad, A. M., Sulieman, H. I., Dawood, H., Rodriguez-Trejo, S., Kassem, M., Naji, K. K., & Dawood, N. (2021). Cost-benefit analysis of BIM-enabled design clash detection and resolution. *Construction Management and Economics, 39*(1), 55–72. https://doi.org/10.1080/01446193.2020.1802768

7. Demirdöğen, G., Işık, Z., & Arayici, Y. (2020). Lean management framework for healthcare facilities integrating BIM, BEPS and big data analytics. *Sustainability, 12*(17), 7061. https://doi.org/10.3390/su12177061
8. Evjen, T. Å., Hosseini Raviz, S. R., Petersen, S. A., & Krogstie, J. (2020). Smart facility management: future healthcare organization through indoor positioning systems in the light of enterprise BIM. *Smart Cities, 3*(3), 793–805. https://doi.org/10.3390/smartcities3030040
9. Elghaish, F., Abrishami, S., Hosseini, M. R., Abu-Samra, S., & Gaterel, M. (2019). Integrated project delivery with BIM: an automated EVM-based approach. *Automation in Construction, 106*, 102907. https://doi.org/10.1016/j.autcon.2019.102907
10. Gerrish, T., Cook, M., & Ruikar, K. (2016). BIM for the management of building services information during building design and use. *Science and Technology for the Built Environment, 22*(3), 249–251. https://doi.org/10.1080/23744731.2016.1156947
11. Iddon, C. R., & Firth, S. K. (2013). Embodied and operational energy for new-build housing: a case study of construction methods in the UK. *Energy and Buildings, 67*, 479–488. https://doi.org/10.1016/j.enbuild.2013.08.041
12. Kalman, N., Hammill, B., Schulman, K., & Shah, B. (2015). Hospital overhead costs: the neglected driver of health care spending? *Journal of Health Care Finance, 41*(4).
13. Kunz, J., Fischer, M., Haymaker, J., & Levitt, R. (2002). Integrated and automated project processes in Civil engineering: experiences of the center for integrated facility engineering at Stanford University. In *Computing in Civil Engineering Proceedings* (pp. 96–105). Reston, VA: ASCE.
14. Li, H., Lu, W., & Huang, T. (2009). Rethinking project management and exploring virtual design and construction as a potential solution. *Construction Management and Economics, 27*(4), 363–371. https://doi.org/10.1080/01446190902838217
15. Lu, W., Fung, A., Peng, Y., Liang, C., & Rowlinson, S. (2014). Cost-benefit analysis of Building Information Modeling implementation in building projects through demystification of time-effort distribution curves. *Building and Environment, 82*, 317–327. https://doi.org/10.1016/j.buildenv.2014.08.030
16. Lucas, J., Bulbul, T., & Thabet, W. (2013). An object-oriented model to support healthcare facility information management. *Automation in Construction, 31*, 281–291. https://doi.org/10.1016/j.autcon.2012.12.014
17. MacLeamy, P. (2004). Collaboration, integrated information and the project lifecycle in building design, construction and operation. Accessed 18 July 2020. https://kcuc.org/wp-content/uploads/2013/11/Collaboration-Integrated-Information-and-the-Project-Lifecycle.pdf
18. Marmo, R., Nicolella, M., Polverino, F., & Tibaut, A. (2019). A methodology for a performance information model to support facility management. *Sustainability, 11*(24), 7007. https://doi.org/10.3390/su11247007
19. Muñoz-La Rivera, F., Vielma, J. C., Herrera, R. F., & Carvallo, J. (2019). Methodology for building information modeling (BIM) implementation in structural engineering companies (SECs). *Advances in Civil Engineering 2019*. https://doi.org/10.1155/2019/8452461
20. Rani, N. A. A., Baharum, M. R., Akbar, A. R. N., & Nawawi, A. H. (2015). Perception of maintenance management strategy on healthcare facilities. *Procedia-Social and Behavioral Sciences, 170*, 272–281. https://doi.org/10.1016/j.sbspro.2015.01.037
21. Sacks, R., & Barak, R. (2008). Impact of three-dimensional parametric modeling of buildings on productivity in structural engineering practice. *Automation in Construction, 17*(4), 439–449. https://doi.org/10.1016/j.autcon.2007.08.003
22. Shin, M. H., Lee, H. K., & Kim, H. Y. (2018). Benefit–cost analysis of building information modeling (BIM) in a railway site. *Sustainability, 10*(11), 4303. https://doi.org/10.3390/su10114303
23. Wang, W., Zmeureanu, R., & Rivard, H. (2005). Applying multi-objective genetic algorithms in green building design optimization. *Building and Environment, 40*(11), 1512–1525.

Advances in Construction Materials

Effective Reuse of Concrete Debris in Soil–Column Study

S. V. Sivapriya, Jijo James, M. Naveen Prasath, and Tanishka Priyadharshini Ramesh

Abstract Expansive soils are problematic soils which need to be stabilized for improving their performance and mitigating their damaging effects on structures constructed on top of them. Stone columns technique for stabilization of soft and expansive soils has been proven effective over time. However, the utilization of huge amounts of stone in the formation of columns put undue pressure on natural resources. Hence, in this investigation, an attempt was made to partially replace the stones with concrete debris, which is abundantly generated due to urbanization and modernisation. A model column of diameter 30 mm and height 155 mm was simulated in the laboratory in a CBR mould filled with an expansive soil of consistency 0.4. Load-penetration behaviour of plain soil, soil stabilized with stone column and the modified stone column were studied. 5 mm stone chips were used as the column material. This material was blended in a ratio of 1:1 with concrete debris and the modified column was formed. This was also simulated in a finite element tool. The results of the finite element analysis gave confidence in understanding the penetration behaviour. The model developed was applied to an existing field study using conventional stone columns to increase the bearing of a foundation. The outcome showed that the modified column applied to the conditions of the study showed better settlement behaviour than the conventional column. The column with demolished waste is able to take more load compared to conventional stone which is indicated by settlement along the length of the column. Considering this fact, demolished waste plays its role well in improving the bearing of the soil and is found as the most effective reuse method. With India generating more than 150 million tons of demolished waste, this can be an effective way of reusing/recycling the demolished waste. This also reduces the carbon footprints leading to a better environment.

S. V. Sivapriya (✉) · M. Naveen Prasath · T. P. Ramesh
Department of Civil Engineering, Sri Sivasubramaniya Nadar College of Engineering, Chennai, India
e-mail: sivapriyavijay@gmail.com

J. James
Department of Civil Engineering, College of Engineering - Guindy, Anna University, Chennai, India
e-mail: jijothegreat@gmail.com

A. Vilventhan et al. (eds.), *Advances in Construction Materials and Management*, Lecture Notes in Civil Engineering 346,
https://doi.org/10.1007/978-981-99-2552-0_15

Keywords Demolished waste · Reuse · Recycle · Stone column · Soil stabilization

1 Introduction

Problematic soils are soils which are known to be difficult to deal with in terms of engineering performance. Soft and expansive clays are one such example of problematic soils. Development of infrastructure facilities on such soils proves to be difficult unless properly designed. There are situations wherein designs though well designed, become uneconomical due to the difficult subsoil conditions. Such soft soils need to be improved to ensure that constructed infrastructure facilities perform satisfactorily. Several ground improvement techniques are available for the improvement of soft clays with stone columns being a well-known and well-proven method. This technique, also called a granular column or granular pile to improve the bearing capacity and reduces settlements. The stone column technique was used in European countries in the 1960s [1] and has been successfully used since. In today's modern world, where sustainability is a prime focus, utilization of stone columns can no longer be sustainable. By installing such columns the bearing, consolidation and stability of soil will be improved which can be homogenous or layered [2–4]. Installation of a large number of columns for projects requires significant quantities of crushed stone which results in the increased exploitation of natural stone deposits in the form of mining and quarrying. Sustainability in this technique can be achieved by finding a suitable replacement for the crushed stone used in the column materials either partially or fully without a significant drop in the performance of the columns. There have been some investigations wherein stones have been replaced with other waste materials including shredded tire wastes, coconut shells, fly ash chips, and crushed concrete to name a few. However, the efficacy of the stone column in performing its function will depend on finding an alternative material that is similar to crushed stones. In a developing country like India, the generation of construction and demolition waste is in significant proportions. Concrete debris forms a significant chuck of the construction and demolition waste which, if sent to landfills, can simply result in quicker filling up of the fill. This concrete debris can be reutilized as a replacement for crushed stones in installation of stone columns for construction projects debris and model its performance using a finite element tool. There have been a few investigations into the utilization of concrete debris as a partial replacement of stones in columns [5, 6]. The present study attempts to investigate the performance of stone columns partially replaced with concrete debris and model its performance using a finite element tool.

2 Material and Methodology

Unit cell concept is adopted to understand the load-penetration characteristics of a column embedded in expansive clay. The unit-cell formation is done using California Bearing Ratio (CBR) test setup.

The expansive clayey soil used in the test is fetched from Thaiyur lake, Chennai,India. It possesses 75.8% liquid limit, and 23.5% of plastic with specific gravity as 2.76 [7] which is found using relevant Indian Standard codes [8, 9]. The procured sample is mixed with 44% of water to achieve a consistency index of 0.4 (Soft). It is allowed to consolidate for a period of two days.

After two days, The soil sample is filled in the CBR mould of 150 mm diameter and 125 mm height by placing a filter medium in the top and bottom [10]. To form the column a 20 mm diameter PVC pipe is inserted before filling the clay soil. The size of the stones used in the column is of particle size retained in a 4.75 mm sieve which is poured from a height of 25 cm to achieve a relative density of 70%. The height of the column is 120 mm with filter medium made of sand for a thickness of 2.5 mm at the top and bottom. The test is done for both soaked(immersing the mould with column in water for 96 hours) and unsoaked conditions and loaded at a rate of 1.25 mm/minute (Fig. 1).

a. Column Formation b. Unit -Cell c. Mould under soaked condition

Fig. 1 Unit–Cell

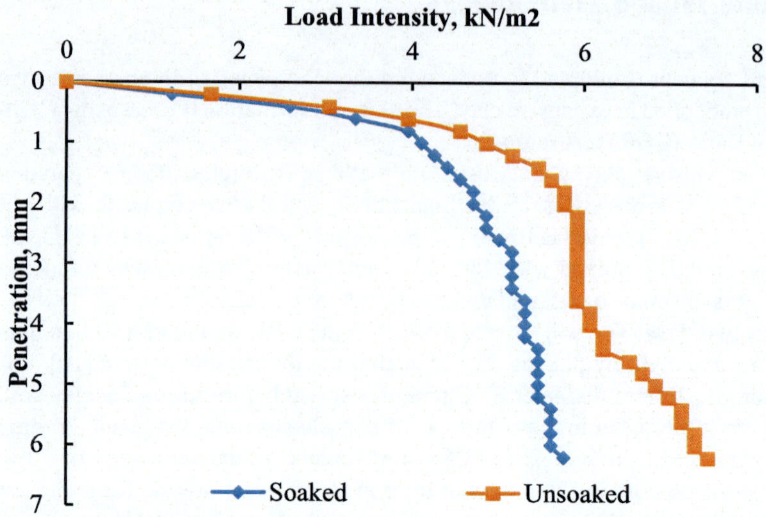

Fig. 2 Experimental Analysis

3 Results and Discussion

3.1 *Experimental Results*

The penetration behaviour of the soaked condition shows less resistance to load compared to unsoaked condition. The load – penetration behaviour shows a similar pattern in the initial phase of loading; this is due to the fact the initial subjected load is considered as seating load. Eventually, with an increase in load the resistance offered by the unsoaked condition is high (Fig. 2). Beyond the initial load, there is a sudden increase in penetration/settlement. This eventually increases for a higher load.

3.2 *Numerical Analysis*

PLAXIS 2D, a finite element tool is used to analyse the load-penetration behaviour of the column (Fig. 2). Similar studies were done previously to understand the behaviour of complex conditions too [11, 12]. To validate the tool, the numerically simulated model is compared with the experimental results under unsoaked conditions. Mohr coloumb analysis was chosen for the same. The load was directly subjected at the rate of 1 N/mm^2. The same parameters that were obtained experimentally or by referring to the design codes, for experimental analysis were taken as the parameters for FEM. Young's modulus of the clay was taken as 4500 kPa and for the stone column as 45000 kPa.The cohesive strength of clay was taken as 15 kPa and Angle of internal

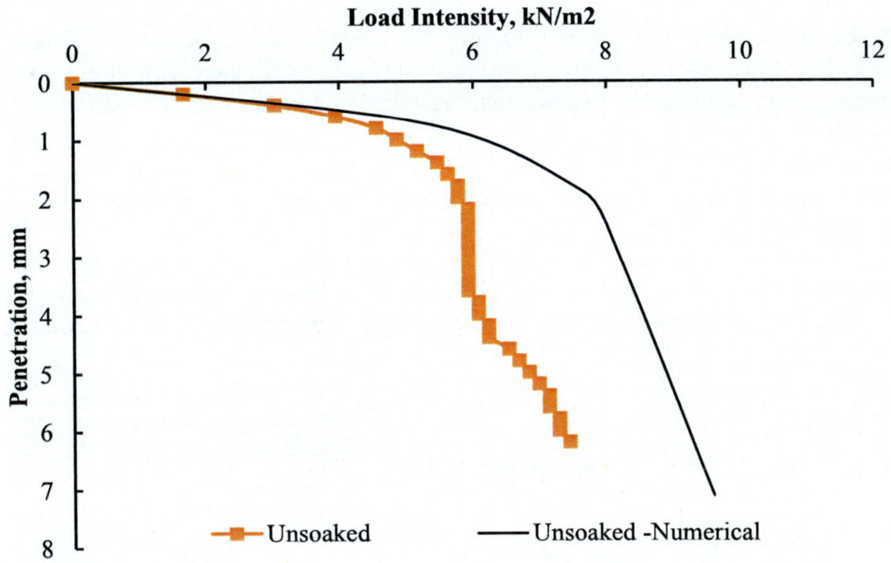

Fig. 3 Validation Graph

friction of the concrete stone column was found to be 30.95° with Poisson's ratio was taken as 0.33 for both the materials. The values are obtained by conducting tests in laboratory from the samples taken from the college permises of M30 grade concrete. The Unsaturated unit weight and Saturated unit weight were taken as 3.490 kN/m^3 and 13.30 kN/m^3 respectively. The initial numerical penetration results agree well with the experimental results. With increase in load intensity the variation increases and the maximum variation observed is 22.72% (Fig. 3).

With this understanding, the stones in stone column are replaced with demolished construction waste and the properties of stone column is replaced with demolished concrete waste in a published paper where 16 stone columns are used in stabilising the weak clay for a maximum depth of 12 m. The angle of internal friction is found using a direct shear test set-up for the M30 grade concrete waste generated as 30.96°, with Youngs Modulus as 20,000 kPa, Poisoon's ratio as 0.21. The crushing value of the concrete is found as 34.83%. The model is generated with the obtained values (Fig. 4) and a fine mesh is generated. The water table is considered at 6.05 m from the ground and in-situ stresses are generated. In the existing study, the stone column material is analysed and drained analysis was also done by replacing the material. The deformation along the length of the stone column and the demolished waste column is extracted and plotted. It is seen the deformation in the initial stage is similar, with increased load intensity the resistance increases. The behaviour of the column with stone shows that the settlement increases with load intensity and the same is observed in the column with demolished waste. The difference between an increase in resistance increases with load intensity (Fig. 5a). The initial difference is due to the seating of stone for the applied load, which eventually increased due to the increase in stiffness of the column. The modulus of the concrete waste is 10 times higher than the conventional column. To understand the increase in strength the improvement bearing improvement ratio is calculated as the ratio of load intensity of column with stones to a column with a demolished waste of the same settlement. With an increase in settlement the bearing ratio increases, and it increases linearly. The load intensity increases initially by 2.8 and then there is a reduction with 2.1 and 1.8 (Fig. 5b).

A non-dimensional chart is developed for load intensity to cohesive strength in x – axis and settlement vs width of the footing (Fig. 6). It is observed that the behaviour of the column with stones shows a linear dip in its behaviour. Whereas for column

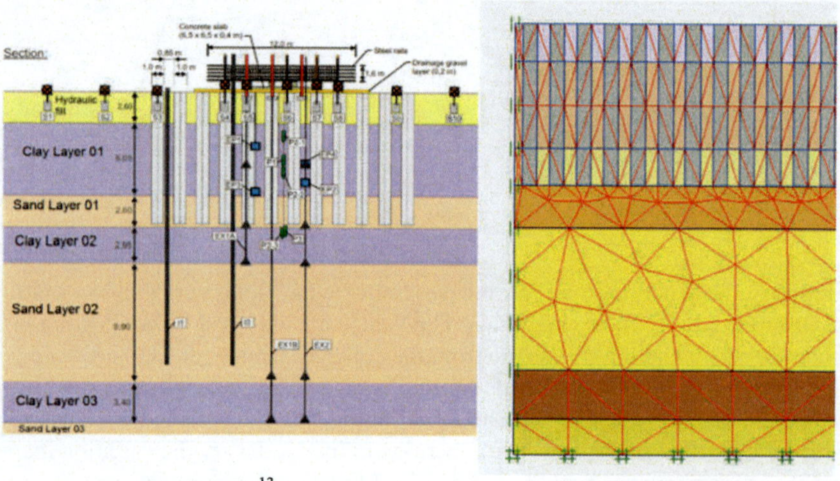

a.Field Model [13]

b. Plaxis Model Generated

Fig. 4 Model generated

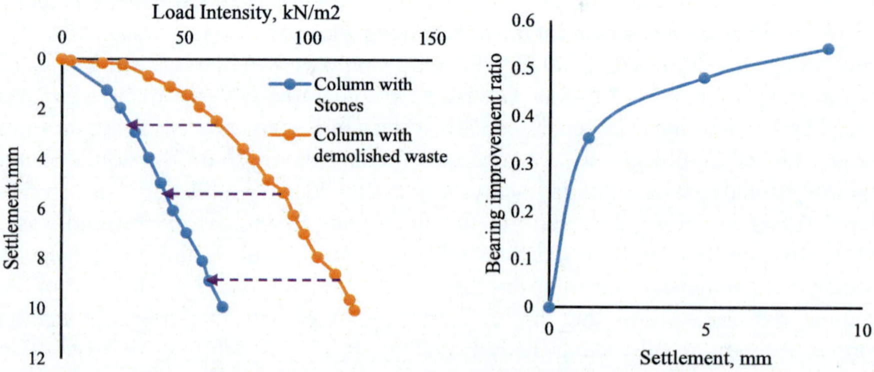

Fig. 5 **a** load-intensity behaviour of columns. **b** bearing improvement ratio

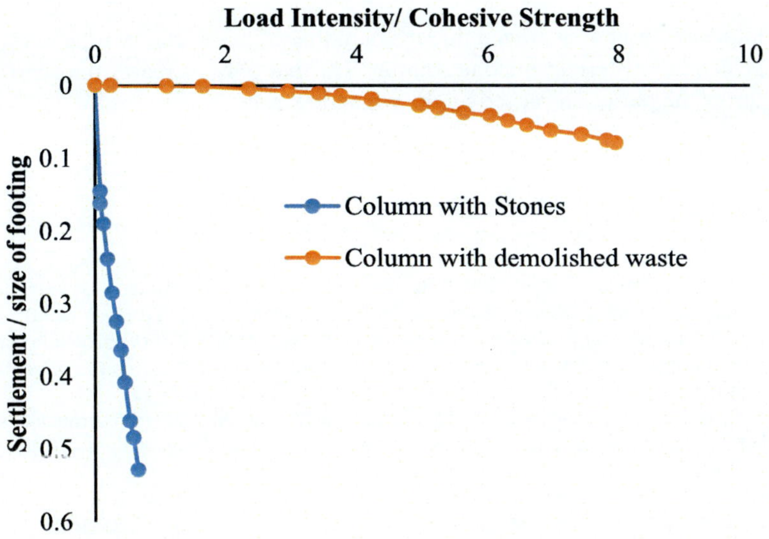

Fig. 6 Non-Dimensional

with demolished waste, it is observed that the increase in modulus plays a role in increasing its stiffness. Already, the stone column is acting as reinforcement for the weak soil and also as a drainage medium. To increase the strength of the soil furthermore, by using construction waste material the resistance increases 2 folds initially and it increases even more than 10 folds later (Fig. 6). It is mainly due to the confinement and the modulus of the material. This also shows the potential use of demolished waste as a reinforcing material.

4 Conclusion

The experimental study is conducted to validate the usage of stone column with the numerical studies where the material is replaced with demolished waste. From the obtained results, the stone materials are replaced with demolished waste in the field to understand the real-time behaviour. The following conclusions have arrived;

1. The unsoaked value of stone column is almost 1.25 times higher than the soaked condition.
2. From the numerical study, it is inferred that the usage of demolished waste increases to a maximum of tenfold due to the high modulus value.
3. The demolished waste column could able to bear the large load. The improvement is due to the confinement and the increase in modulus value.

From the analysis, it is proved that the demolished waste can be effectively used as a replacement material for stone column. This not only increases the bearing but also reduces the pollution caused by the concrete waste.

References

1. Ambily, A. P., & Gandhi, S. R. (2007). Behavior of stone columns based on experimental and FEM analysis. *Journal of Geotechnical and Geoenvironmental Engineering, 133*, 405–415.
2. Basack, S., Indraratna, B., & Rujikiatkamjorn, C. (2016). Analysis of the behaviour of stone column stabilized soft ground supporting transport infrastructure. *Procedia Engineering, 143*, 347–354.
3. Shivashankar, R., Babu, M. R. D., Nayak, S., & Rajathkumar, V. (2011). Experimental studies on behaviour of stone columns in layered soils. *Geotechnical and Geological Engineering, 29*, 749–757.
4. Nayak, S., Balaji, M., & Preetham, H. K. (2020). A study on the behaviour of stone columns in a layered soil system. *Transportation Infrastructure Geotechnology, 7*, 85–102.
5. Salim, N. M., Hasan, S., & Al-Soudany, K. (2020). Effect of replacing crushed stone in stone columns by waste material on soil improvement ratio. *Proceedings of International Structural Engineering and Construction, 7*, 2–7.
6. Kawalec, J., Kwiecien, S., Pilipenko, A., & Rybak, J. (2017). Application of crushed concrete in geotechnical engineering - selected issues. In *IOP Conference Series: Earth and Environmental Science* (Vol. 95).
7. Sivapriya, V., Jijo, J., Yuvaraj, K., & Sushritha, G. (2022). Durability performance of a lime stabilized expansive soil with egg shell ash as a subsidiary admixture. *Gradjevinski materijali i konstrukcije, 65*, 65–71.
8. Bureau of Indian Standard. (1995).*IS 2720 (Part V) Determination of Liquid and Plastic Limit* (pp. 1–17).
9. Bureau of Indian Standard (BIS). (1997).*IS 2720 (Part III/I) Determinationof Specific Gravity of Fine granied Soil.* (pp. 1–10).
10. James, J., & Sivapriya, S. V. (2022). Load-settlement behaviour of stone column with varied spacing. In *Lecture Notes in Civil Engineering* (Vol. 221, pp. 27–31)
11. Sivapriya, S. V., & James, J. (2019). Numerical study on static behaviour of a stone column under uniformly distributed load. In *AIP Conference Proceedings* (Vol. 2161).
12. Houda, H., & Salah, M. (2021). Experimental and numerical study of the behavior of a stone column subject to the loading effect. *Selected Scientific Papers - Journal of Civil Engineering, 16*, 105–114.
13. Almeida, M., et al. (2014). Stone columns field test: Monitoring data and numerical analyses. *Geotechnical Engineering, 45*, 103–112.

Comparative Study to Investigate the Suitability of Sustainable Alternatives in Enhancing Strength Characteristics of Black Cotton Soil

M. N. Asha, B. R. Vinod, R. Anthony, and J. Akshit Jain

Abstract In this paper, the suitability of three different alternatives viz., sisal fiber, brick powder mix, phosphogypsum in stabilizing lime blended BC soil is investigated. BC soil used for the experiments had a maximum dry density of 1.6 g/cc at an optimum moisture content of 22%. For preparing lime blended soil, different proportions of lime have been added viz., 2%, 4%, 6%, 8%, 10%, 12%, 14% and 16%. Since the present study focus on highway application, optimum proportion of lime has been decided based on California Bearing Ratio (CBR value). From the experimental studies, optimum lime content was identified as 10% and to the lime blended BC soil, sisal fiber, brick powder mix, phosphogypsum has been added at different proportions to study their effectiveness. From the experimental results its observed that addition of 10% lime doubles the CBR value of the plain BC soil. Among the different stabilizers, addition of 0.8% of sisal fiber to the lime blended BC soil, increases the CBR value by four times that of the plain BC soil. Based on the CBR value and recommendations given in IRC 37: 2001, pavement cross sections are arrived at for different dosages/proportions of stabilizer.

M. N. Asha (✉)
Department of Civil Engineering, M S Ramaiah Institute of Technology, Bengaluru 560054, India
e-mail: ashamn@msrit.edu

B. R. Vinod · R. Anthony · J. Akshit Jain
Department of Civil Engineering, B M S Institute of Technology and Management, Bengaluru 560064, India
e-mail: vinodbr@bmsit.in

R. Anthony
e-mail: rajanthony@bmsit.in

J. Akshit Jain
e-mail: 1by20cv019@bmsit.in

A. Vilventhan et al. (eds.), *Advances in Construction Materials and Management*, Lecture Notes in Civil Engineering 346,
https://doi.org/10.1007/978-981-99-2552-0_16

1 Introduction

Rapidity in infrastructure development has forced the engineers and practicing professionals to use the in-situ soil for different applications. Highways play a crucial role in infrastructure development since it enables connectivity. Construction and rehabilitation of highways depends greatly on the strength characteristics of in situ soil. Being problematic, black cotton soil (BC soil) is discarded from different construction activities. Since black cotton soils comprise of montmorillonite mineral, it exhibits high swell shrinkage characteristics [1]. Hence, extensive research is being carried out over the years to enhance the engineering properties of BC soil by using pozzolanic materials/admixtures/fibers/geosynthetics [2, 3]. Though lime is a good stabilizer, its effectiveness is observed only over a longer run [4, 5]. Though fibers are effective in increasing strength of soils, for soils with montmorillonite minerals, fibers are generally added along with pozzolanic materials [6, 7]. Suitability of sisal fiber in enhancing strength of soil has been explored by many researchers [8–10]. To ascertain the strength characteristics of the blended soil, unconfined compressive strength has been carried out by many researchers. The length of sisal fiber used for the experiments is generally 10 mm – 20 mm. SEM results from the literature [10] emphasize that there exists strong bonding between sisal fiber and BC soil. Effectiveness of waste materials in stabilizing BC soil has also been explored by few researchers [11, 12]. Owed to pozzolanic action, crushed brick powder has been used as a replacement for cement in making concrete/mortar blocks [13, 14].

According to literature [15] addition of 20% of crushed brick waste enhances flexural strength of stabilized mud blocks. Since pavements are indicative of infrastructural development of country and needs huge quantity of subgrade soil, identification of sustainable alternatives for stabilizing weak soil is important. Phosphogypsum being a synthetic calcium sulphate, is reported to be a good stabilizer [16–18]. 15% phosphogypsum along with lime and GGBS can reduce swell potential characteristics of expansive soil [19].

In the present study use of pozzolanic stabilizer lime in strengthening black cotton soil is investigated. Along with lime, the effectiveness of three other stabilisers viz., sisal fiber, brick powder mix and phosphogypsum is investigated. The broad objectives of the present study can be enumerated as follows:

- Investigate the effect of curing period on strength when BC soil is blended with lime
- Identification of optimum stabilizer content with respect to maximum CBR value
- Estimate the pavement thickness when BC soil is blended with different stabilizers.

2 Methodology

The present study is aimed at stabilizing black cotton soil for highway applications. Since the focus is on application of stabilized BC soil for pavements, California Bearing Ratio value is determined and compared to ascertain the suitability of different stabilizers. Since BC soil has high swell-shrinkage characteristics, pozzolanic additive viz., lime is added to black cotton soil at proportions 2%, 4%, 6%, 8%, 10%, 12%, 14% and 16% by weight of dry soil. On the lime blended BC soil, both unsoaked CBR tests and Unconfined compressive strength tests is carried out. Since the effectiveness of lime as a stabilizer is dependent upon curing period, unconfined compressive strength is determined for 0, 1, 3 and 7 days of curing. Based on CBR value and UCS value an optimum proportion of lime is determined. On to optimum lime blended BC soil (10% lime + 90% BC soil), the effectiveness of three different stabilizers is determined by comparing CBR values. The different stabilizer proportions used for the study are sisal fiber (of 3 cm length and 0.4 mm diameter at proportions 0.2%, 0.4%, 0.6%, 0.8%, 1.0%, 1.2% and 1.4%); brick powder mix (10%, 20%, 30%, 40%, 50%, 60%, 70%, 80% and 90%); phosphogypsum (10%, 20%, 30%, 40%). The methodology adopted is presented in Fig. 1.

To ensure uniform mixing of sisal fiber, it is mixed with optimum lime blended BC soil in dry form. After ensuring uniform mixing only water is added. Photograph of the Unconfined Compression test and CBR test set up is presented in Fig. 2.

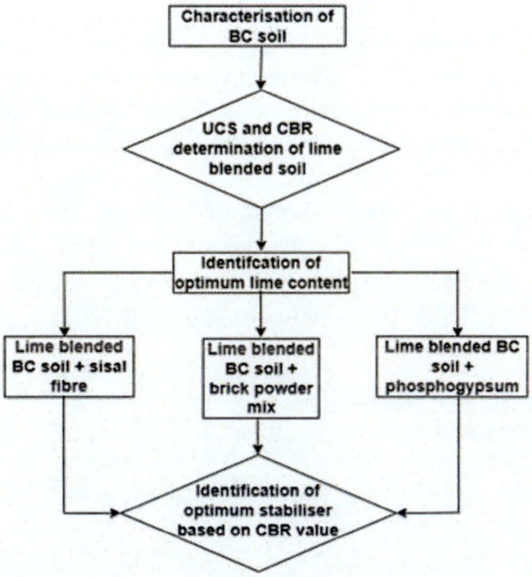

Fig. 1 Methodology adopted for the present study

Fig. 2 Experimental set up for Unconfined compressive strength and California Bearing Ratio determination

3 Materials

Black Cotton soil used for the experimental studies is collected from Laxanatti, Mudhol Taluk, Bagalkot district of Karnataka. The soil exhibited a maximum dry density (MDD) of 1.6 g/cc at an optimum moisture content (OMC) of 22% when densified at standard compaction energy. The CBR value and unconfined compressive strength of the soil was determined at its OMC and MDD and the properties of the soil are summarized in Table 1.

Stabilizers used for the present study are lime, sisal fiber (SF), brick powder mix (BPM) and phosphogypsum (PG). Lime, sisal fiber and phosphogypsum are collected from local vendor. Sisal fiber used for the study had a diameter of 0.4 mm and 3 cm length. In the literature the sisal fiber of length 10–20 mm is used for conducting unconfined compressive strength of soil. In the present study, since CBR results are only compared, the commercially obtained sisal fiber of 30 mm length is used. Brick powder mix is collected from a brick factory situated at Nelamangala. The effectiveness of sisal fiber, brick powder mix and phosphogypsum in stabilizing lime stabilized BC soil has been explored.

Table 1 Properties of the Black Cotton Soil

Property	Value
Specific Gravity	2.6
Liquid limit, %	62
Plastic limit, %	28
IS Soil Classification	CH
Optimum Moisture Content, %	22
Maximum dry density, g/cc	1.6
Unsoaked CBR, %	2.7

4 Experimental Studies

In the present study, to estimate the strength characteristics unconfined compressive strength tests (UCS) and CBR tests are carried out. UCS tests are carried out specimens of 38 mm diameter and 76 mm height. The samples for the tests are extruded from compacted soil, prepared at standard compaction energy and tested at a strain rate of 1.25 mm/min. For curing the samples were kept in air tight desiccators. Unsoaked CBR tests were carried out on specimens prepared at optimum moisture content and maximum dry density.

5 Experimental Results

The subsequent section describes the details of the experimental results. The results are analyzed to judge the effect of pozzolanic stabilizer lime. Also, the effectiveness of different sustainable alternatives in stabilizing lime blended BC soil is also discussed.

5.1 Effect of Lime on BC Soil

The effect of pozzolanic stabilizer on BC soil was ascertained with respect to Atterberg's limits, compaction characteristics, Unconfined compressive strength and CBR values.

5.1.1 Effect on Liquid Limit and Plasticity Index

On adding lime to BC soil, its observed that both liquid limit and plastic limit decreases. Based on liquid limit and plastic limit, plasticity index has been computed and it is observed that plasticity index remains more or less same with the addition of lime. Liquid limit and plastic limit of lime blended BC soil is tabulated in Table 2. A close examination of these limits infers that according to IS soil classification, the blended soil belong to CI (Clay of Intermediate Plasticity) category whereas untreated soil belong CH (Clay of High Plasticity) category.

Table 2 Variation in liquid limit and plastic limit with addition of lime

Property	0% Lime	2% lime	4% lime	6% lime	8% lime	10% lime	12% lime	14% lime	16% lime
Liquid limit, %	62%	41	39	38	37	35	34	32	31
Plastic limit, %	28%	20	18	17	15	15	14	13	12
Plasticity Index	34	21	21	21	22	20	20	19	19

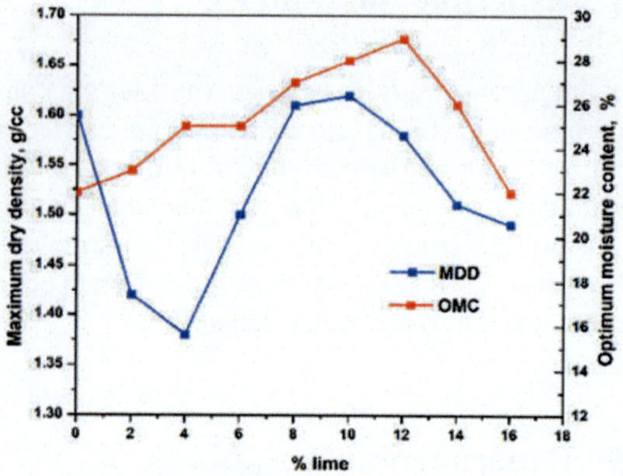

Fig. 3 Variation in OMC and MDD with percentage of lime

5.1.2 Effect of Lime on Compaction Characteristics

To investigate the effect of lime on compaction characteristics, specimens were prepared at standard compaction energy and optimum moisture content and maximum dry density was evaluated. The variation in OMC and MDD when lime is added to BC soil is presented in Fig. 3. When lime is added initially dry density decreases and then increases which could be due to the increase in cations with the addition of lime. The increase in OMC could be attributed to the pozzolanic reaction taking place in the blended soil [4]. With the addition of lime, the thickness of diffused double layer decreases and hence the addition of water enables a denser packing which results in an increase of dry density (beyond 4%). A similar trend is reported in literature [20].

5.1.3 Effect of Lime on Unconfined Compressive Strength and CBR Value

Figure 4 compares the effect of curing period and quantity of lime on unconfined compressive strength on stabilised BC soil. Its observed that maximum improvement in strength is observed at a lime content of 10% and the effectiveness increases with curing period. This infers longevity of strength in lime blended BC soil.

5.1.4 Effect of Lime on CBR Value

Unsoaked CBR tests were carried out on lime blended BC soil and in all cases, CBR value at 2.5 mm penetration was higher than that at 5 mm penetration. It is observed that with addition of lime, CBR value increases (Fig. 5). This emphasizes the fact that addition of lime enhances the strength of the soil which is indicative of its suitability for highway construction.

Fig. 4 Effect of curing period on unconfined compressive strength, kg/cm^2

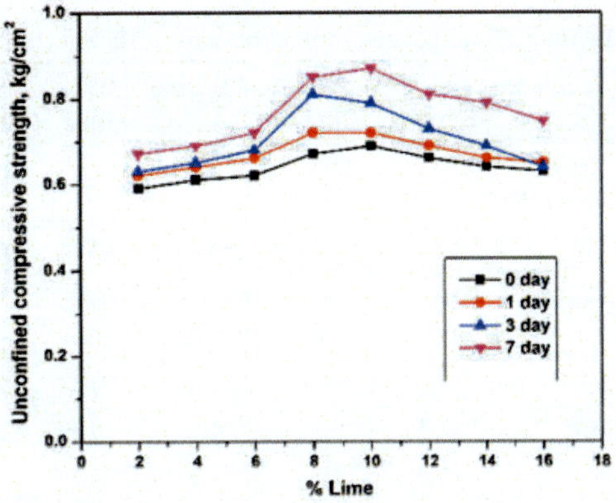

Fig. 5 Effect of lime content on CBR value

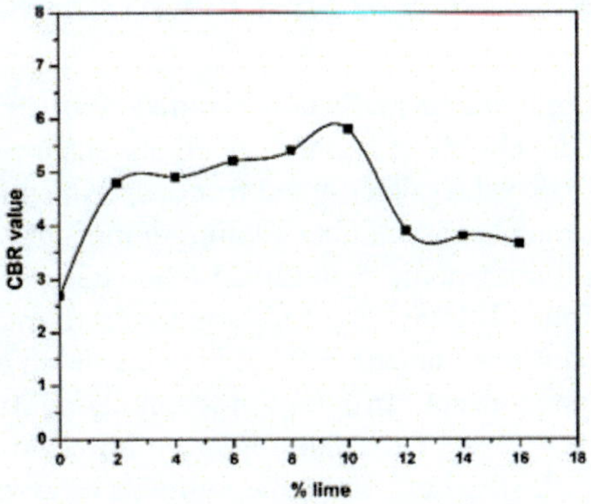

Based on the CBR value pavement thickness is determined for different traffic load intensity in accordance with [21] (measured in terms of million standard axles). Since pavement thickness is dependent upon the CBR value of the subgrade, the total pavement thickness will be the lowest at a lime content of 10%. Table 3 provides the details of total pavement thickness as computed from IRC 37: 2012 for different traffic intensities.

From the experimental results it is observed that at an optimum lime content of 10%, the lime blended BC soil exhibited higher unconfined compressive strength and CBR value. Hence in the further studies lime blended BC soil was treated three different stabilizers to estimate its efficacy.

Table 3 Total pavement thickness for different lime contents

Traffic in msa	0% Lime	2% lime	4% lime	6% lime	8% lime	10% lime	12% lime	14% lime	16% lime
CBR value, %	2.7	4.8	4.91	5.2	5.4	5.8	3.9	3.8	3.65
10	787	668	663.6	651	642	624	706	712	721
20	817	698	693.6	680	670	650	736	742	751
30	837	718	713.6	699	688	666	756	762	771
50	858.5	740	734.5	719	708	686	785	790	797.5
100	888.5	760	754.5	740	730	710	806	812	821
150	915.5	780	774.5	760	750	730	827	834	844.5

5.2 *Effect of Sisal Fiber Content on BC Soil*

Lime blended BC soil exhibited considerable improvement in strength characteristics. However, addition of fibers enhances interfacial friction and, in this section, CBR values determined from experimental studies are compared for seven different contents of sisal fiber. Optimum moisture content and MDD for the conduct of CBR experiments is determined from standard compaction tests and is summarized in Table 4. OMC and MDD is found to increase with addition of sisal fiber upto an optimum content of 0.8% beyond which it decreases. This infers that fibers absorbs moisture upto an optimum percentage and it increases workability of the mix. Figure 6 compares CBR value for seven different contents of sisal fiber in BC soil blended with 10% lime. From the figure its observed that at 0.8% of sisal fiber content, CBR value is maximum which infers that maximum reduction in pavement thickness is obtained at that fiber content. Table 5 provides the total thickness of pavement for 0 and 0.8% of sisal fiber content. Analysis of these results reveal that with the addition of 0.8% sisal fiber, around 14% reduction in pavement thickness is obtained in lime blended (10%) BC soil.

5.3 *Effect of Brick Powder Mix on Lime Blended BC Soil*

Optimum moisture content and maximum dry density is determined for different percentages of brick powder mix and the results are summarized in Table 6. It is observed that with addition of brick powder OMC increases and MDD decreases which could be due to the presence of more fines in the mix. The effectiveness of brick powder mix in stabilizing lime blended BC soil is studied with respect to CBR value and the corresponding pavement thickness. Variation in CBR value with percentage of brick powder mix is presented in Fig. 7.

Table 4 OMC and MDD for different percentages of sisal fiber in lime blended soil

	Plain BC soil	BC soil + 10% lime	BC soil + 10% lime + 0.2% SF	BC soil + 10% lime + 0.4% SF	BC soil + 10% lime + 0.6% SF	BC soil + 10% lime + 0.8% SF	BC soil + 10% lime + 1% SF	BC soil + 10% lime + 1.2% SF	BC soil + 10% lime + 1.4% SF
OMC	22	28	29	31	31	33	32	30	27
MDD (g/cc)	1.6	1.62	1.7	1.85	1.88	1.92	1.89	1.82	1.69

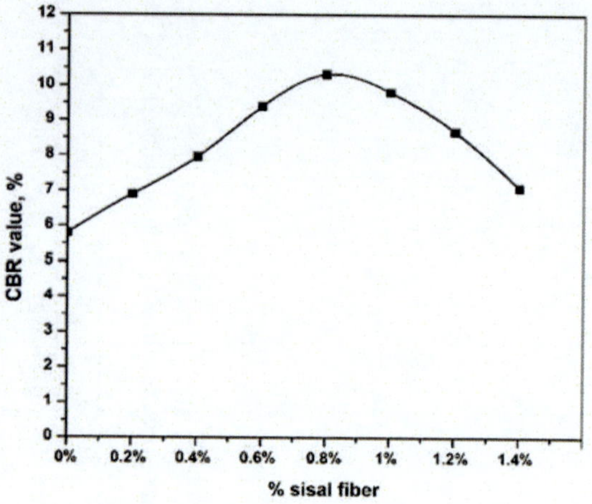

Fig. 6 Effect of sisal fiber content on CBR value of lime blended BC soil

Table 5 Total pavement thickness for different sisal fiber proportion

Traffic in msa	0% sisal fiber + BC soil + 10% lime	0.8% sisal fiber + BC soil + 10% lime
CBR value, %	5.8	10.3
10	624	540
20	650	570
30	666	585
50	686	585
100	710	610
150	730	625

From the figure it is observed that with addition of 10% of brick powder mix, there is an increase in CBR value. But with further addition of brick powder, CBR value went on decreasing. This could be due to the more fines present in the brick powder mix (BPM). The initial increase in the CBR value could be due to the high silica content present in the BPM. However, when compared to the lime blended BC soil, there is a considerable reduction in total pavement thickness for different traffic intensities when 10% brick powder mix is added. The pavement thickness for different traffic intensities for brick powder mix stabilized lime blended BC soil is presented in Fig. 8.

Table 6 OMC and MDD for different percentages of brick powder mix in lime blended soil

	Plain BC soil	BC soil + 10% lime	BC soil + 10% lime + 10% BPM	BC soil + 10% lime + 20% BPM	BC soil + 10% lime + 30% BPM	BC soil + 10% lime + 40% BPM	BC soil + 10% lime + 50% BPM	BC soil + 10% lime + 60% BPM	BC soil + 10% lime + 70% BPM	BC soil + 10% lime + 80% BPM	BC soil + 10% lime + 90% BPM
OMC	22	28	14	16	17	23	24	24	29	30	31
MDD (g/cc)	1.6	1.62	1.9	1.89	1.88	1.88	1.86	1.84	1.83	1.79	1.77

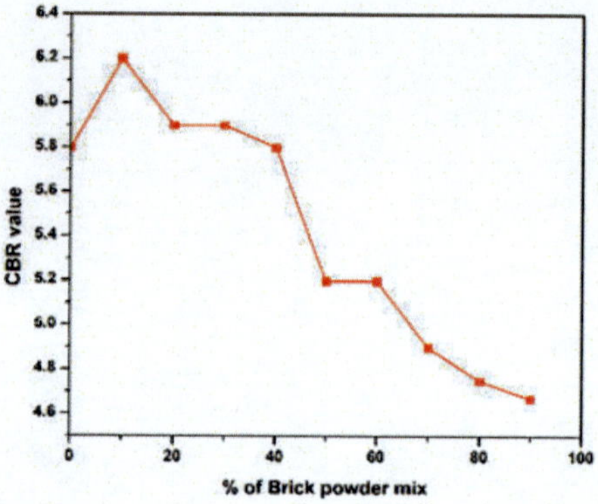

Fig. 7 Effect of brick powder mix on CBR value of lime blended BC soil

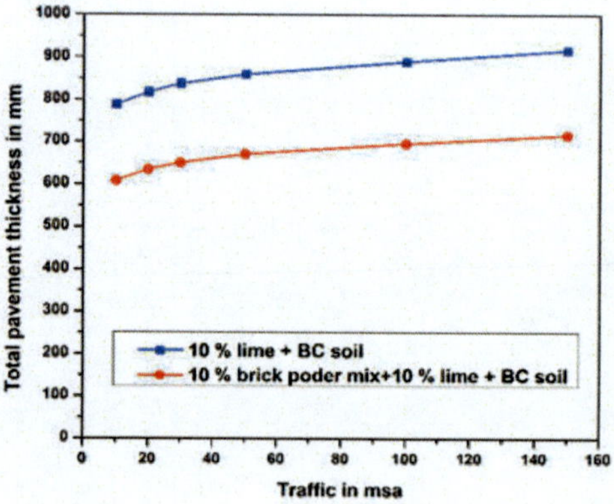

Fig. 8 Total pavement thickness for soil stabilized with optimum brick powder mix

5.4 *Effect of Phosphogypsum on Lime Blended BC Soil*

Phosphogypsum being a synthetic calcium sulphate hydrate, its effectiveness on lime blended BC soil is studied with respect to CBR value and the results are summarized in Table 7. For the studies the weight proportions of phosphogypsum used are 10%, 20%, 30% and 40%. From the experimental results it is observed that there is no improvement in CBR value with the addition of phosphogypsum to lime blended BC soil. However, with respect to plain BC soil, addition of 20% of phosphogypsum increases CBR value by 81%. This infers that effectiveness of phosphogypsum is more or less similar to that of lime and hence there is no additional improvement in strength.

Table 7 CBR value when phosphogypsum is used as stabilizer in lime blended BC soil

Description	Plain BC Soil	0% PG + 10% lime + BC soil	10% PG + 10% lime + BC soil	20% PG + 10% lime + BC soil	30% PG + 10% lime + BC soil	40% PG + 10% lime + BC soil
CBR value, %	2.7	5.8	4.7	4.89	4.12	3.8

6 Discussion on Test Results

The present study compares the effectiveness of different stabilizers in enhancing the strength of BC soil. The optimum dosage of different stabilizer identified with respect to highest CBR value is summarized in Table 8. From the table it is observed that sisal fiber along with lime can enhance strength of BC soil. The increase in CBR value in comparison to plain BC soil is 282%. This infers that pozzolanic action along with interfacial shear provides better improvement in strength. Comparison of total pavement thickness for different optimum dosages of stabilizers is provided in Table 9. The total thickness of pavement decreases with increase in CBR value and the lowest pavement thickness is for BC soil blended with 10% lime and 0.8% sisal fiber.

Table 8 CBR values at optimum dosages of different stabilizers

Description	CBR value, %
Plain BC soil	2.7
BC soil + 10% lime	5.8
BC soil + 10% lime + 0.8% sisal fiber	10.32
BC soil + 10% lime + 10% brick powder mix	6.2
BC soil + 10% lime + 20% phosphogypsum	4.89

Table 9 Total pavement thickness for optimum dosages of different stabilizers

Traffic in msa	Plain BC soil	BC soil + 10% lime	BC soil + 10% lime + 0.8% sisal fiber	BC soil + 10% lime + 10% brick powder mix	BC soil + 10% lime + 20% phosphogypsum
10	787	624	540	608	664.4
20	817	650	570	634	694.4
30	837	666	585	650	714.4
50	858.5	686	585	670	735.5
100	888.5	710	610	695	755.5
150	915.5	730	625	715	775.5

7 Conclusions

The present study compares the effectiveness of different stabilizers in strengthening black cotton soil. Since black cotton soil has montmorillonite mineral it exhibits high swell – shrinkage characteristics. In this study the effect of pozzolanic stabilizers, sisal fibers, waste from brick factory and synthetic calcium sulphate on BC soil is explored. From this study the following conclusions are drawn.

- Effectiveness of pozzolanic stabilizers on strength of weak soil increases with age
- 10% of Lime when added to BC soil increases its CBR value by around 1.14 times
- 0.8% sisal fiber when added to BC soil blended with 10% lime can increase the CBR value by around 2.8 times
- Addition of 0.8% sisal fiber to BC soil blended with 10% lime can decrease the overall pavement thickness by 30%.

8 Scope for Further Work

The present study investigates the effectiveness of different stabilizers in enhancing strength of BC soil. Here mainly CBR tests are used to assess the strength of soil. However, to analyze the behaviour of subgrade soil for pavement applications, resilient modulus is more indicative and also effect of curing period should be investigated for longer durations to ensure its durability.

References

1. Sharma, A., & Sharma, R. K. (2019). Effect of addition of construction–demolition waste on strength characteristics of high plastic clays. *Innovative Infrastructure Solutions, 4*, 1–11.
2. Sharma, A. K., & Sivapullaiah, P. V. (2016). Ground granulated blast furnace slag amended fly ash as an expansive soil stabilizer. *Soils and Foundations, 56*, 205–212.
3. Bahadori, H., Hasheminezhad, A., & Taghizadeh, F. (2019). Experimental study on marl soil stabilization using natural pozzolans. *Journal of Materials in Civil Engineering, 31*, 04018363.
4. Majumder, M., & Venkatraman, S. (2022). Utilization of the lime as subgrade stabilizer in the pavement construction. *Arabian Journal for Science and Engineering, 47*, 4929–4942.
5. Tamassoki, S., et al. (2022). Performance evaluation of lateritic subgrade soil treated with lime and coir fibre-activated carbon. *Applied Sciences, 12*, 8279.
6. Uday, R. A., Kiran, D., Arun Kumar, G. S., Prakash, K. G., & Maddodi, B. S. (2022). Effect of polypropylene macro fiber on geotechnical characteristics of black cotton soil: an experimental investigation and correlation analysis. *Engineered Science, 21*, 1–20. https://doi.org/10.30919/es8d775
7. Rokade, S. (2017). Effect of inclusion of fly-ash and nylon fiber on strength characteristics of black cotton soil. *Electronic Journal of Geotechnical Engineering, 22*, 1941–1950.
8. Almajed, A., Khodadadi, H., & Kavazanjian, E. (2018) Sisal fiber reinforcement of EICP-treated soil (pp. 29–36). https://doi.org/10.1061/9780784481592.004.
9. Wu, Y. K., Li, Y. B., & Niu, B. (2014). Investigation of mechanical properties of randomly distributed sisal fibre reinforced soil. *Materials Research Innovations, 18*, S2953–S2959.

10. Jairaj et al. (2021). Strength and deformation characteristics of lime-admixed black cotton soil reinforced with sisal fibres. In *Lecture Notes in Civil Engineering. LNCE*, (Vol. 136, pp. 753–762).
11. Bhavsar, S. N., & Patel, A. J. (2016). Effect of waste material on swelling and shrinkage properties of clayey soil. *International Journal of application or innovation in Engineering & Management (IJAIEM), 3*, 200–206.
12. Al-Khafaji, Z. S., Al-Naely, H. K., & Al-Najar, A. E. (2018). A review applying industrial waste materials in stabilisation of soft soil. *Electronic Journal of Structural Engineering, 18*, 16–23.
13. Zhao, Y., et al. (2021). Long-term hydration and microstructure evolution of blended cement containing ground granulated blast furnace slag and waste clay brick. *Cement and Concrete Composites, 118*, 103982.
14. Hwang, C. L., Yehualaw, M. D., Vo, D. H., Huynh, T. P., & Largo, A. (2019). Performance evaluation of alkali activated mortar containing high volume of waste brick powder blended with ground granulated blast furnace slag cured at ambient temperature. *Construction and Building Materials, 223*, 657–667.
15. Kasinikota, P., & Tripura, D. D. (2021). Evaluation of compressed stabilized earth block properties using crushed brick waste. *Construction and Building Materials, 280*, 122520.
16. Anamika, B., & Debabrata, G. (2022). Utilisation of phosphogypsum along with other additives in geo-engineering-a review. *Materiales de Construccion, 72*, e288.
17. Kumar, S., & Dutta, R. K. (2014). Unconfined compressive strength of bentonite-lime-phosphogypsum mixture reinforced with sisal fibers. *Jordan Journal of Civil Engineering, 8*, 239–250.
18. Bian, X., Zeng, L., Ji, F., Xie, M., & Hong, Z. (2022). Plasticity role in strength behavior of cement-phosphogypsum stabilized soils. *Journal of Rock Mechanics and Geotechnical Engineering, 14*, 1977–1988.
19. Özkan, İ. (2015). Improvement of expansive soils by using phosphogypsum. Master of Science Thesis (Vol. 13). (Graduate school of natural and Applied Sciences of Middle East Technical University, 2015).
20. Hussain, M., & Dash, S. K. (2016). The influence of lime on the compaction behaviour of soils. *Environmental Geotechnics, 3*, 346–352.
21. IRC 37. (2012). Guidelines for the design of flexible pavements. Indian Roads Congress, New Delhi.

Effect of Compressive Strength and Reinforcing Bar Diameter on Tensile and Cracking Aspects of Reinforced Concrete Prisms

Venkateswarlu Mangalapuri and Durga Gunneswara Rao Thippabhotla

Abstract The purpose of this research is to determine the impact of concrete compressive strength (concrete grade), bar diameter, and reinforcing ratio on the tensile (tensile stress and tension stiffening) and cracking (crack spacing) properties of reinforced concrete (R.C.) prisms under uniaxial tension. In this study, two different concrete grades (20 MPa and 40 MPa) and three different High Yield Strength Deformed (HYSD) steel bars (with a yield stress of 500 MPa) of 8-, 10-, and 12-mm diameter were used as tension chords in prismatic concrete members. From the experimental results, tensile stress and tension stiffening of the members increased as the compressive strength (grade of concrete) of the concrete increased. From this, tensile stresses of the concrete member are affected by the concrete's compressive strength (grade of concrete), but not significantly by using different reinforcing bar diameters. According to cracking properties, crack spacings were influenced by the compressive strength of concrete and the reinforcing ratio (bar diameter). Since concrete's compressive strength increased, crack spacing decreased. Also, as reinforcement ratio (bar diameter) increases, crack spacing decreases. Furthermore, the obtained crack spacings are consistent with Eurocode 2.

Keywords Compressive Strength · Bar Diameter · Tensile Stress · Tension Stiffening · Crack Spacing

V. Mangalapuri (✉) · D. G. R. Thippabhotla
Department of Civil Engineering, National Institute of Technology, Warangal 506004, Telangana, India
e-mail: mv718113@student.nitw.ac.in

D. G. R. Thippabhotla
e-mail: tdg@nitw.ac.in

A. Vilventhan et al. (eds.), *Advances in Construction Materials and Management*, Lecture Notes in Civil Engineering 346,
https://doi.org/10.1007/978-981-99-2552-0_17

1 Introduction

A clear understanding of tensile behaviors of plain and reinforced concrete is essential for the ultimate behavior analysis of the structure under normal and extreme conditions. Likewise, cracks in a structure can have a significant impact on serviceability, durability, aesthetics and energy transfer. A. K. Gupta and S. R. Maestrini proposed an analytical formula for the phenomenon of tensile stiffness in cracked reinforced concrete bars and concludes that the concrete's tensile strength significantly decreases at larger deformations [1]. The Model Code as well as Eurocode-2, must be used as the foundation for the computation of crack spacing and widths for practical applications. The formulas offered by these sources are, however, only semi empirical also were created primarily for beam elements [2, 6, 9]. Tension stiffening effect in cracked reinforced concrete members was mainly caused due to bond effects [4, 5]. After the crack is fully established, the tension stiffening bond factor is not constant and it continues to deteriorate (although more slowly) while continues during the stabilised cracking phase [3].

The tension stiffening effect of cracked concrete is thought to be responsible for the increased stiffness of reinforced concrete slabs [7]. Salah Khalfallah revealed that the cracking and tensile stiffness quality have the greatest influence on the numerical results of flexural RC members under short-term stress conditions [8]. Lee et al revealed that in reinforced concrete members subjected to uniaxial tension, shear or bending, a tension hardening model is proposed that allows estimation of average tensile stress in concrete after the reinforcement has been yield [10]. Simple mechanics-based models were created for various idealized bond properties, to predict crack spacing, crack width, and loads for primary, secondary, and future crack initiation under short-term loading [11]. S. Khalfallah and D. Guerdouh concluded that the shrinkage may drastically lower the members' cracking resistance and stiffness [12]. Shukri et al incorporating tensile stiffening effects to create a mechanical model for simulating R.C joints under reverse cyclic loading [13]. The tensile stiffness increases with the decrease of reinforcement ratio, and not greatly affected by axial force, unless the reinforcement ratio is low and the diameter of the small circular section is high [14].

Although there have been several attempts to provide suitable formulations for estimating crack widths and spacing, there is still a substantial gap between the techniques' consistency and understanding. Most researchers focused their work on beam and slab members in shear, flexure, short term loading, but they do not mention much about how the grade of concrete, size of the member, reinforcing bar diameter, and reinforcement ratio affect the tensile and cracking properties. Therefore, more studies are needed to understand the tensile and cracking properties of structural members in uni-axial tension. Therefore, the present paper focused on how the compressive strength/concrete grade, reinforcing bar diameter, and reinforcing ratio affect the tensile and cracking properties of reinforced concrete prisms under uni-axial tension.

2 Experimental Program

In the present study two concrete grades 20 MPa, 40 MPa are considered. The following variables were considered in the study, they are different concrete grades, reinforcement ratio and bar diameter. Six test series were casted and tested, each series of specimen consisting of different grade of concrete, reinforcement ratio and different bar diameters. Each Reinforced Concrete test member having a concrete prism, 100 × 100 mm cross section and 500 mm long, single HYSD steel bar of length 700 mm cast around in such a way that 100 mm of the bar extended from each end for gripping. HYSD bar sizes varied from 8 mm, 10 mm and 12 mm. Geometric configurations of the samples are presented in Table 1.

2.1 *Materials Used*

In the present study, 53 grade Ordinary Portland Cement (OPC) used. Its specific gravity and specific surface areas are 3.13 and 335 m^2/kg. Fly ash content is partially substituted for cement. Fly ash was obtained from the Ramagundam thermal power plant in Telangana, and the specific gravity and specific surface area of the fly ash used were 2.11 and 450 m^2/kg respectively. Coarse aggregates of nominal size of 20 mm were used and its specific gravity value is 2.83, the nominal size of fine aggregate is 4.75 mm and its specific gravity value is 2.62. SP430 sulfonated naphthalene-based superplasticizer was used and potable water used to mix the concrete. Cement and Fly ash chemical composition mentioned in below Table 2.

Table 1 Geometric configurations of the sections

S.No	Bar Diameter (mm)	Geometric Configurations				Reinforce ment Ratio (%)
		Cross Section	Gross C/S Area (mm^2)	Steel Area (mm^2)	Net C/S Area of Concrete (mm^2)	
1	8	100 × 100	10,000	50.27	9949.73	0.51
2	10	100 × 100	10,000	78.54	9921.46	0.79
3	12	100 × 100	10,000	113.1	9886.9	1.14

Table 2 Cement and Fly ash chemical composition

Binder	CaO	SiO2	Al2O3	Fe2O3	SO3	MgO	Na2O	LOI
cement	60–62%	20–22%	5–6%	4–5%	2–3%	0.5–1	0.12	2.52
Fly ash	1.82	62.32	26.53	4.22	0.37	1.03	0.21	0.86

Table 3 Mix proportions and Compressive Strength values

Mix Designation	Mix Proportions					Compressive Strength MPa
	Cement	Fly ash	Fine Aggregate	Coarse Aggregate	w/c	
20 MPa	360	–	694	1196	0.5	23.46
40 MPa	398	44	736	1063	0.4	42.63

Fig. 1 Specimens casting

2.2 *Mix Proportioning, Casting and Curing of Specimens*

In the present study two concrete grades are considered. They are 20 MPa, 40 MPa. The mix proportioning details presented in Table 3.

To confirm the concrete grade/compressive strength, 100 × 100 × 100 mm size cubes used. And to observe tensile and cracking behaviour of reinforced concrete, cast 100 × 100 × 500 mm size prisms with central rebar (tension chord) of size varying from higher diameter to lower diameter used. Specimens are demoulded from the moulds after 24 h of casting. After demolding, the samples were immersed in water for curing. Table 4, represented the number of samples cast and tests carried on in each test series. Figures. 1 and 2 represented the casting details of specimen and test set up of R.C prism.

2.3 *Tests Conducted*

After casting, concrete cubes are tested for their compressive strength after curing for 28 days, grade confirmation was done from the obtained values.

Fig. 2 Test setup for prism specimen under uni-axial tension

Table 4 Number Samples and tests carried out

Concrete Grade	Specimen	Geometric Configurations	Bar Diameter (mm)	Specimen	Tests Carried out
20 MPa	Cubes	100 × 100 × 100	–	6	Compression test
	Prisms	100 × 100 × 500	8, 10 & 12	9	Uni-axial tension test
40 MPa	Cubes	100 × 100 × 100	–	6	Compression test
	Prisms	100 × 100 × 500	8, 10 & 12	9	Uni-axial tension test

2.4 Uni-axial Tension Test on RC Prisms

In this study, tensile and cracking behaviour of RC (Reinforced Concrete) prisms were determined under uni-axial tension by using 200 kN capacity universal testing machine. This test played a very important role in this study. To measure the axial deformation in the members, two LVDTs were used. The LVDTs are fixed at the corresponding paces of the prisms with a guage length of 400 mm. During the test, cracks in member and their locations were observed and marked. Similarly, the first crack load and yield loads were recorded during the test. Various stages were observed while the test was going on. The deformations associated with the entire test were recorded from the LVDTs and the load values from the UTM.

3 Results and Discussion

3.1 Tensile Behavior

Figures 3 and 4, showed the member load-member strain responses of the 20 MPa and 40 MPa concretes. The recorded values of first crack load and yield load tabulated in Table 5, as the reinforcement ratio increases there is very slight change in tensile stress observed.

From the Figs. 5 and 6, when different reinforcement ratios (steel bars of different diameters) were used as tension chords in 20 MPa grade reinforced concrete prisms, all the specimens showed very slight variations and almost similar maximum tensile stress values. For 20 MPa concrete, the obtained average maximum tensile stress value is 2.62 MPa. The same behavior was observed in 40 MPa grade specimens also. For 20 MPa concrete, the obtained average maximum tensile stress value is 3.18 MPa. From this, it is understood that the tensile properties of concrete are

Fig. 3 Member Load vs Member Strain

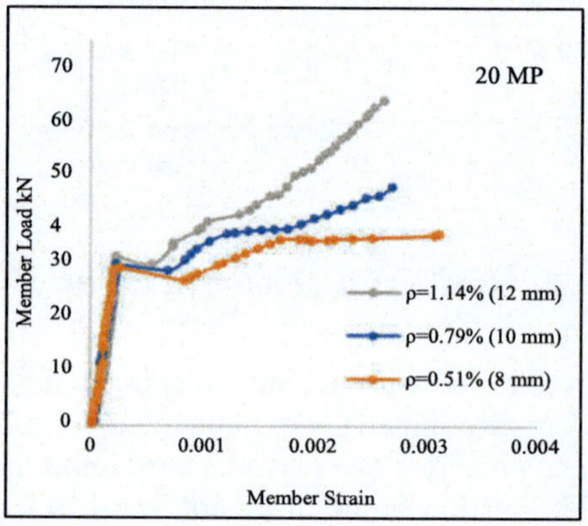

Fig. 4 Member Load vs Member Strain

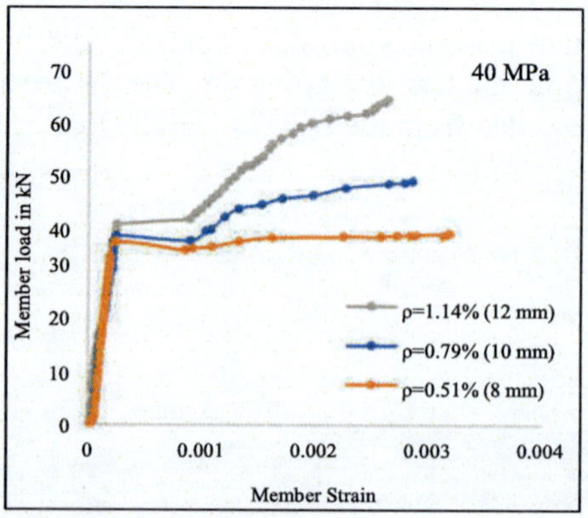

Table 5 First crack load, Yield load and Maximum Tensile strength values

Grade of Concrete	Bar Diameter (mm)	Reinforcement Ratio (%)	First Crack Load (kN)	Yield Load (kN)	Maximum Tensile Strength (MPa)
20 MPa	8	0.51	28.70	35.00	2.63
	10	0.79	29.40	43.60	2.63
	12	1.14	30.80	59.60	2.61
40 MPa	8	0.51	34.30	35.30	3.17
	10	0.79	34.50	45.00	3.16
	12	1.14	36.90	60.00	3.21

dependent on the concrete grade or concrete compressive strength. But, not on the diameter of the reinforcement bar and reinforcing ratio. Since as concrete grade/compressive strength concrete increases tensile strength is also increases. Whereas, in the same grade of concrete, tensile strength of concrete not changed even though the reinforcement ratio increased (bar diameter increased from 8 mm to 10 mm and 12 mm).

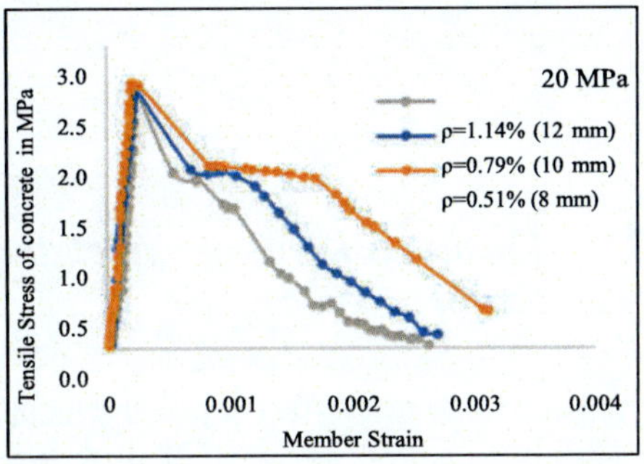

Fig. 5 Tensile Stress of concrete vs Member Strain

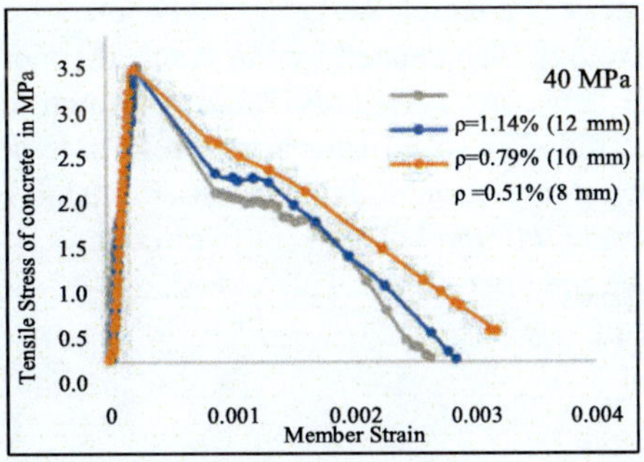

Fig. 6 Tensile Stress of concrete vs Member Strain

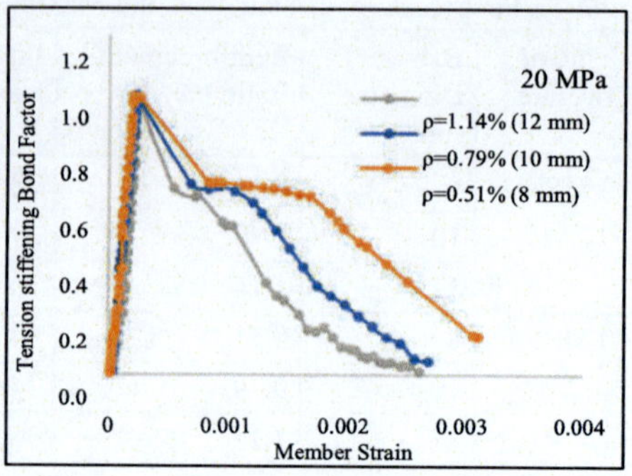

Fig. 7 Tension Stiffening bond factor vs Member Strain

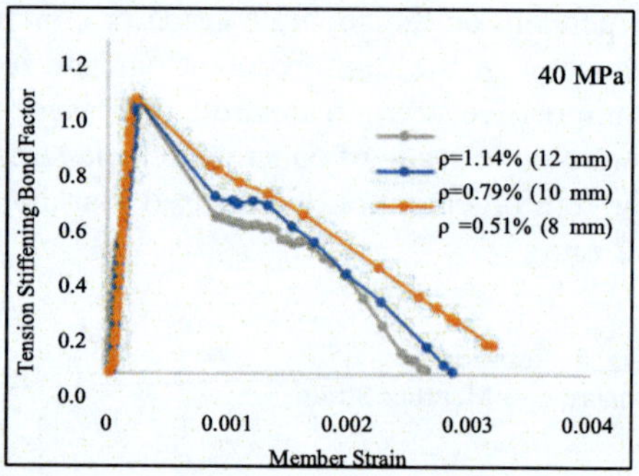

Fig. 8 Tension Stiffening bond factor vs Member Strain

3.2 *Tension Stiffening Effect and Tension Stiffening Bond Factor*

"Tension stiffening effect refers the tension carrying capacity of concrete between cracks in structural member". This tension stiffening effect obtained from member load-member strain response. Tension stiffening indicates the bonding occurred between concrete and steel. "The tension stiffening bond factor obtained by dividing average load carried by the cracked concrete with the force carried by the concrete at first crack." Figures 7 and 8, shown tension stiffening bond factor with respect to member strain for various reinforcement ratios (different bar diameters). From the figures, concluded that better tension stiffening bond factors in 40 MPa concrete grade (compressive strength of concrete) compared to 20 MPa grade concrete over the member strain. This is because of the better bond existed in 40 MPa grade of concrete.

3.3 Cracking Behavior

In this study, maximum crack spacings were measured and compared them with CEB-FIP Model code [9] and some previous studies [15] in crack spacings. Figures 9 and 10 shown the maximum crack spacings over different reinforcement ratios in 20 MPa and 40 MPa concretes.

From the Figs. 9 and 10, it is concluded that as the reinforcement ratio increases the crack spacings are decreased which is happened in both concrete grades. Similarly, as compressive strength/concrete grade increases, cracking spacings are decreased. It indicates that crack spacings are dependent on concrete grade/ compressive strength and reinforcing ratio (different bar diameters). The obtained experimental crack spacings values are compared with Eurocode 2 [6] and previous studies [15]. The experimental values 80-90% agreed with code provision but underestimates the previous studies values. Figure 11, shown the crack mode observed after specimen failure.

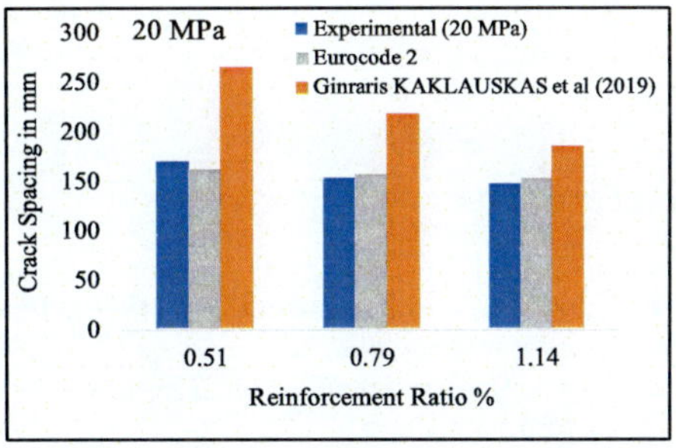

Fig. 9 Crack spacings vs reinforcement ratio 20 MPa

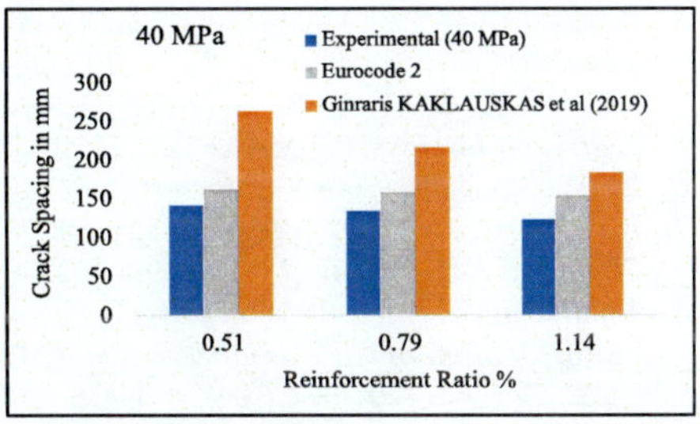

Fig. 10 Crack spacings vs reinforcement ratio 40

Fig. 11 A crack mode observed after specimen failure

4 Conclusions

In this paper, experimental study was conducted to determine effect of compressive strength/concrete grade, reinforcing bar diameter and reinforcing area on the tensile and cracking properties of concrete. Some important points from this study are noted below.

- The compressive strengths of 20 MPa and 40 MPa concrete mixes were achieved at 23.46 MPa and 42.63 MPa, respectively.
- Tensile stresses in concrete are dependent on the concrete grade or compressive strength based on the tensile behaviors. but not on the bar diameter and reinforcing ratio.
- From cracking behaviors, it was understood that crack spacings depend on compressive strength (concrete grade), reinforcing ratio, and bar diameter.
- Finally, the obtained crack spacing is in good agreement with the Eurocode-2 provisions but underestimates the previous study values.

References

1. Gupta, A. K., & Maestrini, S. R. (1990). Tension-stiffness model for reinforced concrete bars. *Journal of Structural Engineering, 116*(3), 769–790.
2. Comité Euro-International du Béton. (1993). *CEB-FIP model code 1990: Design code*. Thomas Telford Publishing.
3. Monti, G., & Spacone, E. (2000). Reinforced concrete fiber beam element with bond-slip. *Journal of Structural Engineering, 126*(6), 654–661.
4. Bischoff, P. H. (2001). Effects of shrinkage on tension stiffening and cracking in reinforced concrete. *Canadian Journal of Civil Engineering, 28*(3), 363–374.
5. Lackner, R., & Mang, H. A. (2003). Scale transition in steel-concrete interaction. I: Model. *Journal of Engineering Mechanics, 129*(4), 393–402.
6. British Standard. (2004). *Eurocode 2: design of concrete structures. Part 1-1* (p. 230).
7. Ian Gilbert, R. (2007). Tension stiffening in lightly reinforced concrete slabs. *Journal of Structural Engineering, 133*(6), 899–903.
8. Khalfallah, S. (2008). Tension stiffening bond modelling of cracked flexural reinforced concrete beams. *Journal of Civil Engineering and Management, 14*(2), 131–137.
9. Code, C. F. M. (2010). *CEB-FIB model code 2010–final draft*. Thomas Thelford, Lausanne, Switzerland.
10. Lee, S.-C., Cho, J.-Y., & Vecchio, F. J. (2011). Model for post-yield tension stiffening and rebar rupture in concrete members. *Engineering Structures, 33*(5), 1723–1733.
11. Muhamad, R., et al. (2012). The tension stiffening mechanism in reinforced concrete prisms. *Advances in Structural Engineering, 15*(12), 2053–2069.
12. Khalfallah, S., & Guerdouh, D. (2014). Tension stiffening approach in concrete of tensioned members. *International Journal of Advanced Structural Engineering, 6*(1), 1–6.
13. Shukri, A. A., et al. (2016). Mechanics model for simulating RC hinges under reversed cyclic loading. *Materials, 9*(4), 305.

14. Morelli, F., et al. (2017). Influence of tension stiffening on the flexural stiffness of reinforced concrete circular sections. *Materials, 10*(6), 669.
15. Kaklauskas, G., Ramanauskas, R., & Ng, P.-L. (2019). Predicting crack spacing of reinforced concrete tension members using strain compliance approach with debonding. *Journal of Civil Engineering and Management, 25*(5), 422–430.

Shear Strength and Settlement Analysis of Stabilized Soil with GGBS and Cement

Darshan C. Sekhar, B. R. Vinod, Anthony Raj, and S. T. Shashank

Abstract Construction on weak soils is difficult as it has poor bearing capacity and especially black cotton soil has high swelling or shrinking characteristics. Hence there is a need for the improvement in the soil properties, which can be addressed using stabilization of the soil. In this study, two soil types i.e., red soil and black cotton soil were replaced by different percentages of GGBS with small amount of cement. The improvement in shear properties, both in cohesion and friction of the soil was observed with all the combinations from laboratory testing. The settlements were analysed in PLAXIS 2D and was observed that within 7 days the settlement reduces by more than 60% in both the soil types with additives when compared to natural soils.

Keywords Bearing Capacity · GGBS · PLAXIS 2D · Settlement · Shear strength Parameter

1 Introduction

One of the soils which is predominantly problematic especially for engineers is the Black cotton soil. In India, the expansive soil is referred to as black cotton soil and covers almost 1/5th of the country. Even though for agricultural practices this soil is

D. C. Sekhar (✉)
Department of Civil Engineering, BMS College of Engineering, Bangalore, India
e-mail: darshancs.civ@bmsce.ac.in

B. R. Vinod · A. Raj · S. T. Shashank
Department of Civil Engineering, B M S Institute of Technology and Management, Bangalore, India
e-mail: vinodbr@bmsit.in

A. Raj
e-mail: rajanthony@bmsit.in

S. T. Shashank
e-mail: 1by20cv035@bmsit.in

A. Vilventhan et al. (eds.), *Advances in Construction Materials and Management*, Lecture Notes in Civil Engineering 346,
https://doi.org/10.1007/978-981-99-2552-0_18

very much useful but because of its swelling and shrinking characteristics very tricky for an engineer. It is indeed tricky as the swelling and shrinkage is not consistent and varies with seasonal changes. This volume instability induces damage to lightly loaded structures. Since the damage is repetitive, repairing and maintenance cost is very high [8, 9]. Because of the heaving and settling problem associated with the foundation soil, it is often replaced with murrum. Though stabilization cannot be effectively implemented at site conditions, but it is one of the alternatives. Since physical stabilization in these soils becomes difficult, chemical stabilization is the go to method. Again they can be categorized into traditional, by product and non-traditional additives. Since the pozzolanic materials are less within soils, addition of GGBS usually promotes the pozzolanic reactions along with cement and thus provides better durability characteristics [5, 7].

Cement is widely used as stabilizer because of its easy availability and high strength within less time. The process consists of hydration, cation exchange, flocculation and the pozzolanic reaction. The formation of CSH gel develops because of the inherent C3S and C2S compounds. The bonding between the soil particles increases with time due to the increased cementitious products [4]. The amount of cement varies upto 15% depending on the amount of clay content. High plastic clays require more cement compared to low plastic clays. But preferably less than 8% to avoid shrinkage cracks [1]. With the addition of GGBS the particle size increases with an immediate reduction in the swell potential and improvement in strength [3, 11, 13]. Addition of GGBS and Cement provides cementitious hydration products and the pozzolanic reactions, which can be prolonged to a longer curing period but this also demands more water content [6]. Furthermore, for strength and workability, GGBS and Cement are used together. In this study we look into using both traditional additive as cement and byproduct as GGBS.

2 Materials and Methods

The natural soil was obtained at a depth of 1 m below the ground surface to avoid organic content from Gulbarga, Karnataka. The soil was later dried and stored for conduction of experiments (Table 1).

Table 1 The properties of the soil obtained are provided below

Soil properties	BC soil	Red soil
Liquid limit	54%	30%
Plastic limit	34%	22%
MDD	1.46 g/cc	1.78 g/cc
OMC	21%	14%
Soil Classification	MH	SP

Table 2 The various combination with GGBS and cement and was tested up to 7 days

Combinations
Soil Alone
Soil + 20% GGBS + 4% cement (3 days)
Soil + 20% GGBS + 4% cement (7 days)
Soil + 40% GGBS + 2% cement (3 days)
Soil + 40% GGBS + 2% cement (7 days)
Soil + 40% GGBS + 4% cement (3 days)
Soil + 40% GGBS + 4% cement (7 days)

GGBS is a by-product obtained from the manufacture of steel and iron industries. It contains high amount of calcium silicate. Since the improvement rate from GGBS in initial stages is less, we have added cement to aggravate the process and to ensure the hydrated products remain for a longer time. The cement used was of 43 grade (Table 2).

The samples are tested for their shear strength parameters from direct shear test. They were prepared using the mould shown below of standard dimensions and the box samples obtained are shown below (Figs. 1, 2 and 3).

Based on the laboratory data obtained, load-settlement analysis was carried out in PLAXIS 2D software.

Fig. 1 Sample preparation mould

Fig. 2 Direct shear soil samples

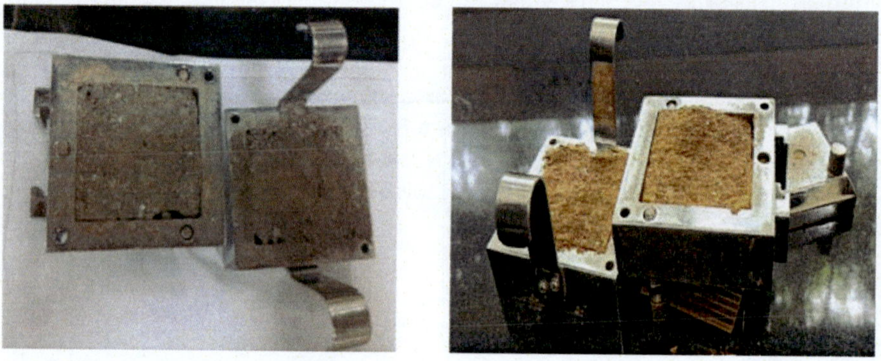

Fig. 3 Soil sample after direct shear test

Load were increased from 10 kPa, 50 kPa, 100 kPa, 150 kPa, 200 kPa and 250 kPa (Fig. 4).

Details of the footing:

Width of the footing – 1 m

Depth of footing – 1.2 m below Ground Level

Thickness of footing – 0.5 m

Footing properties:

Modulus of elasticity (E) – 2.2×10^7 kPa

Poisson's ratio (μ) – 0.15

Unit weight (γ) – 25 kN/m^3

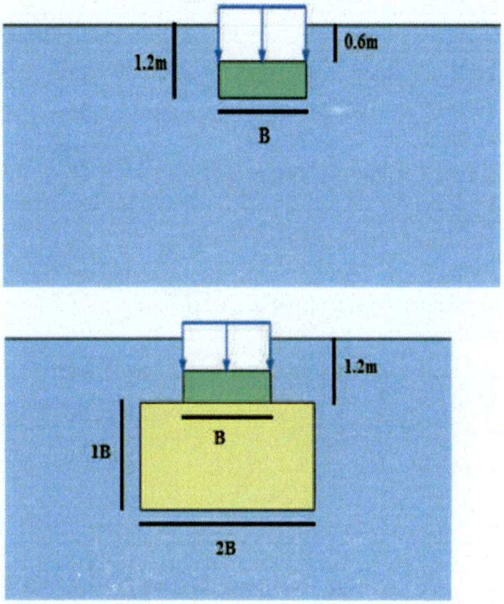

Fig. 4 Mathematical modelling using PLAXIS 2D Software

3 Results and Discussion

The results obtained showed an improvement in both cohesion and friction as seen in Table 3. The improvement in cohesion was lesser compared to the improvement in friction angle. This is due to the fact that even though cohesion improved largely but was reduced due to the testing at submerged condition. However, the friction angle was less affected because of it. The friction angle increase due to the addition of GGBS as it is frictional material initially. Later on with the hydration of cementitious products, the cohesion intercept increases because of the availability of free CaO [12]. But the rate of hydration of these products is lesser, but upon the addition of cement, these reaction rate increases (Fig. 5 and Table 4).

The graphs of the same indicates that with increase in load the settlement increases irrespective of the combinations. But with the addition of GGBS and cement the settlement reduces proving the hardening of the soil mass with time. It is also observed that with increase in GGBS the settlement reduces indicating the formation of cementitious products which have occupied the void space (Burman and Dash, 2022) (Figs. 6, 7 and 8, Tables 5 and 6).

Table 3 BC Soil, GGBS and Cement Combination Properties

Combinations	C (kPa)	Phi (°)	Modulus of elasticity E (MPa)
BC Soil Alone	14	22	1.9
BC Soil + 20% GGBS + 4% cement (3 days)	36	24	3.1
BC Soil + 20% GGBS + 4% cement (7 days)	37	42	5.6
BC Soil + 40% GGBS + 2% cement (3 days)	26	39	3.5
BC Soil + 40% GGBS + 2% cement (7 days)	36	44	4.9
BC Soil + 40% GGBS + 4% cement (3 days)	58	35	5.2
BC Soil + 40% GGBS + 4% cement (7 days)	78	38	6.8

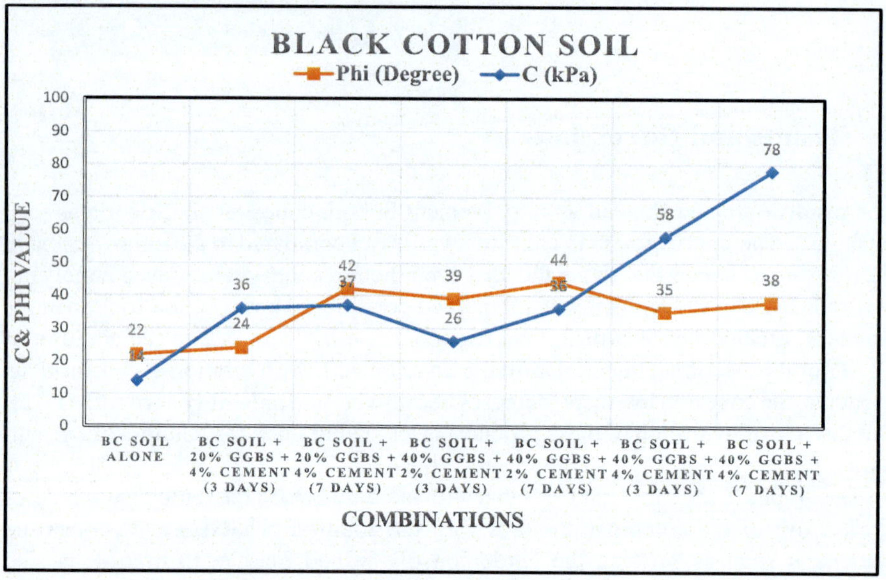

Fig. 5 Shows Black Cotton Soil C (kPa) & Phi (Degree)

Table 4 From the PLAXIS 2D software, the load verses settlement was analyzed upto load intensity of 250 kPa

Settlement (mm)							
Load Intensity (kPa)	Black cotton (BC) soil	BC soil + 20% GGBS + 4% cement (3 days)	BC soil + 20% GGBS + 4% cement (7 days)	BC soil + 40% GGBS + 2% cement (3 days)	BC soil + 40% GGBS + 2% cement (7 days)	BC soil + 40% GGBS + 4% cement (3 days)	BC soil + 40% GGBS + 4% cement (7 days)
10	6.91	4.92	4.15	3.2	2.5	2.1	1.6
50	16.5	15.1	13.5	11.5	8.1	6.9	4.8
100	28.8	24.1	22.5	20.9	15.9	11.6	9.9
150	45	32	30.2	28.5	24.1	19.9	15.2
200	77.1	45.2	40.2	38.01	31.6	26.5	23.9‘
250	152	69.2	57.1	51.2	44.6	35.2	32

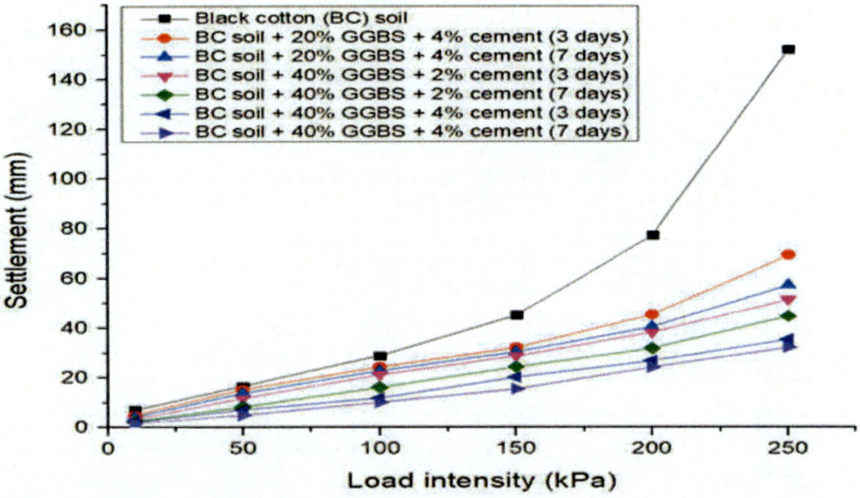

Fig. 6 Shows BC Soil Settlement (mm) verses Load intensity (kPa)

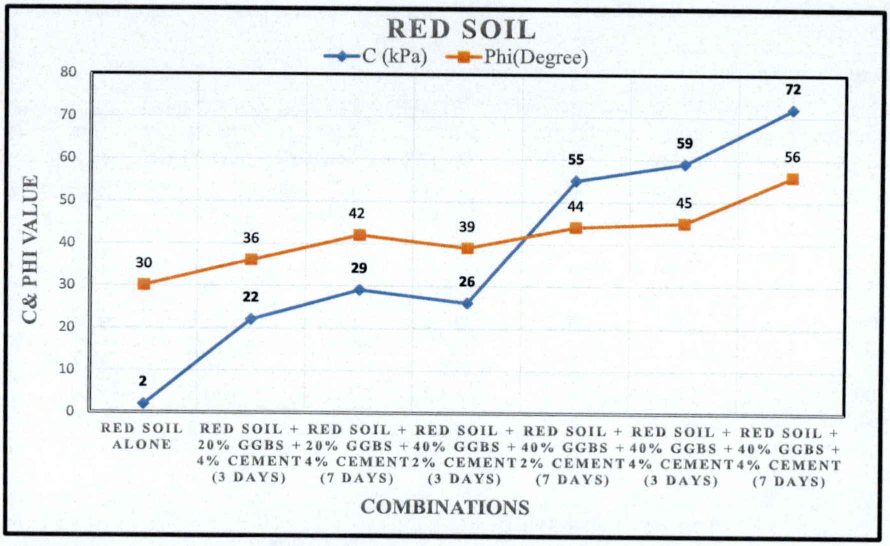

Fig. 7 Shows Red Soil C (kPa) & Phi (0)

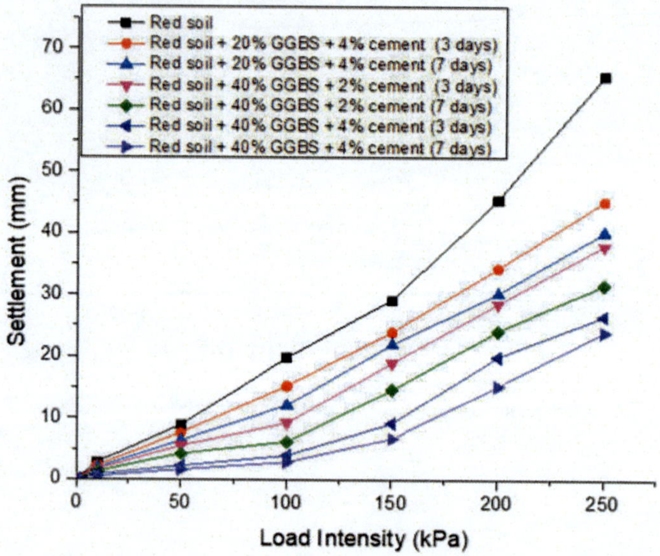

Fig. 8 Shows Red Soil Settlement (mm) verses Load intensity (kPa)

Table 5 Red Soil, GGBS and Cement combination Properties

Combinations	C (kPa)	Phi (^{0})	Modulus of elasticity E (MPa)
Red Soil Alone	02	30	2.6
Red Soil + 20% GGBS + 4% cement (3 days)	22	36	3.9
Red Soil + 20% GGBS + 4% cement (7 days)	29	42	6.2
Red Soil + 40% GGBS + 2% cement (3 days)	26	39	4.6
Red Soil + 40% GGBS + 2% cement (7 days)	55	44	5.9
Red Soil + 40% GGBS + 4% cement (3 days)	59	45	5.8
Red Soil + 40% GGBS + 4% cement (7 days)	72	56	7.9

Table 6 From the PLAXIS 2D software, the load verses settlement was analysed up to load intensity of 250 kPa

Settlement (mm)							
Load Intensity (kPa)	Black cotton (BC) soil	BC soil + 20% GGBS + 4% cement (3 days)	BC soil + 20% GGBS + 4% cement (7 days)	BC soil + 40% GGBS + 2% cement (3 days)	BC soil + 40% GGBS + 2% cement (7 days)	BC soil + 40% GGBS + 4% cement (3 days)	BC soil + 40% GGBS + 4% cement (7 days)
10	2.8	2.2	1.9	1.6	1.3	0.8	0.5
50	8.9	7.6	6.4	5.5	4.2	2.2	1.6
100	19.8	15.2	12.1	9.2	6.2	3.9	2.8
150	29.2	23.9	21.9	18.9	14.6	9.2	6.7
200	45.4	34.3	30.2	28.5	24.1	19.9	15.2
250	65.6	45.2	40.2	38.01	31.6	26.5	23.9

4 Conclusion

1. With the addition of GGBS the friction angle increases at a higher rate than the c cohesion.
2. With increase in curing periods the strength development will also increase with higher friction angle.
3. The load settlement analysis showed that with increase in GGBS, Cement and Curing days the settlement reduces by more than 75 Percentage.

References

1. Bell, F. (1993). *Engineering treatment of soils*. CRC Press.
2. Barman, D., & Dash, S. K. (2022). Stabilization of expansive soils using chemical additives: a review. *Journal of Rock Mechanics and Geotechnical Engineering, 14*, 1319–1342.
3. Cokca, E., Yazici, V., & Ozaydin, V. (2009). Stabilization of expansive clays using granulated blast furnace slag (GBFS) and GBFS-cement. *Geotechnical and Geological Engineering, 27*(4), 489.
4. Goodarzi, A. R., Akbari, H. R., & Salimi, M. (2016). Enhanced stabilization of highly expansive clays by mixing cement and silica fume. *Applied Clay Science, 132*, 675–684.
5. James, R., Kamruzzaman, A. H. M., Haque, A., & Wilkinson, A. (2008). Behaviour of lime slag-treated clay. *Proceedings of Institute of Civil Engineers-Ground Improvement, 161*(4), 207–216.
6. Jongpradist, P., Jumlongrach, N., Youwai, S., & Chucheepsakul, S. (2010). Influence of fly ash on unconfined compressive strength of cement-admixed clay at high water content. *Journal of Materials in Civil Engineering., 22*(1), 49–58.
7. McCarthy, M. J., Csetenyi, L. J., Sachdeva, A., & Dhir, R. K. (2014). Engineering and durability properties of fly ash treated lime-stabilized sulphate-bearing soils. *Engineering Geology, 174*, 139–148.
8. Muttharam, M. (2000). *Engineering behaviour of ash-modified soils of Karnataka*. Ph.D. thesis. Indian Institute of Science, Bangalore, India.
9. Otcovska, T., & Padevet, P. (2016). Microstructure of unburned clay and its shrinkage during drying. In *Applied mechanics and materials* (Vol. 837, pp. 109-112).
10. Parikshith, M. V. (2019). Feasibility of flyash based geopolymer for soil stabilization. *International Journal of Innovation and Technology and Exploring Engineering, 9*, 4348–4351.
11. Sekhar, D. C., Nayak, S., & Preetham, H. K. (2017). Influence of granulated blast furnace slag and cement on the strength properties of lithomargic clay. *Indian Geotechnical Journal, 47*(3), 384–392.
12. Sekhar, D., & Nayak, S. (2019). SEM and XRD investigations on lithomargic clay stabilized using granulated blast furnace slag and cement. *International Journal of Geotechnical Engineering, 13*(6), 615–629.
13. Sharma, A. K., & Sivapullaiah, P. V. (2016). Ground granulated blast furnace slag amended fly ash as an expansive soil stabilizer. *Soils and Foundations, 56*(2), 205–212.
14. Sharma, A. K., & Sivapullaiah, P. V. (2012). Improvement of strength of expansive soil with waste granulated blast furnace slag. In *GeoCongress* (pp. 3920–3928).

A Study on the Effect of Alccofine on the Stability of Soil Slopes

B. R. Vinod, J. Sumalatha, and J. Akshit Jain

Abstract Slope stability issues have existed throughout history, due to human interference with the natural slope balance of soil. As manmade slopes are increasing, the estimation of factor of safety is the important factor to stabilize the soil slopes. This research work consists of laboratory and software evaluation of a stabilized red soil. The soil was treated with Alccofine with varying stabilizer dosages. Unconfined compression tests were performed to establish the stabilizer's optimum dosage. The soil slopes were prepared with the ideal dosage of (4%) for 90, 80, 70 and 60 Degree slopes using cube mould of size 150 mm. The samples were then tested using compression machine corresponding to 7 and 3 days of curing periods. From laboratory studies, it was noticed that stabilization of red soil using Alccofine escalates its strength with increasing curing period. The factor of safety was estimated using Plaxis software and also verified with Taylor's method. The highest values of factor of safety of 2.77 and 2.59 were obtained respectively with Taylor's method and Plaxis 2D software corresponding to a slope angle of 70 Degree.

Keywords Alccofine · Plaxis 2D · Slope Angle · Compression Testing Machine

1 Introduction

The Slope failure occurs due to shear stresses exceeding the shear strength and downward movements of materials due to gravity causing Slope failure. Therefore, factors that tend to reduce the shear strength or increase the shear stresses increase the chances of a slope failure. Various processes can lead to reduction

B. R. Vinod (✉) · J. Akshit Jain
Department of Civil Engineering, B M S Institute of Technology and Management, Bangalore, India
e-mail: vinodbr@bmsit.in

J. Sumalatha
Department of Civil Engineering, M S Ramaiah Institute of Technology, Bangalore, India
e-mail: sumalatha@msrit.edu

A. Vilventhan et al. (eds.), *Advances in Construction Materials and Management*, Lecture Notes in Civil Engineering 346,
https://doi.org/10.1007/978-981-99-2552-0_19

in the shear strength of soil mass. Common factors that reduce shear strength of rock mass are Increased cracking, pore pressure, swelling, strain softening, creep under sustained loads, leaching, decomposition of clayey rock fills, cyclic loading and weathering. The growing need for fill slopes and engineered cut on construction projects has enhanced importance of understanding analytical approaches, investigation tools, and stabilisation procedures for resolving slope stability issues. Alccofine is a micro fine cement-based product used for injection grouting and soil stabilisation in subterranean tunnels. Alccofine is a new era, ultra-fine supplementary cementitious concrete material (SCM) of particle size significantly finer than other SCM's like cement, rice husk ash, fly ash etc. manufactured in India. Impact of Alccofine and lime on geotechnical characteristics for the black cotton soil were investigated by Godayal et.al. [1] and found that there was decrease in liquid limit, plasticity index and plastic limit. The Alccofine was also used in concrete mix as a mineral admixture [7].

Dev Sharma [8] and Soni and Singh [9] studied the use of Alccofine with marble dust for expansive soil stabilization. They found that Alccofine is effective in reducing the volume change behaviour and increasing UCS and CBR of soil. [10], studied the use of Silica Fumes and Alccofine in stabilization of red soil. They noticed that there was a significant reduction in swelling characteristics of the soil and increase in unconfined compressive strength. [11], studied the engineering characteristics of clayey soil with varying amounts of Alccofine and fly ash. [12], studied the potential applications of Alccofine and Rice Husk Ash (RHA) for clayey soil stabilization. Saurav and Ashok [13], studied the application of fly ash and Alccofine as a replacement to cement. They found that Alccofine increases the resistance to segregation, passing ability and filling ability. As well as Alccofine is cost effective and economical as it is cheaper than cement.

The previous research on slope stability analysis using various methods and software tools [2–6, 14] is useful to choose the appropriate method to perform the stability analyses. In the present study, the red soil amended with Alccofine was investigated to ascertain the consequence of amendment on stability of soil slope. Alccofine was added in different proportions and unconfined compression tests were conducted to know the optimum dosage. Later, analysis was conducted corresponding to different angles of slopes using Plaxis software and Taylor's method. From this analysis, it was found that FoS of the slopes enhances on addition of Alccofine to red soil.

2 Materials and Methods

2.1 Red Soil

Red soil used for study was gathered from the fly over construction at MAJ. Sandeep Unnikrishnan Road, dairy circle, Bengaluru, India. The soil is classed as low plasticity clay (CL) by the Indian Standard classification system. The MDD were estimated as 1.8 g/cc and OMC were estimated as 15% for the soil.

2.2 Amended Red Soil

Soil was mixed with 1%, 2%, 3%, 4% and 5% Alccofine to identify the optimum dosage of the amendment. The compaction test results for different mixtures are provided in Table 1 and Fig. 1. It can be observed that MDD reduces and OMC increases with increase of Alccofine content. From the unconfined compression test results, the unconfined compressive strengths corresponding to 0, 1, 2, 3, 4 and 5% Alccofine dosage in soil were estimated as 218, 267, 300, 350, 475 and 375 kPa respectively. The variation in unconfined compressive strength (UCS) with Alccofine content is shown in Fig. 2. The UCS was found high at 4% Alccofine content. This may be due to the formation of pozzolanic compounds formed by the calcium present in Alccofine and silica in soil. Once the silica is in soil is completely utilized for pozzolanic reactions, further addition of Alccofine will not increase the strength as pozzolanic reactions will not be taken place. From the results of unconfined compression tests and compaction tests, 4% is considered as the optimum dosage of Alccofine for this soil.

Table 1 Compaction test results of Alccofine amended soil samples

Specifications	MDD (g/cm^3)	OMC (%)
Soil + 1% Alccofine	1.76	15
Soil + 2% Alccofine	1.66	17
Soil + 3% Alccofine	1.66	18
Soil + 4% Alccofine	1.67	17
Soil + 5% Alccofine	1.63	18

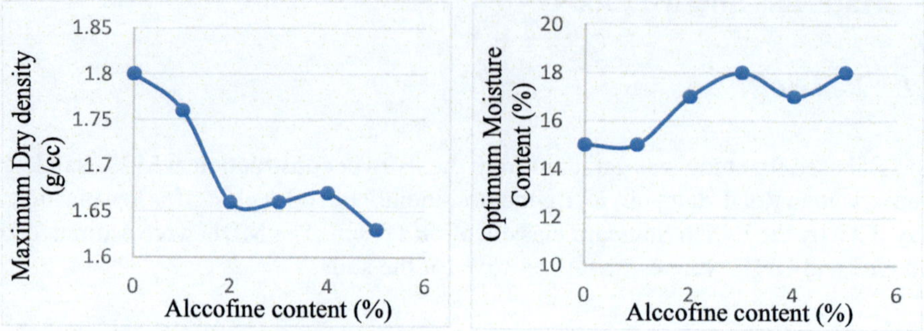

Fig. 1 Compaction characteristics of amended soil

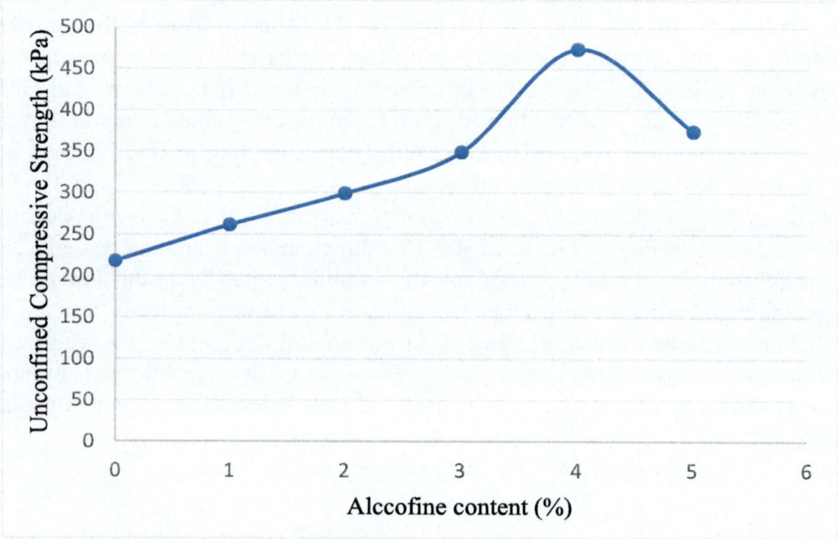

Fig. 2 Variation in UCS with Alccofine content

2.3 *Slope Stability Analyses*

Paxis 2D software and Taylor's method were used to analyse the slope stability. The results of laboratory tests on amended soil (with different percentages of Alccofine) were given as input to the software and the factor of safety were estimated corresponding to various angles of soil slopes.

Table 2 Compression tests result of soil samples with different slope angles

Slope Angle	Period of Curing	Max Load (KN)	Horizontal Deformation (mm)	Vertical Deformation (mm)
90^0	0 Days	1.5	302	1163
	3 Days	35	675	1054
	7 Days	43	765	499
80^0	0 Days	2	212	1049
	3 Days	36	186	600
	7 Days	25.5	729	543
70^0	0 Days	2	326	927
	3 Days	21	149	629
	7 Days	14	124	514
60^0	0 Days	0.8	495	1492
	3 Days	9.5	134	440
	7 Days	12	123	567

3 Results and Discussions

3.1 Amended Red Soil

The cubes of amended soil (with optimum dosage of Alccofine) were prepared using 150 mm cube moulds and they are subjected to curing for 3 and 7 days. Generally, the soil slopes with slope angles less than 60^0 are less prone to failures and hence the slope stability analyses were performed on varying slope angles i.e., 60^0, 70^0, 80^0 and 90^0. The results of compression test conducted on sampled prepared with optimum dosage (4%) is provided in Table 2.

3.2 Results of Undrained Triaxial Tests

Undrained Triaxial tests were performed on Alccofine amended soil samples corresponding to varying curing periods. Shear strength parameters for the amended soil samples corresponding to different curing periods are given in Table 3. The cohesion were estimated as 5.1 kPa and angle of internal friction were estimated 39^0 for the red soil without Alccofine.

Table 3 Shear strength parameters for amended samples

Curing period in days	Cohesion (KPa)	Angle of internal friction
0	11.9	24^0
3	17.9	36^0
7	22.2	28^0

Table 4 Factors of Safety of Soil with 4% Alccofine

Degree	Stability	Height of embankment (cm)	Dry unit weight (kN/m^3)	Cohesion (C) (kN/m^2)	Factor of safety by Taylor's method	Factor of safety by Plaxis 2D Software
90	5.6	15	17.65	17.9	2.43	2.32
80	6.8		17.65	17.9	2.63	2.48
70	8.2		17.65	17.9	2.77	2.59
60	8.7		17.65	17.9	2.23	2.18

3.3 *Factors of Safety*

From the obtained test results of various moulds with different proportions of samples, the Factors of Safety were calculated by using Taylor's Formula (Eq. 1) and Plaxis 2D. The Taylor's stability number (Sn) can be used to estimate the factor of safety (F_C) from the following equation (Eq. 1). The Sn values can be taken from the Taylor's stability chart where the stability numbers were estimated using Friction circle method.

$$Sn = \frac{C}{F_C \gamma H} \tag{1}$$

where 'C' is the cohesion of soil, 'γ' is the unit weight of soil and 'H' is the height of the slope.

The estimated values of factors of safety using both the methods are given in Table 4. From the results obtained, it can be observed that the factor of safety is maximum corresponding to a slope angle of 70^0.

4 Conclusion

The experimental investigation conducted on the Alccofine amended soil samples revealed enhanced geotechnical properties of soil on addition of Alccofine. The optimum dosage of Alccofine for the red soil studied was observed to be 4%. The compression tests and undrained Triaxial tests were conducted on the amended samples prepared with optimum dosage of Alccofine. The results of these tests were given as input to the Plaxis 2D software and an embankment was modelled to know the factors of safety corresponding to different slope angles. From the analyses conducted on the Alccofine amended soil slopes, it was found that the highest values of factor of safety was obtained with 70^0 slope which are 2.77 and 2.59 respectively with Taylor's method and Plaxis 2D software.

References

1. Godayal, A., Kapoor, A., & Garg, P. (2018). Effect of alccofine, lime on geotechnical properties of cohesive soil. *Int. J. Creative Res. Thoughts (IJCRT), 6*(2).
2. Kumar, V., & Parkash, V. (2015). A model study of slope stability in mines situated in south India. *Advances in Applied Science Research, 6*(8), 82–90.
3. Umrao, R.K., Singh, R., Ahmad, M., & Singh, T.N. (2011). Stability analysis of cut slopes using continuous slope mass rating and kinematic analysis in Rudraprayag district, Uttarakhand. *Geomaterials, 2011.*
4. Mahdi Rasouli, M., Mohammad, M., & Kambiz, M. (2011). Design of overall slope angle and analysis of rock slope stability of chadormalu mine using empirical and numerical methods. *Engineering, 2011.*
5. Jacob, A., Thomas, A.., Nath, A.G., & Arsinq, M.P. (2018). Slope stability analysis using Plaxis 2D. *International Research Journal of Engineering and Technology (IRJET), 5*(4), 3666–3668
6. Bidisha, C., & Shivananda, P. (2017). Two-dimensional slope stability analysis by plaxis-2d. *International Journal for Research in Applied Science and Engineering Technology, 5*(9), 871–877.
7. Gupta, S., Sharma, D. S., & Sharma, D. D. (2013). A review on alccofine, a supplementary cementitous material. *International Journal of Modern Trends in Engineering and Research, 3*(2), 148–153.
8. Dev, S., & Sharma, N. (2017). Stabilization of expansive soil with marble dust and alccofine. *International Journal of Advanced Research in Science Engineering, 6*, 1212–1219.
9. Soni, M. K., & Singh, S. (2019). Statistical interpretation of marble dust and alccofine for soil stabilization. *International J Innovative Technology and Exploring Engineering, 8*(7), 1609–1613.
10. Mir, S.H., & Sharma, N. (2019). Amendment of Geotechnical properties of clayey soil using brick kiln dust and alccofine 1101. *International Research Journal of Engineering Technology, 6*
11. Sambyal, L.S., & Sharma, N. (2018). Utilizing fly ash and alccofine for efficient soil stabilization. *International Journal of Science Engineering Research, 9*(3).
12. Rather, S. A., Sharma, N., & Najar, I. A. (2019). Effects of rice husk ash (RHA) and alccofine-1101 on stabilization of clay soil. *International Research Journal of Engineering Technollogy, 6*, 474–478.
13. Saurav, G., & Ashok. K.G. (2014). Experimental study of strength relationship of concrete cube and concrete cylinder using ultrafine slag Alccofine. *International Journal of Scientific and Engineering Research, 5.*
14. Sumalatha, J. (2022). Stability Analysis of Slopes at a Landslide Prone Area: A Case Study on the Landslide at Madikere, India. In *Stability of Slopes and Underground Excavations* (pp. 105–114). Springer, Singapore.
15. Sumalatha, J. (2022). Slope Stability Analysis Using In Situ Ground Reinforcement Techniques for a Landslide Prone Area at Madikere, India. In *Geohazard Mitigation* (pp. 173–183). Springer, Singapore.

Evaluation of Axial Load Carrying Capacity of CFST Columns for Geometrical Cross-Sections

B. Ravi, K. Haribabu, and T. Chandrasekhar Rao

Abstract CFST columns typically have thinner or smaller cross sections of steel and concrete than steel-only columns in terms of structural advantages and synergistic effects on composite construction. This paper focuses on an experimental and FEM analysis of the load carrying capacity of CFST columns under pure axial load. Variables considered for the study are length/diameter (L/D), diameter/thickness ratio (D/t) and shape of the cross-section (rectangle, square, and circular). The experimental values of the circular CFST column's ultimate load (P_u) are compared using the formulas given in two codal provisions: Euro Code-4 (EC4:2004); and AISC (2005). The CFST column elements are also modelled and their strength is evaluated using SAP-2000. It is found that the load carrying capacity of the in filled columns increased with increasing D/t ratio and decreased with increase in L/D ratio. The circular sections outperformed the other two geometrical sections in terms of load carrying capacity for various D/t and L/D ratios. In terms of predicting column capacity, the codal provisions were often very accurate. The experimental and numerical analysis for predicting P_u values are closely related.

Keywords Concrete in-filled steel tube (CFST) · Geometrical cross-sections · L/D · D/t · Ultimate axial load carrying capacity

1 Introduction

CFST structures are composite steel–concrete structure presently employed in civil constructions. They are steel tubes with a concrete core inside. Because of high static and seismic resistant qualities, including strength and ductility, hollow structural steel and concrete are combined to create cross-sections with circular, rectangular, and square dimensions. The concrete core of a short, concentrically loaded concrete-in-filled steel tube experiences more confining stress than regular RCC columns

B. Ravi (✉) · K. Haribabu · T. Chandrasekhar Rao
Civil Engineering Department, Bapatla Engineering College, Bapatla, A.P., India
e-mail: bayyavarapuravi@gmail.com

A. Vilventhan et al. (eds.), *Advances in Construction Materials and Management*, Lecture Notes in Civil Engineering 346,
https://doi.org/10.1007/978-981-99-2552-0_20

confined by lateral ties and a spiral. The reason for this could be that the core of concrete can withstand more stress in a tri-axial state.

For last two decades, experimental research carried out on composite CFST columns. (Shakir-Khalil and Mouli [7]; Zeghiche and Chaoui [1]; Shanmugam and Lakshmi [3]; Zeghiche [2]; Neogi et al. [14] Shakir-Khalil and Zeghiche [8]).The key factors that most significantly affect the behavior of composite columns, according to a review of the literature, are material properties and shape of cross section. The D/t ratio, column geometry, and L/D are the crucial geometrical characteristics that affect the strength.

There are many studies on in filled steel tubes. Some of the relevant studies are mentioned here. The concrete core bears the axial load, preventing or prolonging local tube buckling. Lu and Zhao [4] and Roeder et al. [5] studies shows that interface characteristics affect the axial load capacity of the circular CFST columns, which have concrete strengths in the range of 28.6 MPa–47.2 MPa. According to the test results, large diameter tubes and large D/t ratios has lower bond capacity for circular CFST columns.

A 65 MPa cylinder concrete strength is used in a study on high strength CFST members by Johansson and Gylltoft [15]. Based on the test results, simultaneous loading of the steel and concrete sections has no effect on the behaviour of strength. The bond strength, on the other hand, had a significant impact on the confinement effects and, as a result, the mechanical behaviour effected for those columns if only concrete core is loaded. Tsuda et al. [9], Yamamoto et al. [10], You Q et al. [11], and Yu Z et al. [12, 13] investigated different shapes of CFST columns for different cases.

Fifteen circular columns with different compressive strengths are tested by Giakoumelis and Lam [16]. Unlike the results of Johansson and Gylltoft [15], it demonstrates that while the difference in axial load capacity caused by bonding loss is insignificant for low or normal strength concrete, it is significant for high strength concrete.

Taking into account the various steel tube thicknesses, Guler et al. [17] investigated the behaviour of square CFST columns with a concrete strength of 115 MPa. The test findings demonstrate that for square, high-strength CFST columns, the change in ultimate load caused by bonding loss is significant.

2 Design Codes

The major codes in practice for predicting ultimate load of CFST columns are American code (AISC-2005) and Euro Code (EC-4) apart from Australian, Japan and Chinese codes.

2.1 Euro Code-4(2004)

The EC4 (2004) [5] design code is the latest worldwide standard for composite construction. EC4 covers CFST sections with or without reinforcing, and also fully and partly encased steel sections in concrete. The EC4 provisions are only relevant to CFST members of concrete strength below 50 MPa. According to the EC4, the square CFST columns capacity is given by

$$p_u = A_c f_{ck} + A_s f_y$$

The EC4 considers the confining effect of the steel tube while evaluating the axial load capacity of the circular columns. To calculate the axial load capacity of circular CFST columns, following formula is provided:

$$p_u = \left(1 + \eta_c \frac{t}{D} \frac{f_y}{f_{ck}}\right) A_c f_{ck} + \eta_a A_s f_y$$

$$\eta_c = 4.9 - 18.5\lambda + 17\lambda^2 \geq 0$$

$$\eta_a = 0.25(3 + 2\lambda) \leq 1$$

$$\lambda = \sqrt{\frac{N_{pl,R}}{N_{cr}}} \leq 0.5$$

$$N_{pl,R} = A_c f_{ck} + A_s f_y$$

$$N_{cr} = \frac{\pi^2 (EI)_e}{l^2}$$

$$(EI)_e = A_s E_s + 0.6 E_{cm} A_c$$

where η_c is magnification factor for concrete core and η_a is the enhancement factor for tube

λ = The relative slenderness
l = Effective length of column
E_{cm} = Secant elastic modulus of concrete
$(EI)_e$ = Composite flexural stiffness;

2.2 AISC Code (2005)

According to American code for compact sections.

Table 1 Geometrical properties for all type of cross-sections

L(mm)	D(mm)	t(mm)	D/t	L/D
700	33	3.92	8.42	21.21
700	42	3.5	12	16.67
700	50	3.25	15.38	14
700	60	3.3	18.18	11.67
700	76	3.8	20	9.21

$$P_u = C_2 A_c f_c + A_s f_y$$

$$C_2 = 0.85 \text{ Rectangular sections}$$

$$= 0.95 \text{ Circular sections}$$

3 Experimental and Analytical modelling of CFST Columns

3.1 *Materials*

A total of 15 steel tubes were divided into three sets, i.e., each set consists of five specimens based on geometrical cross-sections, namely circular, square, and rectangular. The variables diameter-thickness (D/t = 8.42 to 20) and length-diameter ratio (L/D = 9.2 to 21.2) are used to analyse each set of CFST. The materials used for the casting of CFST columns are M30 grade concrete and mild steel (Fe250) (Table 1).

3.2 *Experimental Set Up*

Figure 2 depicts a 28 day cured CFST specimen utilised for testing. The CFST columns are tested on a capacity 1000kN computerised universal testing machine, which can give stress–strain plots and axial displacements. All CFST specimens are placed carefully to avoid eccentric loading. Base plates are installed on top and bottom of the sample to provide uniform loading to the concrete and steel. By applying displacement control method, axial deformation values of all columns are noted down (Figs. 1, 3 and 4).

Fig. 1 Casted specimens

Fig. 2 Testing of CFST (D/t=18.18 & L/D=11.67)

Fig. 3 During testing (D/t=18.18 & L/D=11.67)

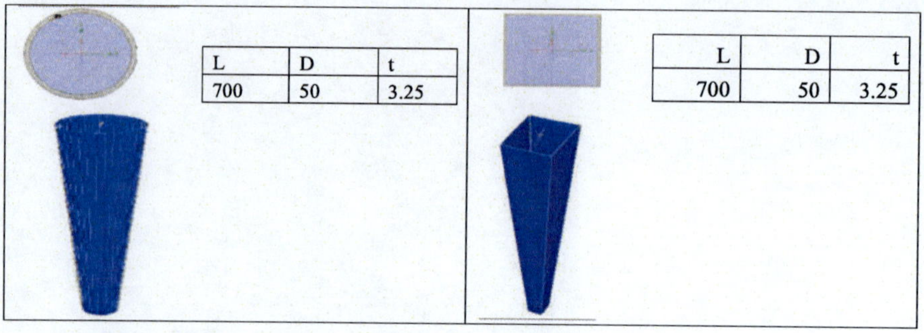

Fig. 4 FEM modeling of Circular & Rectangular columns

3.3 FEM Modeling

SAP2000 is used to model all 15 specimens in this study. Section designer is used to create the core of concrete and surrounded steel tube for the analysis. In the analysis, the end conditions for all the specimens are considered as pinned. The ultimate loads in SAP for all specimens are predicted with known axial displacements obtained from experimental program conducted for the same columns.

3.4 Discussion and Outcomes

The test and FEM results are presented in Table 2. A comparison between ultimate load of all type CFST are shown in the Fig. 5

From Table 2, it is evident that circular columns show better performance compared to the other two types of columns. The reason is that in circular sections, the hoop stress enhances the confinement ratio compared to square and rectangular sections. In square and rectangular sections, the corner experiences stress concentration, resulting in outward buckling of the tube. P_u increases by 15 to 18% for circular columns and 20 to 22% for square and rectangular columns. The comparison between

Table 2 Ultimate loads obtained from Test and FEM modeling

D/t	L/D	P_u -Exp(kN)			P_u -SAP(kN)		
		Circular	Square	Rectangle	Circular	Square	Rectangle
8.42	21.21	101.05	86	78.82	120.5	94.6	89.85
12.00	16.67	146.45	122	111.30	164.3	137.86	126.88
15.38	14.00	174.45	142.35	136.07	187.5	161.57	155.12
18.18	11.67	252.65	210.4	204.65	267.3	231.44	233.3
20.00	9.21	294.95	247.8	236.25	302.2	275.06	269.3

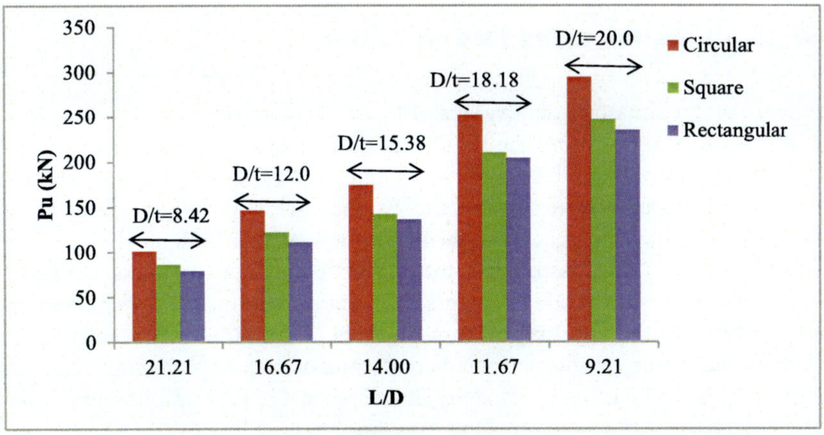

Fig. 5 Variation of ultimate load with Respect to D/t and L/D

ultimate values from experimental and numerical analysis shows good agreement. In the analysis a perfect bond is assumed between concrete and steel, which is not truly existed between the composites, leads to lower experimental values. The circular sections are considered for further study as they are better at resisting applied loads (Fig. 6).

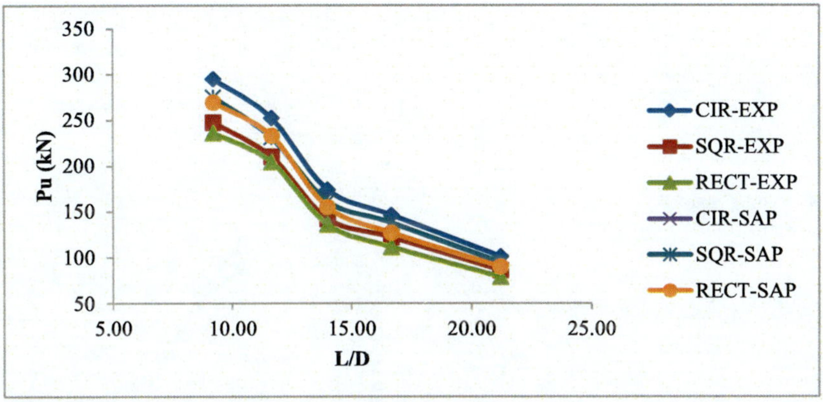

Fig. 6 Comparison of ultimate loads with respect to L/D

4 Data Validation using Design Codes

The experimental and numerical values of P_u of circular sections are validated with EC-4 and AISC (2005). From the Figs. 7 and 8 it is noted that at D/t = 15 and L/d = 15 shown close match between the P_u values of experimental and two design codes. The EC-4 code overestimates the P_u compare with experimental ultimate loads, the reason may be enhancement factors η_{a} & η_c for both steel and concrete are considered in EC-4 on assumption that perfect bond is existing between concrete and steel. Whereas AISC code considers the enhance factor only for concrete there by underestimates ultimate loads. From the Figs. 7 and 8, it is observed that L/D and D/t has influence on ultimate loads determined from experimental, numerical and design codes. The ultimate capacity decreases as the L/D ratio increases, as the column behaviour shifts from crushing (yielding) to buckling (Table 3).

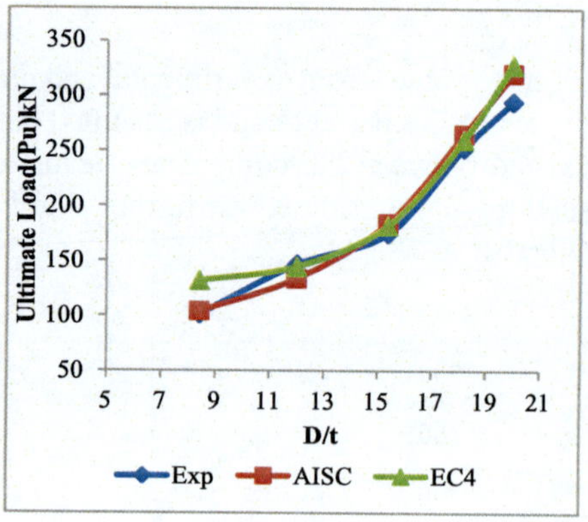

Fig. 7 Validation of P_u values w.r.t D/t

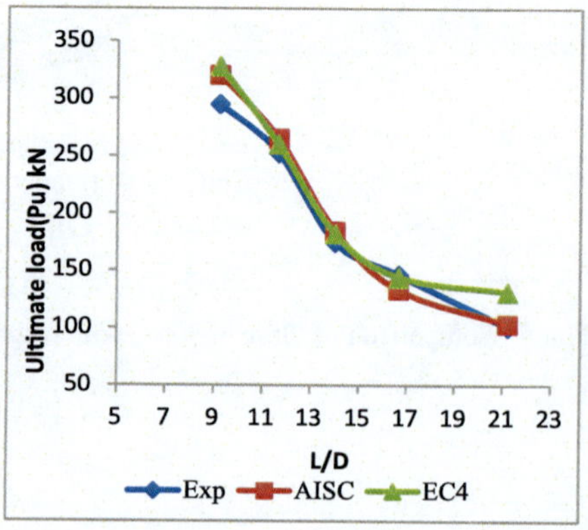

Fig. 8 Validation of P_u values w.r.t L/D

Table 3 Failure loads of circular columns as per design codes

L(mm)	D(mm)	t(mm)	D/t	L/D	Pu-Exp(kN)	Pu-AISC(kN)	Pu-EC4(kN)
700	33	3.92	8.42	21.21	101.05	103.75	131.52
700	42	3.5	12.00	16.67	146.45	133.31	144.19
700	50	3.25	15.38	14.00	174.45	184.08	181.76
700	60	3.3	18.18	11.67	252.65	265.08	259.92
700	76	3.8	20.00	9.21	294.95	320.35	327.98

5 Conclusions

1. CFST circular sections are better in load bearing than square and rectangular sections for different ratios of L/D and D/t considered. The increase in load carrying ability of circular sections is 15 to 22% more than that of square and rectangular sections.
2. The experimental and numerical predictions of ultimate loads for the various geometrical shapes investigated show a fair correlation. The average % variation between experimental and SAP values differ only by 10–15.
3. The determined values from experimental, EC4 and AISC codes are well in agreement for the case L/D = 15 and D/t = 15. The deviation between the experimental and codal values is observed after the ratio of 15.

References

1. Zeghiche, J., & Chaoui, K. (2005). An experimental behaviour of concrete-filled steel tubular columns. *Journal of Constructional Steel Research, 61*, 53–66.
2. Zeghiche, J. (1988). Concrete-filled composite columns. MSc Thesis, *University of Manchester*
3. Shanmugam, N. E., & Lakshmi, B. (2001). State of the art report on steel-concrete composite columns. *Journal of Constructional Steel Research, 57*, 1041–1080.
4. Lu, Z.H. and Zhao, Y.G. (2008) Mechanical behavior and ultimate strength of circular CFT columns subjected to axial compression load. In The 14th World Conference on Earthquake Engineering, Beijing, China
5. Roeder, C. W., Cameron, B., & Brown, C. B. (1998). Composite action in concrete filled tubes. *Journal of Structural Engineering, ASCE, 125*(5), 477–484.
6. Eurocode 4 (2004). Design of composite steel and concrete structures. Part 1.1, General rules and rules for buildings. *European Committee for Standardization*: British Standards Institution
7. Shakir-Khalil, H., & Mouli, M. (1990). Further tests on concrete-filled rectangular hollow-section columns. *The Structural Engineer, 68*(20), 405–413.
8. Shakir-Khalil, H., & Zeghiche, J. (1989). Experimental behaviour of concrete- filled rolled rectangular hollow section columns. *The Structural Engineer, 67*, 346–353.
9. Tsuda, K., Matsui, C., & Ishibashi, Y. (1995). Stability design of slender concrete filled steel tubular columns. In Proceedings of Fifth Asia-Pacific Conference on Structural Engineering and Construction (EASEC-5)
10. Yamamoto, T., Kawaguchi, J., & Morino, S. (2002). Experimental study of the size effect on the behaviour of concrete filled circular steel tube columns under axial compression. *Journal of Structural and Construction Engineering, 561*, 237–244. (in Japanese).

11. Yu, Q., Tao, Z., & Wu, Y. X. (2008). Experimental behaviour of high performance concrete filled steel tubular columns. *Thin-Walled Structures, 46*, 362–370.
12. Yu, Z., Ding, F., & Lin, S. (2002). Researches on behavior of high-performance concrete filled tubular steel short columns. *Journal of Building Structure, 23*(2), 41–47. (in Chinese).
13. Yu, Z. W., Ding, F. X., & Cai, C. S. (2007). Experimental behavior of circular concrete-filled steel tube stub columns. *Journal of Constructional Steel Research, 63*, 165–174.
14. Neogi, P. K., Sent, H. K., & Chapman, J. C. (1969). Concrete-filled tubular steel columns under eccentric loading. *The Structural Engineer, 47*(5), 187–195.
15. Johansson, M., & Gylltoft, K. (2002). Mechanical behavior of circular steel-concrete composite stub columns. *Journal of Structural Engineering, ASCE, 128*(8), 1073–1081.
16. Giakoumelis, G., & Lam, D. (2004). Axial capacity of circular concrete-filled tube columns. *Journal of Constructional Steel Research, 60*(7), 1049–1068.
17. Guler, S., Lale, E., & Aydogan, M. (2013). Behaviour of SFRC filled steel tube columns under axial load. *International Journal of Advanced Steel Construction, 9*(1), 14–25.

Upgrading Recycled Aggregates in Concrete by Using Waste Plastics

Nagala Pavani Pujitha, Sabbisetti Vamsi, Kandisa Vikas, Gadi Sangeetha, and Kumar Raja Vanapalli

Abstract The current study investigates the potential upgradation of recycled coarse aggregates using plastics to be used in concrete. The study investigates the properties of plastic-coated recycled aggregates (PRA) in comparison to the properties of recycled aggregates (RA) and conventional coarse aggregates (CA) using different tests including specific gravity, impact strength, hardness, shape, elongation, and water absorption. The plastic-coated recycled aggregates (PRA) were produced from adding melted PP waste (~180 °C) to demolished aggregates at 5% (w/w) followed by mixing for homogenous coating on the surface. The PRA exhibited 10–24% higher specific gravity, impact strength and hardness with reduced water absorption (by 50%) as compared to RA. The study further investigated the strength and durability properties of concrete cubes produced using PRA as compared to RA and CA. Although the RA exhibited failures for M15, M20 and M25 at compressive strengths of 13.8, 18.6, 20.4 N/mm^2 respectively after 28 days curing, cubes made of PRA exhibited around 21–40% higher values relative to RA and insignificant difference as compared to that of CA. Moreover, the cubes made of PRA also exhibited relatively better results in slump test, Vee Bee test, and Water absorption test. However, chemical leachability and toxicity studies are necessary before commercial application.

Keywords Waste plastics · Recycled Aggregates · Sustainable concrete · Waste Management

N. P. Pujitha · S. Vamsi · K. Vikas · G. Sangeetha · K. R. Vanapalli
Vignan's Institute of Information Technology, Visakhapatnam 530046, Andhra Pradesh, India

K. R. Vanapalli (✉)
National Institute of Technology Mizoram, Aizawl 796012, Mizoram, India
e-mail: kumar.ce@nitmz.ac.in

A. Vilventhan et al. (eds.), *Advances in Construction Materials and Management*, Lecture Notes in Civil Engineering 346,
https://doi.org/10.1007/978-981-99-2552-0_21

1 Introduction

An unprecedented shortage in construction materials amid growing demands warrants a need for sustainable materials [1]. The coarse aggregates obtained from construction and demolition waste which can be reused as a construction material are referred to as recycled coarse aggregate. The use of these recycled aggregates in concrete has been extensively studied in previous literature [2, 3]. The characteristics of recycled aggregate influence various properties of concrete including the strength, distribution size, and water absorption capabilities and so its performance [4]. Moreover, recycling of construction and demolished waste helps in resource conservation and reducing environmental pollution thereby promoting sustainable development [5]. Compared to the conventional natural aggregates recycled aggregates exhibit relatively higher water absorption and lower specific gravity. The recycled aggregates also have relatively higher porosity. However, the variability in meeting the strength and durability requirements has been perceived as a major limitation.

With the rise in the production of plastic waste, a need for sustainable waste management techniques is essential to reduce their environmental impact [6]. So, use of waste plastics in building materials has been lately promoted to reduce the load of conventional materials while providing a solution as opposed to irregular waste disposal. One of the techniques include upgradation of aggregates with plastic coating which was observed in previous studies to improve the durability of flexible pavements [7, 8]. However, there is limited literature available on the use of plastic-coated aggregates in concrete. Moreover, the effects of plastic-coated recycled aggregates in concrete have not been extensively studied before.

The objective of the current study is to investigate the potential upgradation of these recycled aggregates using plastics to be used in concrete. The study investigates the properties of plastic-coated recycled aggregates (PRA) in comparison to the properties of recycled aggregates (RA) and conventional coarse aggregates (CA) using different tests including specific gravity, impact strength, hardness, shape, elongation, and water absorption.

2 Materials and Methods

2.1 *Waste Plastic (Polypropylene)*

Waste polypropylene was collected from the solid waste stream of Vignan's Institute of Information Technology, Visakhapatnam college campus. Polypropylene was selected for its abundance in the waste stream. The plastic was initially shredded into smaller pieces (size <10 cm) and washed thoroughly to remove any contamination. The samples were oven dried (60 °C–24 h) to remove any additional moisture and stored before further processing.

Fig. 1 Process of making plastic coated recycled aggregates

2.2 *Recycled Aggregates (RA)*

Recycled aggregates are the material recovered from construction and demolition waste (C&D waste) after crushing the waste concrete into a material with specified size and quality. Recycled aggregates in the current study have been recovered after processing the C&D waste obtained from the surrounding areas of Duvvada, Visakhapatnam region. After crushing and sieve separating (>4.75 mm) the aggregates from the concrete, they were manually shaken thorough mechanical processes to remove any excess cement attached and stored for further processing.

2.3 *Preparation of Plastic-Coated Aggregates (PCA)*

Approximately 50 gm of shredded polypropylene plastic was initially melted at around 180 °C to obtain melted plastic. To the melted plastic, around 1 kg of recycled aggregate was mixed (5% (w/w) of plastic in RA) thoroughly following which the heat is reduced. The coated aggregates were removed from the bowl and spread to dry in the open air. After complete drying, the plastic-coated recycled aggregates are stored for further processing. The process of making PCA is depicted in Fig. 1.

2.4 *Aggregate Analyses*

To understand the difference between the recycled aggregates used in the study to conventional aggregates, the relative properties of normal aggregates, demolished aggregates, plastic coated aggregates were analysed using impact test and Los Angeles abrasion test. These tests are done to know the relative strength and properties of the plastic-coated recycled aggregates and recycled aggregates in comparison to the conventional aggregates. Although these tests are specifically designed for testing highway materials, for convenience purposes to know relative strength and

Fig. 2 Aggregate analyses techniques **a** Impact test **b** Los Angeles test

durability properties, the properties were adopted in this study even for testing in concrete. Both the tests were conducted in triplicates to obtain consistent results as per the standard procedures specified by IS: 2386 1963. Figure 2 (a)–(b) depicts the testing methods for impact and Los Angeles tests.

2.5 Analyses of Concrete Cubes

The concrete mixtures were prepared following the design of normal concrete method as per Building Research Establishment (BRE) [9]. In this research three different grades of concrete – M15, M20, M25 were selected. A total of 18 different samples were prepared using the combinations of three grades of concrete (M15, M20, M25), three types of aggregates (normal, recycled, and plastic-coated recycled). The cubes were casted and tested after 7 days and 28 days of curing. The formworks used were cubical of 150 mm × 150 mm × 150 mm. All the samples were triplicated for consistent values. The samples were tested for compression test, slump cone test, vee bee test, and water absorption test using standard procedures. Generally, the plastic-coated aggregates surface is smooth and when we mixed plastic coated aggregates there is no bonding in concrete. So, to avoid that issue we are mixing 25% of plastic-coated aggregates and 75% of normal aggregates. This proportion was determined based on experimental trial-and-error basis. Figure 3 (a)–(b) depicts the testing methods for compressive, Vee-bee and slump cone tests.

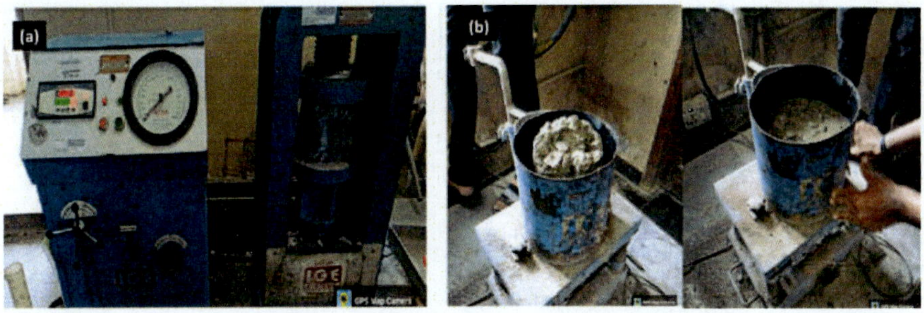

Fig. 3 Cube testing techniques **a** Compressive test **b** Vee-Bee test

3 Results and Discussion

3.1 Relative Aggregate Properties

The properties of normal coarse aggregates, recycled aggregates (RA), and plastic-coated recycled aggregates (PCA) analysed according to the IS 2386 are presented in Table 1. The specific gravities of plastic-coated aggregates were 10–24% higher than the normal and recycled aggregates. Since the specific gravity of a material is indicative of its strength properties, it can be deduced that the relative strength of the recycled aggregates has increased with the coating of plastics. It could be due to the filling of voids in the recycled aggregates with molten plastics resulting in lower porosity and thereby higher bulk density. Similarly, the plastic-coated aggregates exhibited significantly better results in impact test and hardness test as compared to the conventional and recycled aggregates. While the aggregate impact value reflects the toughness of the aggregates to face the impacts, the los angeles abrasion value represents the ability of the aggregates to sustain wear and tear. While the lowest aggregate impact value of 2.19% and hardness value of 15.6% were exhibited by PCA, RA performed the worst with 11.2% and 36.8%, respectively. In the shape and elongation test, normal aggregates exhibited significantly better results as compared to the recycled and plastic-coated recycled aggregates. The hydrophobicity of the plastic has resulted in the lowest water absorption of PCA (0.56%) as compared to that of RA (1.2%) and normal coarse aggregate (0.81%) which should enhance the quality of the concrete.

Table 1 Properties of different types of aggregates (Mean ± S.D.)

S.No	Type of test	Procedure	Normal Coarse Aggregates	Recycled Aggregates	Plastic Coated Recycled Aggregates	Standard Values
1	Specific Gravity	IS: 2386 (Part 3) 1963	2.63 ± 0.12	2.33 ± 0.09	2.89 ± 0.1	2.5–3
2	Aggregate Impact Value	IS: 2386 (part 4) 1963	6.12 ± 0.35%	11.2 ± 0.76%	2.19 ± 0.41%	<30%
3	Los Angels Abrasion Value	IS: 2386 (part 4) 1963	25.8 ± 1.97%	36.8 ± 2.65%	15.6 ± 1.5%	<35%
4	Flakiness Index	IS: 2386 (part 1) 1963	25.1 ± 2.2%	53.6 ± 3.8%	51.7 ± 4.6%	–
5	Elongation Index	IS: 2386 (part 1) 1963	15.3 ± 0.4%	8.2 ± 0.8%	9.9 ± 0.8%	–
6	Water Absorption	IS: 2386 (Part 3) 1963	0.81 ± 0.07%	1.2 ± 0.06%	0.56 ± 0.04%	<1%

3.2 Relative Concrete Cubes Performance

3.2.1 Compressive Strength

Table 2 shows the results for compressive strength in 7 days of curing and 28 days of curing for different types of grades of concrete by mixing of normal aggregates, recycled aggregates, and plastic-coated recycled aggregates. The results of the compressive strength depicted an increase in the compression strength of the cubes by 5–26% (7-day curing), 21–40% (28-day curing) with the use of PRA as compared to conventional recycled aggregates. Moreover, the PRA cubes exhibited relatively no significant difference with respect to the cubes made from normal aggregates. On the other hand, the cubes made from RA failed as substantially lower compressive loads. Similar results were obtained at both 7 day and 28 day curing periods. Although the highest 28-day compressive strengths 21 MPa (M15), 26.5 MPa (M20), 33.6 MPa (M25) were observed with the use of normal aggregates, no significant difference was observed while using PRA. The enhanced quality of the recycled aggregates with the coating of plastics is also reflected in the increased compressive strength of the cubes.

Table 2 Compressive strength testing of samples of varying concrete grades and aggregate mixtures

S.No	Compressive Strength (MPa)	Grade of Concrete	Normal Coarse Aggregates	Recycled Aggregates	Plastic Coated recycled Aggregates
1	7 days curing	M15	10.1 ± 0.5	9.1 ± 0.6(Failure)	11.5 ± 0.9
		M20	13.3 ± 0.9	12.3 ± 0.9 (Failure)	13.2 ± 1.1
		M25	17.8 ± 1.1	16.2 ± 0.8 (Failure)	17.6 ± 1.3
2	28 days curing	M15	21 ± 0.8	13.8 ± 1.0 (Failure)	20.3 ± 1.2
		M20	26.5 ± 1.3	18.6 ± 0.9 (Failure)	26 ± 1.5
		M25	33.6 ± 1.3	20.4 ± 1.2 (Failure)	31.6 ± 1.3

3.2.2 Slump Cone, Vee Bee, and Water Absorption Tests

The results of the slump cone, vee bee, and water absorption tests are presented in Table 3. Slump cone test is performed to determine the workability of concrete, while Vee bee test indicates the mobility and compactibility of the freshly mixed concrete [10]. The concrete made from PRA exhibited the highest slump values of 40 mm, 40 mm, 50 mm for M15, M20, M25 concretes respectively. However, from the slump values for normal and recycled aggregates it can be determined that these concretes have very low workability as compared to the PRA concrete which depicted low workability. The reason could be the hydrophobic nature of the PRAs and irregular surface of the coated aggregate leading to better grip resulting in higher slump values [11]. It was also observed that there were no significant differences between the concretes made from different aggregates as per the Vee Bee test.

The results of the water absorption test indicated that the recycled aggregates had the highest water absorption percentages (0.76–0.98%) which is representative of their high porosity. With the addition of plastic coating the water absorption of

Table 3 Slump cone, Vee Bee, and Water absorption test values of samples of varying concrete grades and aggregate mixtures (Mean ± S.D)

S.No	Type of Test	Grade of Concrete	Normal Coarse Aggregates	Recycled Aggregates	Plastic Coated Recycled Aggregates
1	Slump Cone Test	M15	20 ± 3 mm	30 ± 3 mm	40 ± 5 mm
		M20	20 ± 4 mm	10 ± 2 mm	40 ± 3 mm
		M25	20 ± 2 mm	20 ± 3 mm	50 ± 6 mm
2	Vee-Bee Test	M15	6 ± 1 s	6 ± 0 s	5 ± 1 s
		M20	7 ± 1 s	9 ± 1 s	7 ± 1 s
		M25	10 ± 0 s	8 ± 1 s	8 ± 0 s
3	Water Absorption Test	M15	0.81 ± 0.02%	0.87 ± 0.05%	0.79 ± 0.02%
		M20	0.91 ± 0.05%	0.98 ± 0.07%	0.93 ± 0.03%
		M25	0.5 ± 0.04%	0.76 ± 0.09%	0.59 ± 0.08%

the concrete has significantly reduced by 5–22% as depicted in PRA concrete which reduces the chances of failure. The differences between the PRA concrete and normal conventional concrete were mostly insignificant and inconsistent. This is consistent with the other test results observed in the study. However, relatively lower hardness can be a limitation that can affect fracture toughness of concrete mix.

4 Conclusion

The current study investigates the possibility of upgrading recycled aggregates by using waste plastic for increasing the quality of the concrete. In the current study, the relative properties of recycled aggregates were studied in comparison with the recycled aggregates and plastic-coated aggregates. It was observed that the plastic-coated aggregates had higher specific gravity, impact strength, and abrasion resistance, and lower water absorption value as compared to its precursor (recycled aggregate). The coating of plastic on the aggregates might have reduced the porosity of the voids and increased the strength of the aggregates possibly due to improved binding of materials. The concrete produced from PRAs also exhibited significantly better compressive strength, workability and water resistivity as compared to the concrete made from recycled aggregates. Moreover, the PRA cubes exhibited almost similar properties with no significant difference as compared to the cubes made from conventional aggregates. This indicates the coating the recycled aggregates with plastic might compensate the strength and durability lost from using the recycled aggregates in concrete. So, the use of plastic-coated recycled aggregates in concrete can be an environmentally and economically sustainable alternative to conventional concrete where the benefits of construction & demolition waste and plastic waste management can be utilised without compromising on the quality of the concrete. Future research can be concentrated on optimising studies on proportion of plastic, recycled aggregate percentages for improving the cost dynamics of concrete. Also, long term chemical leachability and environmental risk assessment studies are required as future research before proposing for commercial application.

References

1. Mirza, F. A., & Saif, M. A. (2010). Mechanical properties of recycled aggregate concrete incorporating silica fume. In *Proceedings of the 2nd international conference on sustainable construction materials and technologies, Coventry university and the university of Wisconsin Milwaukee centre for by-products utilization* (pp. 28–30).
2. Abdelfatah, A. S., & Tabsh, S. W. (2011). Review of research on and implementation of recycled concrete aggregate in the GCC. In *Advances in civil engineering 2011*.
3. Tempest, B., Cavalline, T., Gergely, J., & Weggel, D. (2010). Construction and demolition waste used as recycled aggregates in concrete: solutions for increasing the marketability of

recycled aggregate concrete. In *Proceedings of the concrete sustainability conference, Tempe, AZ, USA* (Vol. 1315)
4. Malešev, M., Radonjanin, V., & Marinković, S. (2010). Recycled concrete as aggregate for structural concrete production. *Sustainability, 2*, 1204–1225.
5. Parekh, D. N., & CD, D. M. (2011). Assessment of Recycled Aggregate Concrete" publish by JERS. Vol II/Issue I/January-March.
6. Vanapalli, K., Samal, B., Dubey, B., & Bhattacharya, J. (2019). Emissions and environmental burdens associated with plastic solid waste management (pp. 313–342). https://doi.org/10.1016/B978-0-12-813140-4.00012-1.
7. Chavan, M. A. J. (2013). Use of plastic waste in flexible pavements. *International Journal of Application or Innovation in Engineering and Management, 2*, 540–552.
8. Hake, S. L., Damgir, R. M., & Awsarmal, P. R. (2020). Utilization of plastic waste in bitumen mixes for flexible pavement. *Transportation Research Procedia, 48*, 3779–3785.
9. Teychenné, D. C., Franklin, R. E., Erntroy, H. C., & Marsh, B. K. (1975). Design of normal concrete mixes. HM Stationery Office London, UK.
10. Phul, A. A., Memon, M. J., Shah, S. N. R., & Sandhu, A. R. (2019). GGBS and fly ash effects on compressive strength by partial replacement of cement concrete. *Civil Engineering Journal, 5*, 913–921.
11. Shahidan, S., Pullin, R., Holford, K. M., Bunnori, N. M., & Noorsuhada, M. N. (2013) Quantitative evaluation of the relationship between tensile crack and shear movement in concrete beams. In *Advanced materials research* (vol. 626, pp. 355–359). Trans Tech Publ.

Mechanical Performance of Rice Husk Ash Made Geopolymer Concrete with Partially Replaced Steel Slag as Fine Aggregate

Prabu Baskar, Shalini Annadurai, Dineshkumar Gopalakrishnan, and Erukulla Kethan Chandra

Abstract This paper describes the experimental investigation on the production of Rice Husk Ash (RHA) based geopolymer concrete and partially replacing steel slag as fine aggregate. The research examines the properties of the geopolymer concrete (GPC) that has been activated with an alkaline solution, both when it is fresh and hardened. Different RHA content (R10, R15, R20, R25, R30, and R35) were used, and here R represent Rice Husk Ash, and numerical represent the % of RHA used, as well as the molarity of the NaOH was 10. In addition, other industrial by-products were also used as cementitious materials, such as Ground Granulated Blast Furnace Slag (GGBS), Fly ash with RHA. The GGBS content is fixed at 10%, and different Fly ash content (F80, F75, F70, F65, F60, and F55) were used. These proportions were arrived at to determine the optimum mix proportion. Mechanical properties such as flexural strength, compressive strength, split tensile strength, and bond strength were tested at three, seven, and 28 days. Additionally, the sorptivity, water absorption, and acid attack characteristics of durability were also examined. The test result reveals that on the 28th day, there is an increase in mechanical characteristics by adding 15% of RHA. After getting the optimum mix proportion, it has been used as a controlled mix proportion to replace steel slag as fine aggregate partially. The steel slag was replaced in different percentages (5%, 10%, 20%, 30%, 40%, and 50%) with fine aggregate. In geopolymer concrete, RHA can be used as source material. The partial replacement of steel slag by up to 30% resulted in enhanced mechanical and durability properties. It is suggested that 15% of RHA with 30% of steel slag as a fine aggregate replacement can be used to produce geopolymer concrete with a strength of 25 MPa.

Keywords Geopolymer · Steel Slag · Rice Husk Ash · Strength · Durability

---

P. Baskar (✉) · S. Annadurai
Sona College of Technology, Salem, Tamil Nadu, India
e-mail: prabu.civil@sonatech.ac.in

D. Gopalakrishnan · E. K. Chandra
Vaagdevi College of Engineering, Warangal, Telangana, India

A. Vilventhan et al. (eds.), *Advances in Construction Materials and Management*,
Lecture Notes in Civil Engineering 346,
https://doi.org/10.1007/978-981-99-2552-0_22

1 Introduction

Probably the most often used construction material is concrete. The production of Portland concrete significantly supports carbon dioxide emissions. Global warming is caused by releasing ozone-depleting compounds into the atmosphere, such as carbon dioxide, responsible for around 65 percent of the earth's global temperature increase. Numerous efforts are being undertaken to reduce the use of Portland cement in concrete. These endeavors incorporate using valuable establishing materials, likely fly ash, GGBS, and rice husk ash, and discovering elective fasteners for Portland concrete. The geopolymer breakthrough could reduce cement's CO_2 emissions to the environment by 80% in terms of reducing the risky atmospheric concentration.. A significant portion of cropland is taken up by steel slag, which poses a serious risk of environmental damage. The utilization of steel slag as a resource has grown in importance with the growth of the metallurgy sector to ensure the long-term health of the environment and the steel industry [1, 2].

Steel slag, one of the most promising waste materials, has been the subject of numerous studies [3–6]. Lower compressive strength can be achieved by using steel slag as a fine aggregate replacement while considering size distribution [7]. Currently, steel slag is used as fine and coarse aggregates for concrete, as an asphalt pavement material for highways, as a raw material for clinker, as ballast for railroads, and as a filler material in various excavations. Because it was once released into the environment, steel slag posed a significant environmental threat. Using steel slag in these applications has been found to partially address environmental issues brought on by waste while also slowing down the quick consumption of our limited natural resources. A range of 10-15% of crude steel is thought to be produced as steel slag [8]. In Europe alone, 15.7 million tonnes of steel slag are produced as a by-product, of which 11.5 million tons are recycled, according to World Steel Association data [9]. In China, which makes the most steel worldwide, around 30% of the steel slag is reused. As well as to put it another way, China produces 80 million tonnes of steel slag yearly [10]. Steel production in the world's second-largest country is India, and its integrated steel factories generate about 19 million tonnes of steel slag annually as solid waste. Considering the anticipated expansion projects in Indian steel mills to produce 300 million tonnes of steel by FY 2030, steel slag generation is predicted to reach 60 million tonnes by that year [11]. After metal recovery, most steel slag is dumped or used as fill near steel mills, occupying sizable land areas and polluting air, land, and water.

Recent research has also revealed the usage of steel slag as a fine aggregate in concrete. Local, unprocessed steel slag was used in concrete mixtures by Qasrawi et al. [12], who discovered that doing so improved the concrete's compressive and tensile strengths. According to Mo et al. [1], using steel slag aggregates in place of limestone and sand improved the compressive strength of concrete that had undergone CO_2 curing. Manjunath and Narasimhan [13] reported that alkaline solutions could effectively activate slag. Additionally, using steel slag as a fine aggregate can create masonry materials with a standard strength [14]. Currently, Most steel slag is employed as soil conditioners or embankment fillers. It is also used to make cement and asphalt concrete [15–17]. Pavements made of cement and asphalt perform better when steel slag is used, according to a literature assessment [18–20, 22]. Steel slag is produced in small-sized aggregates to build a framework of asphalt/cement fine fillers, increasing the microscopic qualities of a good binder. Up to the soil subgrade layer, the surface course has all attested to its versatility in application. Pavement development has a significant impact on reducing the stockpiles of steel slag as infrastructure demand rises. Further systematic research is required for the steel slag-modified geopolymer with tertiary blended binder materials. Milled steel slag fine particles were partially replaced for river sand in this study. Then, the mechanical characteristics of RHA and steel slag-based specimens were established. In this research, the investigation was made to examine the geopolymer concrete's strength and long-term properties with RHA and steel slag.

2 Materials Used

2.1 Ground Granulated Blast Furnace Slag (GGBS)

A by-product of the production of iron and steel, molten iron slag, is quenched to create glassy, granular material that is then dried and processed into a fine powder to create GGBS. Depending on the raw materials used in the iron-making process, the chemical composition of slag varies greatly. CaO (30% to 50%), SiO_2 (28% to 38%), Al_2O_3 (8% to 24%), and MgO are the most common components of blast furnace slag (1% to 18%). The expansion of GGBS in geopolymer concrete increases the solid's strength, and further recovery of geopolymer concrete at room temperature is possible [23]. GGBS comprises primary compound ingredients that are comparable to those in conventional Portland concrete but in different quantities. The chemical compositions of GGBS are presented in Table 1.

Table 1 Oxide Composition of Fly ash, GGBS, and RHA

Elements as Oxides	Class F Fly ash %	Requirement as Per IS 3812 - 2003	GGBS %	Requirement as per IS 12089 – 1987	RHA %
SiO_2	55.99	$SiO_2 > 35\%$	41.24	$(CaO + MgO + Al_2O_3)/SiO_2 \geq 1.0$	88.64
Al_2O_3	15.23	Total – 70%	20.64		1.23
Fe_2O_3	21.78		7.28		1.19
CaO	0.17	-	25.45		1.09
MgO	2.45	$<5\%$	2.93	$<17\%$	1.76
LOI	0.62	$<5\%$	Nil	-	$<6\%$

Fig. 1 SEM Image of FA, RHA, and GGBS [25]

2.2 *Rice Husk Ash (RHA)*

Rice husk contains SiO_2 when burned at a controlled temperature, most of which is in amorphous form. Rice husk makes up the world's paddy by 20% output, which totals 500 MT [24]. Organic volatile matter accounts for 75% of the total, with the remaining 25% being inorganic. As a result of the burning, rice husk ash is produced. Ash is made from the husk, which loses 25% of its weight. RHA is amorphous and has high silica content. RHA's amorphous silica will perform pozzolanic action by reacting with cementitious binders. Geopolymer is a high-strength, high-hardness, and chemically stable alumino silicate material. Table 1 provides the oxide compositions of RHA.

2.3 *Fly Ash*

In coal-fired power plants, fly ash is the main product of the coal-burning process. Fly ash is divided into Class C and Class F categories. Geopolymer binders made from fly ash have exceptional durability.

Among these are high protection from salt silica response, better corrosive opposition and high obstruction than fire, low carbonation, and restricted sulfate attack. Fly ash production will rise in the future, particularly in countries like China and India. A 45-micron filter was used to sieve the fly ash. The Blaine fineness of fly ash is 2351 cm^2/g with 2.46 specific gravity, both of which provided chemical evidence of the presence of iron. SEM photographs of the Fly Ash showed that the particles were spherical and ranged in size from 0.5 to 300 m, as evident from Fig. 1. Table 1 provides the oxide compositions of fly ash.

2.4 *Steel Slag*

Typical components of slag include metal oxides and silicon dioxide. On the other hand, metal sulfides and elemental metals can be found in steel slag. Steel slag is

Table 2 Steel Slag's Physical properties

Property	Values
Crushing strength	29.1
Specific Gravity	3.5
Water absorption	2%
Unit Weight (kg/m^3)	1600 - 1920

frequently used to remove trash during the metal smelting process, but it may also be utilized for other things, including helping to regulate temperature and reducing any re-oxidation of the liquid metal that is removed from the furnace that is used to create solid metal [26]. Steel slag can be used as fine aggregates with low fineness. More than 80% of the chemical components of steel slag are made up of CaO, SiO_2, Fe_2O_3, and MgO. The steel slag's physical properties are illustrated in Table 2. Basic oxygen furnace (BOF) slag from single-smelting steel was used in this investigation and was supplied by a local supplier, Salem.

2.5 *Alkaline Liquid*

Due to their lower cost than potassium-based activators, sodium-based activators were chosen for the experiment. Economical suppliers in Salem are where can buy sodium silicate and sodium hydroxide. The sodium hydroxide solids were of a research facility grade in pellets structure with 99% virtue, acquired from neighborhood providers. The pellets were dissolved in water to create the sodium hydroxide (NaOH) arrangement. Depending on molarity level, a solution's mass of sodium hydroxide solids fluctuated. For instance, a solution of sodium hydroxide with an 8 M concentration contains $8 \times 40 = 320$ grammes of solid sodium hydroxide (in pellet form) per liter of distilled water, where 40 is the molecular weight of NaOH. The sodium silicate arrangement (water glass) was utilized, which was obtained from local suppliers. In terms of mass, the sodium silicate arrangement's synthetic composition was $Na_2O = 8\%$, $SiO_2 = 28\%$, and water $= 64\%$. The interaction of the sodium silicate and sodium hydroxide arrangements shapes the soluble fluid.

2.6 *Super Plasticizer (SP)*

It acts as a water reducer equipped to diminish water substance by around 30%. The productivity of the SP is reduced when it is expanded to a firm, solid blend. Adding 1 to 3% by wt. of concrete is volume, depending on how strong the combination is. This research uses a superplasticizer, specifically Conplast SP 430, to get functional binder material at low w/c proportion. The system comprises massive particles (colloidal size) that break down in the water to create particles with an anion.

Table 3 Properties of fine and Coarse aggregate

Description	Coarse Aggregate	Fine Aggregate
Type	Uncrushed	Uncrushed (natural)
Specific gravity	2.6	2.54
Bulk Density	1765 kg/m3	1668 kg/m3
Fineness modulus	6.45	2.76
Grading zone	-	Zone II
Maximum size	20 mm	-

2.7 *Fine and Coarse Aggregates*

Aggregates act as the concrete's fundamental parts, which extraordinarily change the geopolymer concrete's strength, thickness, and properties. The fine aggregate utilized in the research was river sand conforming to zone II according to IS: 383:1970. In addition to river sand, steel slags were replaced with different proportions from 10 to 50%. In this research, a locally available coarse aggregate of size 10 - 20 mm was utilized. The properties of the FA and CA are given in Table 3.

3 Mix Proportion and Test Set up

3.1 *Mix Proportions*

The mix proportions for geopolymer concrete made with various combinations of FA, RHA and GGBS are shown in Table 4. The percentage of RHA particles that were substituted for fly ash ranged from 10 to 35, with 5% increments. The R15 mix with the controlled percentage of binders was produced from several trial mixes, providing a compressive strength of 21 MPa. In the R15 combination, the FA was partially replaced by steel slag (SS) to a proportion of 10% to 50%. Table 5 presents the mix proportions for RHA-based geopolymer concrete made with steel slag as fine aggregate. The manufacturing process of GPC with steel slag is illustrated in Fig. 2. Alkaline solutions (sodium based) are kept ready a day before making concrete. NaOH solution is diluted with one liter of water to make the required level. Polymerization occurs when NaOH and Na_2SiO_3 solutions blend, liberating high heat. The solution is utilized after 24 h. The liquid component is prepared during cast by mixing alkaline liquid, superplasticizer, and extra water (if required).

Table 4 Mix Proportion for different mixes of Geopolymer Concrete

Mix Id	RHA	GGBS	Fly Ash	CA	FA	NaOH	Na_2SiO_3	Water	SP
R10	39.43	39.43	315.44	1293.60	554	45.10	112.60	59.14	11.83
R15	59.15	39.43	295.72	1293.60	554	45.10	112.60	59.14	11.83
R20	78.86	39.43	276.01	1293.60	554	45.10	112.60	59.14	11.83
R25	98.57	39.43	256.30	1293.60	554	45.10	112.60	59.14	11.83
R30	118.29	39.43	236.58	1293.60	554	45.10	112.60	59.14	11.83
R35	138.00	39.43	216.87	1293.60	554	45.10	112.60	59.14	11.83

Table 5 Mix proportion for partially replaced steel slag as fine aggregate

Mix Id	Fly Ash	GGBS	RHA	FA	Steel Slag	CA	NaOH	Na_2SiO_3	Water	S.P
R15	295.725	39.43	59.145	554	0	1293.6	45.1	112.6	59.14	11.83
GP1	295.725	39.43	59.145	498.6	55.4	1293.6	45.1	112.6	59.14	11.83
GP2	295.725	39.43	59.145	443.2	110.8	1293.6	45.1	112.6	59.14	11.83
GP3	295.725	39.43	59.145	387.8	166.2	1293.6	45.1	112.6	59.14	11.83
GP4	295.725	39.43	59.145	332.4	221.6	1293.6	45.1	112.6	59.14	11.83
GP5	295.725	39.43	59.145	277	277	1293.6	45.1	112.6	59.14	11.83

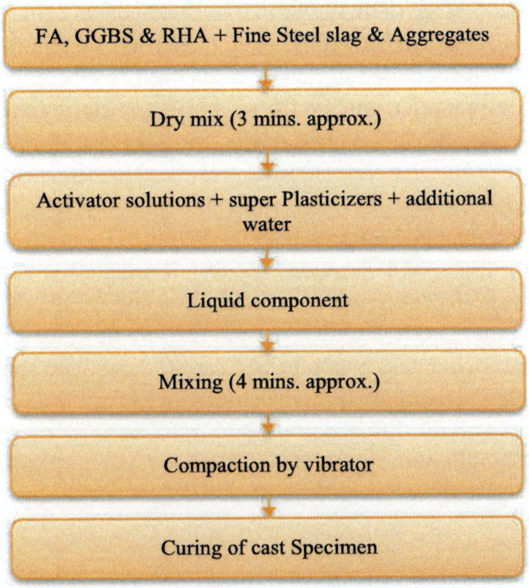

Fig. 2 Manufacturing process of GPC with steel slag

3.2 Test Specimens and Methods

The test specimens for the compressive strength test were made of a cube having a size of 100 mm × 100 mm × 100 mm. Three cubes were cast and examined for each mix percentage after three, seven, and twenty-eight days. Cast PVC pipe moulds with a 200 mm height were used to create the specimens for the split tensile strength test, which were made of 100 mm diameter cylinders. Three cylinders were made and tested for each mix proportion at 7 and 28 days. Cast iron steel moulds were used to create the test specimens for the flexural strength test, which were prisms with dimensions of 100 mm × 500 mm by 100 mm. Three quantities of prisms were cast and tested for each mix proportion at 7 and 28 days.

A dry specimen's ability to absorb water is measured by its absorption rate. Therefore, it is determined how much water is present overall in the permeable pore space. ASTM C1585 - 13 states that the test entails drying the specimen to a constant weight, weighing it, submerging it in water for the designated time, and then weighing it again. Absorption is the change in weight expressed as a percentage of the starting weight, and the size, position, and continuity of the pores define permeability. Concrete with low water permeability prevents water intrusion. The ability to freeze and thaw is not present. The sorptivity test measures the speed at which water is absorbed. Concrete's sorptivity is tested using a specimen cylinder with dimensions of 100 mm in height and 50 mm in diameter that has been curing for 28 days. Before usage, the specimen must be dried at 110 °C in an oven for a day, then cooled for 24 h in a dry atmosphere. When hardened concrete is allowed to come into contact with water at a depth of 5 mm, water will be drawn in by capillary action. Every half-hour, the amount of water absorbed is noted. Concrete is naturally alkaline.

pH is the parameter that is utilized to indicate its nature. The pH value for excellent concrete should range from 9 to 11. Acidic materials are those whose pH value is less than 7. Concrete may be damaged by acids, rendering it unusable. Clause 8.2 of IS 456 - 2000 provides the requirements for concrete's durability. 100 mm of size standard cubes were manufactured and used as specimens. By measuring the weight of the 28-day-cured specimen and submerging it in 0.1 normal sulphuric acid, an acid resistance test is conducted for 90 days. The same test is carried out by immersing several samples in hydrochloric acid for 90 days. After the acid immersion period, the weight difference is again measured using hydrochloric acid. The compressive strength of the specimen is assessed and contrasted with the compressive strength of a sample not immersed in acid.

4 Results and Discussion for RHA-Based Geopolymer Concrete

4.1 Compressive Strength

Table 6 presents the compressive strength development of RHA-based geopolymer concrete in different curing periods. The objective of the present research is to achieve M20-grade concrete. The test results revealed that adding RHA content in a greater % decreases the compressive strength. The ultimate strength of 25.2 MPa was reached in the R25 mix. Due to the larger proportion of RHA used, there was a variation in compressive strength. Because of loose and porous microstructure during ambient curing, the increase in RHA lowered compressive strength and slowed geopolymerization. The more reactive silica was added to the RHA, which increased the concentration of Si–O-Si linkages in the geopolymer matrix. Another factor is that excessive RHA has a larger surface area than other source materials, producing extremely dense products by occupying the pores. The Si–O-Si bonds were stronger than the Si–O-Al and Al–O–Al ones. The compressive strength of 21 MPa was attained in 15% of the replacement of RHA itself. So the R15 blend is considered the controlled proportion for partially adding steel slag with fine aggregate.

Table 6 Strength development of RHA-based Geopolymer concrete

Mix Id	Compressive Strength (MPa)			Split Tensile Strength (MPa)		Modulus of rupture (MPa)		Density (kg/m^3)		
	3d	7d	28d	7d	28d	7d	28d	3d	7d	28d
R10	5.26	13.68	20.0	1.49	2.40	2.37	2.86	2492	2458	2468
R15	6.10	14.37	21.0	1.61	2.46	2.43	2.93	2468	2598	2480
R20	5.95	15.46	22.6	1.59	2.56	2.52	3.04	2434	2216	2460
R25	6.63	17.24	25.2	1.67	2.70	2.66	3.21	2452	2492	2408
R30	5.80	14.57	21.3	1.57	2.48	2.44	2.95	2312	2312	2411
R35	5.50	14.03	20.5	1.52	2.43	2.40	2.90	2354	2202	2490

4.2 Split Tensile Strength and Modulus of Rupture

The development of modulus of rupture and split tensile also follows a similar trend to compressive strength. Table 6 shows the split and flexural strength achieved curing at 7 and 28 days. It is seen from the test results that up to the replacement of 25% of RHA to fly ash increases the strength, but beyond that, there are decreases in the split and flexural strength in both curing periods. Because it was challenging to break the Si and Al bonds of the monomer from the fly ash particles to initiate the reaction, the tests' results reveal the amount of increased tensile and flexural strength in the fly ash mix was less during ambient curing. The highest split and flexural strength values were 2.70 MPa and 3.21 MPa, respectively.

4.3 Density

The density at various ages of RHA-based GPC specimens during ambient curing is presented in Table 6. At seven days, the RHA specimens' density ranged from 2492 to 2354 kg/m^3, and at 28 days, it ranged from 2468 to 2490 kg/m^3. The geopolymer concrete density increased as the proportion of RHA with fly ash and GGBS increased. The greater surface area of RHA particles that filled the pores in the concrete was considered to be the cause of the increase in density.

5 Results and Discussion for RHA-Based Geopolymer Concrete with Steel Slag

5.1 Compressive Strength

The compressive strength development of RHA-based geopolymer concrete with different proportions of steel slag as fine aggregate is illustrated in Table 7. The maximum compressive strength of 25.2 MPa was attained with 30% partial replacement of SS as fine aggregate. The formation of a calcite layer on the surface of the fine aggregates, which hinders the bonding of the paste with the aggregates, may cause a drop in compressive strength when steel slag is used as fine aggregate. Upon visual inspection of failed samples, instances of aggregate debonding from the paste were found. The binding strength between the paste and particles significantly influences the properties of concrete. Geopolymer concrete mixtures' mechanical and durability characteristics are adversely affected when the bond between the coating layer and paste is stronger than the bond between the coating layer and aggregate surface.

Table 7 Strength development of RHA-based Geopolymer concrete with SS as FA

Mix Id	Compressive Strength (MPa)			Split Tensile Strength (MPa)		Flexural Strength (MPa)		Density (kg/m^3)		
	3d	7d	28d	7d	28d	7d	28d	3d	7d	28d
R15	5.26	13.68	20.0	1.61	2.46	2.43	2.93	2498	2456	2480
GP1	6.10	14.37	21.0	2.53	2.73	2.99	3.52	2309	2397	2415
GP2	5.95	15.46	22.6	2.74	2.89	3.08	3.63	2223	2155	2407
GP3	6.63	17.24	25.2	3.15	3.43	3.19	3.75	2241	2431	2355
GP4	5.80	14.57	21.3	3.43	3.67	3.31	3.89	2101	2251	2358
GP5	5.50	14.03	20.5	3.97	4.11	3.40	4.15	2498	2456	2480

5.2 *Split Tensile Strength and Flexural Strength*

Split tensile and flexural strength development differs from compressive strength development in some ways. The split and flexural strengths attained at the 7 and 28-day curing periods are displayed in Table 7 below. The test findings show that adding more steel slag to the fine aggregate in concrete with an RHA foundation boosted the concrete's split and flexural strengths. The geopolymerization reaction proceeds quickly, increasing the split and flexural strengths. The split strength value was 4.11 MPa, and the flexural strength value was 4.15 MPa.

5.3 *Water Absorption*

Figure 3 depicts the % of water absorption and strength loss due to the different specimens submerging in water for a while. After 90 days of immersion, the specimen containing fly ash, GGBS, and RHA with 50% of steel slag had the highest water absorption value, at 6.15%. According to the test results, the RHA-based geopolymer concrete's ability to absorb water was found to be increased by adding more RHA to fly ash and GGBS. Compared to the other two pozzolanic materials, RHA had a very high degree of fineness. This is explained by a denser geopolymer matrix that causes fewer pores in the concrete. But the increase in the % of steel slag resulted in increased water absorption properties. The compressive strength also drastically decreased due to the more porous and increased water absorption capacity of RHA and steel slag after immersing in water for 90 days.

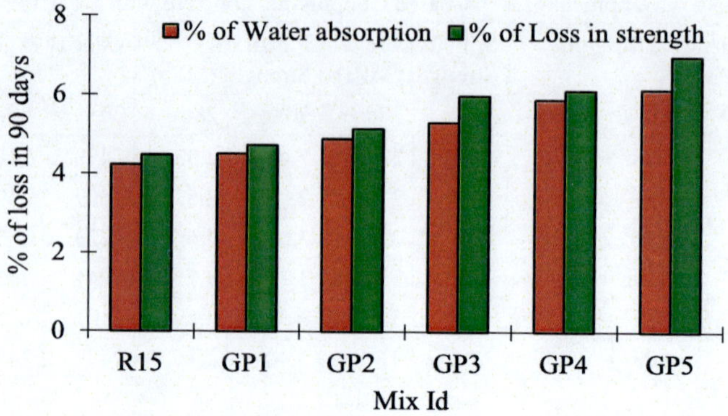

Fig. 3 % of Water absorption and Strength Loss

5.4 Sorptivity

Figure 4 shows the sorptivity value for replacing RHA-based geopolymer concrete with a mixture of steel slag for different percentages. Compared to concrete made from the controlled mix, every replacement mix exhibits an increased sorptivity value. The R15 mixture's maximum sorptivity value was found to be 0.1113 mm/min$^{1/2}$.

The replacement mix's combination of sorptivity has a minimum value of 0.1189 mm/min$^{1/2}$ for 10% steel slag replacement. 0.0418 mm/min$^{1/2}$ is the variance between the maximum and least sorptivity value. When its value is less, it indicates little water absorption by capillary action, which results in less porosity [20]. The findings demonstrated that the sorptivity value increases for replacement of geopolymer concrete when it is partially added with SS, and it was therefore concluded that as the % of substitution with steel slag rises, the porosity increases because the fine aggregates in steel slag are very coarse and resulted in increased porosity due to more absorption of water.

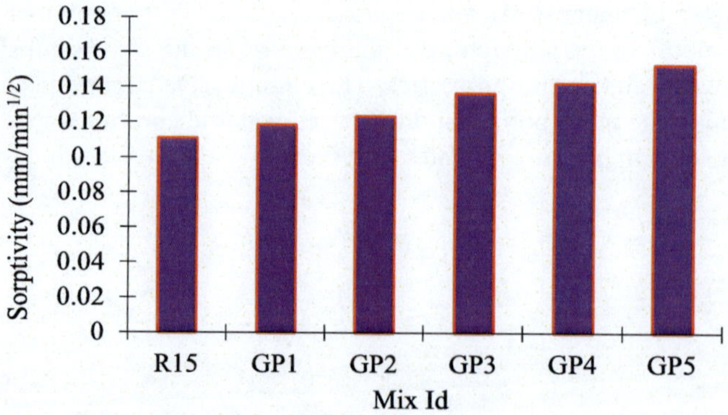

Fig. 4 Sorptivity values of different proportions of steel slag

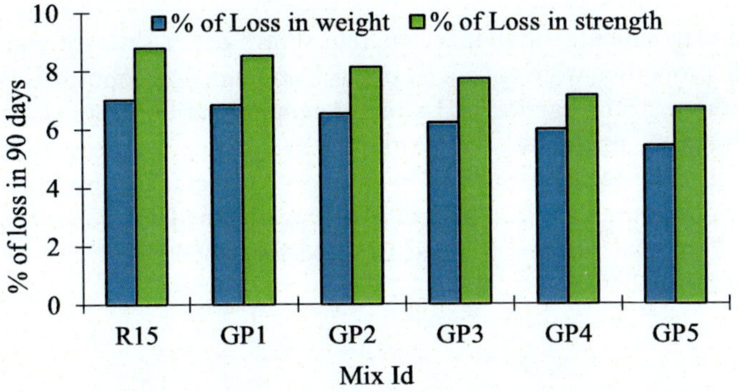

Fig. 5 % of weight loss and Strength Loss when immersed in HCl solution

5.5 *Acid Attack (HCl Solution)*

The weight of the specimen is decreased by the increasing percentage of replacement of steel slag when it is submerged in HCl solution for 90 days. Figure 5 displayed the weight loss % and strength loss % for various specimens.

Also discovered after and before immersion in HCl solution was the strength for each percentage of replacement specimens. When exposed to HCl for 90 days, the strength of the concrete with SS substitution decreases. However, the percentage of compressive strength loss was lower for a smaller amount of partial replacement of SS than for steel slag replacement % increases in the R15 mix. The mix GPC1 has a larger compressive strength reduction than the specimen not submerged in acid, with a compressive strength reduction of 8.53%. The strength of the specimen GPC5 was reduced by 6.73% when submerged. The control mix's compressive strength decreased 8.79% when immersed in an acid medium. The loss of strength is nearly the same for all specimens generated from SS. The compressive strength of GPC1 is shown to be lost at a much lower percentage than that of R15. This is so that steel slag can withstand weak acids and bases without losing their chemical resistance [27, 28]. The structures are dissolved by extremely potent acids or bases that cause ion exchange processes. So, steel slag-added geopolymer concrete exhibits superior acid resistance compared to RHA-based geopolymer concrete.

6 Conclusion

Geopolymer concrete is identified as an eco-friendly construction material with superior mechanical and durability properties. The investigation results indicate that the GPC made with 65% fly ash, 10% GGBS, and 25% RHA shows good mechanical and durability properties in the different curing periods. It is possible to attain the M20 grade of geopolymer concrete using the above mix proportions. In addition to the above, steel slag-added geopolymer concrete shows better strength and durability. The strength properties were enhanced by 26% compared to controlled geopolymer concrete without steel slag. The RHA-based geopolymer concrete made with 30% of partial replacement of steel slag exhibits better strength properties, but beyond that, the addition of steel slag resulted in an adverse effect on strength and durability properties. The optimum replacement of steel slag obtained was 30%. The durability test results show that steel slag added GPC performance better up to 30% of the replacement level of steel slag.

References

1. Mo, L., Zhang, F., Deng, M., Jin, F., AlTabbaa, A., & Wang, A. (2017). Accelerated carbonation and performance of concrete made with steel slag as binding materials and aggregates. *Cement and Concrete Composites, 83*, 138–145. https://doi.org/10.1016/j.cemconcomp.2017.07.018
2. Yi, H., Xu, G., Cheng, H., Wang, J., Wan, Y., & Chen, H. (2012). An Overview of utilization of steel slag. *Procedia Environmental Sciences, 16*, 791–801. https://doi.org/10.1016/j.proenv.2012.10.108
3. Brand, A. S., & Roesler, J. R. (2015). Steel furnace slag aggregate expansion and hardened concrete properties. *Cement and Concrete Composites, 60*, 1–9. https://doi.org/10.1016/j.cemconcomp.2015.04.006
4. Dash, M. K., Patro, S. K., & Rath, A. K. (2016). Sustainable use of industrial-waste as partial replacement of fine aggregate for preparation of concrete – a review. *International Journal of Sustainable Built Environment, 5*, 484–516. https://doi.org/10.1016/j.ijsbe.2016.04.006
5. Ferrerira, V. J., et al. (2015). Evaluation of the steel slag incorporation as coarse aggregate for road construction. *Journal of Cleaner Production, 130*, 175–186. https://doi.org/10.1016/j.jclepro.2015.08.094
6. Pang, B., Zhou, Z., & Xu, H. (2015). Utilization of carbonated and granulated steel slag aggregate in concrete. *Construction and Building Materials, 84*, 454–467. https://doi.org/10.1016/j.conbuildmat.2015.03.008
7. SubathraDevi, V., & Gnanavel, B. K. (2014). Properties of concrete manufactured using steel slag. *Procedia Eng., 97*, 95–104. https://doi.org/10.1016/j.proeng.2014.12.229
8. Nassar, Nedal T., Brainard, Jamie., Gulley, Andrew., Manley, Ross., Matos, Grecia., Lederer, Graham., Bird, Laurence R., Pineault, David., Alonso, Elisa., Gambogi, Joseph., Fortier, Steven M.: Evaluating the mineral commodity supply risk of the U.S. manufacturing sector. Science Advances, 6(8), eaay8647 (2020). doi:https://doi.org/10.1126/sciadv. aay8647.
9. World Steel Association. (2019). *Steel Statistical Yearbook 2019—Concise Version*. Belgium.
10. Kang, L., Du, H. L., Zhang, H., & Ma, W. L. (2018). Systematic research on the application of steel slag resources under the background of big data. *Complexity, 2018*, 1–12. https://doi.org/10.1155/2018/6703908
11. https://www.csir.res.in/csirblog/steel-slag-road-sustainable-valorization-steel-slag-road-making-aggregates
12. Qasrawi, H., Shalabi, F., & Asi, I. (2009). Use of low CaO unprocessed steel slag in concrete as fine aggregate. *Construction and Building Materials, 23*, 1118–1125. https://doi.org/10.1016/j.conbuildmat.2008.06.003
13. Manjunath, R., & Narasimhan, M. C. (2018). An experimental investigation on self-compacting alkali activated slag concrete mixes. *J. Build. Eng., 17*, 1–12. https://doi.org/10.1016/j.jobe.2018.01.009
14. Santamaría-Vicario, I., Rodríguez, A., Gutiérrez-González, S., & Calderón, V. (2015). Design of masonry mortars fabricated concurrently with different steel slag aggregates. *Construction and Building Materials, 95*, 197–206. https://doi.org/10.1016/j.conbuildmat.2015.07.164

15. Pasetto, M., Baliello, A., Giacomello, G., & Pasquini, E. (2017). Sustainable solutions for road pavements: a multi-scale characterization of warm mix asphalts containing steel slags. *Journal of Cleaner Production, 166*, 835–843. https://doi.org/10.1016/j.jclepro.2017.07.212
16. Rosales, J., Cabrera, M., & Agrela, F. (2017). Effect of stainless steel slag waste as a replacement for cement in mortars. Mechanical and statistical study. *Construction and Building Materials, 142*, 444–458. https://doi.org/10.1016/j.conbuildmat.2017.03.082
17. Biskri, Y., Achoura, D., Chelghoum, N., & Mouret, M. (2017). Mechanical and durability characteristics of High Performance Concrete containing steel slag and crystalized slag as aggregates. *Construction and Building Materials, 150*, 167–178. https://doi.org/10.1016/j.conbuildmat.2017.05.083
18. Liu, W., Miao, P., & Wang, S.-Y. (2017). Increasing microwave heating efficiency of asphalt-coated aggregates mixed with modified steel slag particles. *Journal of Materials in Civil Engineering, 29*, 1–9. https://doi.org/10.1061/(asce)mt.1943-5533.0002026
19. Li, W., Lang, L., Wang, D., Wu, Y., & Li, F. (2018). Investigation on the dynamic shear modulus and damping ratio of steel slag sand mixtures. *Construction and Building Materials, 162*, 170–180. https://doi.org/10.1016/j.conbuildmat.2017.12.026
20. Goli, H., Hesami, S., & Ameri, M. (2017). Laboratory evaluation of damage behavior of warm mix asphalt containing steel slag aggregates. *Journal of Materials in Civil Engineering, 29*, 2–10. https://doi.org/10.1061/(asce)mt.1943-5533.0001832
21. Furlani, E., et al. (2018). Synthesis and characterization of geopolymers containing blends of unprocessed steel slag and metakaolin: the role of slag particle size. *Ceramics International, 44*, 5226–5232. https://doi.org/10.1016/j.ceramint.2017.12.131
22. NetingerGrubeša, I., JelčićRukavina, M., & Mladenovič, A. (2016). Impact of high temperature on residual properties of concrete with steel slag aggregate. *Journal of Materials in Civil Engineering, 28*, 1–13. https://doi.org/10.1061/(asce)mt.1943-5533.0001515
23. Mehta, A., & Siddique, R. (2018). Sustainable geopolymer concrete using ground granulated blast furnace slag and rice husk ash: Strength and permeability properties. *Journal of Cleaner Production, 205*, 49–57. https://doi.org/10.1016/j.jclepro.2018.08.313
24. Wang, Y., Hu, S., & He, Z. (2019). Mechanical and fracture properties of fly ash geopolymer concrete addictive with calcium aluminate cement. *Materials, 12*(18), 2982. https://doi.org/10.3390/ma12182982
25. Annadurai, S., Rathinam, K., Kanagarajan, V.: Development of eco-friendly concrete produced with Rice Husk Ash (RHA) based geopolymer. Adv. Concr. Constr. **9**, 139–147 (2020). https://doi.org/10.12989/acc.2020.9.2.139.
26. Bai, T., Song, Z. G., Wu, Y. G., Hu, X. D., & Bai, H. (2018). Influence of steel slag on the mechanical properties and curing time of metakaolin geopolymer. *Ceramics International, 44*, 15706–15713. https://doi.org/10.1016/j.ceramint.2018.05.243
27. SubathraDevi, V. (2018). Durability properties of multiple blended concrete. *Construction and Building Materials, 179*, 649–660. https://doi.org/10.1016/j.conbuildmat.2018.05.056
28. Wasim, M., Ngo, T. D., & Law, D. (2021). A state-of-the-art review on the durability of geopolymer concrete for sustainable structures and infrastructure. *Construction and Building Materials, 291*, 123381. https://doi.org/10.1016/j.conbuildmat.2021.123381

Study on Influence of Secondary Treated Wastewater on Mechanical Properties of Concrete

T. N. Guruprasad and N. Uday Shankar

Abstract This research is to study the influence of secondary water-treated wastewater on mechanical parcels of concrete and compare it with mechanical parcels of freshwater concrete. Concrete is the most commonly used construction material in the world and is one of the largest water users. Approximately 150 L of water are consumed per cubic meter of concrete mix, not including other water consumption in concrete production. But water is a critical environmental issue, with a water supply and quality increasingly constrained around the world. The wastewater samples for our study (secondary treatment wastewater) were collected from the wastewater treatment plant at SSIT Tumakuru. The characteristics of secondary treated wastewater (STWW) were determined. In the M20 concrete mix, potable water (PW) was replaced by secondary treated wastewater (STWW), and the performance of fresh and hardened concrete was compared. The workability (slump value) of the concrete was reduced by 5 mm when STWW was used instead of PW. By replacing PW with STWW, the compressive strength and splitting strength of concrete increased by 1.13% and 1.06%, respectively.

Keywords Secondary treated wastewater (STWW) · Potable water (PW) · Freshwater (FW)

1 Introduction

In the construction industry, cement concrete is widely used as a foundation material, superstructure material, and roof material.

T. N. Guruprasad (✉) · N. U. Shankar
Department of Civil Engineering, Sri Siddhartha Institute of Technology, Tumkur 572105, karnataka, India
e-mail: guruprasadtn@ssit.edu.in

N. U. Shankar
e-mail: shankarudaycool@gmail.com

A. Vilventhan et al. (eds.), *Advances in Construction Materials and Management*, Lecture Notes in Civil Engineering 346,
https://doi.org/10.1007/978-981-99-2552-0_23

Sulfates in ground water cause concrete structures to deteriorate over time. It is vital that all living things have access to water. In the early days, water was used primarily for domestic purposes. Besides daily needs for water, modernization has increased the demand for water in construction, decoration, and sewage. The construction industry is required to consume a large amount of fresh water. Concrete mix water is used for pouring, but only 3% of it exists in the ocean, 1% exists in the ground, and 97% exists in the oceans. There are about 5 billion cubic meters of concrete used each year [6]. One cubic meter of concrete requires approximately 150 L of water, without considering other uses of water [1].

Seawater and alkaline water have been investigated for use in concrete mixes, as well as mine, mineral, and sewage waste waters, as well as wastewater from ready-mixed concrete factories. Since the impurities in each type of water vary, it is challenging to come to a consensus about their usage in concrete mixes. Only comparative tests have been devised now to evaluate the acceptability of combining water. Concrete created with ambiguous water is generally compared to concrete made with suitable water in comparison tests [7].

Freshwater (FW) and treated wastewater were used to test the properties of cement mortar (TWW). In order to compare the fresh and hardened characteristics of the materials, TWW was used instead of fresh water in the M20 concrete mix. By substituting FW for TWW, concrete's compressive strength, split tensile strength, and flexural strength are reduced [5]. STWW satisfies the requirements of IS 456-2000.

By combining STWW, the compressive strength is increased. At 28 days, the compressive strength of concrete with STWW is equal to that of concrete with potable water, the split tensile strength is nearly the same, and the flexural strength of concrete with STWW has improved [2].

The laboratory analysis of the wastewater samples' chemical characteristics was done under the designations of primary treated wastewater (PTWW) and secondary treated wastewater (STWW). Comparing compressive, tensile, and flexural strength of concrete specimens with PTWW and STWW. In spite of taking less time with PTWW, cement paste using potable water had identical start and ultimate setting times to cement paste using STWW. In STWW, residential wastewater, the compressive strength rose over time, and the findings of the tensile and flexural strength tests were the same. Consequently, STWW did not increase compressive or tensile strength in this study. However, STWW does boost compressive strength in STWW concrete after it has cured for 60 days [3].

Experimental analysis was done to determine if utilizing treated effluent to mix concrete was appropriate. Tap water (TW), first-stage wastewater (PTWW), second-stage wastewater (STWW), and third-stage wastewater (TTWW), all acquired from the nearby Raqqa wastewater treatment facilities, were used to create concrete cube examples. Concrete slump and density were not affected by the kind of water utilized during mixing. However, it was shown that setting times grew longer as water quality declined. For ages up to one year, PTWW and STWW showed poorer strength development. The strength of TW-made concrete at the 3 and 7-day early concrete ages. The usage of STWW and PTWW enhanced the risk of steel corrosion. As a result, both the initial and ultimate setting times were affected by how water was mixed and set times. As the type of mixing water deteriorated, the setting time increased. PTWW and STWW were found to affect the setting time most. In this case, concrete with TTWW was stronger than concrete with tap water [8].

2 Materials

2.1 Cement

A cement is generally referred to as an adhesive, a substance that sets up on its own and is capable of binding materials together. The word "cement" dates back to the Romans, who used the term "opus caementicium" the term refers to a masonry structure resembling concrete and made from crushed stone and quicklime. Cements are tested for specific gravity, fineness, consistency, and initial and final setting times. Pozzolan and powdered brick additives are added to quicklime to provide a hydraulic binder.

2.2 Fine Aggregate

Well graded river sand passing 4.75 mm in order to remove any foreign particles, then mixed the sand with a suitable fine aggregate. We are currently testing aggregate specific gravity, fineness modulus, and coarseness distribution.

2.3 Coarse Aggregate

The coarse aggregate used was the aggregate with a maximum size of 20 mm. Tests were conducted to evaluate the water absorption, specific gravity, flakiness, and elongation index of the aggregate as well as its grading.

2.4 Water

Water is regarded as a key component in concrete strengthening for acceptable quality concrete that meets specified criteria. Since the time of the ancient Romans, concrete has been used in buildings, and it is typically mixed with drinkable water. By 2025, half of the world's population will live in regions with insufficient access to fresh water, according to estimates from the World Meteorological Organization (WMO) and the United Nations (UN) [4]. Since freshwater is virtually universally accessible, the concrete industry uses it unrestrictedly for all purposes. But as freshwater supplies become more limited, the scenario is changing.

2.4.1 Secondary Treated Waste Water

According to "conventional knowledge" in the literature on concrete technology, water can be used to mix and cure concrete if it is safe for human consumption. The use of treated wastewater in concrete has, however, been demonstrated to be beneficial in several studies. The mechanical properties and durability of concrete mixed with biologically treated wastewater have been studied to a limited extent. Utilization of treated wastewater also mitigates the negative environmental effects associated with disposing of treated wastewater.

In secondary wastewater treatment, microorganisms are used to remove pollutants from wastewater through biological means. There are two types of secondary biological processes, aerobic and anaerobic. Under specific conditions, coupled anaerobic–aerobic processes may also be used.

3 Experimental Tests and Results

The mix design is prepared for M20 grade concrete, we got an average value slump of 95 and 90 mm for potable water concrete and secondary treated wastewater concrete respectively. The 18 cubes of 150 mm × 150 mm × 150 mm are cast for testing out of this 9cubes using PW (portable water) and 9cubes using STWW (*Secondary treated wastewater) had undergone compression test. The 18 cylinders of dimensions 150 mm diameter (D) and 300 mm height (L) were cast Out of these 9 cylinders using PW (portable water) and 9 cylinders using STWW (*Secondary treated wastewater) had undergone the Split tensile test.

3.1 Tests and results of PW and STWW

We conducted tests on Total Hardness, Calcium Hardness, Chloride, Dissolved Oxygen, Alkalinity, pH, Total Dissolved Solids, and Turbidity for both PW and STWW. The results are the below Table 1.

Table 1. Tests results on the water with IS limits

Sl. no	Properties	Results of PW	Result of STWW	IS limits
1	Total Hardness	132 mg/l as CaCO3	136 mg/l as CaCO3	<200 mg/l
2	Calcium Hardness	134 mg/l as CaCO3	160 mg/l as CaCO3	<200 mg/l
3	Chloride	60.07 mg/l	150 mg/l	<250 mg/l
4	Dissolved Oxygen	5.51 mg/l	7 mg/l	<8 mg/l
5	Alkalinity	200 mg/l as CaCO3	210 mg/l as CaCO3	<250 mg/l
6	pH	6.85	7.45	>6
7	Total Dissolved Solids	440 ppm	1600 ppm	<2000 ppm
8	Turbidity	3 NTU	4.8 NTU	<5 NTU

3.2 *Compression Test*

Compression tests are conducted on both PW and STWW cubes using a 1000 kN UTM (Universal Testing Machine) in Fig. 1. The test results of both PW cubes and STWW are below Table 2 and Table 3 respectively. The average compressive of PW concrete and STWW concrete is almost same in the both. So, there are no effects on compressive strength we can use STWW concrete to save water. As we can see from the results, A 7-day cure of STWW reduced its compressive strength by 2.08% of PW. A 14-day cure increased STWW's compressive strength by 1.39% of PW. A 28-day cure increased STWW's compressive strength by 1.13% of PW are shown in Fig. 2. We can conclude that Potable water can be replaced by Secondary Treated Wastewater in Construction Industry.

Fig. 1 Compression test using CTM

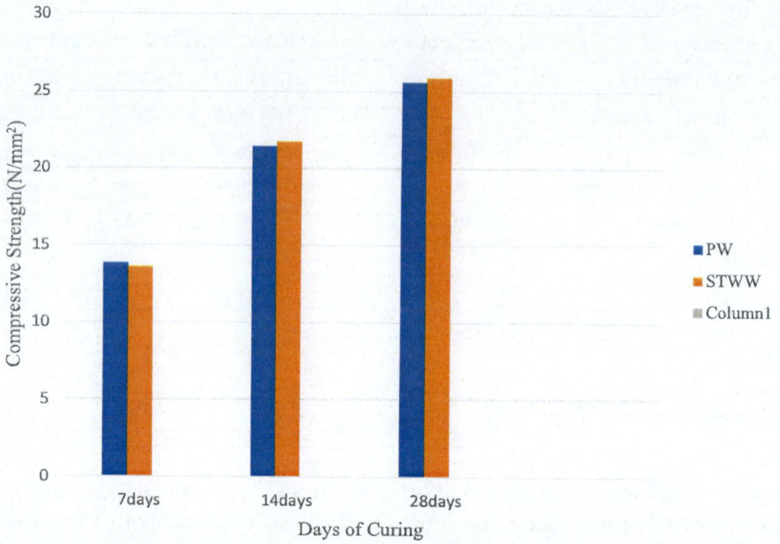

Fig. 2 Comparison between Compressive strength of (PW & STWW)

Table 2. Compressive strength test results of potable water

Age of M20 grade concrete	Trial	Date of curing	Date of testing	Load = P (KN)	Compressive strength = P/A (N/mm^2)	Average strength (N/mm^2)
7 Days	1	13/04/2022	20/04/2022	350	15.56	13.92
	2			310	13.77	
	3			280	12.44	
14 Days	1	13/04/2022	27/04/2022	470	20.88	21.48
	2			510	22.67	
	3			490	21.78	
28 Days	1	13/04/2022	11/05/2022	550	24.47	25.63
	2			600	26.67	
	3			580	25.77	

3.3 Split Tensile Strength

A splitting tensile test is generally carried out to obtain the tensile strength of concrete. The 18 cylinders of dimensions 150 mm diameter (D) and 300 mm height (L) were cast Out of these 9 cylinders using PW (portable water) and 9 cylinders using STWW (*Secondary treated wastewater) for this test as the Fig. 3. The test results of both PW cubes and STWW are the below Table 4 and Table 5 respectively.

Table 3. Compressive strength test results of secondary treated wastewater

Age of M20 grade concrete	Trial	Date of curing	Date of testing	Load = P (KN)	Compressive strength = P/A (N/mm^2)	Average strength (N/mm^2)
7 Days	1	12/04/2022	19/04/2022	290	12.89	13.63
	2			250	11.11	
	3			380	16.89	
14 Days	1	12/04/2022	26/04/2022	510	22.67	21.78
	2			480	21.33	
	3			520	23.11	
28 Days	1	12/04/2022	11/05/2022	580	25.77	25.92
	2			540	24	
	3			630	28	

Fig. 3 Split tensile Test

Table 4. Split tensile strength test results of potable water

Age of M20 grade concrete	Trial	Date of curing	Date of testing	Load = P (KN)	Split tensile strength = (2P)/(πDL) (N/mm^2)	Average strength (N/mm^2)
7 Days	1	19/04/2022	26/04/2022	110	1.56	1.75
	2			140	1.98	
	3			120	1.70	
14 Days	1	19/04/2022	04/05/2022	160	2.26	2.40
	2			200	2.83	
	3			150	2.12	
28 Days	1	13/04/2022	11/05/2022	250	3.53	3.77
	2			260	3.68	
	3			290	4.10	

Table 5. Split tensile strength test results of secondary treated wastewater

Age of M20 grade concrete	Trial	Date of curing	Date of testing	Load = P (KN)	Split tensile strength = (2P)/(πDL)(N/mm^2)	Average strength (N/mm^2)
7 Days	1	20/04/2022	27/04/2022	110	1.56	1.70
	2			130	1.84	
	3			120	1.69	
14 Days	1	20/04/2022	04/05/2022	170	2.40	2.52
	2			160	2.26	
	3			210	2.90	
28 Days	1	12/04/2022	11/05/2022	270	3.82	3.81
	2			**290**	**4.70**	
	3			**250**	**3.53**	

As we can see from the results, after 7 days of curing the Split Tensile strength of STWW decreased by 2.85% of PW, after 14 days of curing the Split Tensile strength of STWW increased by 5% of PW, and after 28 days of the curing the Split Tensile strength of STWW is increased by 1.06% of PW are shown in Fig. 4. Therefore, Secondary treated waste water can be used instead of Potable Water without affecting Split Tensile strength.

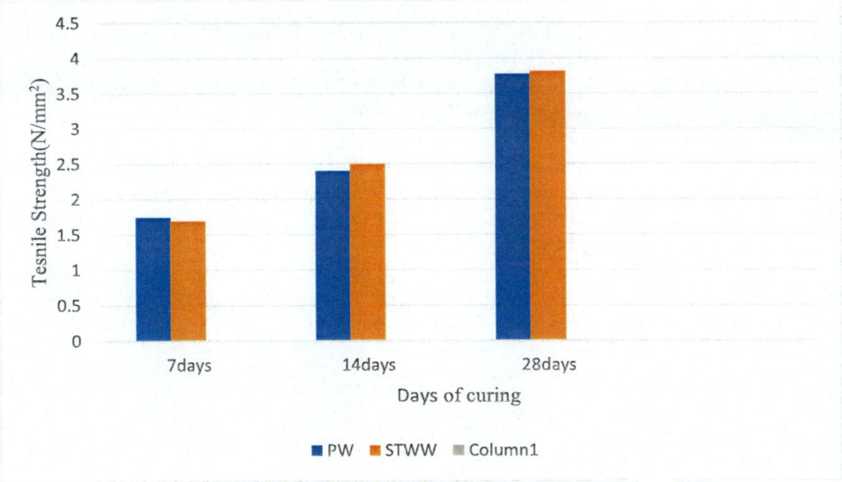

Fig. 4 Comparison between Split Tensile strength of (PW &STWW)

4 Conclusions

- Using the pH determination test, we found that STWW has a pH of 7.45, so it is safe to use it as construction water because it is slightly basic and has no detrimental effects.
- From the Turbidity test we got the value of 5.7 NTU which is more than the IS Limits (i.e., > 5 NTU), because of a slight difference in the value it won't affect many concrete properties.
- The tests which are conducted on water shows that STWW contains fewer impurities and it is fit for Construction Purpose as per IS Code provision. The compressive strength of STWW is increased by 1.13% of the Compressive Strength of Potable Water after 28 days of curing which shows that STWW can be used instead of Fresh Water in the Construction Industry and thereby reduces Water Scarcity.
- The Split Tensile Strength of STWW is increased by 1.06% of Split Tensile Strength Potable Water after 28 days of curing hence; the Reuse of Treated water is considered the best option in Construction Industry to save fresh water.

References

1. Detal, G.B. (2018). Effect of Treated Wastewater on Mechanical Properties of Concrete.
2. Narkhede, K.L., & Chavan, F.I. (2017) Effect of Treated Wastewater on the Strength of Concrete.
3. Ramkar, A.P., & Ansari, U.S. (2016). Effect of Treated Wastewater on Strength of Concrete (2016).

4. Shaik, A.R., & Inamdar, V.M. (2016). Study of utilization of wastewater in concrete. *Journal of Mechanical and Civil Engineering* 105–108.
5. Narender Reedy, A., Naveen, K., & Siva Rani, N. (2015). Use of treated domestic wastewater as mixing water in cement mortar. *International Journal of Science and Engineering Invention* 23–31.
6. Bardhan, S. (2011). *Assessment of Water Resource Consumption in Building Construction in India.* WIT Press.
7. Rama Mohan Rao, P., Moinuddin, S.M.K., & Jagadeesh, P. (2013) Effect of Treated Wastewater on the Properties of Hardened Concrete.
8. Terro, M., Ghusain, I.A.I. (2003) Mechanical Properties of Concrete Made with Treated Wastewater at Ambient and Elevated Temperatures. *Kuwait J Sci Engineering* 213–288.

Strength Properties of Concrete with Rice Husk Ash and Quarry Dust for Sustainable Construction Applications

P. V. R. K. Reddy and D. Ravi Prasad

Abstract The economic and environmental difficulties related with the utilization of traditional construction material for construction projects, such as use of cement and river sand have risen in parallel with the human population growth. As a result, researchers are focussing alternate construction materials that are cost effective and sustainable. The aim of current investigation is to develop a cost effective and sustainable concrete composite using rice husk ash and quarry dust. The conventional concrete with a characteristic strength of 30 MPa, was developed with cement and river sand. Initially, rice husk ash was proposed as a substitute for cement in varying percentages of 10 to 30%. Quarry dust was then utilized as a substitute for sand in mixes comprising rice husk ash with percentages ranging from 10 to 30% to develop non-conventional concrete. The slump of concrete and the strength characteristics such as compressive, split tensile and flexural strength were measured at 7 and 28 days, respectively. The strength characteristics of the specimens decreased with increasing percentage replacements of quarry dust in comparison with conventional concrete. Based on the experimental findings, the concrete having 10% rice husk ash and 20% quarry dust having superior strength properties. Hence rice husk ash and quarry dust could be employed as partial substitutes for cement and river sand, respectively, in the development of sustainable construction materials.

Keywords Concrete · Strength properties · Rice husk ash · Quarry dust · Sustainable constructions

---

P. V. R. K. Reddy (✉) · D. Ravi Prasad
Civil Engineering Department, National Institute of Technology, Warangal, India
e-mail: pvrkr@student.nitw.ac.in; pvrkreddy.ce@anits.edu.in

D. Ravi Prasad
e-mail: raviprasad@nitw.ac.in

P. V. R. K. Reddy
Civil Engineering Department, Anil Neerukonda Institute of Technology and Sciences, Bheemunipatnam (Mandal), Visakhapatnam 531162, India

A. Vilventhan et al. (eds.), *Advances in Construction Materials and Management*, Lecture Notes in Civil Engineering 346,
https://doi.org/10.1007/978-981-99-2552-0_24

1 Introduction

The progress of sustainable construction materials from locally obtained industrial and agricultural waste has the ability to valorize these wastes by converting them into wealth [1]. Since the need for affordable and sustainable new construction continues to rise in practically every country. Use of traditional building materials have not solved the housing deficit problem; rather, they have resulted in global warming through the emission of carbon dioxide into the environment from the cement manufacturing process [2]. For sustainable construction, it is necessary to develop durable and environmentally friendly construction materials from waste materials that have beneficial properties [3]. The substitution of natural or waste materials for traditional construction materials such as cement is critical to maintain sustainability and providing environmentally friendly construction materials [21–26]. Investigators have effectively used several biomass and waste materials for the creation of construction materials [1–3].

Historically, rice has been a popular staple food and an agricultural crop in most of the countries. In the process of milling rice, around 78% of paddy grains is converted into bran, rice, and broken rice, while other 22% is converted into husk [4]. Introducing rice husk ash in the development of concrete composites was investigated, and it has been reported that fine rice husk ash improved the mechanical characteristics, representing that cement can be substituted with rice husk ash to manufacture durable concrete [5]. Moreover, the findings indicated that coarse rice husk ash is detrimental to strength gain of concrete composites. Another study examined the impact of rice husk ash on microstructure and strength characteristics of autoclaved aerated concrete (AAC), and they discovered that while a lighter-weight AAC was produced when sand was substituted for rice husk ash, the water demand increased as a result, which contributed to the diminution in compressive strength [6].

The issue of landslides and erosion has been exacerbated by ongoing river and water channel dredging in quest of river sand for construction. In the construction industry, finding a substitute to the usage of river sand is critical to minimise the continual depletion of river sand. Researchers studied the application of quarry dust as supplement for river sand or fine aggregate. Quarry dust is a waste generated during blasting of rock or stone at quarry sites. It is primarily composed of quartz minerals, which accounts for its usage in the development of durable construction materials due to the hardness of quartz. It has been proven to be useful in the development of sustainable cementitious composites [7]. It was discovered that adding quarry dust improved the composite's ability to shrink, and the workability rose as the replacement amount of quarry dust increased. Recycling quarry dust for use as supplemental cementitious material in low-carbon cement manufacturing observed that the use of quarry dust reduced the average pore size distribution, as a result densified microstructure and enhanced compressive strength at early-age [8].

Bheel et al. [9] investigated the strength characteristics of concrete developed using wheat straw ash and rice husk ash and demonstrated that the increase in proportion of cement replaced with wheat straw ash and rice husk ash, the drying shrinkage of the specimens decreased. The study found that as curing age increased, the mechanical characteristics were enhanced. For the construction of hollow sandcrete blocks, quarry dust was employed as a 100% substitute for river sand, together with other minerals additives [10]. According to the findings of this investigation, 100% quarry dust substitute resulted in reduced compressive strength in comparison with the mix developed using partial substitute of quarry dust with pozzolanic materials. The impact of recycled aggregate and rice husk ash on concrete characteristics was also studied [11]. The findings demonstrated that a concrete mix having recycled concrete coarse aggregates at 100% and rice husk ash at 10–15% met the design standards of the construction sector.

Despite prior efforts to create durable and environmentally acceptable construction materials from agricultural and industrial waste. There has been very few or no research on using rice husk ash and quarry dust in combination to substitute cement and river sand, respectively. According to this study, incorporating rice husk ash as a substitute for cement reduces the emission of CO2 during cement manufacturing process, while introducing quarry dust as an alternate for river sand reduces environmental issues created by the continual dredging of rivers to acquire sand. As a result, the study aims to create a sustainable concrete using rice husk ash and quarry dust as partial substitutes for cement and river sand, respectively.

2 Materials and Methods

2.1 Materials

The materials used in this study are cement, sand, coarse aggregate, rice husk ash, and quarry dust. Ordinary Portland cement of 53 grade according to IS:269, river sand and crushed granite stones according to IS:383. The quarry dust was sourced from the granite crushing process. The raw rice husk (RH) was sourced from the rice mill. Rice husk ash was produced by calcining RH in an electric furnace at 650 degrees Celsius. The rice husk ash was sieved through a 75-μm to get smaller pieces that could substitute part of the cement in the concrete composite.

2.2 Concrete Mix and Specimen Preparation

In this investigation, standard-grade concrete was designed to achieve 30 MPa after 28 days. The mix was designed in accordance with IS:10, 262–2019. Conventional concrete specimens were produced using cement and river sand as references. Non-conventional concrete mixes were produced with several proportions of rice husk ash (10%, 20%, and 30%) and quarry dust (10%, 20%, and 30%) for the replacements of cement and river sand, respectively. The concrete mix proportions are

Table 1 The concrete mix proportions per m^3

Mix ID	Rice husk ash (kg)	Quarry dust (kg)	Water/cement ratio	Cement (kg)	River sand (kg)	Coarse aggregates (kg)	Superplasticizer (%)
CC	–	–	0.44	380	860	1050	0.6
R10	38	–	0.44	342	860	1050	0.6
R20	76	–	0.44	304	860	1050	0.6
R30	114	–	0.44	266	860	1050	0.6
R10Q10	38	86	0.44	342	774	1050	0.6
R10Q20	38	172	0.44	342	688	1050	0.6
R10Q30	38	258	0.44	342	602	1050	0.6

presented in Table 1. A total 126 composite specimens were developed, with 18 conventional composite specimens and 108 non-conventional composite specimens. The composite mixes were used to prepare cube, cylinder, and prismatic specimens and cured for 7 and 28 days, respectively. For each batch three similar specimens were prepared.

2.3 *Slump Test*

The workability of fresh concrete was tested in accordance with IS 1199:1959, and the designated slump value for the conventional concrete was 100 mm. The slump cone test was used to gauge the concrete's workability after it had been poured into the mold in four levels, each measuring 1/4 of the mold's height (top 100 mm, bottom 200 mm, and height 300 mm) and having been tamped 25 times. The mold is then instantly detached from the concrete by slowly and cautiously elevating it vertically. The slump is instantly determined by evaluating the height between the mold and the highest point of the specimen. Slump value for conventional and non-conventional composites was measured.

2.4 *Compressive Strength*

The test was performed on a 100 mm cubes using 200 Ton capacity compression testing machine (CTM) in accordance with code IS 516–1959 for each curing period (7 and 28 days). A constant increase in load was made at a rate of 14 N/mm^2/min until no more load could be taken. The load was introduced gradually to avoid shock. For each sample, the highest load applied on the specimen was noted, and the average of three specimens were considered for each test.

2.5 Tensile Strength

The tensile strength test was conducted in accordance with code IS 516–1959 on cylinders of size 100 mm in dia and 200 mm in height for each curing age of 7 and 28 days on a 200-ton capacity CTM. Without applying any shock, the load was gradually increased till failure at a nominal rate between 1.2 and 2.4 $N/mm^2/min$. The maximum load given to each specimen was noted, and the average of three specimens was considered for each test.

2.6 Flexural Strength

In accordance with IS 516–1959, the standard procedure for flexural strength was conducted on beams measuring 100 × 100 × 500 mm on universal testing machine for each period of 7 and 28 days curing. The load was applied steadily at a rate of 0.7 $N/mm^2/min$. The specimen is tested in standard four-point bending. The highest load applied to the specimen was noted. For each test, the averages of three similar specimens were considered.

3 Results and Discussions

3.1 Chemical Composition of Raw Materials

The chemical composition of the materials was evaluated using the conventional Thermo scientific X-ray fluorescence (EDXRF) technique. EDXRF has been employed to evaluate the oxide contents of the samples. The chemical composition of materials is shown in Table 2. Silicon dioxide (SiO2) was discovered to be the most abundant ingredient in the rice husk ash and quarry dust. The sum of silicon, iron, and aluminium oxides in the rice husk ash sample was 87.78%, which is higher than the 70% for pozzolana materials recommended by ASTM C618 [12], showing that rice husk ash could be a pozzolana material. The 87.78% SiO2 + Fe2O3 + Al2O3 is consistent with other research findings of 86.01% [13], 82.97% [14], and 93.12% [6]. However, the sum of the silicon, iron, and aluminium oxides for the quarry dust sample was 77.44%, which is greater than 70%, suggesting that quarry dust is not a pozzolanic material since it is an inert material. In contrast, the percentage of CaO in quarry dust is larger than in rice husk ash. When this CaO is combined with pozzolanic materials, cement hydrates of calcium-aluminum- silicate-hydrate (C-A-S-H) and calcium-silicate-hydrate (C-S-H) are generated. This CaO is accountable for the hydration process [2]. Unlike the rice husk ash, the quarry dust contains significant amount of strontium oxide (SrO) of 5.06% as well as a tiny amount of BaO of 0.05%. Because of its high refractive index, SrO in quarry dust falls under

Table 2 Chemical compositions of raw materials

Oxides	Rice Husk Ash	Quarry Dust	Cement	River sand
SiO2	86.74%	62.73%	21.54%	78.59%
Al2O3	0.84%	12.31%	5.32%	10.52%
MgO	1.12%	3.73%	1.95%	0.85%
K2O	1.57%	2.88%	0.71%	0.52%
CaO	0.54%	2.65%	63.61%	1.95%
P2O5	0.64%	0.43%	–	0.65%
MnO	0.41%	0.31%	–	1.21%
Fe2O3	0.20%	2.40%	3.61%	1.85%
TiO2	0.06%	0.36%	–	–
SO3	0.16%	0.05%	2.65%	1.03%
SrO	5.06%	6.53%	–	0.54%
BaO	0.05%	0.03%	–	–
ZnO	0.05%	0.02%	–	–
CeO2	2.49%	0.88%	–	–

the category of ceramic fluxing oxides and may offer superior interface for appropriate binding of the constituting compounds in the composite, whereas BaO in tiny amounts will offer shielding for various kinds of radiation in quarry dust containing composites [15]. As seen in Table 2, the most common oxides found in Portland cement are CaO, SiO2, and Al2O3.

3.2 *Slump Test Results for the Concrete Composites*

Six different types of non-conventional mixes were developed, as previously mentioned. Figure 1 depicts the slump test results for conventional and non-conventional concrete. The slump values indicates that all the mixes had good workability and reached the desired slump of minimum 100 mm. Introducing rice husk ash as a replacement for cement reduced workability slightly. The slump observed for conventional concrete was 175 mm. Incorporating rice husk ash the slump value observed as 170 mm, 165 mm, and 160 mm for 10%, 20%, and 30% replacements, respectively. Furthermore, rice husk ash along with quarry dust addition decreased the slump of the mixes even further. Quarry dust substitute with river sand reduced the slump value from 170 to 160 mm, 150 mm, and 140 mm for 10%, 20%, and 30% replacement, respectively. Previous investigations have found similar trend [16]. The decrease in slump could be ascribed to the increased dust and silt concentration in the quarry dust, as higher amount of fine particles provides a large surface area and consequently increases the water requirement in cement composites [16].

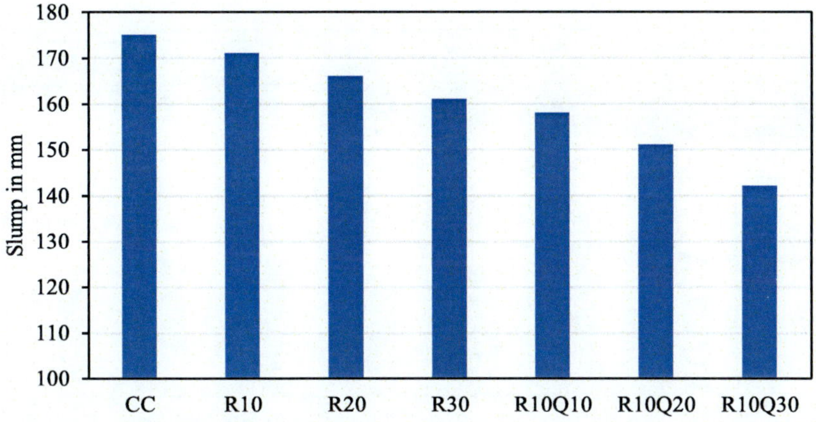

Fig. 1 Slump values of concrete composites

3.3 Strength Properties of the Concrete Composites

Compressive strength was one of the most significant mechanical characteristics for hardened concrete to be evaluated, along with other strength characteristics such as flexural and split tensile strength. The development of strength properties was observed at various curing periods. Compressive, split tensile, and flexural strength developments for the conventional and non-conventional concrete composites at 7 and 28 days are shown in Fig. 2, 3, and 4, respectively. When the cement was partially substituted by rice husk ash, compressive strength of the R10 mix was improved by 14.6% and 10.4% at 7 and 28 days, respectively, in comparison with conventional concrete. Similarly, split tensile and flexural strengths of R10 mix were improved by 12.2% and 9.2% at 7 days, and by 13.8% and 9.8% at 28 days, respectively, compared with conventional concrete. When the rice husk ash percentage replacement is increased, a further drop in strength properties was observed. Then again with the partially substitution of sand with quarry dust, the strength properties were reduced. Similar trend was observed [17]. When compared with the respective R10 mix, the 28 days compressive, split tensile and flexural strength of the concrete mix containing 20% quarry dust was reduced by 9.8%, 8.6%, and 9.1%, respectively. Based on the concrete strength properties, the non-conventional concrete which contains 10% rice husk ash and 20% quarry dust, was selected as the superior mix among all the non-conventional mixes considered. This evaluation was primarily based on the strength characteristics at 28 days.

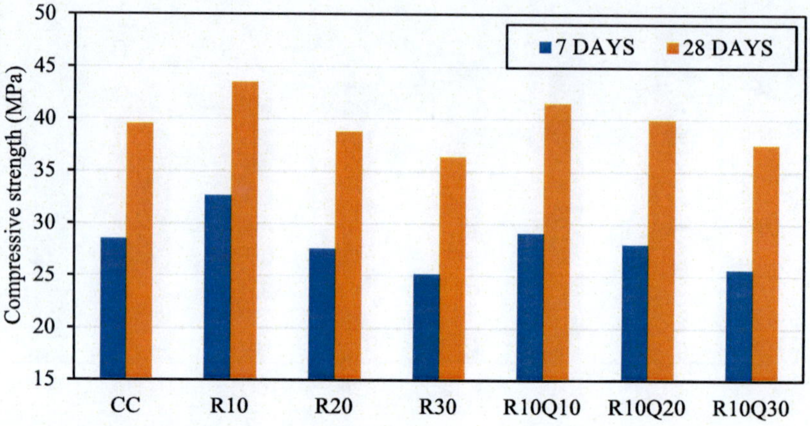

Fig. 2 Compressive strength of concrete composites

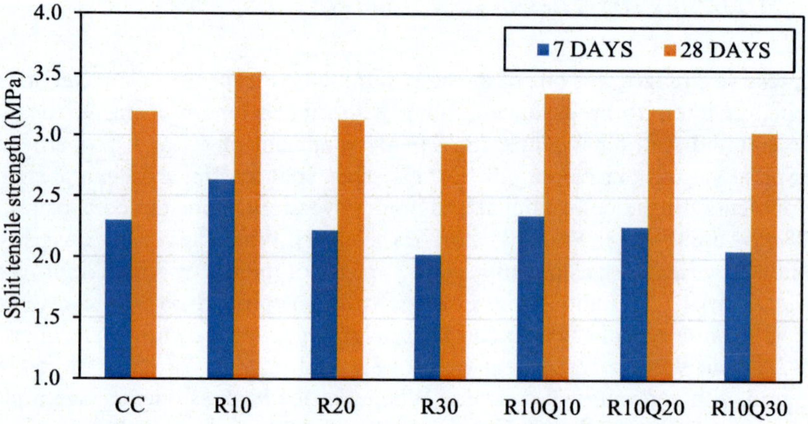

Fig. 3 Split tensile strength of concrete composites

3.4 Strength Mechanism of the Concrete Composites

As the replacement percentages of rice husk ash and quarry dust increased, it was found that the strength characteristics of non-conventional mixes were decreasing, as presented in Figs. 2, 3 and 4. The chemical composition shows that rice husk ash comprises less CaO (lime), which may account for the reduced strength properties found in mixes having a higher percentage substitution of rice husk ash and quarry dust in comparison with high CaO concentration of cement employed in this investigation. Reduced CaO content will lead to lower strength properties because CaO is required for pozzolanic reaction which will cause the hydration process of concrete matrix. Because rice husk ash is a hydrophilic substance, the decline in strength of

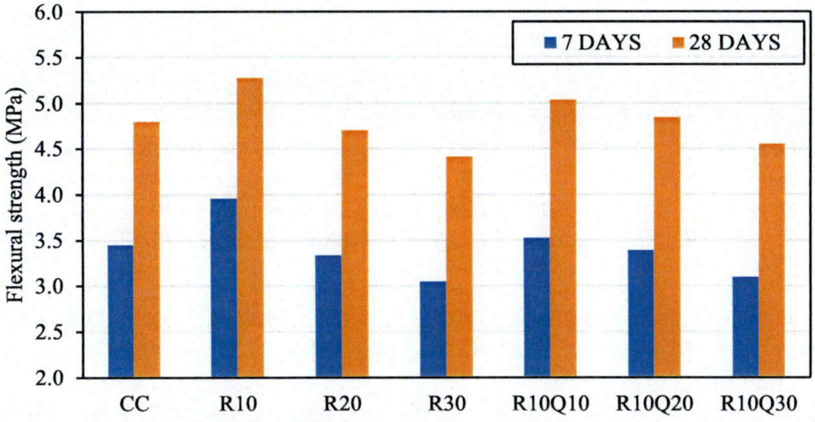

Fig. 4 Flexural strength of concrete composites

concrete composites with high quantities of rice husk ash might be due to a lack of accessible water for cement hydration [18]. The high silica (SiO2) in the rice husk ash caused decreased strength properties resulting in the chemical reaction among the rice husk ash and other components of the concrete matrix, such as cement, river sand, and quarry dust. When calcium hydroxide from cement hydration interacts with Al2O3 and SiO2 from rice husk ash and quarry dust, C-A-S-H and C-S-H are formed [19]. The excessive silica in rice husk ash does not contain enough $Ca(OH)_2$ or CaO to react with or generate hydration products such as C-A-S-H and C-S-H, resulting in inferior strength characteristics in the concrete composite with a high content of rice husk ash. It has been found a drop in strength properties as a result of rice husk ash as a substitute [13, 18]. However, the economic and environmental advantages of employing rice husk ash in construction enable it a feasible alternative. It has the potential to minimize greenhouse gas emissions due to the manufacture of rice husk ash requires less embodied energy than cement manufacturing. The process of rice husk agricultural waste to usable rice husk ash for construction materials demonstrates the transformation of trash into wealth as well as a reduction in the enormous expense of cement in construction projects. In contrast to conventional concrete mixes, the improvement in strength properties of non-conventional concrete mixes with increasing curing period is because of a further reaction among the pozzolanic compounds and CaO from cement and quarry dust, as a result, the development of stronger cement hydrates [20].

4 Conclusions

The need for environmentally friendly construction materials has grown in recent years, as a result of global warming. An attempt was made in order to reduce CO2 emission and environmental issues, this work investigates the utilization of rice husk ash from agricultural waste and quarry dust from industrial waste as partial substitutes for cement and river sand, respectively. Following are significant results of the investigation:

- The slump values of non-conventional concrete mixes decreased as the replacement percentages of rice husk ash and quarry dust increased when compared to conventional composites.
- The non-conventional composites had the best strength properties at 10% rice husk ash replacement when compared to conventional composites. The proportion of rice husk ash replacement was increased further, a drop was observed in strength.
- The strength parameters of non-conventional composites declined with the partial substitute of sand with quarry dust when compared with the R10 mix.
- The strength properties of a non-traditional composite, including 10% rice husk ash and 20% quarry dust, were considered as the optimum composition.

The present investigation suggested the feasibility of developing environmentally acceptable and sustainable construction materials by combining agricultural and industrial wastes. It helps to address the environmental problems brought on by the excessive use of cement and river sand if these waste products are employed in construction materials as a suitable proportion.

References

1. Arancon, R. A. D., Lin, C. S. K., Chan, K. M., Kwan, T. H., & Luque, R. (2013). Advances on waste valorization: new horizons for a more sustainable society. *Energy Science Engineering, 1*, 53–71.
2. Obianyo, I.I. et al. (2021) Performance of lateritic soil stabilized with combination of bone and palm bunch ash for sustainable building applications. *Cogent Engineering 8*, 1921673
3. Sandanayake, M., Bouras, Y., Haigh, R., & Vrcelj, Z. (2020). Current sustainable trends of using waste materials in concrete—a decade review. *Sustain, 12*, 1–38.
4. Nagrale, S. D., Hajare, H., & Modak, P. R. (2012). Utilization of rice husk ash. *International Journal of Engineering Research and Applications, 2*, 2–7.
5. Ahsan, M. B., & Hossain, Z. (2018). Supplemental use of rice husk ash (RHA) as a cementitious material in concrete industry. *Construction and Building Materials, 178*, 1–9.
6. Kunchariyakun, K., Asavapisit, S., & Sombatsompop, K. (2015). Properties of autoclaved aerated concrete incorporating rice husk ash as partial replacement for fine aggregate. *Cement and Concrete Composites, 55*, 11–16.
7. Ban, C.C., Sern, L.J., & Ramli, M.B. (2019). The mechanical strength and durability properties of ternary blended cementitious composites containing granite quarry dust (GQD) as natural sand replacement. *Construction and Building Materials, 197*, 291–306.

8. Zhao, Y., Qiu, J., Xing, J., & Sun, X. (2020). Recycling of quarry dust for supplementary cementitious materials in low carbon cement. *Construction and Building Materials, 237*, 117608.
9. Bheel, N., Awoyera, P., Shar, I.A., Sohu, S., Abbasi, S.A., Prakash, A.K. (2021) Mechanical properties of concrete incorporating rice husk ash and wheat straw ash as ternary cementitious material. *Advances in Civil Engineering, 2021*, 2977428.
10. Sojobi, A. O., Awolusi, T. F., Aina, G. B., Oke, O. L., Oladokun, M., & Oguntayo, D. O. (2021). Ternary and quaternary blends as partial replacement of cement to produce hollow sandcrete blocks. *Heliyon, 7*(6), e07227.
11. Subhasmita, R., Kumar, R., Bhusan, B., & Dey, T. (2018). Influence of incorporation of rice husk ash and coarse recycled concrete aggregates on properties of concrete. *Construction and Building Materials, 173*, 289–297.
12. Pourkhorshidi, A. R., Najimi, M., Parhizkar, T., Jafarpour, F., & Hillemeier, B. (2010). Applicability of the standard specifications of ASTM C618 for evaluation of natural pozzolans. *Cement and Concrete Composites, 32*, 794–800.
13. Begum, R., Habib, A., & Mostafa, S. (2014). Effects of rice husk ash on the non-autoclaved aerated concrete concrete. *International Journal of Engineering Innovations and Research, 3*, 116–121.
14. Onyelowe, K.C. et al. (2021) Swelling potential of clayey soil modified with rice husk ash activated by calcination for pavement underlay by plasticity index method (PIM). *Advances in Materials Science and Engineering, 2021*, 6688519.
15. Zezulova, T., Opravil, S.T. (2017). Influence of barium oxide additions on Portland clinker. *Ceram. -Silik´aty 61*, 20–25
16. Raman, S.N., Ngo, T., Mendis, P., & Mahmud, H.B. (2011). High-strength rice husk ash concrete incorporating quarry dust as a partial substitute for sand. *Construction and Building Materials, 25*, 3123–3130.
17. Safiuddin, M., Raman, S.N., & Zain, M.F.M. (2007). Flowing concretes with quarry waste fine aggregate. *Journal of Civil Engineering Research and Practice, 4*, 17–25
18. Sandhu, R.K., & Siddique, R. (2017). Influence of rice husk ash (RHS) on the properties of self-compacting concrete: a review. *Construction and Building Materials, 153*, 751–764
19. Kufre, R., Ufot, D., Christopher, I., & Chibuzor, K. (2021). Effect of micro-sized quarry dust particle on the compaction and strength properties of cement stabilized lateritic soil. *Cleaner Materials, 2*, 100023
20. Onyelowe, F., Onyelowe, D.B., Van, I., Obianyo, E.K. (2020) Overview of ash as supplementary cementitious silicate-based composite and construction material. *Epitoanyag Journal Silicate Based Composite Materials Difference, 72*, 80–85
21. Reddy, P.V.R.K., & Prasad, D.R. (2022). A study on workability strength and microstructure characteristics of graphene oxide and fly ash based concrete. *Materials Today: Proceedings, 62*, 2919–2925. https://doi.org/10.1016/j.matpr.2022.02.495
22. Reddy, P.V.R.K., & Ravi Prasad, D. (2023). Static and Dynamic Mechanical Properties of Graphene Oxide and Fly Ash Based Concrete, Springer Nature Singapore, 279–288. https://doi.org/10.1007/978-981-19-8979-7_23
23. Reddy, P.V.R.K., & Prasad, D.R. (2022). Investigation on the impact of graphene oxide on microstructure and mechanical behaviour of concrete. *Journal of Building Pathology and Rehabilitation, 7*(1). https://doi.org/10.1007/s41024-022-00166-1
24. Reddy, P.V.R.K., & Prasad, D.R. (2023). The role of graphene oxide in the strength and vibration characteristics of standard and high-grade cement concrete. *Journal of Building Engineering, 63*, 105481. https://doi.org/10.1016/j.jobe.2022.105481
25. Reddy, P.V.R.K., & Prasad, D.R. (2023). Graphene oxide reinforced cement concrete – a study on mechanical durability and microstructure characteristics. *Fullerenes Nanotubes and Carbon Nanostructures, 31*(3), 255–265. https://doi.org/10.1080/1536383X.2022.2141231
26. Reddy, P.V.R.K., & Prasad, D.R. (2022) Synergetic Effect of Graphene Oxide and Fly Ash on Workability, Mechanical and Microstructural Properties of High-strength Concrete. *Jordan Journal of Civil Engineering, 16*(3), 507–517.

An Experimental Study on Red Mud Based Geo Polymer Mortar with GGBS, M Sand and Metakaolin

T. Sowmyashree, S. Muralidhara, Ahamed Sharif, and Vali Mohammed

Abstract The current research work is to find the possibility of replacing the conventional binder by Red mud completely in the production of geopolymer technology based mortar mix. Commercial garbage products have a negative environmental impact as a result of storage problems, waste disposal problem and causing hazardous impact on the environment. By assessing the effectiveness of utilisation of red mud in order to resolve these issues in addition of GGBS, M sand and Metakolin as partial quantities in the geopolymer mortar production. The mix is chemically activated with alkaline activator to influence geopolymerisation with NaOH of considerably lesser concentration of 2Molarity. The produced mortar was checked against compressive strength and water absorption as per IS 3495:1992. Red mud, GGBS and Metakaolin are used as a binder material. The compatibility of raw materials are checked in order to get flowable mix as it is need not be compacted. The Process employed an alkaline activator. Red mud used in varied percentage from 30 to 60%. Metakaolin is used upto maximum value of 10% in the composition to make the mix economically viable. GGBS was used for strength enhancement of produced mix. Research proved that geopolymer mortar mix can be produced with CS from 14.1 Mpa to 38.1 Mpa by using redmud content from 40 to 60% with maximum water absorption of 18.9%. Optimum percentage of raw materials combination is obtained and to study the strength of mortar mix, masonry prism test was conducted. The results proved that Red mud, GGBS, M sand, and Metakaolin can be geopolymerized and used as a sustainable alternative to traditional mortar. This would be the solution for waste disposal issues of many industrial by-products across the world, simultaneously fulfil the current increased demand of construction materials due to rapid urbanisation in the world.

T. Sowmyashree (✉) · A. Sharif · V. Mohammed
Presidency University, Bengaluru 560064, India
e-mail: sowmyashreet@bmsce.ac.in

A. Sharif
e-mail: ahamedsharif@presidencyuniversity.in

S. Muralidhara
BMS College of Engineering, Bengaluru 560019, India
e-mail: murali.civ@bmsce.ac.in

A. Vilventhan et al. (eds.), *Advances in Construction Materials and Management*, Lecture Notes in Civil Engineering 346,
https://doi.org/10.1007/978-981-99-2552-0_25

1 Introduction

Nowadays, it is normal practice to look for recycling options for various industrial wastes or byproducts. This is done to comply with regulations, but it is also done to reduce disposal costs and prevent land and water contamination [12].

The Bayer chemical process is used to treat bauxite to create high-grade aluminium, and the waste product produced during this process is red mud, [11] which is intractable. Storage in impounded dike deposits close to alumina processing facilities is the most typical method of disposal. Red mud deposits' high alkali concentration presents numerous environmental and health problems. Restoring the biological values of the deposit sites is extremely difficult because of how perilous it is. Few attempts have been made in the past to grow plant cover over the red mud residue deposits and bauxite mining wastes.

Since ancient times, mortar has been used extensively in construction; the oldest known specimen was discovered in Israel and is considered to be 10,000 years old. The durability of most of the old mortar that is still in use today is proof of how durable the substance is. In actuality, a mortar joint serves as a gap-filling adhesive, a sealer, a bearing pad, and the glue that binds the units together while keeping them apart.

Mortar can be produced either on-site in a concrete mixer or off-site by specialised vendors in a factory-controlled environment. In both cases, the manufacturing of the cement involves premium cement, sand, and occasionally lime. Commercial use of Portland cement in the 1920s, which greatly increased the possibility for masonry construction, is what gives today's mortars their high quality.

In comparison to earlier methods that used diverse kinds of lime, new mortar materials offered far more consistent strength growth and a significantly faster rate of strength rise. This made it possible to plan and carry out construction much more quickly. Although the working characteristics of the limes used in mortar prior were satisfactory, the rate of strength increase was slow, especially during colder months. This meant that even a high-quality lime with a decent ultimate strength would provide significant difficulties when used in the winter. With the clear exception of instances of really severe winter weather with heavy precipitation or subfreezing temperatures, the availability of the new Portland cements changed this and allowed construction to continue year-round.

2 Objectives

An effort is made to lessen the effect of industrial waste on the ecosystem by employing it in the construction industry and to solve the issues caused by the depletion and obsolescence of raw materials used to manufacture traditional building materials. The goals of the current effort are listed below.

1. To study the use of red mud as an additional cementitious material (CM) and how it affects mortar's compressive strength
2. To examine the impact on strength parameters of red mud (as a cement substitute) and used manufactured sand (as a fine aggregate substitute) in the mortar mix.
3. To check the possibility of producing mortar by substituting industrial wastes for traditional raw materials, environmental contamination may be reduced and our natural resources can be preserved.

3 Methodology

The following steps should be taken for the preparation of Geo-polymer Mortar

1. Procuring materials mentioned in Sect. 3.1.
2. Characterization of raw materials.
3. Determining the mix quantity for all combinations of Red mud and GGBS to produce geo-polymer masonry mortar.
4. Producing geo-polymer cubes using different proportion of Red mud, M sand, Metakaolin and GGBS using NaOH and Na2SiO3 as alkaline activators.
5. Determination of strength properties after curing.

3.1 Materials Used

Geopolymers are still not widely employed around the world despite having significant advantages in terms of economic, environmental, and structural performance. Fly ash is typically the chosen material for the synthesis of geopolymers, and significant research has been done on geopolymers based on fly ash.

- Red mud: India is said to produce 4.71 million tonnes of red mud, or 6.7% of the approximately 70 million tonnes of red mud produced worldwide. [7]. Currently used disposal techniques include dry stacking, lagooning, ocean discharge, and dry disposal. [9].
- Metakaolin: poly (sialate) geopolymers, which are low Si/Al geopolymer systems, are formed using a chemical process. The aluminosilicate minerals rely on metakaolin's dissolution and hydrolysis [8] The aluminate and silicate species' condensation processes were investigated [2]
- Ground granulated blast furnace slag

Table 1 Properties of Redmud & GGBS

SI. No	Particular	Red Mud	GGBS
1	Colour	Red	White
2	pH	10.52	–
3	Fineness	–	17%
5	Specific Gravity	3.16	2.9

- Manufactured sand
- Chemical Activator
    - Sodium Silicate
    - Sodium Hydroxide (2Molarity).

3.2 Characterization of Raw Material

Red Mud
The Hindalco refinery in Belgaum, Karnataka, India, provided the red mud. It first appeared as rocks that were coarsely processed to break down the larger lumps so that research could be done on them. A digital pH metre was used to evaluate the sample's pH (Table 1).

Metakaolin
When kaolin, a mineral found in china clay, heated until a certain temperature between 600 and 800 deg C, metakaolin is produced as a byproduct. It is the best pozzolanic component for concrete application. Metakaolin's particles are not as fine as silica fume but are smaller than cement particles.

Ground Granulated Blast Furnace Slag
A leftover from the iron industry. To help the red mud-based geopolymer in the current study set up quickly at room temperature, GGBS is added.

Manufactured Sand
To make it, hard stones are crushed into sand-sized, angular fragments, which are then cleaned and carefully graded for use as construction aggregate. It is a superior option for building than river sand.

3.3 Production of Geopolymer Mortar

The preliminary research on mix proportioning, processing, and behavior of red mud-based geopolymer composites is presented in this part. In contrast to the manufacturing of traditional concrete materials, where the water cement ratio is based on both the amount of water needed for hydration and the requirements for workability, the water content for geopolymer materials shall be based only on those needs [1]. The flow test is typically thought of as a standard process to determine the water content [6]. Consequently, early research through trial flow experiments is crucial to the creation of geopolymers. It should be highlighted that this also impacts molarity. Test specimens in the shape of 50 mm and 70 mm cubes were manufactured as well as a variety of trial mixtures of red mud-alkali based composites.

3.4 Geopolymer Combinations Used

The investigation of strength growth and water absorption by the polymerization of various combinations of red mud, m sand, metakaolin, and GGBS is the main goal of this work, as previously said. Studies on compressive strength and water absorption must be completed in order to determine whether the method for making mortar may be used in other situations. Initial red mud-alkali based composite experimental mixes were prepared in large quantities (Table 2).

Table 2 Test results for all mixes of Geopolymer cubes

SI. No.	Mix Composition in %				Test results			
	Mix	Red mud (%)	Metakaolin (%)	M sand (%)	GGBS (%)	Qty of alkaline solution (%)	Water Absor ption in %	Compress ive Strength in MPa
1	M1	40	5	35	20	24	8.77	18.0
2	M2	30	10	40	20	23	9.56	16.9
3	M3	35	10	30	25	24	10.10	18.9
4	M4	50	10	25	15	27	9.15	14.1
5	M5	30	10	30	30	25	10.4	33.7
6	M6	30	10	20	40	25	18.9	37.9
7	M7	30	10	10	50	24	10.06	38.1
8	M8	50	10	20	20	27	9.68	14.8
9	M9	60	10	10	20	28	10.8	16.5

4 Production of Geopolymer Mix

The investigation of red mud-based geopolymer binder determines its potential application in the building sector [3]. A crucial component of building constructions are brick, mortar. The study to synthesise brick mortar using red mud and examine their mechanical properties was motivated by the high demand for masonry materials (Fig. 1).

4.1 *Mix Calculation for Geopolymer Mortar Mix*

Chemical activator contains NaOH / Na2SiO3 of 1: 2.5. 1Lit of water is mixed with 80 g of NaOH.

For 70 mm Cubes
For cube volume of $3.43\text{x}10^{-4}\text{m}^{3}$, Mass of Soil added is 0.755 kg, Quantity of solution = 755x 0.32 = 241.6 g.

4.2 *Mixing*

The trial mixes were dry mixed followed by wet mixing by adding chemical activator then the mixtures became cohesive. Geopolymer composite is poured into cubes and kept for drying. The prepared samples of geopolymer cubes were demoulded after 24hrs, the produced cubes are kept in ambient temperature for 7 days then they are checked for absorption of water and saturated mechanical properties after being submerged in water for 24 hours. For each mix proportion, the aforementioned process was repeated Fig. 2 and 3.

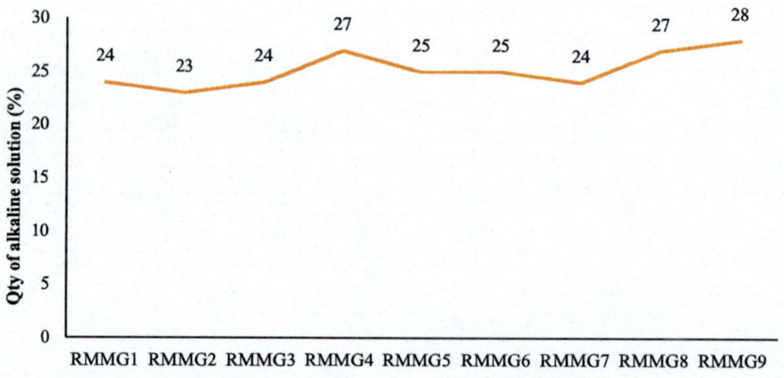

Fig. 1 Flow Test Results of all mix composites found by Flow table

Fig. 2 Produced Geo polymer cube specimens for testing

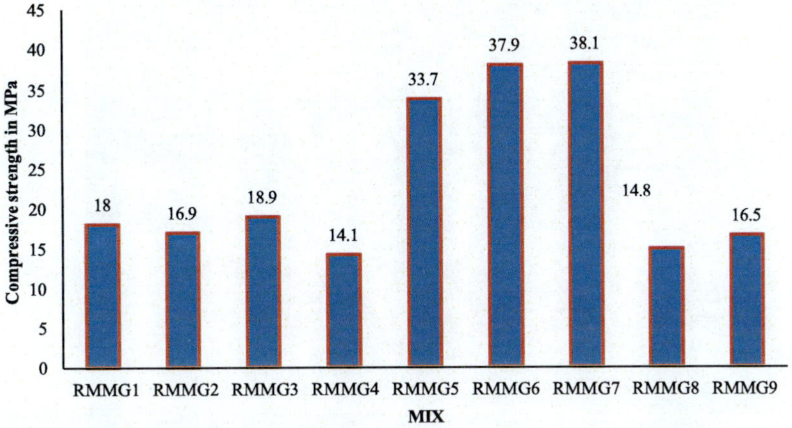

Fig. 3 Compression Strength of geopolymer mix composites

According to Indian Standard IS 3495 (Part 2):1992 the average wet CS of mortar is more than 3.5 Mpa (Fig. 4).

The prism was cast using four bricks that almost all had mortar that was 5 mm thick. The sample was studied seven days after ambient sun curing. The vertical fracture moved first from top brick to the second and third bricks, as seen in the image. The prism failed as a result of cracks that started in the mortar and spread to the bricks (Fig. 5).

Fig. 4 Brick masonry prism

Fig. 5 Failure pattern of brick prism

4.3 Discussion

Depending on the bauxites used and the process parameters, red mud is a very complex product. Therefore, red mud should be seen as a class of substances with certain characteristics, such as being created during the refining of bauxite and being extremely alkaline. Various elements and mineralogical phases in iron oxides make up the majority of the material. Significantly higher specific surface. The size distribution of fine particles (Table 3).

The greater surface energy of the particles and stronger flocculability are caused by the greater number of tiny particles. Red mud is hence highly retentive and viscous.The increase in Metakaolin (from 5% to 10%) shows increase in compressive strength. When we mix Red mud + M sand + Metakaolin + GGBS the strength of the cube will be more comparing to the mixing of only Red mud +Metakaolin + GGBS.

All specimens with red mud percentage exhibited excellent strength properties as compared to conventional mix. Moreover, all variations of red mud demonstrated better strength, where admixtures contribute at the higher level. Presence of GGBS aided in the setting of red mud, which was otherwise difficult to set at ambient temperature.

It is important to note that the presence of red mud impacted the strength for all variations. This was rather unprecedented because it was the fly ash that contained much greater percentage of reactive silica than red mud. Hence, it was feasible to conclude that red mud's iron content was what differentiated it from fly ash because fly ash had a larger percentage of alumina. Red silt geopolymeric's potential for use in the construction industry is determined by the inquiry. Bricks are a key element in building construction. Due to a shortage of masonry units, an experiment was conducted to develop mortar from red mud and evaluate their mechanical properties.

Table 3 Compression strength test results for brick prism with geopolymer mortar mix

Mix Combination				Compressive Strength in MPa	Standard values as per IS Code 3495
Red mud(%)	Metakaolin(%)	M sand(%)	GGBS(%)		Should be greater than 3.5 Mpa
50	10	25	15	5.9	

5 Conclusion

The following is a list of results reached as a result of the research on geopolymer composites for civil engineering applications.

1. This study, primarily aimed at production of geopolymer mortar using varying proportions of Red mud, M sand, Metakaolin and GGBS as raw materials presents the various trial mixes carried out prior to the production of actual geopolymer mortar.
2. Use of red mud, metakaolin, M sand and GGBS based in geo polymer technology to play vital role in mechanical properties of mortar.
3. With red mud content of 30 to 60% and alkaline activator content with 2 M concentration, compressive strength obtained in range of 14.1–38.1 N/mm^2, water absorption gets 8.77 to 18.9% which is within the conforming range as per standards. According to the experimental findings, geopolymerization of red mud, metakaolin, M sand, and GGBS can be utilised as a long-lasting replacement for conventional mortar.
4. The molarity of the NaOH and the ratio of $NaOH / Na_2SiO_3$ plays a key role in the workability of the mix and also impacts on the durability of the specimens. Red Mud can play a major role in replacing huge amount of fertile clay and cement that is being used in mortar.

References

1. Duxson, P., Fernández-Jiménez, A., Provis, J.L., Lukey, G.C., Palomo, A., & van Deventer, J.S. (2007). Geopolymer technology: the current state of the art. *Journal of Material Science*, 42(9), 2917–2933
2. Sagoe-Crentsil, K., & Weng, L. (2007). Dissolution processes, hydrolysis and condensation reactions during geopolymer synthesis: Part II—High Si /Al ratio systems". *Journal of Materials Science, 42*, 3007–3014.
3. Zhang, G., He, J., & Gambrell, R.P. (2010). Synthesis, Characterization, and mechanical properties of red mud–based geopolymers. *Journal of the Transportation Research Board, Washington*
4. Ribeiro, D.V., & Morelli, M.R. (2011). Use of red mud as addition for Portland cement mortars Brazilian congress on ceramics, Porto de Galinhas, PE, Brasil
5. Rai, S., Wasewar, K. L., Mukhopadhyay, J., Chang, K. Y., Hasan, U., & Bandopadhyay, A. (2012). Neutralization and utilization of red mud for its better waste management. *Arch Environ Sci, 6*, 13–33.
6. He, J., Jie, Y., Zhang, J., Yu, Y., & Zhang, G. (2013). Synthesis and characterization of red mud and rice husk ash-based geopolymer composites. *Cement and Concrete Composites, 37*, 108–118.
7. Ahmari, S., & Zhang, L. (2013). Utilization of cement kiln dust (CKD) to enhance mine tailings-based geopolymer bricks. *Construction and Building Materials, 40*, 1002–1011.
8. Reddy, G.N., & Chandra, K.S. (2014). Characterization and comprehensive utilization of red mud - an overview. *International Journal for Scientific Research & Development*, 2, 2321–0613
9. Ashok, P., & Sureshkumar, M.P. (2014). Experimental studies on concrete utilizing red mud as a partial replacement of cement with hydrated lime

10. Mudduli, S.D., Nayak, B.D., & Mishra, B.K. (2014). Geo-polymer GGBS building bricks by atmospheric curing, Sadguru Publication, Udaipur
11. Sowmyashree, T., Arpitha, D. J., Manjunath, C. S., & Siddu Karthik, C. S. (2016). Study on mechanical properties of red mud as a partial replacement of cement with hydrated lime for M40 grade concrete with superplasticizer. *International Journal of Research in Engineering and Technology, 5*(4), 2319.
12. Sowmyashree, T., Manjunath, C. S., Arpitha, D. J., & Panditharadhya, B. J. (2016). Comparison on durability properties of red mud as a partial replacement of cement with hydrated lime for different grades of concrete with and without superplasticiser. *Int J Eng Res Technol, 5*(5), 572–577.

Service Life Estimation of Reinforced Concrete Bridges

B. Kavitha and M. B. Anoop

Abstract The objective of the present study is to estimate of service life of reinforced concrete structural members subjected to chloride-induced corrosion of steel reinforcement. Based on the review of literature, a procedure for service life estimation of reinforced concrete structural elements has been identified. Though the procedure used in this thesis is applicable for different types of structural components, a T-beam girder of 18 m span bridge was selected to illustrate the procedure. The bridge is modelled using a commercial finite element analysis package (SAP 2000) and analysed for IRC load cases. The service lives of the bridge girder with respect to minimum required load carrying capacity are determined for different exposure conditions as defined in IS456-2000. Two approaches are considered in the present study for meeting the service life requirement in the very severe and extreme conditions: i) Use of different maintenance methods, and ii) Use of High-Performance Concrete. Four different maintenance options (silane surface coating, cathodic protection, rebuilding and cover replacement) are considered, and their relative performance is studied using life-cycle costing. It is noted that the procedure for estimation of service life will be useful for scheduling the maintenance activities. It is also found that for very severe and extreme exposure conditions, use of High-Performance Concrete is a feasible alternative for satisfying the service life requirement.

B. Kavitha (✉)
NIT Warangal, Hanamkonda, India
e-mail: kavithab@nitw.ac.in

M. B. Anoop
CSIR- SERC, Chennai, India
e-mail: anoop@serc.res.in

A. Vilventhan et al. (eds.), *Advances in Construction Materials and Management*, Lecture Notes in Civil Engineering 346,
https://doi.org/10.1007/978-981-99-2552-0_26

1 Introduction

The chloride-induced corrosion of the reinforcement steel in concrete is one of the significant mechanisms which reduce the capacity of RC structures located in coastal areas and other severe environments. The reduction in area of steel reinforcement will lead to the reduction of resistance, referred from Anoop et al. 2003 and 2012. The corrosion problems in the reinforced concrete (RC) bridges has increased significantly in the past few decades. The corrosion protection measures, corrosion-inhibiting admixtures and coated reinforcing steel will help in reducing corrosion in the new RC bridges. The present study proposes a procedure for estimating service life of reinforced concrete (RC) flexural elements.

A literature review was conducted on the experimental studies and analytical procedures related to service life prediction of RC structural elements.○

[4] presented a probabilistic model to estimate the costs for repair and replacements. Their model can be used to estimate life-cycle costs for RC structures in marine environments. [7] developed the life-cycle maintenance plan for deteriorating bridges. They considered the lifetime condition, safety levels and life-cycle maintenance amount as objective functions in their model. Later, [11] formulated a model that estimates the time taken from corrosion initiation to corrosion cracking by developing a relationship between the internal radial pressure and the steel mass loss caused by the expansion of corrosion products. The long-term corrosion process of RC beams is studied by [10]. The RC members were kept in a chloride environment for 17 years under service loading and cracking maps were drawn for those beams. The total chloride contents were measured along with that few mechanical tests were also performed. They proved that the bending cracks and their widths have their effect on service lives. [1] observed the effect of reinforcement steel corrosion on flexural behaviour of RC beams. He observed that the product of corrosion current density and corrosion period ($I_{corr}T$) is the important factor which affects the flexural strength of RC beam [8]. They used two RC beams under sustained load in a chloride environment for 14 and 23 years respectively. In 2010, [6] modelled the chloride penetration rate for concretes in Tidal zone. They have calculated the efficiency factors of mineral admixtures by considering the basic parameters as type and content of mineral admixtures and w/c ratio.

From the review of literature, a procedure for estimating the service life of RC structural elements has been identified. Fick's second law is selected for determining time for corrosion initiation, Rodriguez et al. model is selected for determining rate of corrosion and Tuutti's schematic model is selected for estimating service life.

2 Service Life Estimation of the Bridge

The service life of any RC structure can be spilt in two different phases. The starting phase is the corrosion initiation phase and the second phase is propagation of corrosion.

Initiation Period (t_i):
During initiation phase aggressive substances (co_2, chloride ions) that can depassivate the steel penetrates from the surface into deep concrete.

The length of initiation period is controlled by the rate of diffusion of the chloride ions in the concrete. The diffusion process follows Fick's second law of diffusion mentioned with Eq. 1 below (Adolf Fick, 1855).

$$\frac{\partial c_f}{\partial t} = D\frac{\partial^2 c_f}{\partial x^2} \quad (1)$$

where, D = diffusion coefficient (m^2/s); t = time (s); x = distance from concrete to the reinforcement (m). The initiation period is usually much longer than the propagation period.

Propagation Period (t_p):
The second phase is the propagation period, t_p, the time between corrosion initiation and cracking. This phase lasts until an unacceptable degree of corrosion damage has occurred.

Time for Corrosion Initiation:
The Fick's second law of diffusion is used to compute the time for corrosion initiation in Eq. 2 (Adolf Fick, Non-steady state diffusion equation, 1855).

$$t_i = \frac{c^2}{4D}[erf^{-1}\left(\frac{C_s - C_{cr}}{C_s}\right)]^{-2} \quad (2)$$

where,

c- clear cover (m); D - diffusion coefficient (m^2/s); c_{cr} - critical chloride concentration (ppm); c_s - surface chloride concentration (ppm).

Rate of Corrosion Process:
Using Rodriguez et al. model, the rate of corrosion can be estimated in Eq. 3 as

$$r_{corr} = 0.0115 I_{corr}\alpha \quad (3)$$

where

r_{corr} is the rate of corrosion (mm/year);

I_{corr} is the average value of corrosion current density, ($\mu A/cm^2$);

α is the factor for incorporating the effect of highly localised pitting corrosion (4 to 8); and 0·0115 is for converting $\mu A/cm^2$ to mm/year.

Remaining Diameter:
From [2], the remaining diameter of the steel reinforcement at any time t, $\Phi(t)$ is considered to be in Eq. 4.

$$\Phi(t) = \Phi(0) - r_{corr}(t - t_i) \tag{4}$$

where
$\Phi(t)$ - the remaining dia of the steel reinforcement (mm);
$\Phi(0)$ - the initial dia of the steel reinforcement (mm);
t_i - time taken for corrosion initiation (years).

2.1 *Procedure*

The procedure to estimate the service life of RC flexural members as follows:

i. Select a reinforced concrete structural element for estimating durability-based service life. For this study a 18 m T-Beam bridge girder is selected from Standard Plans for Highway Bridges [9].
ii. Select the exposure condition of the member.
iii. Estimate the service life, t_{SL} of the particular member based on the codes of practice. For this study required service life is fixed as 60 years.
iv. Design the flexural member for satisfying safety and serviceability requirements. (Design from Standard Plans for Highway Bridges [9] adopted in the present study)
v. Calculate the service load moment, Mo, at a particular section
vi. Choose C_s, C_{cr} and D values based on grade and type of concrete, w/c ratio and exposure condition [12]
vii. Estimate t_i, using Fick's second law of diffusion.
viii. Choose values for α and I_{corr} based on different exposure conditions and w/c ratio.
ix. Determine the rate of corrosion of reinforcement, r_{corr}. The Eq. (3) can be used to estimate the rate of corrosion of steel reinforcement.
x. For t > ti, estimate the remaining dia from Eq. (4).
xi. Calculate the area of remaining steel reinforcement.
xii. Estimate the moment carrying capacity of the particular section with the remaining area of reinforcement.
xiii. Using provided moment carrying capacity and required moment carrying capacity, find capacity ratio (υ).
xiv. The service life of the member can be considered as finished when υ is falling below unity.

3 Finite Element Modelling

SAP Models and Results:

From the Standard Plans for Highway Bridges, by the Indian Roads Congress, Ministry of Surface Transport, New Delhi, a plan of bridge with 18 m span is selected. The length of the bridge is 12 m with 4 longitudinal and 3 cross girders.

Description:

1) The considered design loads are:

- Higher value of Carriage way load-One lane of IRC class 70R or two lanes of class A
- Wearing coat load is considered as 2kN/sq.m
- Footpath load is taken as 5kN/sq.m

2) All reinforcing steel should be of HYSD Bars (Grade designation S 415) conforming to IS: 1786 (except for mesh reinforcement which shall be MS bars grade designation S 240 conforming to IS: 431 part-1 mild steel). The general plan of T-Beam bridge is shown in the Fig. 1 and the detailing of longitudinal girder is shown in Fig. 2. The section details are adopted as such and modelled using finite element package SAP2000. The model is analysed for IRC load combinations and the maximum required moment capacity is found.

Capacity Ratio (υ)

Capacity ratio is the ratio between provided moment carrying capacity at 't' and the required moment carrying capacity at time 't', shown below as Eq. 5.

$$\text{Capacity ratio } (\upsilon) = \frac{\text{PMC (t)}}{\text{RMC (t)}} \tag{5}$$

where, PMC- provided moment carrying capacity in kNm and RMC-required moment carrying capacity in kNm.

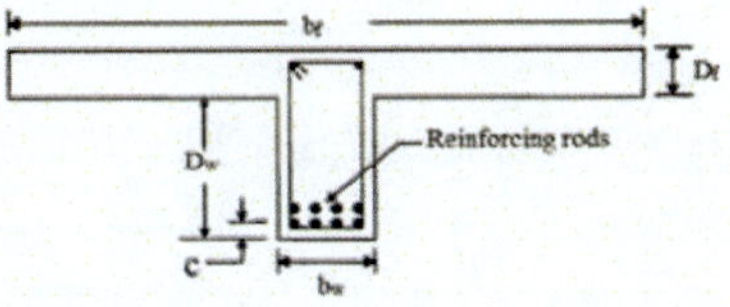

Fig. 1 T- beam cross-section

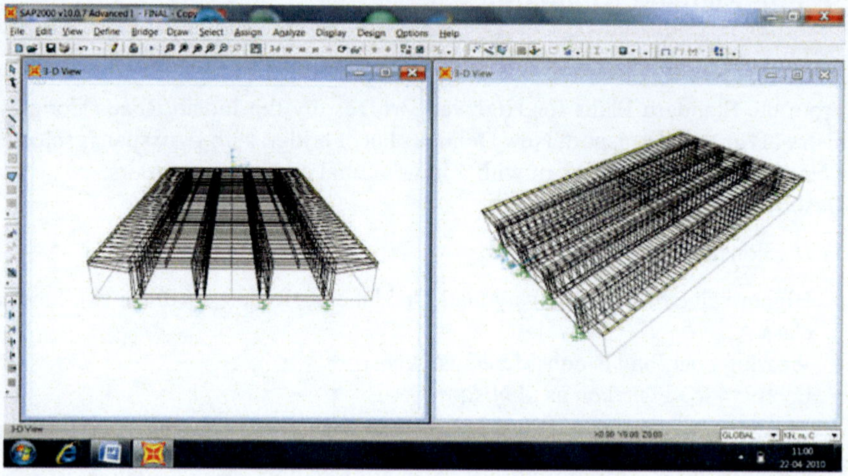

Fig. 2 Detailing of longitudinal girder of a T-beam bridge in SAP 2000

4 Service Life Estimation

The parameters used to find service lives in various exposure conditions are given in Table 1, (parameters referred from *ACI Committee 20, a guide to durable concrete, Chapter IV, 1994 and chloride diffusion coefficient of concrete using open-circuit potential measurements by [13].

The degradation in capacity ratio with time 't', due to chloride induced corrosion of steel reinforcement in bridge girder exposed to different environmental conditions (referred from [3]), are shown in Fig. 3.

The required service life of the beam is fixed as 60 years. As noted from Fig. 3 for moderate and severe exposure conditions the section adopted for bridge girder found to be adequate. However, in very severe and extreme exposure conditions, the capacity ratio decreasing below unity i.e. the service life of bridge girder is below

Table 1 Considered parameters for various exposures

Sl.no	Exposure condition			
	Moderate	Severe	Very Severe	Extreme
D* (cm^2/s)	8×10^{-8}	6.5×10^{-8}	6.5×10^{-8}	6.5×10^{-8}
C_s *(%)	0.25	0.5	0.75	0.9
C_{cr}* (%)	0.13	0.13	0.13	0.13
A	6.52	6.52	8	8
I_{corr}	0.591	2.063	12.88	41.13
Grade of concrete	M25	M30	M30	M30
Water to cement ratio	0.45	0.4	0.4	0.4

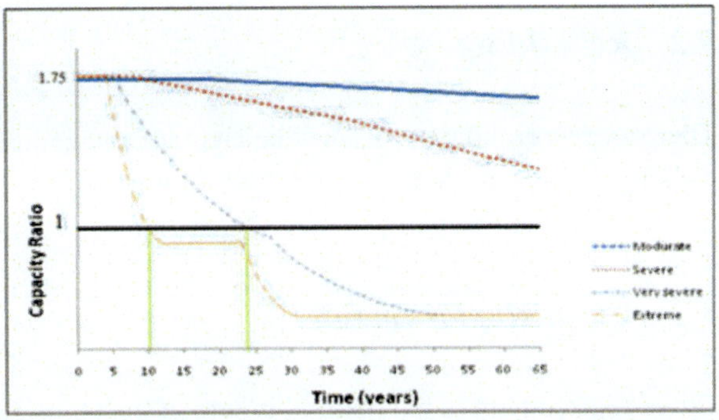

Fig. 3 Service life of bridge girder

Table 2 The values of service life for the bridge girder in different exposure conditions

S. No	Exposure Conditions	Service Life (years)
1	Moderate	> 60
2	Severe	> 60
3	Very severe	24
4	Extreme	10

60 years. The values of estimated service life for different exposure conditions are given in Table 2.

It can be observed from the Table 2 that the estimated service life is very lower than 60 years in very severe and extreme exposure conditions, suitable measures need to be taken to ensure the service life in these exposure conditions. Two such measures that are being increasingly used are considered in the present study, they are i) Use of different maintenance methods and ii) use of high-performance concrete.

5 Maintenance Options

5.1 Surface Coating

The surface coating is used to improve the durability of RC structures by increasing its resistence to carbonation, chloride ingress, sulphate ingress, corrosoin of embedded steel,etc. The silane films are consisting of dense, well-packed and homogeneous Si–O rich layers.

5.2 Rebuilding

There are few circumstances in which it is advantageous to rebuild the whole structure instead of going for some other repair options.

5.3 Cover Replacement

Cover replacement is the method in which the chloride ions can be prevented from reaching the reinforcements embedded in the concrete by replacing the cover. This can be done by chipping off the concrete cover to sufficient thickness and applying fresh concrete over the chipped surface. Bonding agent is also used prior to applying the cover concrete for a good bonding between the new concrete and the old concrete.

5.4 Cathodic Protection

Cathodic protection (CP) will stop corrosion in existing RC structures. It makes the metal a cathode via a direct current, or it may connect it to a sacrificial anode. The cathodic areas in the electrochemical cell will not get corroded.

5.5 Life Cycle (LC) Cost Analysis

The LC cost of an alternative option anticipates all the costs that are may likely to occur over the service life. It includes costs of the initial investment, inspection and maintenance, repair, rehabilitation/replacement charges.

5.6 Cumulative Life-Cycle Maintenance Cost

Any maintenance option can be selected by considering the direct and indirect costs. Equation 6 gives the cumulative life cycle cost.

$$C_{pv} = \Sigma_{i=1}^{N} \frac{c_{t_i}}{(1+u)^{t_i}} \tag{6}$$

where, C_{pv} = the cumulative life cycle cost (present value)
C_{ti} = cost for maintenance option selected and applied at time t_i;
υ = discount rate; N = number of maintenance cycles needs.

5.7 Comparison of Maintenance Options

Number of maintenance interventions required for bridge girder in very severe and extreme exposure conditions are calculated based on the service life and duration of maintenance effect and is tabulated as below in Table 3:

The life cycle cost required in terms of initial cost for bridge girder by all the four maintenance options are calculated for very severe and extreme exposure conditions and are represented in the Fig. 5 and 6. By comparing life cycle costs of maintenance options namely surface coating, cathodic protection, rebuilding and cover replacement, it is found to adopt surface coating as an economical maintenance method in both very severe and extreme exposure conditions.

Table 3 Different maintenance options and number of cycles

Exposure Conditions/ Maintenance Options	Number of Maintenance Interventions			
	Surface Coating	Cathodic Protection	Rebuilding	Cover replacement
Very Severe	6	5	2	9
Extreme	6	5	5	11

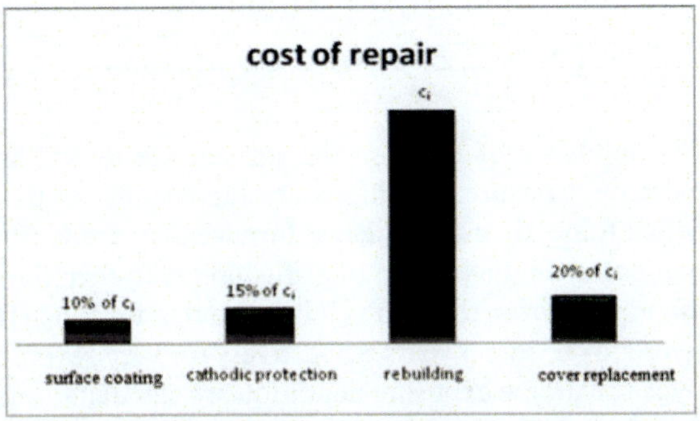

Fig. 4 The cost required for each maintenance option

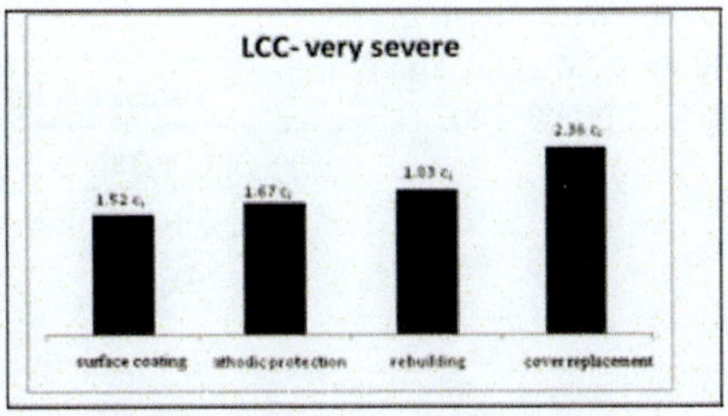

Fig. 5 LLC for very severe condition

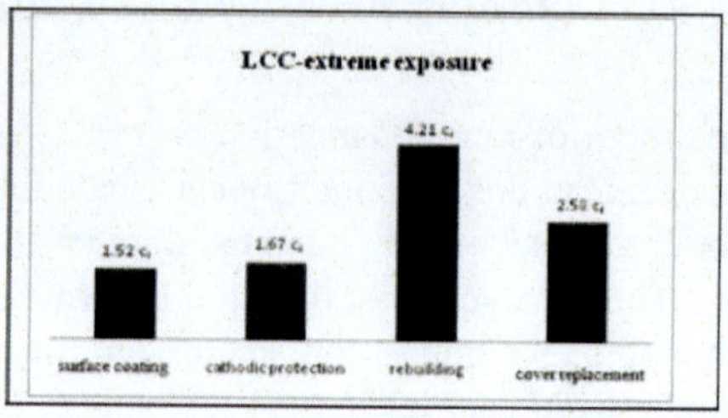

Fig. 6 LLC for extreme condition

6 High Performance Concrete

High performance concrete is required to exhibit enhanced strength, durability, or serviceability in order to meet requirements of specific projects. In establishing those specified properties careful consideration needs to be given to design requirements, properties of locally available materials and the construction method. Slag, Silica fume and Fly ash have been accepted as supplementary cementitious materials in HPC. For mix design, ACI Committee method is adopted. 1:1.64:2.12:0.39 proportion is taken as a basic proportion for M30 and appropriate mix proportions for 30% replacement with mineral admixtures are given in Table 4:

6.1 Different Exposure Conditions

The influence of HPC on the service life of RC bridge girder in very severe and extreme exposure conditions are represented in the Fig. 7 and 8 respectively. The Silica fume, fly ash and Blast furnace slag are used as a mineral admixture in HPC and based on the change in diffusion coefficient the service lives are calculated and compared. From the results it is noted that for Silica fume, fly ash and slag the service lives are 46, 56, 53 years respectively for very severe exposure condition. The service lives in extreme exposure condition are calculated and are compared. For silica fume, fly ash and slag the service lives are 31, 38, 35 years respectively.

Table 4 Mineral Admixture vs Mix proportion

Mineral Admixture	Mix proportion
Fly Ash	1: 1.55: 2.12: 0.39
Silica Fume	1: 1.53: 2.12: 0.39
Slag	1: 1.62: 2.12: 0.39

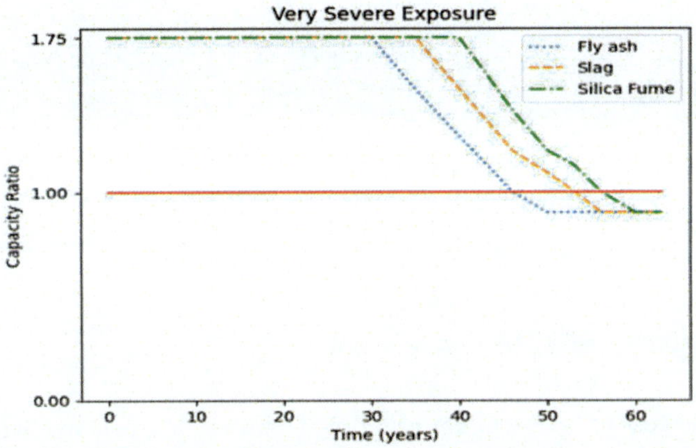

Fig. 7 Service life of RC bridge with HPC in very severe exposure

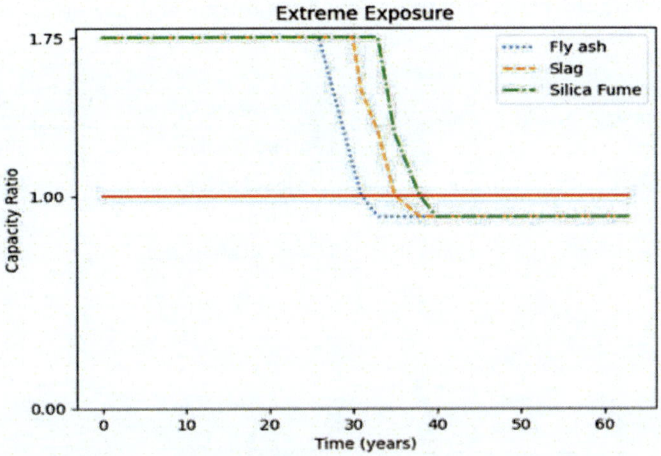

Fig. 8 Service life of RC bridge with HPC in extreme exposure

6.2 Comparison

The service lives of bridge girder in very severe and extreme exposure conditions using HPC with mineral admixtures namely Silica Fume, Slag and Fly Ash are compared with that of ordinary concrete and listed in Table 5. From the results it is noted that usage of HPC will increase the service life more than two times.

Table 5 Various mineral admixtures in different exposure conditions

Mineral Admixture Used	Service Life (years)	
	Very severe	Extreme
OPC	24	10
Fly Ash	46	31
Silica Fume	56	38
Slag	53	35

7 Comparison of Life-Cycle Cost

The life cycle cost of bridge girder using high performance concrete with different maintenance options are estimated and noted in Table 6 and are illustrated in Fig. 9.

Where C_2 is the Initial Cost. From the above figure it is noted that the cover replacement method is an economical maintenance option for the bridge girder considered in the very severe and extreme exposure conditions using high performance concrete.

Table 6 Different maintenance options

S. No	Maintenance Options	C2
1	Surface Coating	1.49
2	Cathodic protection	1.63
3	Rebuilding	1.94
4	Cover Replacement	1.19

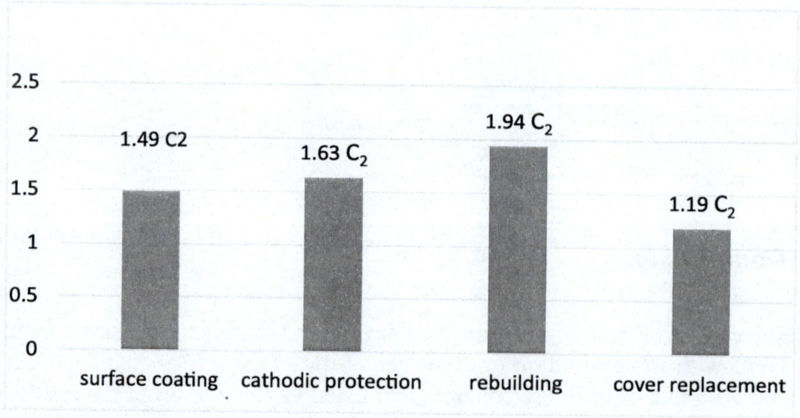

Fig. 9 Comparison of life cycle cost

8 Conclusions

Equal emphasis should be placed on developing methods to prevent corrosion in all new RC bridges to be built in the future. As IS456 [5] is updated on a durability basis, standard bridge plans should be modified accordingly. The obtained results show that the condition of a large number of bridges in coastal cities is concerning. By comparing the life-cycle costs of four different maintenance options, it is discovered that surface coating is an economical maintenance method in both very severe and extreme exposure conditions. According to the findings of this study, using HPC will increase the service life of structures by more than twofold. It can also be seen that the cover replacement method is a cost-effective maintenance option when using high-performance concrete in both very severe and extreme exposure conditions.

References

1. Azad, A.K., Ahmad, S., & Azher, S.A. (2007). Residual strength of corrosion-damaged reinforced concrete beams. *ACI material journal*, 104-M05, 40–47
2. Anoop, M. B., Balaji Rao, K., & Appa Rao, T. V. S. R. (2003). A methodology for durability-based service life design of reinforced concrete flexural members. *Magazine of Concrete Research, 55*(3), 289–303.
3. Anoop, M. B., Raghuprasad, B. K., & Balaji Rao, K. (2012). A refined methodology for durability-based service life estimation of reinforced concrete structural elements considering fuzzy and random uncertainties. *Computer-Aided Civil and Infrastructure Engineering, 27*(3), 170–186.
4. Val, D. V., & Stewart, M. G. (2003). Life-cycle cost analysis of reinforced concrete structures in marine environments. *Structural Safety, 25*, 343–362.
5. Indian Standard- Code of Practice for Plain And Reinforced Concrete- Fourth Revision. IS: 456–2000
6. Bermúdez, M.A., & Alaejos, P. (2010). Models for chloride diffusion coefficients of concretes in tidal zone. *ACI Materials Journal*, 3–11
7. Liu, M., & Frangopol, D.M. (2005). Multiobjective, Maintenance planning optimization for deteriorating bridges considering condition, safety and life-cycle cost. *Journal of Structural Engineering, ASCE*, 833–842
8. Zhang, R., Castel, A., & François, R. (2009). The corrosion pattern of reinforcement and its influence on serviceability of reinforced concrete members in chloride environment. *Cement and Concrete Research, 39*, 1077–1086.
9. Standard Plans For Highway Bridges, Ministry Of Surface Transport, Road Wing, New Delhi, R.C.C T-Beam Bridge And Slab Superstructure
10. Vidal, T., Castel, A., & François, R. (2007). Corrosion process and structural performance of a 17 year old reinforced concrete beam stored in chloride environment. *Cement and Concrete Research, 37*, 1551–1561.
11. El Maaddawy, T., & Soudki, K. (2007). A model for prediction of time from corrosion initiation to corrosion cracking. *Cement & Concrete Composites, 29*, 168–175.
12. ACI Committee 20, guide to durable concrete, Chapter IV (1994)
13. Erdogˇdua, S.ß., Kondratovab, I.L., Bremnerc, T.W. (2003). Determination of chloride diffusion coefficient of concrete using open-circuit potential measurements. Cement & Concrete Research

An Experimental Investigation on Fresh Characteristics of Self Compacting Waste Plastic Fiber Reinforced Concrete

Khanapuram Anand Goud, B. Bhavani, and S. Manideepa Sai

Abstract Self-compacting concrete (SCC) is a concrete which is having high filling ability, passing ability, high segregation resistance without applying any external force. Because of its flow properties, concrete will pass through congested reinforcement with full compaction. At present time domestic plastic waste causing high damage to the environment which damages the eco system and reduces the yield of the soil. Hence an attempt has been made to determine whether they can be used successfully in self-compacting concrete (SCC) to increase the properties. Waste plastic fiber reinforced self-compacting concrete is a composite material made up of cement, fine aggregates, and coarse aggregates and plastic fibers (PET-Polyethylene Terephthalate). The waste plastic fibers are replaced with fine aggregate to reduce the utilization of natural sand. The current study focuses on the fresh properties of SCC with polyethylene terephthalate (PET) fibers. Various percentages of waste plastic fibers like 2.5%, 5%, 7.5%, and 10% are added by weight of fine aggregate. The fresh properties of self-compacting concrete, such as flow ability, passing ability, segregation resistance, and filling ability, were studied at each percentage of plastic fibres. The findings of this study show that the fresh properties of self-compacting concrete with waste plastic fibre reinforcement meet the workability requirements as per the EFNARC guidelines.

Keywords Self-compacting concrete · Cement · PET (Poly ethylene terephthalate) · Waste plastic fibers

1 Introduction

In 1988, Japan was the first country to achieve a long-lasting concrete structure. Prior to 1998, and for a long time beginning in 1983, Japan was particularly interested in addressing the issue of concrete toughness.

K. Anand Goud (✉) · B. Bhavani · S. Manideepa Sai
Institute of Aeronautical Engineering, Dundigal, Hyderabad 500043, Telangana, India
e-mail: anandgoud.k3@gmail.com

A. Vilventhan et al. (eds.), *Advances in Construction Materials and Management*, Lecture Notes in Civil Engineering 346,
https://doi.org/10.1007/978-981-99-2552-0_27

Self-compacting concrete constantly anticipated that would be more grounded and tougher than before while being expense and strength effective. The significant favorable circumstances that concrete has over the constructional substance must be preserved. The chance of being manufacture relatively in any region, the capacity to make the frame forced by the state of a shape and an ease part.

The aspect of self-compacting Concrete is primarily determined by the materials used. The primary characteristics that influence self-compacting concrete in its fresh state, [1] The most vital is that self-compacting concrete properties can be allowed to fluctuate with in substantially stricter restrictions.

Self-Compacting Concrete is one of the main reasons to change the aspect of building construction. SCC does not require any skill labor. In view of vibration being pointless, the commotion is diminished and the danger of creating issues because of the utilization of vibrating hardware is lessened. The SCC properties are considered from EFNARC.

The mix design procedure mainly focused on flow properties under its own weight without applying external forces. Only self-compacting concrete can be categorized as a concrete sample [2, 5]. It has the various attributes.

1. **Filling ability**:
   The ability of concrete to completely fill a form work without leaving voids is known as filling ability. It involves the material's capacity to flow freely under its own weight and, if necessary, both horizontally and vertically upward.
2. **Passing ability:**
   The ability of concrete capacity to flow through clogged reinforcement, like a space between steel reinforcing bars, while bearing its own weight. Testing methods such as L-box, Fill box, U-box, and J-ring can be used to determine passing ability.
3. **Segregation resistance:**
   The aggregate particle distribution is still coherent in both the vertical and horizontal directions.

The properties of self-compacting concrete have been tested using a variety of methods. Slump flow test, V-Funnel test, T 50 test, L-Box test, U-box test, and J-Ring test are among the various tests used to ascertain workability.

2 Materials Used

In this study, Portland pozzolana cement was utilized. Fine aggregate of size 2.36 mm is used for preparing SCC sample. Coarse aggregates of size 10 mm and 16 mm were used. Potable water is used as per IS 456:2000. Polycarboxylate ether (PCE) was used as water reducing agent [6] (Figs. 1, 2, 3, 4 and 5).

Fig. 1 Cement

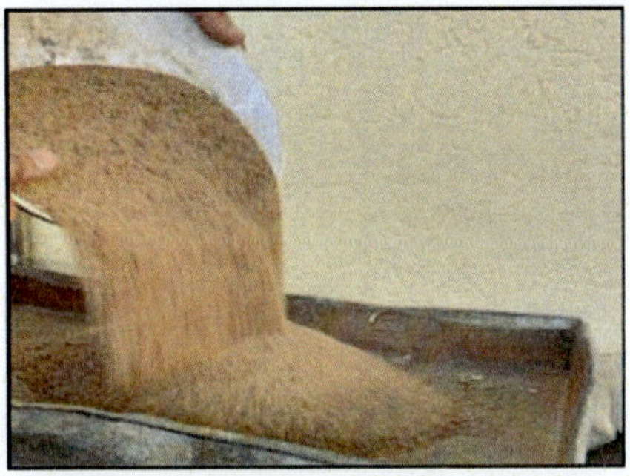

Fig. 2 Fine Aggregate

Fig. 3 Coarse Aggregate

Fig. 4 Water

Fig. 5 Polycarboxylate ether (super plasticizer)

Waste Plastic Fibres (polyethylene terephthalate)
The Waste plastic fibers are collected from used water bottles [3]. These bottles were crushed in the mechanical crushers [4, 7]. The size of the waste plastic fibers is not uniform (Fig. 6 and Table 1).

Implementation and Methodology

1) **Slump flow test:**

   In the absence of obstructions, the slump flow test is used to evaluate the horizontal free flow of self compacting concrete. This method is mainly based on the flow of the slump concrete from cone. In the absence of obstructions, the slump flow test is used to evaluate the horizontal free flow of self-compacting concrete. This method is mainly based on the flow of the concrete from slump cone.

Fig. 6 Waste Plastic Fibres (polyethylene terephthalate)

Table 1 Properties of plastic waste

Thickness (mm)	0.6
Length (mm)	5–7
Bulk density (kg/m^3)	1320

2) **T50 Test:**

The T50 test determines the amount of time required for self-compacting concrete to cross a 50 cm diameter in the absence of obstructions.

3) **J-ring test:**

This j-ring test is employed to assess both passing and filling ability. The j-ring test quantifies the flow time, flow spread, and blocking step. This flow spread indicates that the deformability of SCC is limited due to the blocking effect of reinforcement bars and flow time represents the rate of determination within a flow distance. Blocking step indicates the blocking effect.

4) **V-Funnel Test:**

This V- funnel test is also conducted to determine the passing ability of the concrete. In V- funnel test v is the flow time in which the prepared concrete mix can easily passing through a closed opening in a v funnel and signals the filling limit due to no obstacles or any restrictions.

5) **L - Box Test:**

This test evaluates the flow of the concrete as well as the extent to which it is blocked by reinforcement. L box test determines the passing ability of SCC.

6) **U - Box Test:**

The U box test is performed to determine the filling ability of self-compacting concrete. The sample is passed through U-shaped box which is provided with reinforcement at the bottom of the U-box base.

Mix Proportion

In Self compacting concrete for M30 grade of concrete 8 different proportions are mixed and different trails conducted for this experimental procedure [6]. For M30 grade concrete the nominal mix ratio is 1: 1.57: 1.45: 0.45 (Table 2).

Table 2 Mix Proportions for M30 grade of concrete

S. No.	Waste Plastic (%)	Waste Plastic (kg)	Cement (kg)	Fine Aggregate (kg)	Coarse Aggregate (kg)	Water (Liters)	Super Plasticizer (Liters)
1	0	0	530	836	771	191	4.77
2	2.5	20.9	530	815.1	771	191	4.77
3	5	41.8	530	794.2	771	191	4.77
4	7.5	62.7	530	773.2	771	191	4.77
5	10	83.6	530	752.4	771	191	4.77

Table 3 Slump Flow Diameter Value (mm)

S. No.	Plastic (%)	Slump Flow diameter (mm)
1	0	665
2	2.5	675
3	5.0	630
4	7.5	615
5	10	485

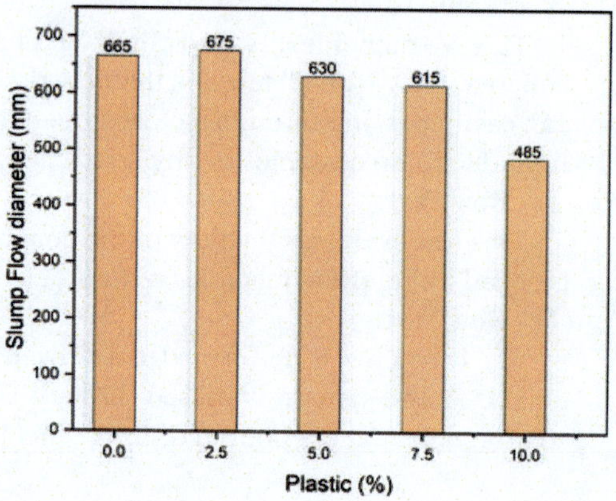

Fig. 7 Slump flow diameter

3 Results

3.1 Slump Flow Test

The slump flow test of sample has been performed for 0%, 0.25%, 5%, 7.5% and 10% of partial substitution of fine aggregates [8] with waste plastic fibres.

The test results are mentioned in Table 3 and Fig. 7 (Fig. 8).

With increase in the plastic percentage from the above result, it is observed that at 2.5% of plastic replacement the slump flow diameter is increasing. Further increase in plastic percentage decreases [9, 10] the horizontal flow of sample.

3.2 T_{50} Test

The T_{50} test of sample has been performed for 0%, 0.25%, 5.0%, 7.5%, and 10% of partial replacement of fine aggregate with waste plastic fibers [11].

The test results are mentioned in Table 4 and Fig. 9.

Fig. 8 Slump flow test

Table 4 T_{50} Slump Flow Test

S. No.	Plastic (%)	Time (sec)
1	0	6
2	2.5	5
3	5.0	4.5
4	7.5	5
5	10	16

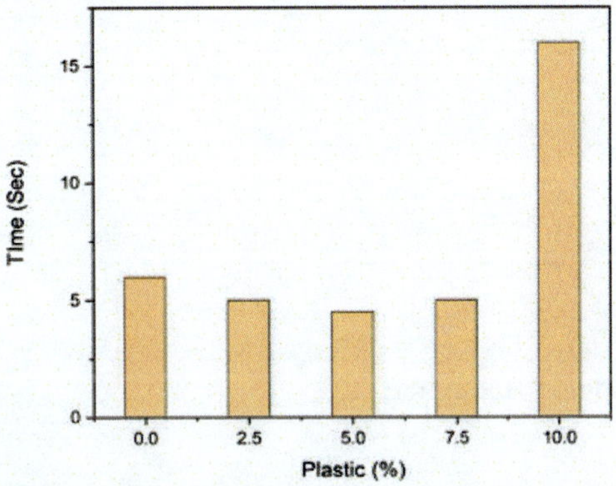

Fig. 9 T_{50} time

From the above results it is observed that upto 7.5% of plastic replacement, there is a decrease in time to cross the 50 cm diameter [12]. Further increase in plastic percentage increase in the time to flow the sample.

3.3 *J – Ring Test*

The J-Ring test on sample has been conducted for 0%, 0.25%, 5%, 7.5% and 10% of partial substitution of fine aggregate with waste plastic fibers [13].

The test results are mentioned in Table 5 and Fig. 10.

From the above results it is observed that, up to 2.5% of plastic replacement shows good results [14]. Further increase in plastic percentage increasing the height of sample which is not an accepted value as per EFNARC guidelines (Fig. 11).

Table 5 J-Ring Height Value (mm)

S.No.	Plastic (%)	Difference (h2-h1) in mm
1	0	7
2	2.5	8
3	5	12
4	7.5	22
5	10	20

Fig. 10 J Ring test results in mm

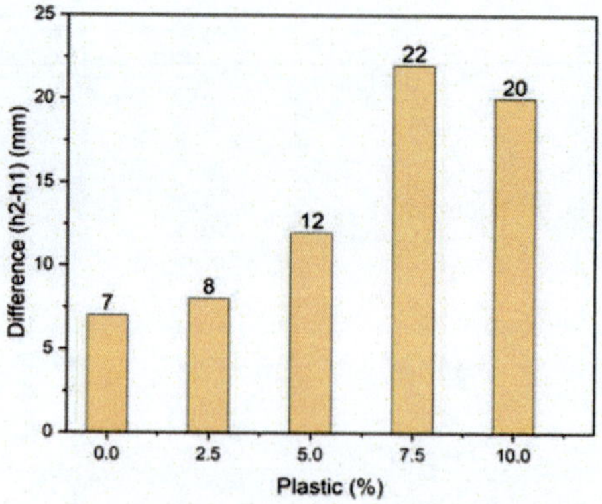

Fig. 11 J Ring Test

Table 6 V-Funnel Flow Time (sec)

S.No.	Plastic (%)	Time (sec)
1	0%	7
2	2.5%	7.9
3	5.0%	8.5
4	7.5%	9
5	10%	17

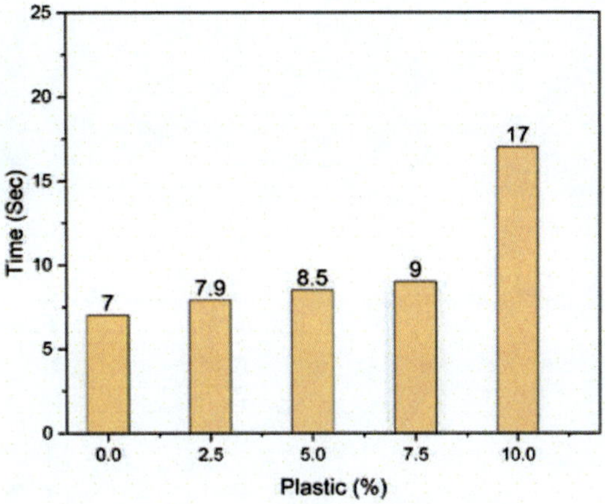

Fig. 12 V-Funnel Test Time

3.4 V- Funnel Test

The V-funnel test of sample has been performed for 0%, 0.25%, 5.0%, 7.5%, and 10% of partial replacement of fine aggregate with waste plastic fibres.

The test results are mentioned in Table 6 (Figs. 12 and 13).

From the above mentioned findings, it can be inferred that up to 7.5% plastic replacement, the results are within acceptable limits as per EFNARC guidelines. Further increase in waste plastic effecting the concrete's flowability.

3.5 L – Box Test

The L-Box test has been performed for 0%, 0.25%, 5.0%, 7.5%, and 10% of partial replacement of fine aggregate with waste plastic fibres (Table 7 and Figs. 14, 15).

From the above result it is observed that up to 7.5% of plastic replacement the blocking ratio is within the limits. An increase in the percentage of plastic is causing sample blockage near the reinforcement.

Fig. 13 V-Funnel Testing

Table 7 L-Box Height Ratio

S.No.	Plastic (%)	h_2/h_1(mm)
1	0%	0.89
2	2.5%	0.92
3	5.0%	0.93
4	7.5%	0.88
5	10%	0.76

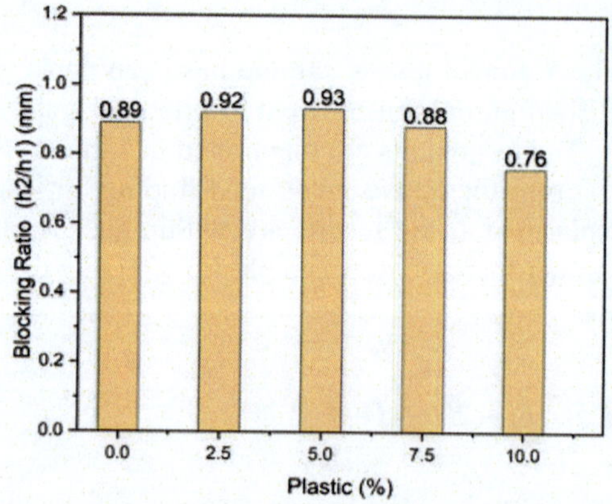

Fig. 14 L-Box Test

3.6 U-Box Test

The U-box test on sample has been performed for 0.25%, 5.0%, 7.5%, and 10% of partial replacement of fine aggregate with waste plastic fibres [15].

The results are mentioned in Table 8 (Figs. 16 and 17).

Fig. 15 L-Box Testing

Table 8 U-Box Height Values (mm)

S.No.	Plastic (%)	Difference (h_2-h_1)mm
11	0	14
2	2.5	15
3	5	13
4	7.5	30
5	10	27

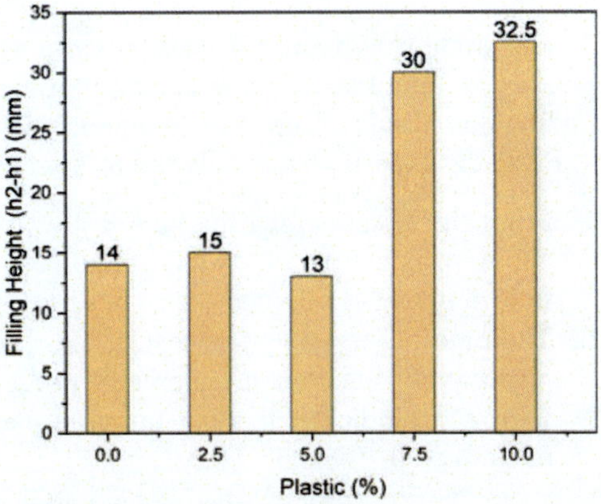

Fig. 16 U- Box Test results

U-Box test has been conducted and results obtained are within limits up to 7.5% plastic replacement, further increase in plastic % causing sample to block near the reinforcement which is not a good sign for SCC.

Fig. 17 U- Box Test

4 Conclusions

Present study is about use of plastic waste as a replacement material of fine aggregate in concrete. Plastic waste is ubiquitous in the environment that causes damage. Adding this plastic waste to the concrete decreases the damage to environment.

From the experimental study the following observations were made,

1. From the observations the slump flow values are in the range of 665 mm to 485 mm. as per the EFNARC guidelines at 2.5% plastic replacement, the slump flow diameter is satisfied.
2. The time observed in T50 test is 5 s, 4.5 s, 5 s for 2.5%, 5%, 7.5% of plastic respectively, which is in acceptable range as mentioned by EFNARC.
3. J-Ring test indicates that use of waste plastic up to 5% gives satisfied results as mentioned in Table 5.
4. By increasing waste plastic in concrete V-Funnel time also increases. Up to 7.5% plastic replacement the time is in acceptable range.
5. It can be observed that all the mixes from 0 to 10% of waste plastic satisfy the flowability measured by L-Box and U-Box tests.
6. The compressive strength is decreasing with increase in plastic percentage. Higher compressive strength is achieved at 2.5% plastic.

Acknowledgements I would like to express my gratitude to Mr. K. Anand Goud, Assistant Professor, who helped me with my research and guided me in the right direction.

References

1. Hajime, O., & Masahiro, O. (2003). Self-compacting concrete. *Journal of Advanced Concrete Technology, 1*(1), 5–15. https://doi.org/10.3151/jact.1.5
2. Siddique, R. (Ed.). (2019) *Self-compacting concrete: materials, properties and applications.* Woodhead Publishing.
3. Al-, A. I., Noaman, A. T., & Mosleh, W. K. (2019). Mechanical properties and impact behavior of PET fiber reinforced self-compacting concrete (SCC). *Composite Structures, 224*, 111021.
4. Mounika, K., Goud, K. A., & Malagavelli, V. (2021). Study on workabilty and strength of M35 grade self compacting concrete by using plastic waste and alccofine materials. *Elementary Education Online, 20*(6), 27–37. https://doi.org/10.17051/ilkonline.2021.06.004
5. Al-, A. I., & Hilal, N. N. (2016). The possibility of enhancing some properties of self-compacting concrete by adding waste plastic fibers. *Journal of Building Engineering, 8*, 20–28.
6. Frhaan, W. K. M., et al. (2022). Relation between rheological and mechanical properties on behaviour of self-compacting concrete (SCC) containing recycled plastic fibres: A review. *European Journal of Environmental and Civil Engineering, 26*(10), 4761–4793.
7. Kohistani, A. S., & Singh, K. (2018). A review – utilization of plastic waste and alccofine in self - compacting concrete. *International Research journal of Engineering and Technology (IRJET), 5*, 56–72.
8. Hama, S. M., & Hilal, N. N. (2017). Fresh properties of self-compacting concrete with plastic waste as partial replacement of sand. *International Journal of Sustainable Built Environment, 6*(2), 299–308.
9. Al-, A. A., & Dalal, T. R. (1997). Concrete containing plastic aggregates. *Concrete International, 19*(8), 47–52.
10. Ghernouti, Y., et al. (2015). Fresh and hardened properties of self-compacting concrete containing plastic bag waste fibers (WFSCC). *Construction and Building Materials, 82*, 89–100.
11. Hilal, N., & Hadzima-Nyarko, M. (2021). Improvement of eco-efficient self-compacting concrete manufacture by recycling high quantity of waste materials. *Environmental Science and Pollution Research, 28*(38), 53282–53297.
12. Shahidan, S., Isham, I., & Jamaluddin, N. (2016). A review on waste minimization by adopting in self compacting concrete. In *MATEC Web of Conferences* (Vol. 47). EDP Sciences.
13. Dadzie, D. K., Kaliluthin, A. K., & Kumar, D. R. (2020). Exploration of waste plastic bottles use in construction. *Civil Engineering Journal, 6*(11), 2262–2272.
14. Foti, D. (2013). Use of recycled waste pet bottles fibers for the reinforcement of concrete. *Composite Structures, 96*, 396–404.
15. Yaakob, M. Y., et al. (2016). A review on different forms and types of waste plastic used in concrete structure for improvement of mechanical properties. *Journal of Advanced Research in Applied Mechanics, 28*(1), 9–30.

Mechanical Properties of Concrete with Micro Level Reinforcement Using Natural and Synthetic Fibres

J. Philips, R. L. Lija, and V. Vandhana Devi

Abstract The effects on engineering properties such as fresh and hardened characteristics of various fibres dispersed into the normal concrete have been investigated. In this research, the design mix M30 grade concrete is blended with 0.5% of various natural and synthetic fibres. The fibres used are Jute, Steel, Sisal, and Glass. Adding various fibres into the control mix concrete can dramatically increase the engineering strengths like flexural, split tensile and compressive strength. Superplasticizer is added to fibre-blended concrete in the proportion of 2% of the cement mass to increase workability. Comparing the results of fibre blended concrete with M30 grade plain concrete, the positive influence of adding 0.5% of various fibres increases the engineering properties during a 7-day and a 28-day cure period has been studied. According to the testing results, concrete with glass fibres added increases compressive strength by 35.82% when compared to control mix concrete after 28 days of curing. The split tensile strength of glass fibre blended concrete is 20.26% more than the nominal mix concrete. From this investigation, it is noticeable that glass fibre mixed concrete has shown optimum values in all strength aspects in comparison with natural fibres.

Keywords Fibre-blended concrete · Steel fibre · Glass fibre · Sisal fibre · Jute fibre · Mechanical properties

1 Introduction

This occurrence can be linked to its superior strength, durability and cost-effectiveness over other forms of construction materials [1, 2]. However, its application has been severely limited due to its poor susceptibility to crack opening, limited fracture strain capacity, and inability to bear stress. Concrete may have

J. Philips (✉) · R. L. Lija · V. V. Devi
KCG College of Technology, Chennai, India
e-mail: philipsjeyaseelan@gmail.com

A. Vilventhan et al. (eds.), *Advances in Construction Materials and Management*, Lecture Notes in Civil Engineering 346,
https://doi.org/10.1007/978-981-99-2552-0_28

various materials added to it in an effort to improve its lacking mechanical characteristics [3–5]. These concrete's intrinsic flaws may be greatly improved by using the idea of micro-level reinforcement, i.e. the inclusion of different discontinuous chopped fibres. By bridging micro-cracks in the matrix, glass or polypropylene fibre can disperse stress and stop it from propagating at the crack points [1]. The goal of creating fibre concrete is to increase the material's toughness, resistance to impact loads, and bending strength. Despite the fibres being randomly distributed, the fibre concrete behaves consistently under stress [6].

Fibre-reinforced concrete's structural integrity is improved by using small linear fibres that are equally scattered and randomly oriented. It has numerous advantages, including increased structural strength, less steel rebar, and improved freezing–thawing resistance. Its behaviour is influenced by the type, dose, density, and dispersion of the fibres necessitating a thorough investigation of each instance. Steel, glass, polymer, synthetic, and other forms of fibres are frequently utilized in girders, slabs, and infrastructure. By replacing steel fibres in concrete reinforcement with recycled plastic fibres, such as PET fibre, corrosion-related problems, excessive weight, and high initial costs may be avoided, and the environmental impact of aggregation of plastic may be reduced [7].

Since a very long time ago, fibres are created into building materials. The combinations of synthetic and natural fibres provide amazing results since the combination significantly improves the mechanical properties of the composite material. Due to the fact that natural fibre reinforcement requires a substantially lower fibre ratio than conventional reinforcement, it is significantly more beneficial than standard reinforcement bars with regard to energy and financial advantages. Recently, the potential of natural and synthetic fibres as reinforcing materials in a building, particularly concrete, has been examined. They found that adding fibres to concrete boosted mechanical performance by up to 1.0%, but that adding more fibres lowered mechanical performance due to its impracticability. In order to boost the concrete's tensile strength, it's advised to add 1.0% fibre [8].

Steel fibre is the one of these that is most frequently utilized in concrete. Steel fibre's susceptibility to corrosion, however, limits its use. Synthetic fibres are frequently suggested as a solution to this issue. Synthetic fibre manufacture is time and resource-intensive. In these circumstances, natural fibres are frequently taken into account as a viable substitute fibre to create fibre-reinforced concrete. Natural fibres are favourable for the environment, economics, and conservation of energy and resources. Based on their origins, natural fibres are sometimes categorized into three categories: mineral, plant, and animal. Reinforcing cementitious materials with naturally derived fibre can considerably improve their properties [9, 10]. Cellulose-based plant fibres may provide a chance to add cement to concrete composites as a useful reinforcing agent because of their low density, high tensile strength, and unique microstructure [11]. Due to their demonstrated effectiveness, to improve concrete's tensile properties, synthetic fibres have been used. Steel and propylene, two typical synthetic fibres, are pricey and may increase the project's cost. On the other hand, natural fibres make it possible to improve the tensile qualities of concrete at a lower cost while also being sustainable and ecologically sound. The trunks, skins, leaves,

Table 1 Physical Properties of OPC53 Grade

Physical characteristics	Results	Requirement as per IS12269-1987[16]
Standard consistency	36%	—
Fineness	5%	Maximum 10%
Specific gravity	3.15	Minimum 2.65
Soundness	2 mm	Maximum 10 mm
Initial setting time	50 min	Minimum 30 min
Final setting time	280 min	Maximum 600 min

and tissue of plants are potential sources of natural fibres. The primary components of jute, flax, oil palm, banana, sugarcane, sisal, coconut, and pineapple are cellulose, hemicelluloses, and lignin [11]. These fibres can be found easily and just need minimal preparation. Natural fibres derived from plants are vulnerable to degradation in cement-based matrixes because water and alkaline pore solutions may erode these fibres [12].

The combination of steel and artificial fibres enhances the structural concrete components' capacity for strength, flexibility, and post-peak durability. Steel fibre is being utilized to offer a bridging effect over thermal stress-related cracking. A bridging effect will then be provided by the steel fibre against heat cracks [13].

2 Materials Used for Fibre-Blended Concrete

2.1 Ordinary Portland Cement

In this study, cementitious material of Grade 53 of Ordinary Portland cement was employed to create fibre-blended concrete. According to IS: 4031-1988 [14, [15], the physical parameters of OPC in Grade 53 were evaluated, and the findings were tabulated in line with IS: 12,269-1987 [16]. Table 1 lists the physical properties of the Grade 53 OPC utilized in this research study.

2.2 Fine Aggregate

In this analysis, natural river sand with granules smaller than 4.75 mm was utilized as the fine aggregate. A test was conducted on the fine aggregate in line with IS 383-1970 [17]. Tables 2 and 3 demonstrate, respectively, the gradation of fine aggregates and their physical properties. Figure 1 depicts the sieve analysis gradation curve for the fine particles.

Table 2 Sieve analysis test results of fine aggregate

IS sieve size (mm)	Weight retained (Kg)	Cumulative weight (Kg)	Cumulative % retained	Cumulative % retained
4.75	0.00	0.00	0.0	100.0
2.36	0.08	0.08	8.0	92.0
1.18	0.16	0.24	24.0	76.0
0.60	0.35	0.59	59.0	41.0
0.30	0.26	0.85	85.0	15.0
0.15	0.15	1.00	100.0	0.0

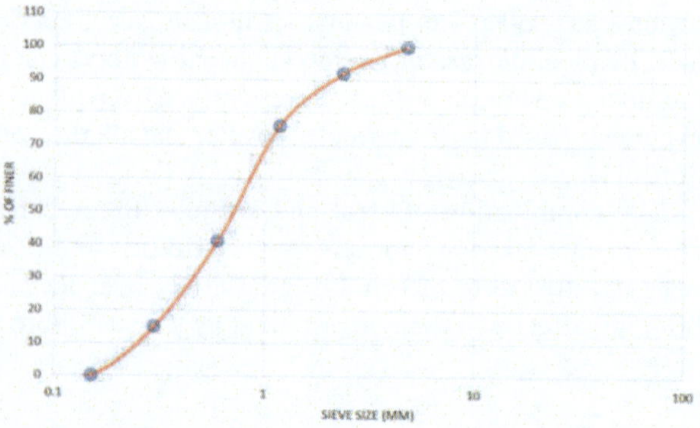

Fig. 1 Sieve analysis gradation curve of fine aggregate

2.3 *Coarse Aggregate*

In this study, angular crushed aggregates with particle sizes ranging from 20 to 10 mm were locally accessible coarse aggregate utilized to make a fibre-blended concrete. Its properties in accordance with IS 383-1970 [17] decide the strength of the concrete and it should be free from the minerals and chemicals impurities. The coarse aggregates were examined confirming to IS: 2386-1963 [18, 19] and the outcomes are presented in Table 3.

Table 3 Physical properties of fine and coarse aggregates

Characteristics	Fine aggregate	Coarse aggregate
Size (mm)	Less than 4.75	10–20
Bulk density (Kg/m^3)	1.6	1.15
Specific gravity	2.65	2.84
Fineness modulus	2.76	7.16

Table 4 Physical parameters of admixture

Parameters	Values
Type	Liquid
Color	Dark brown
Specific gravity	1.2 ± 0.3

2.4 *Chemical Admixture*

Ceraplast300 is a high-grade naphthalene-based super plasticizer that is highly recommended for improved concrete workability and early and final strengths. It was collected from Cera—Chem Private Limited, Kilpauk. Because of its excellent dispersion qualities, Ceraplast300 increases the workability of concrete without causing air entrainment and also coats the particles. The uniform distribution guarantees that the mix is more homogeneous, reducing the likelihood of segregation during placement. The properties of ceraplast300 conforming to IS 9103-1999 [20] is listed in Table 4.

2.5 *Fibre*

Fibre is a filament that is used to make vegetable tissue, mineral material, or textile. The fibres were collected from Go-Green Private Limited, Triplicane. The fibres used in this investigation were Jute, Sisal, Steel and Glass fibres. Steel and glass fibres are synthetic whereas jute and sisal are natural fibres. Table 5 lists the physical parameters of the fibres employed in this research investigation.

2.5.1 Jute Fibre

A long, glossy, soft vegetable fibre called jute may be twisted into strong, coarse strands. The major plant that produces jute fibre, which is a member of the Malvaceae family, is *Corchorus Clitoris*. In terms of production volume and the range of uses, jute is among the most economical natural fibres, ranking second only to cotton.

Table 5 Physical parameters of fibres

Parameters	Steel	Glass	Sisal	Jute
Density (g/cm^3)	7.8	2.6	0.9	1.3
Tensile strength (MPa)	>1450	2000–4000	375–570	230
Elongation (%)	<4	2.5–4.8	3–5	1.7
Modulus of elasticity (GPa)	200	70–80	16–26	26.5
Aspect ratio	50	150	100	25

Lignin and cellulose are combined to create jute fibres. The fibres are between off-white and brown in colour and between one to four meters long (3–13 ft).

2.5.2 Sisal Fibre

Sisal, which is produced from the plant Agave Sisalana, is native to southern Mexico but has been widely planted and naturalized in various countries. It produces a rigid fibre that is used to make a variety of items. Sisal has historically been the main component of agricultural twine in terms of strength, endurance, ability to extend, preference for specific dye materials, and resistance to degradation in salt-water (binder twine and baler twine). Sisal has been utilized as an environmentally friendly enhancing agent in composites to replace asbestos and fibre-glass in a variety of industries including the automotive industry.

2.5.3 Steel Fibre

Single steel fibres short pieces of steel may be readily and randomly disseminated in a new concrete mix using conventional blending techniques. They have an aspect ratio of 20 to 100. Unlike reinforcing bars or wires, which are often employed just when necessary, steel fibres are typically dispersed across a cross-section. Comparing steel fibres to continual reinforcement bars or wires, steel fibres, are narrower and much more closely spaced. Steel fibres have been shown to lessen plastic shrinkage cracks and are frequently incorporated in concrete in low quantities (can be less than 1%). Although steel fibres typically have minimal impact on the free shrinkage of concrete, when used in sufficient amounts; they can increase fracture resistance and decrease crack breadth.

2.5.4 Glass Fibre

Glass fibre is a material comprised of many, incredibly thin glass fibres. When employed as a thermal insulating layer, glass fibre is constructed using a bonding agent to trap several tiny air cells. Other fibres with material characteristics usually comparable to glass fibre include polymers and carbon fibre. Although it is less stiff and robust than carbon fibre, it is significantly cheap and fragile when used in composites. Glass-fibre polymeric composite rebar has the potential to replace steel rebar in areas where steel corrosion is a problem.

Table 6 Mix design proportions of fibre blended concrete

Mix ingredients (Kg/m^3)						
Mix	Cement	Fine aggregate	Coarse aggregate	Water	Fibre	Admixture
CONTROL	400	668	1218	186	0.00	0.00
STFBC 0.5%	398	668	1218	186	2.00	2.00
GLFBC 0.5%	398	668	1218	186	2.00	2.00
SIFBC 0.5%	398	668	1218	186	2.00	2.00
JTFBC 0.5%	398	668	1218	186	2.00	2.00

3 Experimental Program

3.1 Mix Design

The 53grade of ordinary portland cement, fine aggregates, coarse aggregates, admixture (Ceraplast300) and fibres (Jute, Sisal, Steel and Glass) were utilized to mix the FBC. A total of 5 mix proportions were made, one of which was a control mix with no fibre mixing and the other four were concrete mixtures with a fibre concentration of 0.5%. Table 6 displays the various mix ratios. Following their preparation per IS 10262-2009 [21], the FBC was cast in a typical cube, cylinder, and beam moulds. Concrete's compressive strength and split tensile were tested using cubes of 150 mm size and cylinders of 150 mm in diameter and 300 mm in height, respectively. Six cubes were made for compressive strength testing, and six cylinders were cast for split tensile tests over 7 and 28 days for the percentage of every mix. Beams of size 100 × 100 mm cross-sectional area and 500 mm length were cast to measure the flexural strength of FBC for a curing period of 7 days and 28 days. Totally six number of beams were cast for each mix proportion.

3.2 Testing of Concrete Specimens

Before casting the sample in moulds, the fibre-mixed concrete's performance was evaluated by slump testing per IS 1199-1959 [22]. The samples were stored at room temperature for a whole day. The specimens are then removed from the mould and let to cure for 7 or 28 days before being checked for signs of hardening.

The cubes' compressive strength and the cylinders' split tensile strength were measured using a compression testing machine (CTM) with a 2000kN capacity. The specifications defined in IS: 516-1959 were used to measure the compressive strength of the cube [23]. The cylinders' divided tensile strength was determined per the standards outlined in IS: 5816-1999 [24]. The flexural strength of the concrete beams was ascertained using the Universal Testing Machine (UTM), which has a 400kN capability.

3.3 Labelling of Concrete Specimens

To distinguish the concrete mixtures, the samples have been properly labelled.

- STFBC 0.5% indicates concrete blended with steel fibre with a dosage of 0.5%
- GLFBC 0.5% indicates concrete blended with glass fibre with a dosage of 0.5%
- SIFBC 0.5% indicates concrete blended with sisal fibre with a dosage of 0.5%
- JTFBC 0.5% indicates concrete blended with jute fibre with a dosage of 0.5%

4 Experimental Program

4.1 Fresh Properties of FBCs

By including 0.5 per cent of various fibres through the concrete mixtures during testing, the impact of fibres upon the effectiveness of fresh fibre-blended concrete was investigated. Table 7 and Figure 2 display the testing of the slump values of the fibre-blended concrete. The concrete control slump value is greater than the fibre-blended concrete, as seen in Fig. 3.

The addition of fibres therefore significantly affects workability. As previously stated, adding fibre volume to concrete reduces its workability. With the inclusion of Ceraplast-300 chemical admixture, improved workability may be achieved by the w/c ratio for M30 concrete is 0.45. The steel fibre blended concrete has a slump value nearer to the control mix concrete. But the natural fibres have a low slump value compared to control concrete and synthetic fibre-blended concrete due to the absorption of water by natural fibres. The absorption of water percentage is higher for natural fibres compared to synthetic fibres.

Table 7 Fresh properties of Fibre blended concrete

Mix	Water/Cement ratio	Slump value (mm)
CONTROL	0.45	20
STFBC 0.5%	0.45	18
GLFBC 0.5%	0.45	17
SIFBC 0.5%	0.45	15
JTFBC 0.5%	0.45	16

Fig. 2 Testing of slump values of FBCs

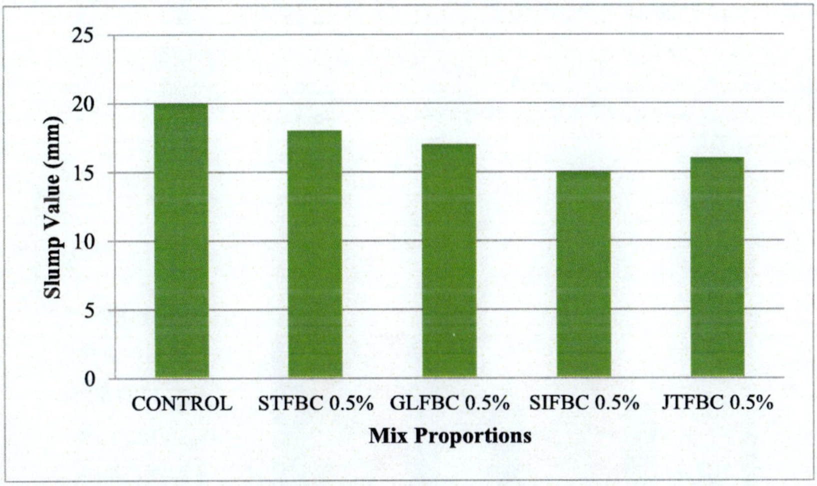

Fig. 3 Slump Values of the Various FBCs

4.2 Hardened Properties of FBCs

4.2.1 Compressive Strength

The mix proportion was created for M30 grade concrete, per IS456-2000 [21] IS 10262-2009 [25]. The results showed that the control mix's compressive strength was 22.22 N/mm2 after 7 days and 31.85 N/mm^2 after 28 days. Table 8 shows several fibre-concrete strengths mixed with different fibres. Figure 4 shows the changes in FBC's strength after 7 and 28 days. The GLFBC 0.5% mix, after curing for 7 days exhibited a 45% increase in compressive strength in comparison with the control mixture, and had the greatest compressive strength of every Glass fibre blended concretes.

The strength of the concrete of JTFBC 0.5% mixes was about equal to that of the control mix after curing for 28 days although it eventually reached M30's characteristic strength. The GLFBC 0.5% mix showed an ideal compressive strength of 35.8% when compared to the control mixture in terms of compressive strength. According

Table 8 Hardened mechanical properties of FBCs

Mix	Compressive strength (MPa)		Split tensile strength (MPa)		Flexural strength (MPa)	
	7 Days	28 Days	7 Days	28 Days	7 Days	28 Days
CONTROL	22.22	31.85	2.60	3.06	2.99	4.00
STFBC 0.5%	29.84	36.70	2.60	3.25	6.62	8.48
GLFBC 0.5%	32.14	43.26	3.20	3.68	6.88	9.71
SIFBC 0.5%	29.92	36.59	2.59	3.02	5.33	7.95
JTFBC 0.5%	23.10	32.87	2.59	3.06	6.18	7.52

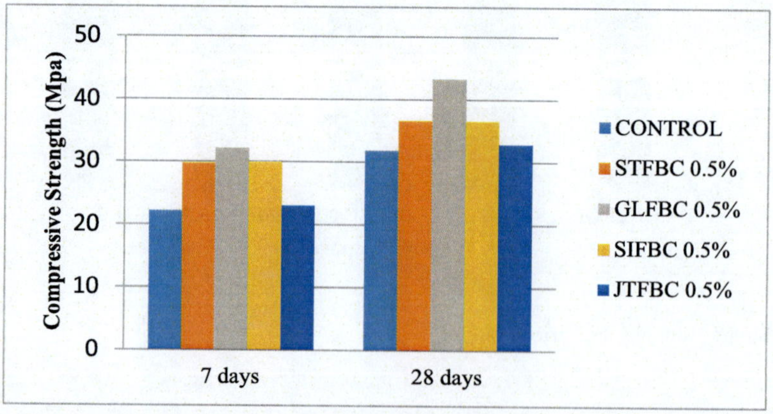

Fig. 4 Compressive strength comparison of the Various FBCs

to the results of cube compressive strength tests, the compressive strength of synthetically mixed concrete rose noticeably higher than that of naturally blended concrete. For Natural fibre blended concrete, SIFBC 0.5% mix showed a good increment in strength like a synthetic fibre STFBC 0.5% mix. The elongation percentage and elasticity of sisal fibre are primarily responsible for the increase in compressive strength relative to other natural jute fibre. The sisal fibre has 3–5% elongation compared to jute fibre. The sisal fibre elongation percentage is nearly equal to synthetic fibres.

4.2.2 Split Tensile Strength

The tensile modulus of the fibre-blended concrete made of various natural & synthetic fibres exhibits a little increase, in comparison with the control mixture's strength. The divided tensile strengths of every fibre-blended concrete mixture are displayed in Table 8. According to Table 8, the divided tensile strength of the majority of the fibre-blended concrete is equal to or slightly higher than that of the divided tensile strength of the control mixture. The specimen GLFBC 0.5% showed a substantial increase in 28 days divided tensile strength compared to 7 days divided tensile strength. The split tensile strength of GLFBC 0.5% concrete increased by 23% and 20% at 7 and 28 days, respectively, in comparison to the Control Mix concrete's split tensile strength. The split tensile strength of FBC at 7 & 28 days is shown in Figure 5. The tensile modulus of glass fibres, which increases bond strength, is responsible for the improvement in split tensile strength in GLFBC 0.5%. The outside failing regions of the examined specimens showed that the mortar paste and fibre embedding had a very strong bond, increasing mortar strength and, consequently, the tensile properties of the fibre-blended concrete.

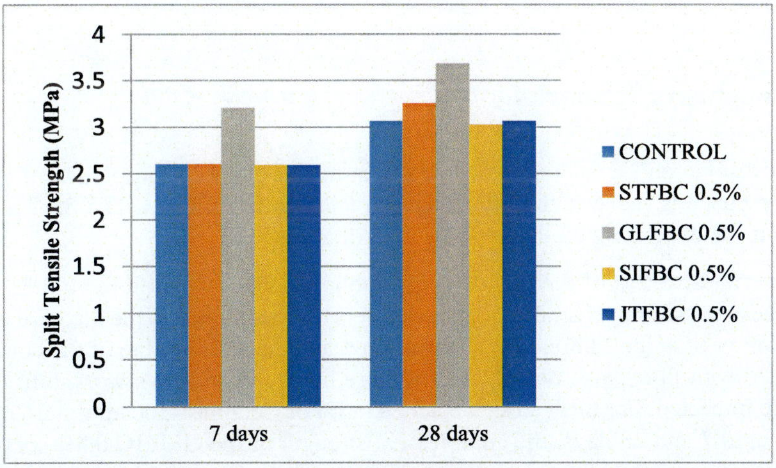

Fig. 5 Split tensile strength comparison of the Various FBCs

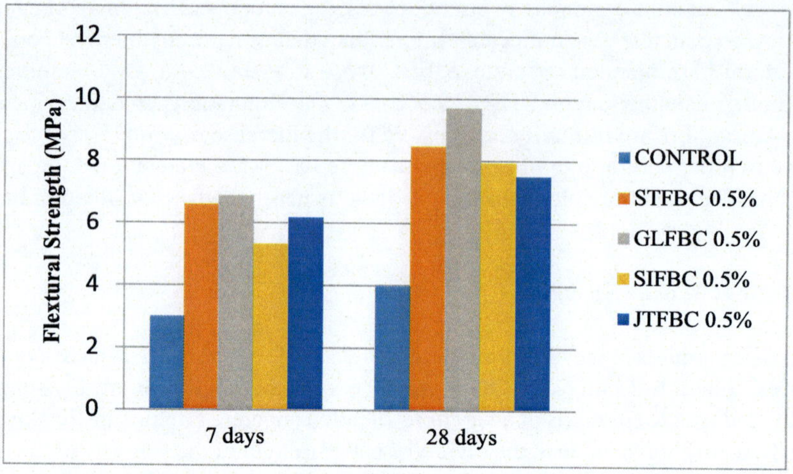

Fig. 6 Flexural strength comparison of the Various FBCs

4.2.3 Flexural Strength

The results showed that the flexural strength of the control mix was 2.99 N/mm^2 after 7 days and 4 N/mm^2 after 28 days. The flexural strength result for fibre-blended concrete is depicted in Table 8. The flexural strength of the GLFBC 0.5% mixture was 6.88 N/mm^2 after 7 days and 9.71 N/mm^2 after 28 days of curing, nevertheless. The combination of blended fibre and mortar produces extremely strong bond strength while viewing the specimens. The flexural strength of fibre-blended concrete at 7 and 28 days after curing is shown in Figure 6.

5 Conclusive Remarks

This investigation's goal focused on examining the physical properties of fibre-blended concrete made with various natural and synthetic fibres. The final results can be made depending on the findings of various tests:

1. The cohesiveness of the blended concrete is reduced by the fibre volume. The cohesiveness of the blended concrete suffers noticeably due to the increase in the quantity of natural fibres that absorb water, and the slump value of the concrete made with fibres also decreases. The fibre-blended concrete's workability was also impacted. For this cause, we added 0.5% of Ceraplast-300admixture.
2. At ages 7 and 28 days, the compressive strength of the GLFBC 0.5% mix was nearly greater than the compressive strength of all other combinations.

3. At ages 7 days and 28 days, the impact strength of the synthetic fibre mixed concrete with the following mixes, GLFBC 0.5% and STFBC 0.5%, was greater than the compressive strength of the natural fibre blended concrete mixture.
4. The majority of the fibre-blended concrete's split tensile strength was practically identical to the control mixture's split tensile strength. However, when glass fibre was added to the concrete, a considerable rise in split-tensile was seen.
5. At a curing time of 7 days and 28 days, fibre-blended concrete exhibits a spectacular increase in flexural strength in comparison with the control concrete mixture.
6. Compared to steel fibre and other natural fibres, it's advised that adding glass fibre to the control mixture will help it reach the desired target strength.
7. The aspect ratio of glass fibre is higher than all other fibres was predicted to be one of the reason for the increment of higher compressive strength.

References

1. Yuan, Z., & Jia, Y. (2022). Mechanical properties and microstructure of glass fiber and polypropylene fiber reinforced concrete: An experimental study. *Construction and Building Materials*. https://doi.org/10.1016/j.conbuildmat.2020.121048
2. Lija, R.L., Beulah Gnana Ananthi, G., Roy, K., & Lim, J.B.P. (2020). Effect of super absorbent polymer on microstructural and mechanical properties of concrete blends using granite pulver. *Structural Concrete*. https://doi.org/10.1002/suco.201900419
3. Lija, R.L., Beulah Gnana Ananthi, G., Roy, K., & Lim, J.B.P. (2020). Influence of super absorbent polymer on mechanical, rheological, durability, and micro-structural properties of self-compacting concrete using non bio-degradable granite pulver. *Structural Concrete*. https://doi.org/10.1002/suco.201900470.
4. Kishor Kumar, V., Rajan, N.K., & Vandhana Devi, V. (2022). Relationship between Gaestel index and rheology of blended bitumens. In *AIP conference proceedings* (Vol. 2452, p. 020006). https://doi.org/10.1063/5.0118092
5. Rekha, Y., Suriya, S., & HamedulIrshad, H. M. (2022). Comparative study on oven curing of geo-polymer concrete over conventional concrete. *Materials Today: Proceedings, 55*, 462–469. https://doi.org/10.1016/j.matpr.2022.02.344
6. Karatas, M., & Gunes, A. (2015). Engineering properties of self-compacting concrete produced by polypropylene and steel fiber. *Periodica Polytechnica Civil Engineering, 59*(2), 95–102. https://doi.org/10.3311/PPci.7348
7. Alani, A.H., Johari, M.A.M., Noaman, A.T., Muhamad Bunnori, N., & Majid, T.A. (2022). Effect of the incorporation of PET fiber and ternary blended binder on the flexural and tensile behaviour of ultra-high performance green concrete. *Construction and Building Materials*. https://doi.org/10.1016/j.conbuildmat.2022.127306
8. Ahmad, J., & Zhou, Z. (2022). Mechanical properties of natural as well as synthetic fiber reinforced concrete: a review. *Construction and Building Materials*. https://doi.org/10.1016/j.conbuildmat.2022.127353
9. Islam, Md.J., Islam, K., Shahjalal, Md., Khatun, E., Islam, S., Razzaque, A.B. (2022). Influence of different types of fibers on the mechanical properties of recycled waste aggregate concrete. *Construction and Building Materials*. https://doi.org/10.1016/j.conbuildmat.2022.127577
10. Ren, G., Gao, X., & Zhang, H. (2022). Utilization of hybrid sisal and steel fibers to improve elevated temperature resistance of ultra-high performance concrete. *Cement and Concrete Composites*. https://doi.org/10.1016/j.cemconcomp.2022.104555

11. Islam, M.S., & Ahmed, S.J.U. (2018). Influence of jute fiber on concrete properties. *Construction and Building Materials*. https://doi.org/10.1016/j.conbuildmat.2018.09.048
12. Onuaguluchi, O., & Banthia, N. (2016). Plant-based natural fibre reinforced cement composites: A review. *Cement and Concrete Composites*. https://doi.org/10.1016/j.cemconcomp.2016.02.014
13. Hussain, I., Ali, B., Akhtar, T., Jameel, M.S., & Raza, S.S. (2020). Comparison of mechanical properties of concrete and design thickness of pavement with different types of fiber-reinforcements (steel, glass, and polypropylene). *Case Studies in Construction Materials*. https://doi.org/10.1016/j.cscm.2020.e00429
14. Methods of Physical test for Hydraulic Cement, IS 4031:1988(Part 3), Bureau of Indian Standard, New Delhi.
15. Fineness test on Cement, IS 4031:1996(Part 1), Bureau of Indian Standard, New Delhi.
16. Specifications for Ordinary Portland cement 53 Grade, IS 12269:1987, Bureau of Indian Standard, New Delhi.
17. Specifications for Coarse and Fine Aggregates from Natural Sources for Concrete, IS: 383:1970, Bureau of Indian Standards, New Delhi.
18. Methods of test for Aggregates on Concrete, IS 2386:1963(Part 3), Bureau of Indian Standard, New Delhi.
19. Fineness Modulus test on Aggregates, IS 2386:1963(Part 5), Bureau of Indian Standard, New Delhi.
20. Specifications for Concrete Admixtures, IS 9103: 1999, Bureau of Indian Standard, New Delhi.
21. Recommended Guidelines for Concrete mix Design, IS 10262:2009, Bureau of Indian Standard, New Delhi.
22. Methods of Sampling and Analysis of Concrete, IS 1199-1959, Bureau of Indian Standards, New Delhi.
23. Methods of test for Strength of Concrete, IS 516: 1959, Bureau of Indian Standard, New Delhi.
24. Methods of test for Splitting Tensile Strength of Concrete Cylinders, IS 5816:1999, Bureau of Indian Standard, New Delhi.
25. Plain and Reinforced Concrete Code of Practice IS: 456:2000, Bureau of Indian Standards, New Delhi.

An Improved Methodology for Making Recycled Concrete Aggregate (RCA) - Sustainable Construction Material

Jagdish Godihal

Abstract Construction industry is one of the topmost sector contributing in the development of a country. However the generation of construction and demolition has an impacts on the environment. This paper highlights the comparative studies of properties of Recycled Concrete Aggregates (RCA) and Natural Aggregates (NA) as per Indian standards. The sources of natural aggregates have been dwindling, wherein the disposal of construction and demolition (C&D) waste has been problem in most of the urban areas, due to space constraints. Hence, this lead to the practice of using RCA as a secondary raw material in construction. The maximum utilization of RCA proportioning with natural aggregate in the preparation of concrete, thereby reduces the environmental impact and conservers natural resources. In this paper an attempt has been made to draw the conclusions based on the studies carried out at the concrete technology laboratory of Presidency University, Bengaluru, regarding the efficient utilization of RCA in preparing RAC. The tests revealed that the performance of RCA is affected by the adhered cement mortar on the aggregates. The adhered mortar on aggregate enhances water absorption and density both of these have a considerable impact on desired properties of concrete. To reduce this impact, the adhered mortar on the aggregate has been removed by the abrasion process. This study has been carried out to ascertain both the physical and mechanical properties of the RCA and its performance were studied and compared with those of natural aggregates (NA). This provides an insight to adopt the methodology to improve the performance of RCA in RAC. The study summarizes sustainability gains of RCA utilization. In addition to this, the optimal proportioning of RCA to prepare RAC has improved its stake as a feasible sustainable alternative to NA in construction. In this study an improved methodology has been proposed to remove the mortar on RCA and to enhance the effectiveness of recycled aggregates with reference to desired properties.

J. Godihal (✉)
Presidency University, Bengaluru, Karnataka 560064, India
e-mail: drjagdishgodihal@presidencyuniversity.in

A. Vilventhan et al. (eds.), *Advances in Construction Materials and Management*, Lecture Notes in Civil Engineering 346,
https://doi.org/10.1007/978-981-99-2552-0_29

Keywords Sustainability · Environment · Recycled Concrete Aggregates (RCA) · Recycled Aggregate Concrete (RAC) · Natural Aggregates (NA) · Natural Aggregates Concrete (NAC)

1 Introduction

Concrete being one of the most versatile construction material used all over the world due to unique and impeccable properties and cost benefits. The consumption of coarse and fine natural aggregates has been increasing tremendously nowadays. About 70% of the total concrete volume in a concrete mix comprises of these aggregates.

The urbanization trends witnessed in recent years lead to the higher rate of demolition of old structures for creating a space for new construction [1]. This generates a lot of C&D waste in urban area, providing an opportunity to use recycled aggregates to prepare concrete, as one of the sustainable approach [2]. It has been recognized that the C&D waste volumes have been increasing making it difficult to dispose appropriately, most of it lands in burying low lying areas or into the sanitary landfills. However, in India after the implementation of C&D Waste management rules 2016, it has provided impetus to reuse, recycle by identifying the stakeholders and assigning the roles and responsibilities in effectively and efficiently managing the C&D wastes in construction.

The increasing cost of natural aggregates production and ever increasing demand for construction material, especially coarse aggregates, has widen the scope of using RCA to partially replace the NA has gained more importance [3]. The RCA are adhered with hydrated cement paste on its surface which reduces the specific gravity where as it increases the porosity [4]. Thus, this research aims to the study the properties of M30 grade concrete with various proportion of NA and RCA to identify at the optimal proportioning of RCA and NA to meet the two dimensions of sustainability, i.e., environment and economy. The results obtained by this study provides insights and necessary process improvements and approaches required to optimal use of the recycled concrete aggregate as a sustainable construction material.

2 Methodology

The steps followed in the methodology adopted for the study (Fig. 1) has been mentioned as below:

- Evaluation of preliminary tests on recycled concrete aggregates and natural aggregates.
- Mix design is done as per IS 10262-2019 [5] and IS 383-2016 [6]
- Removal of adhered mortar on recycled course aggregate by abrasion process.
- Determining the workability using Slump test for all mix proportions.

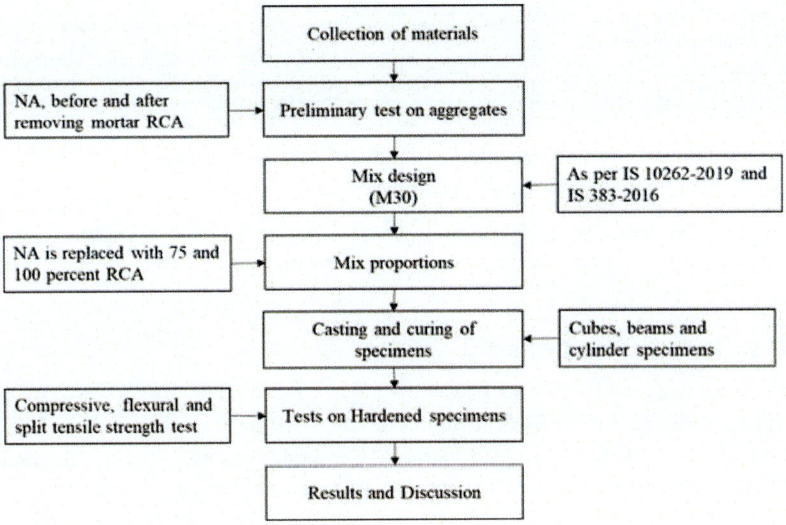

Fig. 1 Methodology adopted for the study

- Casting of cubes, cylinders and beams for both with and without adhered mortar RCA for different concrete strength like Compressive, Split tensile, Flexural.

2.1 *Materials*

Ordinary Portland Cement (OPC), grade 43 as per IS 269-015 [7] and clean portable water is used in the investigation as per the requirement mentioned in IS 456:2000 [8].

2.1.1 Fine Aggregate and Coarse aggregate

The properties of RCA were tested in accordance with tests specified in IS 383:2016 [6]. The tests have been carried out for both RCA, NA. Recycled fine aggregates and coarse aggregates were sourced from rock crystal Kerala crushers in Maranayakanahalli, Bengaluru.

2.1.2 Proportion of Natural Aggregates with RCA

Both recycled coarse aggregates and recycled fine aggregates used to design the concrete mix with proportioning shown in Table 1.

Table 1. Material quantity for mix design

Particulars	Material quantity for mix design for water cement ration 0.46
Sample 1	100% NA: 100% M-sand: Cement
Sample 2	75% RCA+25%NA: Cement
Sample 3	100% RCA: Cement
Sample 4	75% Abrasion RCA+25%NA: Cement
Sample 5	100% Abrasion RCA: Cement

3 Results and Discussion

The experiments for this study done using RCA in different context, one being, with adhered mortar on the (RCA) aggregates and another one without adhered mortar on the (RCA) aggregates.

3.1 Test Result with Adhered Mortar on the (RCA) Aggregates

The following tests have been carried out on recycled concrete aggregates sourced from rock crystal Kerala crushers in Maranayakanahalli, Bengaluru.

3.1.1 Workability

The workability of concrete has been ascertained by slump cone test by replacing natural aggregates with recycled aggregates at varying percentage as per the standards IS1199, 1959 [9]. The results have been depicted in Table 2. The decreasing trend in workability has been observed with increased percentage of RCA.

The graphical representation of variations in the workability are shown in the Fig. 2.

Table 2. Workability

Particulars	Design slump value 75 mm
Sample 1	65
Sample 2	60
Sample 3	59

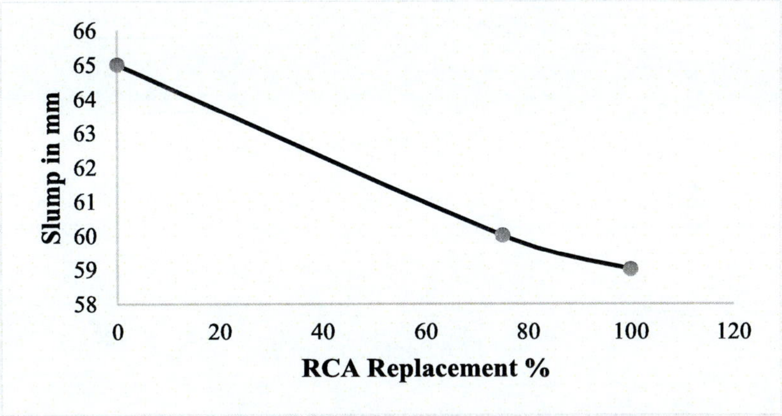

Fig. 2 Slump test with adhered mortar on RCA

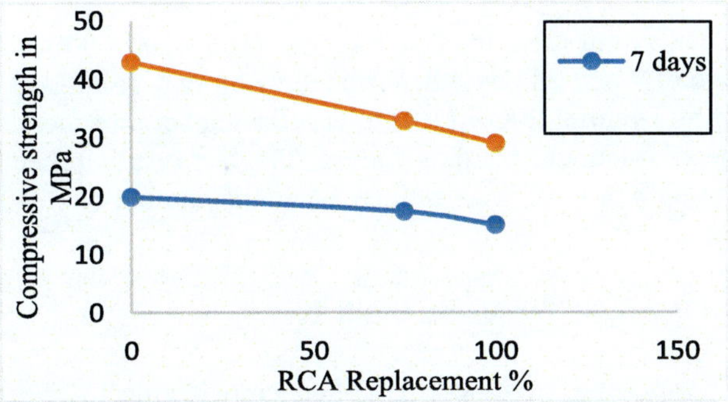

Fig. 3. Compressive strength with adhered mortar on RCA

3.1.2 Compressive Strength Test

This tests relates to quality of concrete. It was conducted on specimens after seventh day and twenty-eight day of curing. The cubes are casted as per the proportion shown in the Table 1 having size 150 mm, under curing for 24 h. The test was conducted as per the standards mentioned in IS:516-1959 [Reprint 2006] [10]. The test results provided in Table 3.

Figure 3 shows the average compressive strength of concrete mixes tested for 7 and 28 days respectively

Table 3. Compressive strength of concrete with adhered mortar on RCA

Particulars	7 days	28 days
Sample 1	19.8 MPa	42.9 MPa
Sample 2	17.3 MPa	32.8 MPa
Sample 3	15.0 MPa	29.0 MPa

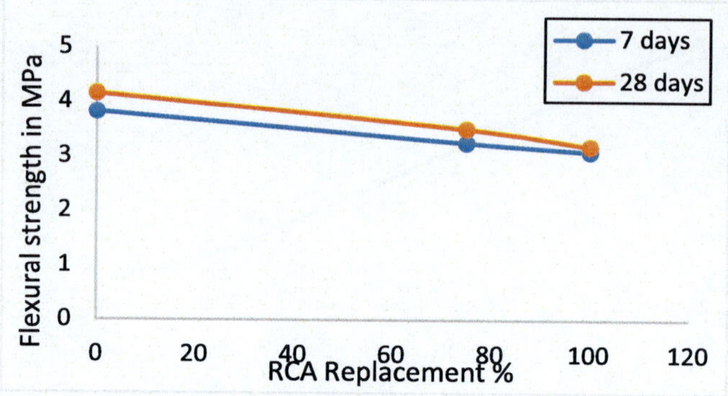

Fig. 4 Flexural strength with adhered mortar on RCA

3.1.3 Flexural Strength Test

The flexural strength of each sample (100 * 100 * 500 mm beams, mix at the concrete age of 7 and 28 days) determined using two-point load bending test as per IS:516-1959 [Reprint 2006] [10]. It tests the capacity of the slab or unreinforced concrete beam to endure bending failure. The test results reflects about the decrease in the strength of RAC with an increase in the amount of RAC, as shown in the Fig. 4.

3.1.4 Split Tensile Strength Test

The splitting tensile strength of concrete having 150 * 300 mm cylinders sample has been carried out with reference to IS 5816: 1999 [11] for 7 and 28 days. Figure 5 represents the results of split tensile test of various samples tested.

3.1.5 Ultra Sonic Pulse Velocity (UPV)

The homogeneity and quality of the concrete is ascertained by UPV, conducted referring IS:13311(part1) 1992 [12]. Increase proportioning of RCA in concrete decreases the UPV as shown in the Fig 6.

3.2 The Results Without Adhered Mortar on the (RCA) Aggregates

The recycled concrete aggregates (RCA) contain hydrated cement paste adhered to the surface. This increases water absorption leading to decrease in strength. The RCA

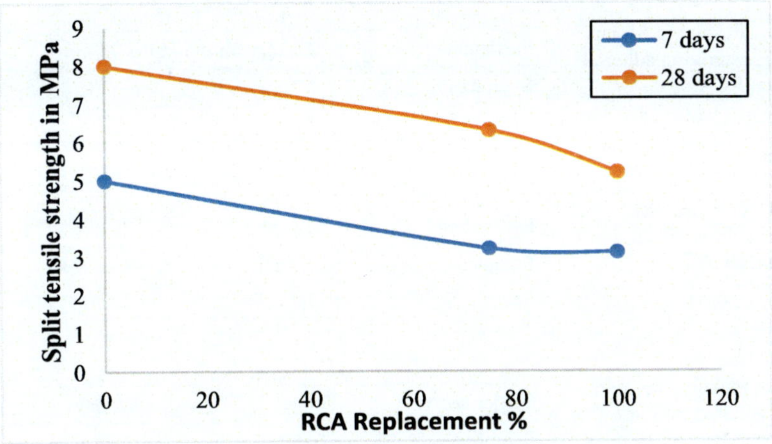

Fig. 5 Split tensile strength with adhered mortar on RCA

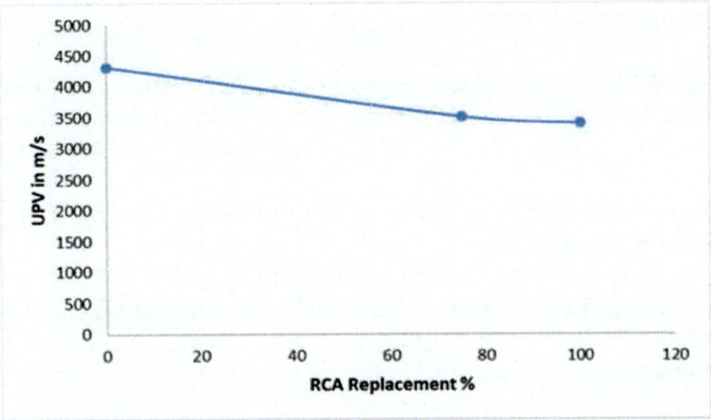

Fig. 6 UPV with adhered mortar on RCA

have been placed in abrasion testing machine for 100 revolutions to remove part of adhered mortar on the recycled coarse aggregate.

3.2.1 Workability

Table 4 reflects the status of the slumps test of the concrete prepared using RCA with and without adhered mortar. The slump seems to be slightly higher for concrete prepared with the RACs without adhered mortar. The adhered mortar removed by abrasion process to lower the water absorption of RCA.

Table 4. Results of Slump Test after removing adhered mortar on RCA

Particulate	Design Slump Value 75 mm
Sample 4	63
Sample 5	61

Table 5. Compressive strength without adhered mortar on RCA

Particulate	7 days	28 days
Sample 4	17.2 MPa	33.6 MPa
Sample 5	15.5 MPa	30.5 MPa

Table 6. Flexural strength without adhered mortar on RCA

Particulate	7 days	28 days
Sample 4	3.45 MPa	3.62 MPa
Sample 5	3.50 MPa	3.55 MPa

3.2.2 Compressive Strength Test

Table 5, depicts the test results after removing the adhered mortar on the RCA. The results reflects the increasing in compressive strength of RAC without adhered mortar on RCA.

3.2.3 Flexural Strength Test

The Flexural strength varies as the removed adhered mortar aggregates have lower water absorption increasing in flexural strength as compared to before removing adhered mortar on the RCA depicted in the Table 6.

3.2.4 Split Tensile Strength Test

This test on RCA without adhered mortar increased as compared with normal recycled concrete aggregates as shown in the Table 7.

Table 7. Tensile strength without adhered mortar on RCA

Particulate	7 days	28 days
Sample 4	3.9 MPa	7.5 MPa
Sample 5	3.5 MPa	6.2 MPa

Table 8. UPV without adhered mortar on RCA

Particulate	28 days
Sample 4	3884
Sample 5	3640

3.2.5 Ultra Sonic Pulse Velocity (UPV)

As the UPV of after removed adhered mortar aggregates are slightly increased with compared to before removing adhered mortar on recycled aggregates. The results of UPV test after removed adhered mortar test are shown in Table 8.

3.3 Disscusion

The comparison of Natural aggregates, before removing adhered mortar and after removing adhered mortar on the recycled coarse aggregates are discussed below:

3.3.1 Workability

The effect on slump as the control slump is 75 mm decreased to 58 mm in case of 100 percent RCA with adhered mortar on the recycled aggregates, but gradually increased to 61 mm with the removal of adhered mortar by putting RCA in abrasion machine for 100 revolutions i.e., after removing adhered mortar content. The effect of processing on workability are shown in the Fig. 7

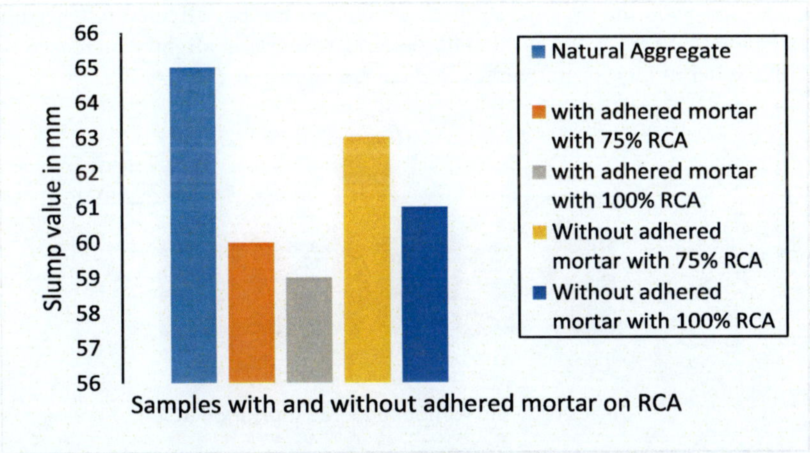

Fig. 7 Slump with and without adhered mortar on RCA

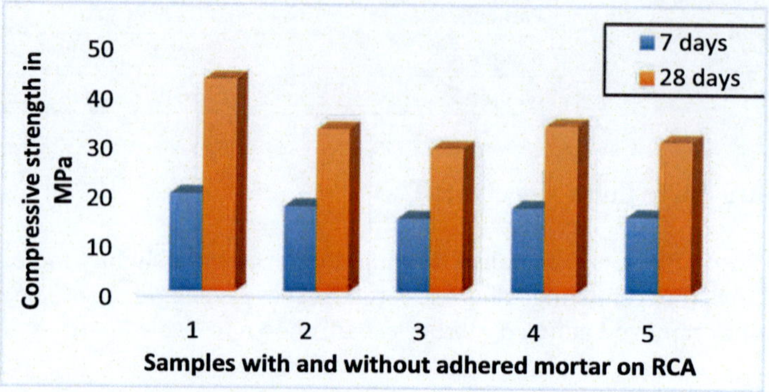

Fig. 8 Compressive strength with and without adhered mortar on RCA

3.3.2 Compressive Strength Test

The strength of RAC observed to be lower than that made with natural aggregate as RCA surfaces adhered with mortar. Fig. 8. Shows the results of 28-days compressive strength of concrete with different proportions of recycled concrete aggregates with and without adhered mortar.

3.3.3 Flexural Strength Test

The reduction in flexural strength with replacement of RCA with natural aggregates at 28 days. The variations in the flexural strength tests conducted for RCA in comparison of natural aggregates, before and after removing adhered mortar on RCA are shown in Fig. 9. However, the flexural strength seemed to be less affected when without adhered mortar on RCA are used. The flexural strength is slightly increased after removing adhered mortar on the recycled coarse aggregates (RCA).

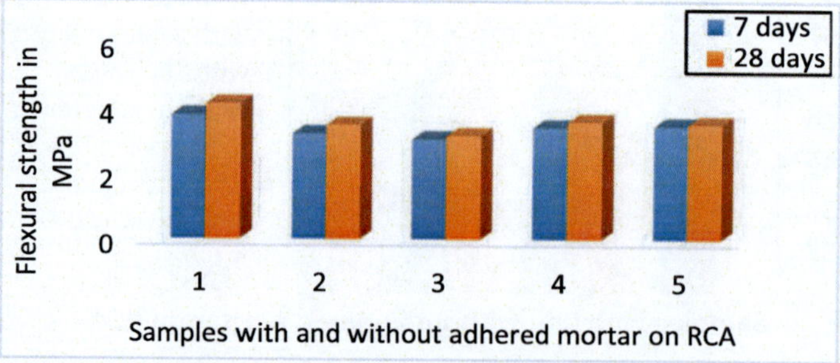

Fig. 9 Flexural strength with and without adhered mortar on RCA

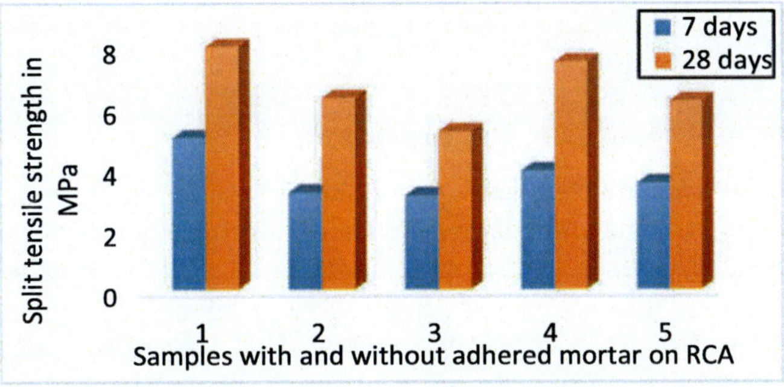

Fig. 10 Split tensile strength with and without adhered mortar on RCA

3.3.4 Split Tensile Strength Test

The comparison of split tensile strength for natural aggregates and RCA, before and after removing adhered mortar content on the recycled coarse aggregates are shown in the Fig. 10.

3.3.5 Ultrasonic Pulse Velocity Test

The UPV test results for all natural aggregates, with and without adhered mortar on the recycled aggregates are shown in the Fig. 11. The results of UPV test on RCA without adhered mortar is improved as compared to with adhered mortar on the recycled aggregates as the recycled aggregates, making the aggregate to bond with the water and cement.

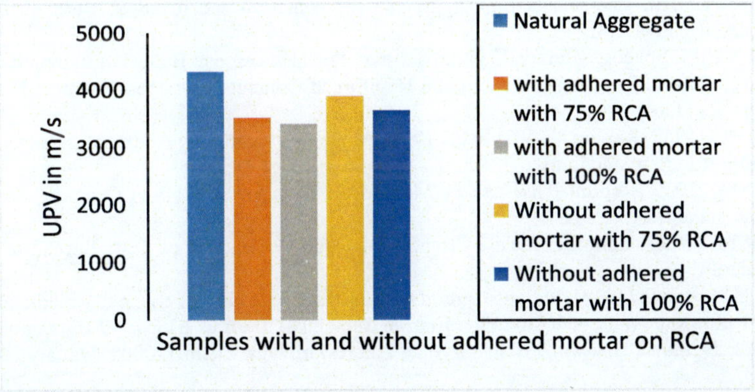

Fig. 11 UPV with and without adhered mortar on RCA

4 Conclusion

The conclusions of the study are as given below:

- The recycled concrete aggregates can be used for non-structural construction. However, for the structural component, RCA alone does not fulfil the specifications as per IS:383, In fact, the blended concrete, RCA with NA up to 50%, meet this criteria.
- The strength of recycled aggregate concrete (RAC) decrease with an increase in RCA percent, but using the RCA without adhered mortar, improves the compressive strength of the RAC about 30%.

References

1. Xie, T., Gholampour, A., & Ozbakkaloglu, T. (2018). Toward the development of sustainable concretes with recycled concrete aggregates: comprehensive review of studies on mechanical properties. *Journal of Materials in Civil Engineering, 30*(9) (2018). https://doi.org/10.1061/(ASCE)MT.1943-5533.0002304. © 2018 American Society of Civil Engineers.
2. Nieto, D., Dapena, E., Alaejos, P., Olmedo, J., & Pérez, D. (2019). Properties of self-compacting concrete prepared with coarse recycled concrete aggregates and different water: cement ratios. *Journal of Materials in Civil Engineering, 31*(2). https://doi.org/10.1061/(ASCE)MT.1943-5533.0002566.
3. Deng, X. H., Lu, Z. L., Li, P., & Xu, T. (2016). An investigation of mechanical properties of recycled coarse aggregate concrete. *Archives of Civil Engineering, 62*(4), 19–34. https://doi.org/10.1515/ace-2015-0107.
4. Mullick, A.K. (2015). [8] - Management of C&D Wastes in India –Current Status. JNTUH, Kukatpally, Hyderabad.
5. IS 10262. (2019). Concrete Mix Proportioning Guidelines. Bureau of Indian Standards.
6. IS 383. (2016). Coarse and Fine aggregate for concrete- specification. Bureau of Indian Standards, New Delhi.
7. IS 269-2015. Specification for 43 grade ordinary Portland cement: Bureau of Indian Standards.
8. IS 456:2000. Indian Standard Plain and Reinforced Concrete - Code of Practice: Bureau of Indian Standards.
9. IS : 1199 - 1959 [Reprint NOVEMBER 1991]. Methods of Sampling And Analysis of Concrete: Bureau of Indian Standards.
10. IS : 516 - 1959 [Reprint JUNE 2006]. Methods of Tests for Strength of Concrete: Bureau of Indian Standards.
11. IS 5816. (1999) Splitting Tensile Strength of Concrete-Method of Test: Bureau of Indian Standards.
12. IS:13311 (part1) (1992). Indian standard non-destructive testing of concrete - methods of test Part 1 ultrasonic pulse velocity. First Reprint September 1996 @ BIS: 1992 Bureau of Indian Standards Manak Bhavan, 9 Bahadur Shah Zafar Maro New Delhi 110002, January 1992.

Experimental Investigation of the Effects of Fly Ash on Functionally Graded Recycled Coarse Aggregate Concrete Beams Incorporating Fibers

N. Mareeswari Andal and N. Kaviya

Abstract The construction industry's continued expansion encourages larger carbon emissions during the production of building materials, which in turn encourages the concrete industry to pursue sustainability. Two distinct grades of concrete, M30 and M20 were employed in this study. M30 grade concrete is used in the compression zone, and engineered concrete is strategically used to replace the tensile portion of the concrete that surrounds the primary longitudinal reinforcement in an RC member. The investigation consists of determining the bending behavior of functionally graded engineered concrete beams. The RCA content (30% RCA) and 0.75% polypropylene fibers was used as one of the criteria for analyzing how the RCA concrete beams behaved structurally under flexure.

Keywords Recycled coarse aggregates · XRD · EDS · SEM · Functionally graded beam · Polypropylene fibers

1 Introduction

The construction industry is the major exploiter of natural resources starting from land to utilization of various resources like metals, fuels, minerals, etc. Besides the environmental effects, large quantities of waste generated from old buildings need attention. In recent days, various experiments and research have been going on regarding the re-usage of aggregates obtained from C&D wastes. The production of cement leads to higher CO_2 emissions of about 5–7 % of the total CO_2 anthropogenic emissions [1]. Fly ash is a pozzolanic material obtained as a by-product of the thermal power plant industry.

N. M. Andal · N. Kaviya (✉)
Government College of Technology, Coimbatore 13, India
e-mail: kaviyanagarajan07@gmail.com

A. Vilventhan et al. (eds.), *Advances in Construction Materials and Management*, Lecture Notes in Civil Engineering 346,
https://doi.org/10.1007/978-981-99-2552-0_30

In our study, optimization of concrete mix containing Recycled Coarse Aggregate and fly ash was done to determine the effective proportion of RCA. Six concrete mixes containing a varying proportion of RCA ranging from 10% to 40% in the Natural Coarse Aggregate were made. For each concrete mix, cubes of dimension 100 × 100 × 100 mm were cast.

The optimum amount of RCA was determined by conducting tests like the slump test and compressive strength test. The initial and final setting time of cement was determined. SEM and XRF for fly ash were carried out. Physical properties like specific gravity, water absorption, particle size distribution analysis, bulk density for aggregate, etc. were also studied and carefully considered for the determination of mechanical properties.

2 Research Methodology

An optimum amount of RCA incorporating class F fly ash as binding material was found in the first stage of the research. The second stage involved the casting of beams in two different groups.

Controlling specimens and testing specimens were two different sets of prepared beam specimens. The control specimens were three beams made up of full-depth M20, M30, and functionally graded with M30 and M20 concrete beams. (See Table 1) Table 2 lists the weight ratios for basic mixtures. The test specimens were built of M30 grade and engineered concrete with fly ash and RCA (see Table 3). In this study, maximum load carrying capacity, maximum deflection, beam stiffness, and ductility data for two groups of beams are compared.

Table 1. Conventional beam-mix proportions

Beam ID	Mix ID	OPC (%)	Fly Ash (F) (%)	RCA (%)	Polypropylene fiber (%)
B2C	M20	100	–	–	–
B3C	M30	100	–	–	–

Table 2. Weight proportions of the basic mix

Mix ID	Grade	Mix ratio	W/C ratio	Cement kg/m^3	Fine aggregate kg/m^3	NCA kg/m^3
M2C	M20	1:2.42:4.25	0.5	314	759.37	1334.52
M3C	M30	1:1.85:3.49	0.43	378.31	714.67	1322.55

Table 3. Functionally graded beam - mix proportions

Beam ID	Mix ID	Proportion details of Engineered Concrete in the Lower portion					
		Binding materials		CA			PP fibre (%)
		OPC (%)	Fly Ash (%)	NCA (%)		RCA (%)	
				12.5 mm	20 mm		
FB2-3C	M20 &M30	–	–	30	70	–	–
FB2-F	M20&M2	70	30	30	70	–	–
FB2-M5	M20&M5	70	30	21	49	21	–
FB2-M5P	M20&M5	70	30	21	49	21	0.75

3 Experimental Program

All the materials used in this study were according to Indian Standards.

3.1 Materials

Binding material OPC of grade 53 having a specific gravity of 3.06 was used as per IS 12269–2013. Fine aggregates were tested for their gradation. Their sieve analysis results and physical properties are given below in Table 4. M-sand confirming Zone II as per Table 4 of IS 383 – 2016 was used. Two types of coarse aggregates namely NCA and RCA were used. The maximum size of CA is 20 mm. NCA of size 20 mm was replaced partially by RCA (see Table 5) for finding the optimum amount of RCA. Both the coarse aggregates were used according to IS 2386 (part 1):1963. Polypropylene fibers were used to improve the performance of engineered concrete in the tension zone. Superplasticizer Master Glenium SKY 8233 which possesses a water-reducing ratio of 16%, was added to the mix in the proportion of 0.75%, and class F Fly ash with specific gravity 2.1 confirming to IS: 3812 – 2013(part I) was used in the proportion of 30% to the weight of cement. As per IS 456 – 2000, potable water free from salt was used during the casting and curing process of concrete.

3.2 Microstructural Characterization

The chemical composition of a sample is determined by X-ray fluorescence (XRF) analysis by measuring the X-rays that the sample emits. Table 6 provides information on the chemical composition of Class F Fly Ash. It revealed that fly ash comprised a total of 77.75% SiO_2, Al_2O_3, and Fe_2O_3 together with 6.25% CaO. Scanner electron microscopy (SEM) analysis was then conducted. Results from SEM can be used to

Table 4. Sieve analysis of natural fine aggregates

S No	Sieve Size	% Passing	Grading Zone from Table 4 (IS:383-2016)
1	75 μ	1.6	100
2	150 μ	5.7	90–100
3	300 μ	19.9	75–100
4	600 μ	40.4	55–90
5	1.18 mm	68	35–59
6	2.36 mm	85.8	8–30
7	4.75 mm	100	0–10
8	10 mm	100	–
9	PAN	1.6	–

Table 5. Optimization mix proportions

Mix ID	Grade	Binding materials		Coarse aggregate		
		OPC (%)	Fly Ash (%)	NCA (%)		RCA (%)
				12.5 mm	20 mm	
M1	M20	100	–	30	70	0
M2	M20	70	30	30	70	0
M3	M20	70	30	30	63	7
M4	M20	70	30	30	56	14
M5	M20	70	30	30	49	21
M6	M20	70	30	30	42	28

examine the particle shapes (see Fig 1). Most particles are under 45μm in size and have a spherical form.

Crystallographic Open Database's COD Inorganic 2015 and Cement 2014 datasets were used for phase identification, and X-ray diffraction (XRD) was used to establish the mineralogical composition of recycled aggregate alone. This process made use of the Match! Software (v.1. 10, Crystal Impact, Bonn, Germany). Along with the aforementioned, X-ray microanalysis (EDS) was also used, which could validate the

Table 6. Composition of fly ash

Oxide	Amount (%)
SiO_2	50.45
Al_2O_3	19.25
Fe_2O_3	8.05
CaO	6.25
MgO	1.77
K_2O	8.4

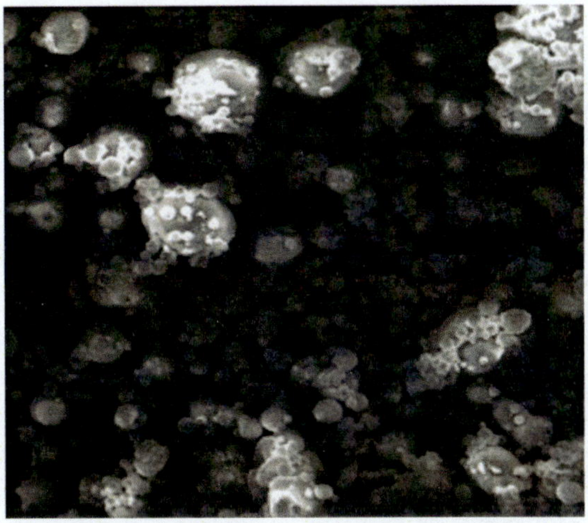

Fig. 1. Scanning electron microscopic images of class F Fly ash

Fig. 2 Curing of beams

XRD results of recycled aggregate. Table 7 and Table 8 show the physical properties of NA, RCA, cement, and fly ash. A pH meter was used on recycled aggregates to take readings at room temperature, which were used to verify the findings of the chemical analysis. For this, 20 mL of distilled water was combined with 10 g of treated and untreated recycled aggregates. After stirring this mixture with a glass rod for a while, the pH readings were taken [2].

Table 7. Physical properties of natural aggregates and RCA

Properties	NCA	NFA	RCA
Specific gravity	2.947	2.736	2.361
Water absorption (%)	0.64	4.17	3.38
Bulk density – loose state (g/cc)	1.618	1.653	1.286
Void ratio – loose state (%)	45.09	39.58	45.53
Bulk density – dense state (g/cc)	1.931	1.965	1.555
Void ratio – dense state	34.48	28.18	34.13
Fineness modulus		2.513	

a Abbreviation: RCA Recycled Coarse Aggregate NCA, NFA, Natural Fine aggregate; Natural coarse aggregate;

Table 8. Physical properties of cement and Fly Ash

Properties	Cement	Fly ash
Specific gravity	3.05	2.13
Soundness expansion	2.6 mm	1.36 mm
Specific surface area (m^2/kg)	310	332
Consistency (%)	31	
Initial Setting time (min)	128	
Final setting time (min)	225	

4 Methods

There is little information available on the study of the combined effects of fly ash and optimum recycled coarse aggregate (RCA) quantity. RCAs typically contain 30–35% original cement paste and 67%–70% original aggregate by volume. The tensile strength and compressive strength of RCA were reduced by 40% and 60%, respectively, compared to regular concrete, while drying shrinkage was reduced by 60% [3]. Because of the decreased tensile characteristics of concrete, cracks were more likely to form when concrete included more than 50% RCA [4]. Due to the adhered mortar, recycled aggregate concrete is deemed to be of lower quality (Table 9).

Table 9. Tests performed and codes followed

S No	Test	Codes
1	Pozzolana Reactivity (Strength Activity Index)	IS: 4031 (part 6) – 1988
2	Compressive strength	(516, 1959) IS: 516–1959

4.1 Fabrication of the Beam Specimens

Three types of concrete mixtures were used to create the FGM group of beams, M20, M30, and Engineered concrete respectively. Vertically, the gradation pattern was made. After pouring, a needle vibrator was used to compact the concrete. The beam specimens were made using steel molds (Fig. 2).

5 Test Results and Discussion

The beams were tested under four-point loading conditions that were simply supported. To further measure the beam deflections, LVDTs were positioned in the midspan and at the loading locations. Utilizing hydraulic jacks, the load was transferred to the beams. The test continued until the beam failed. The XRD results from the recycled coarse aggregate are shown in Fig. 3. The XRD test results revealed that typical multiphase mixtures had peaks and reflection overlap.

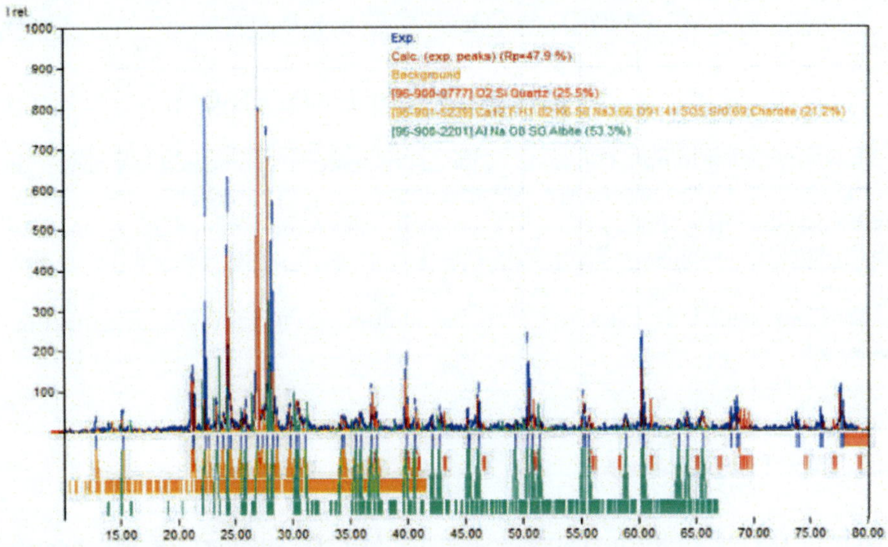

Fig. 3 X-Ray diffraction patterns of Recycled coarse aggregates

5.1 Mineralogical Composition of Recycled Coarse Aggregate

Phase identification was automated by the Match! software. The programme looks for a number of crystalline phases that are connected to the original aggregates and the cement paste that was used to bind them. The majority of the components of concrete are normally made up of aggregates [5], therefore the plentiful minerals in each sample studied belonged to the principal aggregates. The feldspar family (Albite), boggsite, charoite, and okenite were identified by the XRD results (zeolite family). The reason for the broadened feldspar peak morphologies is assumed to be the occurrence of near XRD reflections of homologous crystallographic planes in a variety of feldspar minerals. According to peak intensity, calcium oxygen compounds, like charoite, occur as multiphase. Aggregate made of limestone is not used in this location. Since the majority of the paste-adhering recycled aggregates are made up of portlandite $Ca(OH)_2$ and C-S-H hydrated calcium silicates, the large presence of calcite is principally brought on by the carbonation of cement hydration products in that paste (C-S-H gel). These phases are carbonated, as stated in Eqs. 1 and 2.

$$Ca\,(OH)_2 + CO_2 \rightarrow CaCO_3 + H_2O \tag{1}$$

$$C - S - H + CO_2 \rightarrow CaCO_3 + SiO_2.n\,H_2O \tag{2}$$

It is interesting to note that none of the tested materials contained any peaks for portlandite $Ca(OH)_2$, despite the fact that Liu et al. [12] discovered that calcium hydroxide $Ca(OH)_2$ undergoes rapid carbonation when the samples are ground into fine particles for analysis. Since $Ca(OH)_2$ is released from the cement during this grinding process, some of the CaCO3 in the diffractogram is the carbonate that is created as a result of exposure to CO_2 in the air after grinding for powder diffraction techniques.

5.2 SEM -EDS and pH

Figure 4 and Table 10 provide the results of the SEM-EDS experiment. If the compositional variance matched the XRD results, scan areas and elemental chemical analyses are provided in percentage by weight. Because of the calcium chemical components present, typical concrete has a pH value of roughly 10.67. The concrete mix M5, which contains 30% recycled coarse aggregate in place of natural coarse aggregate, has a pH value of 10.93 after 28 days.

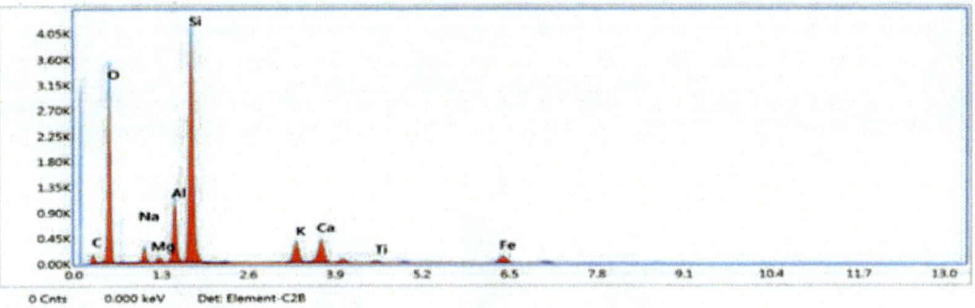

Fig. 4 Results of the chemical analyses by EDS

Table 10. Results of the chemical analysis by EDS of the RCA expressed as a weight percentage

Element		C	O	Na	Mg	Al	Si	K	Ca	Ti	Fe
Weight %		2.68	46.62	3.21	0.8	7.5	25.65	4.14	5.44	0.6	3.37

5.3 *Effect of Fly Ash on Strength Activity Index of Blended Cement*

The Strength activity index of fly ash blended cement mortar specimens is shown in Fig. 5. According to a study, adding fly ash to a mortar diminishes its compressive strength even after just seven days. The strength activity index (SAI) of the mortar is lower than the 80% threshold required by IS 3812 (part 1): 2013 to be categorized as pozzolanic materials. The gradual and constant pozzolanic reaction of fly ash in concrete is responsible for the poor compressive strength value of the fly ash blended mortar at an early age. Calcium hydroxide and calcium silicate hydrate (C-S-H) are the main byproducts of cement hydration. Hardened concrete's primary source of strength is (C-S-H). Because $Ca(OH)_2$ is weak and dissolves in water to generate cavities, it degrades the quality of hardened concrete. The pozzolanic reaction causes $Ca(OH)_2$ to change into secondary C-S-H gel when fly ash is added to the mixture as a cement substitute.

6 Properties of Concrete Mixes

The mix details of various concrete mixes are presented in Tables 1, 2 and 3. Fresh and mechanical properties of these concrete mixes were studied and described below.

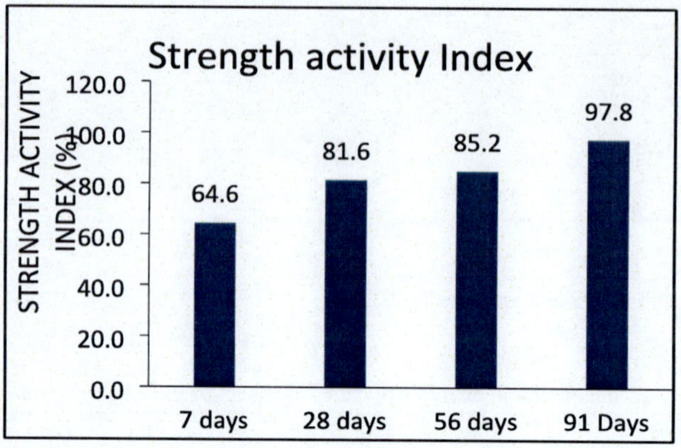

Fig. 5 Strength activity index

Table 11. Slump value of various mixes

Mix ID	M1	M2	M3	M4	M5	M6
Slump (mm)	121	132	98	92	85	75

6.1 Slump Test

Table 11 shows the test results of slump values of all specimens. Control mix M1 exhibits great workability; mix M2, which contains 30% fly ash, has the highest level of workability. The primary disadvantage is the presence of adhered mortar on the surface of coarse recycled concrete aggregate.

6.2 Mechanical Properties

Figure 6 shows the target mean strength for all mixes. The 30% fly ash content of mix M2 resulted in a small increase in compressive strength due to the pozzolanic interaction between the silica in fly ash and the $Ca(OH)_2$ released during the hydration of OPC. With 10% more RCA content, concrete's strength is marginally reduced to 30.92 MPa. On the other hand, compared to a standard mix of M2, the compressive strength is increased with the inclusion of 20% (M4) and 30% (M5) RCA content. In other areas, some researchers demonstrated that a small amount of $CaCO_3$ from limestone filler might replace some of the cement [6]. The main characteristics of recycled coarse aggregate are high water absorption and low specific gravity. Compressive strength decreased when RCA content increased by over 30%. This decrease may be related to adhered mortar on RCA, which prevents appropriate bonding of cement and aggregate.

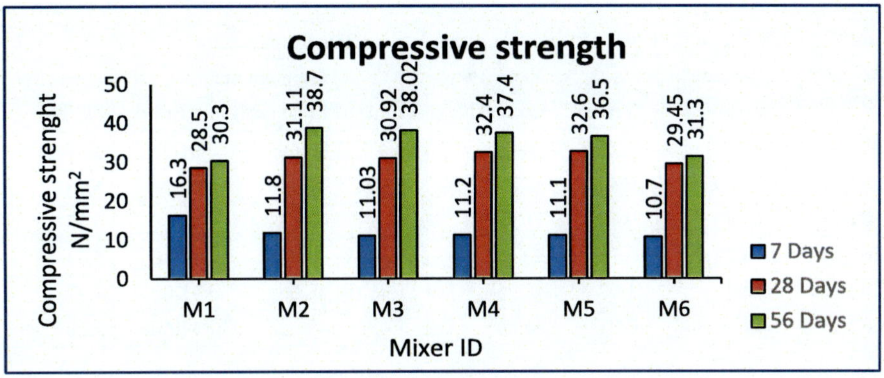

Fig. 6 Compressive strength

6.3 Load Deflection Relationship

Table 12 shows each specimen's first cracking load, yield, and ultimate load. A load cell of 500 kN was used to find the applied load. The load-deflection curves of the beams are shown in Fig. 7. The load value at the tip of the initial linear zone in the load-deflection curve is the first cracking load. According to the test results, controlling mix B3C (M30 grade) has a little higher initial cracking load than controlling mix B2C. The corresponding deflection is 0.84 and 1.26 mm for the conventional mix of B3C and B2C respectively. This might be explained by the increased rigidity of higher-grade concrete.

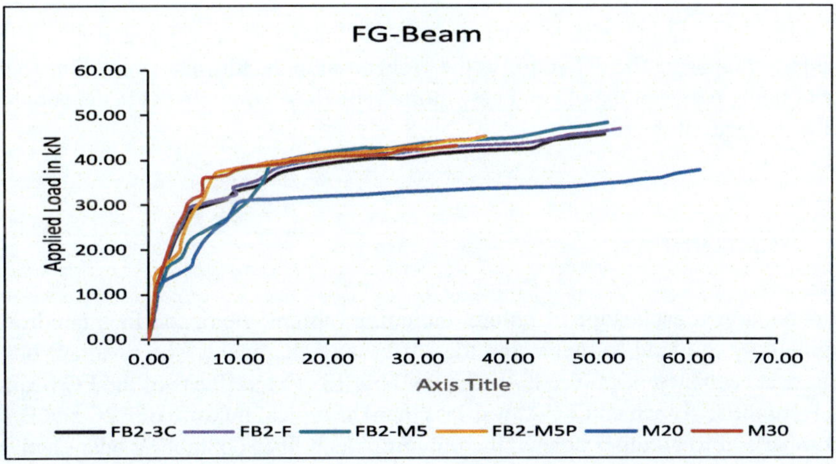

Fig. 7 Load deflection curve

Table 12. Loads and Midspan deflection at rupture, yield, the ultimate state

Beam Specimen	Rupture		Yield State		Ultimate State		Ductility index
	Load (kN)	δ mm)	Load (kN)	δ mm)	Load (kN)	δ (mm)	
B2C	11.29	1.26	30.86	10.4	37.69	61.26	5.89
B3C	12.69	0.84	32.29	5.99	42.92	34.26	5.72
FB2-3C	10.57	0.78	31.59	8.69	45.63	50.69	5.83
FB2-F	10.22	0.76	32.11	9.68	46.83	52.36	5.41
FB2-M5	10.01	1.02	32.96	11.96	48.06	51.02	4.27
FB2-M5P	14.76	0.91	36.86	7.48	45.02	37.66	5.03

Comparing the FG-controlling specimen (FB2-3C) to the standard mix B2C, the controlling specimen performed better in deflection with a lower load-carrying capacity. These results are similar to those in the study by M M A Pratama et al. [7]. The beam FB2-M5 containing 30% fly ash and 30% RCA had the lowest initial cracking load. The interfacial transition zones between the RCAs and new mortar cement paste were associated with the reduced cracking load of the RCA beam. The functionally graded beam containing polypropylene fiber is still more efficient than FG beams that solely carry fly ash and RCA. This is caused by the influence of polypropylene fiber distribution in concrete pores. As a result, concrete has enhanced crack resistance and is more durable than any other mix.

6.4 *Structural Ductility*

The ability of a structure an inelastic reaction is known as ductility, and it is determined by comparing the deflection at the yield point to the ultimate point. Due to the poor bonding between the old and new mortar, the RCA concrete FG beam achieves the lowest level of ductility.

6.5 *Beam Stiffness*

The deflection at each stage of failure, including rupture, yield, and final condition, is divided by the load to determine the stiffness of the beam. The resultant beam stiffness at each step of failure is shown in Table 13. The stiffness of the FG beams, which comprise fly ash and RCA, was less than the typical mixture of B2C and B3C. In comparison to all other beams, the polypropylene fiber-containing beam had the highest stiffness value. This might be because the fibres are distributed uniformly throughout the material, which prevents cracks from forming during the actual rupture stage and gives the material more rigidity than a standard B2C combination.

Table 13. Beam stiffness at Rupture, Yield, and Ultimate condition

Beam ID	Beam stiffness (kN/mm)		
	Rupture	Yield	Ultimate
B2C	9.81	2.97	0.62
B3C	13.57	5.39	1.25
FB2-3C	15.09	3.6	0.9
FB2-F	14.69	3.32	0.89
FB2-M5	10.08	2.76	0.94
FB2-M5P	16.29	4.93	1.2

7 Conclusion

1. Optimization of mixes containing a varying percentage of RCA was conducted. The compressive strength of Fly ash blended RAC increases up to 30% RCA replacement whereas the further increase in RCA proportion leads to a decrease in compressive strength.
2. Being a good pozzolanic material, the usage of fly ash in the concrete mix enhances the compressive strength of RAC. When comparing the stiffness of the conventional M20 grade beam with FG - conventional and Fly ash blended beam, the stiffness of the latter ones is approximately 50% more than that of the former one.
3. Due to the debonding effect of RCA, a considerable amount of decrease in RAC beam stiffness can be found. Polypropylene fibers increase the stiffness of the FG beam by 66% which can be explained by the bridging effect of fibers on the pores.

References

1. Stajanča M., & Eštokova A., (.2012). Environmental Impacts of Cement Production Lviv Polytechnic National University Institutional Repository.
2. Moreno-Pérez, E., et al. (2018). Chemical and mineralogical characterization of recycled aggregates from construction and demolition waste from Mexico City. *Minerals, 8*(6), 237. https://doi.org/10.3390/min8060237
3. Talamona, D., & Tan, K. H. (2012). Properties of recycled aggregate concrete for sustainable urban built environment. *Journal of Sustainable Cement-Based Materials, 1*(4), 202–210. https://doi.org/10.1080/21650373.2012.754571
4. Ikea, T., Yamane, S., & Sakamoto, A. (1988). Strength of concrete containing recycled aggregate concrete. In *Proceedings 2nd RILEM Symp. on Demolition and Reuse of Waste, RILEM Publications*, Bagneux, France. https://doi.org/10.11159/icsect19.124.
5. Omary, S., Ghorbel, E., & Wardeh, G. (2016). Relationships between recycled concrete aggregates characteristics and recycled aggregates concretes properties. *Construction and Building Materials, 108*, 163–174. https://doi.org/10.1016/j.conbuildmat.2016.01.042.

6. Angulo, S. C., Ulsen, C., John, V. M., Khan, H., & Cincotto, M. A. (2009). Chemical-mineralogical characterization of C&D waste recycled aggregates from Sao Paulo Brazil. *Waste Management, 29*, 721–730. https://doi.org/10.1016/j.wasman.2008.07.009
7. Pratama, M. M. A., Vertian, T., Umniati, B. S., & Yoh, W. H. (2019). Flexural behaviour of the functionally graded concrete beams using two-layers and three-layers configuration. *IOP Conference Series: Materials Science and Engineering, 669*(1), 012054. https://doi.org/10.1088/1757-899X/669/1/012054
8. IS: 12269 – 2013. a code of practice for specification for 53 grade ordinary Portland cement.
9. IS: 383-206 (2016). a code of practice for specification for coarse and fine aggregates for concrete.
10. IS: 2386 (part1)1: 1963.; a code of practice for specification for methods of test for aggregates for concrete.
11. IS: 3812.1.2013. a code of practice for specification of fly ash.
12. Liu, Q., Tong, T., Liu, S., Yang, D., & Yu, Q. (2014). Investigation of using hybrid recycled powder from demolished concrete solids and clay bricks as a pozzolanic supplement for cement. *Construction and Building Materials, 73*, 754–763.

Recycled Aggregate Concrete Hollow Block as a Sustainable Walling Material

Darshini Shekhar and Jagdish Godihal

Abstract Ever-increasing construction projects in India and all over the world had a tremendous impact on the natural habitat by exploring the natural resource. For such activities meanwhile, construction practices generate huge proportion of construction and demolition waste. As a result, there is a need to compute the environmental impacts of construction materials (CDW) and to develop minimization strategies. This study has been carried out to ascertain sustainability and to reuse, recycle the CDW constituents like recycled concrete aggregates and coal ash replacing natural coarse aggregates (NCA) and cement respectively. In terms of stability, the compressive strength of circular and rectangular recycled concrete hollow blocks is measured by varying different percentage compositions of recycled coarse aggregates (RCA) and coal ash. The strength, grade properties of the recycled concrete hollow block is evaluated per code 1S2185:2005. To relate with the environmental impacts embodied energy and embodied carbon emission of each hollow block is calculated based on Inventory of Carbon and Energy database.

Keywords Sustainability · Waste reduction · Recycling · Carbon footprint · Green building

1 Introduction

Construction and demolition waste (CDW) is a global issue. India's construction industry is expected to produce 10–12 million tonnes of waste every year. According to a survey 2017 by the Technology Information Forecasting and Assessment Council (TIFAC), new construction waste in India averages 50 kg/m^2, resulting in 50 million tonnes of waste, whereas dismantling of old structures results in 300–500 kg/m^2 of demolished waste. CDW accounts for the majority of municipal solid waste. Landfilling CDW contributes to the depletion of land resources, and unauthorized

D. Shekhar (✉) · J. Godihal
Presidency University, Bengaluru, Karnataka 560064, India
e-mail: darshinis@presidencyuniversity.in

A. Vilventhan et al. (eds.), *Advances in Construction Materials and Management*, Lecture Notes in Civil Engineering 346,
https://doi.org/10.1007/978-981-99-2552-0_31

waste disposal pollutes soil, water, and air. Overcoming the current garbage siege phenomena and reducing CDW is a pressing and difficult problem all over the world [1]. Some of the negative societal consequences include land destruction, rising energy, the formation of solid waste, dust and the degradation of the natural, particularly non-renewable materials. [2]. If we assume that 5% of current buildings are demolished each year, this trash will amount to 288 million tonnes. Construction and demolition debris comprises two types of waste namely recycled concrete aggregate (RCA) and recycled aggregate (RA). Sand, gravel, crushed stone, tile, and wood are examples of inert mineral materials that make up RA. RCA stands for Recycled Concrete Aggregate, which is made from crushed concrete made using NA.

A study found that recycled aggregate could serve as a bridge between aggregate exigency and supply in India according to research by the Ministry of Urban Development (MoUD), with the housing sector estimating a 55,000 million cubic meter aggregate shortage. One approach to all of these problems is to recycle CDW and use as an alternative coarse aggregate for concrete.

Aggregate recycling may be necessary for the creation of "Recycled Concrete" in the future leading to a shortage of natural resources. Every year, coal-fired power plants produce tonnes of combustion waste, which includes flyash and coal bottom ash, and the large percentage with this ash is buried in landfills. Cement concrete hollow blocks due to its maintainability, acoustic and fire proof, thermal insulation, less weight, and faster construction pace, serve an important role in today's construction industry. They are a better and more cost-effective adjunct to burnt mud bricks. [3]. Construction can be done faster since concrete hollow blocks are often larger than traditional clay building bricks and require less mortar. It has a nice look to it and can be readily incorporated into any architectural style. On both the outside and interior walls, it can be finished in a variety of ways. Cement plaster, lime gauged, or plasticizer could also be used to finish it.

2 Literature Review

Aslani, Hou, Nejadi, Sun, and Abbasi (2019) [4] made recycled aggregate concrete (RAC) blocks, replacing natural stone with coarse RCA, river sand with fine RCA. The strength, ecological and cost benefits of RAC block were greater. Fly ash as an active additive and bricks-mortar of the demolished building is used in the construction of smaller hollow concrete blocks. It was found that concrete's performance can meet required strength, that limiting the water-to-cement ratio(w/c) appropriately can effectively boost concrete quality. Hollow concrete blocks' uniformity and service life can be continued to improve by properly adding fly ash, and concrete workability can be upgraded by adequately lowering the w/c [5]. For the design characteristic strength of 45 MPa, three different concrete mixtures with 100% OPC and 10%, 20% OPC replacement with coal ash have been investigated. Concrete with coal ash as a component of the additives dropped 4.7% and 11.8% of its compressive strength, respectively [6]. Supplementary cementitious materials like limestone and coal ash, can be combined with recycled aggregates to produce eco-friendly concretes and mortars with good mechanical characteristics and sulfate resistance. It is suggested that recycled aggregate amounts of about 70% be used in the creation of an ecological mortar, with 30% of coal ash replacing portland cement to ensure the densification of the concrete matrix and interfacial transition zone. These mixtures are advised for non-structural uses like precast that will experience moderate sulfate action [7]. Even though granite aggregate, the second most popular aggregate, is used less frequently than quartzite, it still has similar mechanical properties to concrete made with RCA. CDW can be used to create concrete with a compressive strength(fck) of up to 30 N/mm^2 when natural aggregate is replaced with RCA [8]. The RAC can be completely used in place of natural coarse aggregates in concrete without affecting the slump or strength of M20 grade concrete [9]. The maximum amount of cement that can be substituted with fly ash out of reducing the concrete stability varies depending on the w/c ratio. In low w/c ratio concretes, the ratio of partial cement replacement with fly ash will be higher, and in high w/c ratio concretes, it will be significantly smaller [10]. Construction commodities with a large embodied energy (EE) were discovered to transmit more carbon than those with a low EE [11]. Embodied carbon can account for 2% to 80% of life cycle carbon footprint, relying on variables such as building usage, geographical location, physical parameters, and presumptions about building lifesapn and future supply of energy [12]. The choice of building materials, maintainability, wide availability, and recovering facilities are critical factors in a building's ability to operate efficiently in terms of energy and carbon output [13].

Traditional concrete's main components, cement and aggregates, can be supplemented with more sustainable choices to minimize its environmental impact. [14, 15]. Portland cement, which accounts for 74 to 81% of total emissions from regularly used concrete mixtures, is the most significant source of carbon emissions [16]. SCM (supplementary cementitious materials) like fly ash, granulated blast furnace slag can complement cement, and natural aggregates (NA) can be substituted with recycled aggregates [17–19]. Additionally, RAC helps to protect the world's mineral resources. Due to the long distances involved in shipping, recycled aggregate should only be used when it is regionally reachable. Conversely, their embodied carbon value could be consistently greater than that of traditional materials. Once the distance exceeds 15 to 25 km, the ecological rewards of using RA turn balanced or negative because the environmental effects of RCA are incredibly susceptible to shipping distances [20, 21]. Furthermore, it was outlined that cement has the maximum influence on the environment throughout concrete production, and that increase in the share of RCA in concrete composites has minor effect on CO_2 emissions [22].

An innovative type of concrete masonry unit with good serviceability, environmental protection is being investigated in this article. Two different hollow block (circular and rectangular) specimens of different concrete compositions are evaluated under compressive loads to determine compressive strength. Water absorption and density of a variety of blocks casted has been compared. The Inventory of carbon database is used to investigate the embodied energy and embodied carbon output from conventional and composite concrete hollow block. It is intended that it will serve as a source of inspiration for future scientific study and engineering applications involving recycled concrete hollow blocks made using recycled coarse aggregate and Coal ash. It is a new proposal for the raw material source for wall materials, as well as a new method for reusing concrete waste. It has substantial environmental and practical implications for engineering action in the construction industry.

3 Methodology

The recycled aggregate-coal ash concrete hollow block is casted following IS 10262(2019) and IS 2185 (part 1):2005 [23, 24]. This composite hollow concrete masonry wall specimen functional property test is done and test results are compared with same dimension hollow block made of conventional materials. Environmental impact analysis of recycled aggregate concrete hollow block is done concerning embodied energy and embodied carbon emission in the surrounding by means of equation suggested by Jimenez et al. (2018) and the Inventory of Carbon Energy database. The sequence of the investigation is shown in Fig. 1.

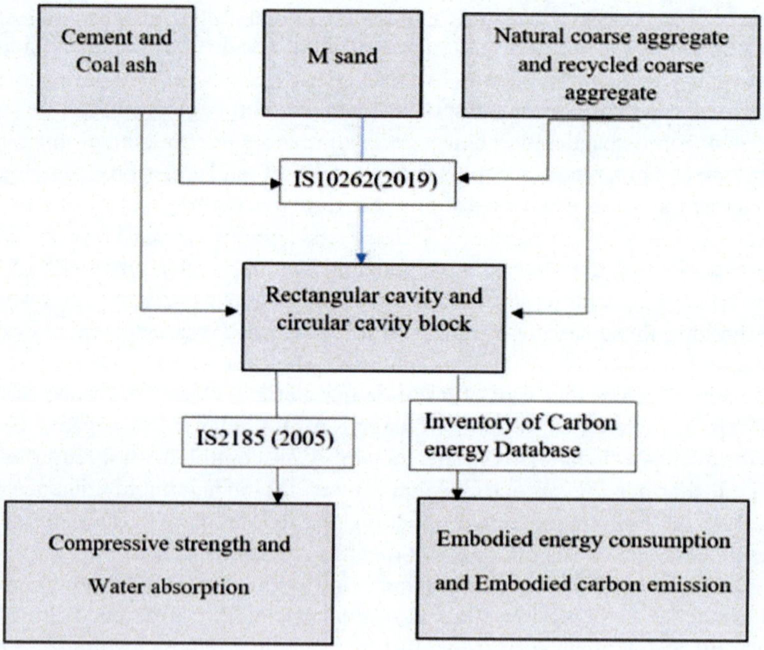

Fig. 1 Methodology of investigation

Fig. 2 Preparation of circular hollow block and rectangular hollow block

3.1 Preparation of Specimen Sample

The specimen is mix designed for the goal strength of M30 grade compressive strength. Concrete materials in this research consist of M-sand, natural coarse aggregate (NA), recycled coarse aggregate (RCA)(<6mm size), OPC cement 53 grade, and coal ash (CA) in the manufacturing of hollow concrete blocks. RCA is bought from a CDW recycling plant located near Yelahanka, Bangalore. Coal ash is obtained by burning coal from a clay brick manufacturing unit where the residue found at the bottom of kiln is collected. Specific gravity water absorption test of all material is done before mix design and casting is done on mould size 500 *100 * 100 mm.

The proportions of concrete hollow blocks in the mix is provided in Table 2. 25 mm bed thickness in both blocks, face shell and web thickness of the hollow blocks, compressive strength, dimension, water absorption, block density is calculated as per the IS2185 (part 1) 2005 code. The casted rectangular hollow and circular hollow concrete blocks is shown in Fig. 2.

4 Evaluation and Results

4.1 Compressive Strength of the Recycled Concrete Hollow Block

Table 1 enlists the basic property test results of hollow block materials. Ten concrete hollow blocks of each 5 rectangular cross-sections and the circular cross-section are casted by varying percentage of RCA, coal ash for NA and cement replacement respectively. Hollow size is 200 * 50 mm and 50 mm diameter is stated in the case of RCB (rectangular hollow block) and CCB (circular hollow block) respectively. Material composition variation in two different hollow block is shown in Table 2.

Table 1 Basic property test of concrete material

Sl no	Particulars	Specific gravity	Water absorption
1	Cement	2.91	-
2	Coal Ash	2.30	-
3	M sand	2.62	2.59
4	NCA	2.75	1.35%
5	RCA	2.32	2.57

Table 2 Material quantity for mix design, W/C = 0.55

Sample number	RCHB	CCHB	Material quantity for mix design for W/C = 0.55
0	R0	C0	100%cement: 100%NA: Msand
1	R1	C1	20%CA + 80%cement:30%RCA + 70%NA: Msand
2	R2	C2	30%CA + 70%cement:50%RCA + 50%NA: Msand
3	R3	C3	40%CA + 60%cement:70%RCA + 30%NA: Msand
4	R4	C4	50%CA + 50%cement:90%RCA + 10%NA: Msand

The distribution of RCA and coal ash raising could be observed in the reduction of compressive strength parallelly. The compressive strength of CCHB is found to be better than RCB for the same mix proportion. CCB(C4) and RCB(R4) specimen where a maximum percentage of RCA, coal ash mixed as achieved 38 percent of target strength. Fig. 3 and Fig. 4 gives a clearer insight of variation in compressive strength of different shapes, mix proportion concrete hollow blocks. Under the compression test, the cracks of RCB specimens are found at the web part of blocks due to their lesser thickness, CCB is mostly crushing at the three-fourth length of the specimen.

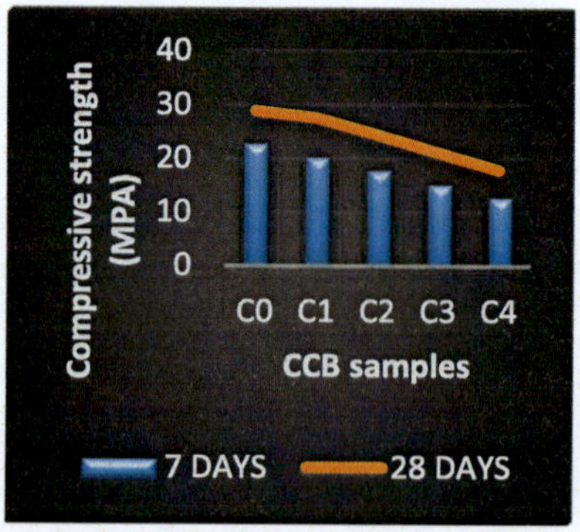

Fig. 3 Circular concrete hollow block compressive strength

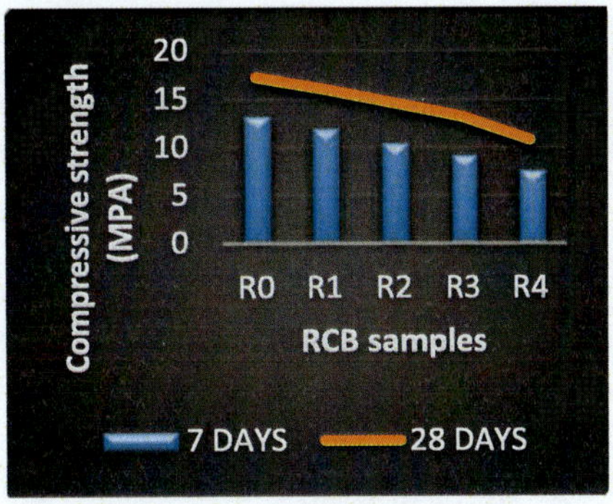

Fig. 4 Rectangular concrete hollow block compressive strength

4.2 *Water Absorption and Density*

For 24 h, the concrete specimens had been totally submerged in room temperature water. The samples were then weighed (C). They must be removed from the water and allowed to drain for one minute before being weighed (A). A damp cloth should be used to remove any visible surface water. (B) is a dry mass of concrete. Water absorption of variable concrete mixtures is displayed in Table 3.

Calculate the water absorption as follows:

Water absorption, $kg/m^3 = \frac{A-B}{A-C} \times 100\%$

Concrete hollow block with water absorption of less than 10% and density more than 1500 kg/m^3 with minimum compressive strength of 5MPa is considered to be grade A block. From Figs. 3, 4 and Table 3 observation recycled aggregate hollow blocks of all mix proportion is outlined to be grade A hollow blocks achieving the characteristics featured in IS 2185 (2005).

Table 3 Water absorption of different concrete mixtures

Sample number	Recycled Concrete hollow block (RCHB)	Density-kg/m^3	Concrete hollow block (CCHB)	Density-kg/m^3	Water absorption (%)
0	R0	1497	C0	1738	6.1%
1	R1	1500	C1	1741	6.9%
2	R2	1518	C2	1742	7.87%
3	R3	1535	C3	1744	8.12%
4	R4	1545	C4	1746	8.65%

4.3 Environmental Impacts of Concrete Mixture

The Inventory of carbon and energy (ICE) database by Geoffrey Hammond, Craig Jones, Fiano Lowrie, and Peter Tse has specified embodied carbon coefficient (ECC) and embodied energy coefficient (EEC) and is expressed in MJ/kg and $kgCO_2/kg$. It has been estimated by UK industrial fuel consumption data for all the materials and is specified from cradle to factory production. Embodied energy and Embodied carbon calculations were made from the following equation

$CO2\text{-}e = \sum Q1F1 + Q2F2 + \cdots + QnFn$ (Jimenez et al. 2018).

Where Q is material quantity in kg, F is embodied coefficient.

Embodied energy and Embodied carbon emission of different concrete mix and hollow shape blocks is shown in Fig. 5 and Fig. 6.

While recycled and waste materials have lower coefficients of embodied energy and carbon than virgin materials, concrete materials embodied energy and carbon emissions decreases as conventional materials are replaced with each other. Due to the larger hollow ratio, rectangular hollow block manufacturing consumes less energy and emits less carbon dioxide into the atmosphere than CCB.

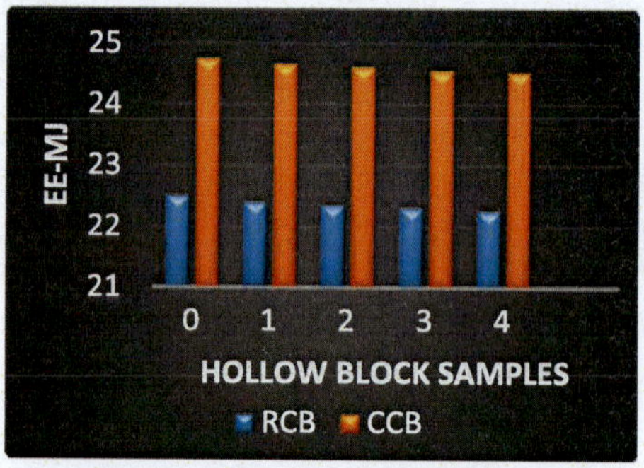

Fig. 5 Embodied energy for different concrete blocks

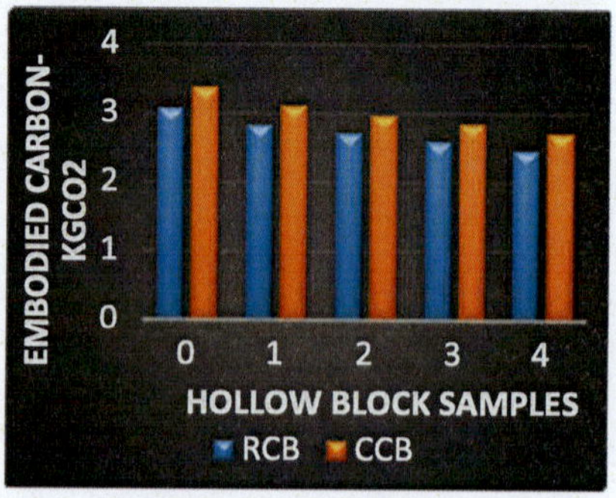

Fig. 6 Embodied carbon for different concrete blocks

5 Conclusion

1. The current study indicates that recycled aggregate concrete can fully meet the strength requirements for concrete hollow blocks to be used in load-bearing buildings, despite having a lower strength than conventional aggregate concrete. For the purpose of manufacturing hollow blocks, substitute coal ash can up to 50% and RCA up to 90% while still achieving compressive strengths greater than 5 MPa.
2. Compared to conventional mix, recycled aggregate-coal ash blocks absorb more water and are slightly denser. However, water absorption is less than the maximum percentage accepted by the 1S2185 code.
3. Embodied energy of 22.53 MJ and 24.63 MJ is consumed for R0 and C0 blocks, respectively; with approximately 1%, energy can be saved for one block when R4 and C4 usage is implemented.
4. One conventional concrete hollow block is estimated to emit 3 to 3.4 kg of CO_2. The carbon emission per unit hollow block could be 2.5 to 2.7 $kgCo_2$ if 50% of the cement was replaced with coal ash and 90% of the natural aggregate was replaced with recycled coarse aggregate.
5. Recycled aggregates combined with coal ash can be used to create grade A hollow masonry units with sufficient characteristic strength, reducing environmental impact and consuming less embodied energy

References

1. Contreras, M., et al. (2016). Recycling of construction and demolition waste for producing new construction material (Brazil case-study). *Construction and Building Materials, 123*, 594–600.
2. Lu, W., & Yuan, H. (2010). Exploring critical success factors for waste management in construction projects of China. *Resources, conservation and recycling, 55*(2), 201–208.
3. Kamal, M. A. (2020). Analysis of Autoclaved Aerated concrete (AAC) blocks with reference to its potential and sustainability. *Journal of Building Materials and Structures, 7*(1), 76.
4. Aslani, F., Hou, L., Nejadi, S., Sun, J., & Abbasi, S. (2019). Experimental analysis of fiber-reinforced recycled aggregate self-compacting concrete using waste recycled concrete aggregates, polypropylene, and steel fibers. *Structural Concrete, 20*(5), 1670–1683. https://doi.org/10.1002/suco.201800336
5. Yang, L., & Li, S. (2019). Analysis of the effect of substitution rate on the mechanical properties of recycled concrete hollow block. *Advances in Materials Science and Engineering, 2019*, 1–12. https://doi.org/10.1155/2019/7546376
6. Bostanci, S. C. (2020). Coal ash use as a cement replacement in concrete production. *IOP Conference Series: Materials Science and Engineering, 800*(1), 012010. https://doi.org/10.1088/1757-899X/800/1/012010
7. Bedoya, M. A., & Tobón, J. I. (2022). Incidence of recycled aggregates and ternary cements on the compressive strength and durability of ecological mortars. *Case Studies in Construction Materials, 17*, e01192. https://doi.org/10.1016/j.cscm.2022.e01192

8. Akhtar, M. F., Naqvi, M. W.,Alam, M. M., & Shariq, M. (2018). Use of Different Types of Aggregate vis-a-vis Demolition Waste as an Alternative Material for Concrete. In: Urbanization Challenges in Emerging Economies: *Resilience and Sustainability of Infrastructure*, pp. 679–690. Reston, VA: American Society of Civil Engineers
9. Prathima, G., Ganesh, B., &Nagaraja, K. P. (2018). Optimum Utilization of Alternate Material for Aggregates—An Approach for Waste Management in Urban Areas for Sustainability. In *Urbanization Challenges in Emerging Economies*: Energy and Water Infrastructure; Transportation Infrastructure; and Planning and Financing, pp. 221–231. Reston, VA: American Society of Civil Engineers
10. Rao, B. K., Reddy, M. A. K., & Rao, A. V. (2021). Effect of flyash as cement Replacement Material and Pore Filling Material in Concrete. In *Materials Today: Proceedings*
11. González, M. J., & Navarro, J. G. (2006). Assessment of the decrease of CO_2 emissions in the construction field through the selection of materials: practical case study of three houses of low environmental impact. *Building and Environment, 41*(7), 902–909.
12. IbnMohammed, T., Greenough, R., Taylor, S., OzawaMeida, L., & Acquaye, A. (2013). Operational vs. embodied emissions in buildings—a review of current trends. *Energy and Buildings, 66*, 232–245.
13. Atmaca, A., & Atmaca, N. (2015). Life cycle energy (LCEA) and carbon dioxide emissions (LCCO2A) assessment of two residential buildings in Gaziantep, Turkey. *Energy and Buildings, 102*, 417–431.
14. Jimenez, L. F., Dominguez, J. A., & Vega-Azamar, R. E. (2018). Carbon footprint of recycled aggregate concrete. In *Advances in civil engineering, 2018*
15. Tosic, N., Marinkovic, S., Dasic, T., & Stanic, M. (2015). Multicriteria optimization of natural and recycled aggregate concrete for structural use. *Journal of Cleaner Production, 87*, 766–776.
16. Flower, D. J., & Sanjayan, J. G. (2007). Greenhouse gas emissions due to concrete manufacture. *The International Journal of Life Cycle Assessment, 12*(5), 282–288.
17. Gursel, A. P., Maryman, H., & Ostertag, C. (2016). A life-cycle approach to environmental, mechanical, and durability properties of "green" concrete mixes with rice husk ash. *Journal of Cleaner Production, 112*, 823–836. https://doi.org/10.1016/j.jclepro.2015.06.029
18. Collins, F. (2010). Inclusion of carbonation during the life cycle of built and recycled concrete: influence on their carbon footprint. *International Journal of Life Cycle Assessment, 15*, 549–556.
19. Xu, B., Bompa, D. V., & Elghazouli, A. Y. (2020). Cyclic stress–strain rate-dependent response of rubberised concrete. *Construction and Building Materials, 254*, 119253.
20. Knoeri, C., Sanyé-Mengual, E., & Althaus, H. J. (2013). Comparative LCA of recycled and conventional concrete for structural applications. *The International Journal of Life Cycle Assessment, 18*(5), 909–918.
21. Kleijer, A. L., Lasvaux, S., Citherlet, S., & Viviani, M. (2017). Product-specific life cycle assessment of ready-mix concrete: comparison between a recycled and anordinary concrete. *Resources, Conservation and Recycling, 122*, 210–218.
22. Sabău, M., Bompa, D. V., & Silva, L. F. (2021). Comparative carbon emission assessments of recycled and natural aggregate concrete: environmental influence of cement content. *Geoscience Frontiers, 12*(6), 101235.
23. IS 10262(2019) Concrete Mix Proportioning — Guidelines (Second Revision), Bureau of Indian standards manak bhavan, 9 bahadur shah zafar marc3 New Delhi 110002
24. IS 2185 (2005) part 1 Indian Standard concrete masonry units — specification part 1 hollow and solid concrete blocks (Third Revision), Bureau of Indian standards manak bhavan, 9 bahadur shah zafar marc3 New Delhi 110002 Embodied Carbon - the Inventory of Carbon and Energy (ICE) (BG 10/2011)

Seismic Fragility of Building Subjected to Pounding Effects with Damper

Rajan L. Wankhade and Amarsinh B. Landage

Abstract In the present work efforts are made to perform seismic analysis of adjacent building frames having relatively closed spacing. These building frames undergo pounding due to seismic forces and ground motion. Effect on behavior of these frames is observed for combination of Storey height and spacing. To mitigate the effect of large deformation in closely spaced frames due to earthquake, dampers are employed further. Hence, the effects of dampers are also considered with the pounding phenomenon. Four different models are prepared considering height of buildings and difference in slab level. Comparison is made between results obtained for these frames with and without dampers. We observed that by the use of dampers to reduce pounding effect, seismic displacement is reduced up to 32%. Similarity axial load, shear forces and bending moments are mitigated to about 51%, 31% and 27% respectively.

1 Introduction

During the severe earthquake; closely spaced infrastructure undergoes pounding effect and damage within seismically active zones. We generally avoid closed spacing in buildings lying in region prone to moderate and strong ground shaking. Many adjacent buildings are designed without considering the pounding effect. It is necessary to consider the pounding effect in closely spaced adjacent buildings while designing in earthquake prone zones. Dampers are efficiently employed in building frames to mitigate earthquake effects [1–4]. Pounding effects and its further influences is properly studied further [5].

Building frames excited by earthquake induced vibrations can be suppressed by tuned mass dampers [6, 7]. Some eccentricities with eccentric seismic pounding of

R. L. Wankhade (✉)
Civil Engineering Department, Netaji Subhas University of Technology, New Delhi, India
e-mail: rajanw04@gmail.com

A. B. Landage
Civil Engineering Department, Government College of Engineering, Karad, Maharashtra, India

A. Vilventhan et al. (eds.), *Advances in Construction Materials and Management*,
Lecture Notes in Civil Engineering 346,
https://doi.org/10.1007/978-981-99-2552-0_32

symmetric Buildings is also highlighted [8]. A system with multiple dampers is optimized for multiple tuned mass dampers considering different parameters [9]. In literature assessment of seismic pounding hazard for Taipei City is also available [10]. An approach for efficient performance based seismic design is required to be developed considering dampers in conjunction with response based damage assessment [11, 12]. Depending upon Near-Fault and Far-Fault Earthquake Motions design procedure is also required to be modified in regards of seismic pounding [13–15]. Some correlation is required to be established between seismic acceleration parameters and damage indices for building frames [16, 17]. Probability analysis is further required with pounding effects and dampers for adjacent buildings based upon a novel optimum criterion [18–20]. In case of passive devices a bidirectional and homogeneous tuned mass damper is found to be more effective [21]. A robust design approach is also developed for dampers and base isolation techniques [22–24]. It is observed that vertically distributed multiple TMD can help in establishing effective performance based design and levels [25].Some comparative response assessment is also given for SMRF buildings [26]. Bracing system is also useful for ductile braced frames [27]. Genetic algorithm and its application for designing optimal multiple TMD is required for mitigating the seismic response of structures [28]. Actual analysis is also required for such frames with fixed base and isolated system [29]. Performance based probabilistic design is very useful for pounding effect [31, 32]. Numerical work is also performed for smart composite beams and plates [33–39]. Pounding for bridges with strong ground motion is recorded [40]. Some experimental, analytical and numerical performance of RC beams with V-shaped reinforcement is available in literature [41]. Seismic fragility is highlighted for building frames subjected to pounding effects considering SSI [42]. Moreover, the influence of using dampers in building frames is suggested as a retrofitting technique to mitigate the intensity of pounding.

2 Methodology and Model

The study is focused on the investigation of the effects of pounding on the limit state capacity of adjacent multi story SMRFs. Figure 1 shows adjacent closed spaced buildings subjected to earthquake induced vibrations. In the present model response modification factor is adopted as, R = 8. Over-strength factor is, Ω = 3. Appropriate floor dead and live loads are considered as given further. Young's modulus of 200 GPa. The defection amplification factor is assumed as Cd = 5.5. To stimulate the pounding force, impact force can be given as,

$$F_{(t)} = K_{imp}\,\delta_{(t)} + C_{imp}\,\dot{\delta}_{(t)} \qquad (1)$$

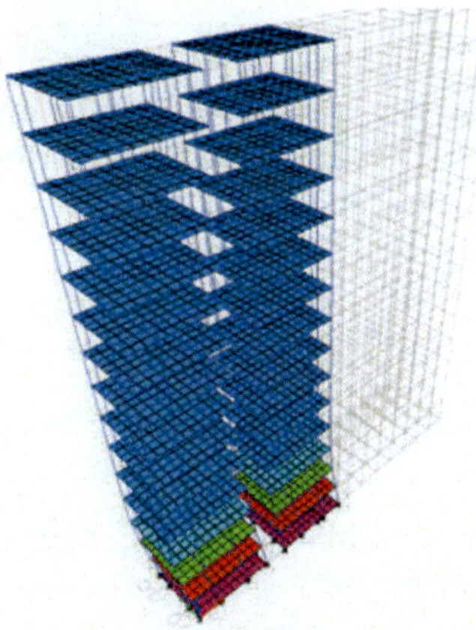

Fig. 1 Closed spaced adjacent buildings subjected to earthquake induced vibrations

In the above equation, $\delta_{(t)}$ and $\dot{\delta}_{(t)}$ represents relative deformation and velocity of building frames undergoing pounding. Whereas $K_{imp}\, and\, C_{imp}$ denotes the element's stiffness and the damping coefficient respectively. Damping coefficient is expressed as.

$$C_{imp} = 2\xi \sqrt{K_{imp} \frac{m_1 m_2}{m_1 + m_2}} \tag{2}$$

Here 'm1' and 'm2 'represents masses of the adjacent buildings. 'ς' is impact damping ratio.

3 Seismic Separation Distance to Reduce Pounding Effect

Seismic pounding is exerted in the case when the separation gap between neighbouring buildings is insufficient to response the relative earthquake ground motion. To reduce pounding effect, different seismic codes have restricted the separation distance between the adjacent buildings. Minimum separation distances can be calculated by absolute sum (ABS) or Square Root of Sum of Squares (SRSS) as given by the international building code. According to ASCE, the adjacent building frames may be separated by a distance, δMT, given as below.

$$\delta_{MT} = 2\xi \sqrt{\left(\frac{C_d\, \delta_m}{I}\right)_1^2 + \left(\frac{C_d\, \delta_m}{I}\right)_2^2} \tag{3}$$

where, Cd is defection amplification factor. 'I' and 'δM' are taken as importance factor and the maximum inelastic response displacement respectively.

Also in the capacity spectrum approach, it is suggested to design building frames in such a way that it should perform some specific performance. Target displacement is specified to evaluate performance where the intersection of the capacity spectrum with the suitable demand spectrum (the displacement at the performance at the performance point is equivalent to the target displacement in the coefficient method).

3.1 Details of the Building Frames

The building frames considered in the study is having first one as 15 storey and second having different storey for comparison. The floor height of both the building is varied to create different cases. The material used are M25 concrete and fe500, fe415 steel bars for columns and beams. Different checks like column PMM interaction ratio, beam/column capacity ratio, column/beam capacity ratio, and drift storey checks were performed on the building to check the stability of structure and wherever necessary the size of column is increased to pass the checks. Columns are 300x450mm and beams are of 230x300mm sizes. Slab is considered as 150mm thick used in both the building models. Figure 2 shows model of closely spaced adjacent building frames with dampers having same storey level and equal height.

Further Fig. 3 demonstrates models of closely spaced adjacent building frames provided with dampers. This model is different storey level and different height of buildings.

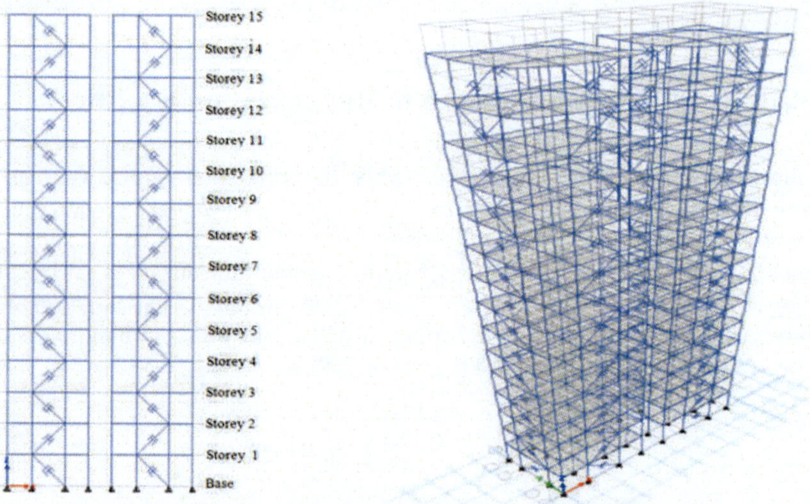

Fig. 2 Closely spaced adjacent building frames with dampers having same storey level and height

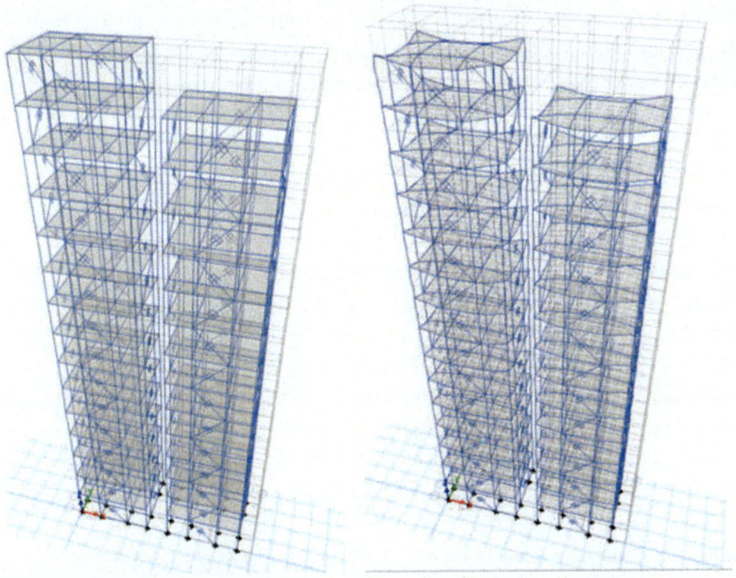

(a) Unreformed Shape (b) Deformed Shape

Fig. 3 Closely spaced adjacent building frames with dampers having different storey level and height

3.2 Loads and Load Combinations

All the structural elements of building frame contribute to the gravity load. It consists of the self-weight of beams, columns, slabs, walls and other permanent members. The self-weight of beams and columns (frame members) and slabs (area sections) is automatically considered by the program itself. The wall loads have been calculated and assigned as uniformly distributed loads on the beams.

Wall load = Unit weight of brickwork x thickness of wall x (height of floor – depth of beam) x (1 – opening percent).

=19.2*0.3*(2.8448-0.5) *0.7

=13.50*0.7

= 9.454 kN/m

Live loads is applied in the form of UDL on uniform area on the slab elements as per IS 1893 (Part 1) 2002.

Live load on roof 2.0 KN/m^2

Live load on all other floors 3.0 KN/m^2

Different load combinations are employed according to IS 1893 (Part 1) 2002 considering the limit state design of RC structures.

e) Frame section property reinforcement data

We have adopted concrete material for column as M25 grade. The depth column is 450 mm and the width is 300 mm. Design type p-M2-M3 Design [column] is adopted. We are using different types of rebar materials as follow.

Rebar material for Longitudinal bar: HYSD500 and for Confinement bar ties: HYSD 415.

Longitudinal bar size and area: 12 mm^2

Corner bar size and area: 16 mm^2

Confinement bar size and area: 8 mm^2

f) Frame section property data- Beam formation.

The material use for beam is M25. The depth of beam section is 300 mm and width of beam section is 230 mm. Design Type: M3 design only is adopted for Beam,

Rebar material for longitudinal bars isHYSD500 and rebar material for Confinement bars (ties) is HYDS 415.

g) Slab Properties.

Material for slab is adopted as M25 grade concrete and the thickness of slab is taken as 150 mm.

Hence Table 1 provides loading patter employed with respect to the analysis carried out.

Sesmic zone factor, Z= 0.36 ……. from IS 1893 (part 1):2002 Table 2 Clause 6.4.2}

Table 1 Defined Load Patterns

Load	Type	Self wt multiplier	Load	Auto lateral load
Dead	Dead	1	Dead	-
Live	Live	0	Live	-
WL	Super dead	0	WL	-
PWL	Super dead	0	PWL	-
FF	Super dead	0	FF	-
Eqx	Seismic	0	Eqx	IS 1893 2002
Eqy	Seismic	0	Eqy	IS 1893 2002

Table 2 Mass Source Data

Load Pattern	Multiplier
Live	0.25
Wind	1
PWL	1
FF	1

Special RC Moment Resisting Frame:

Response reduction factor R= 5…from IS 1893(part 1): 2002 {Clause6.4.2}

Following Table 2 shows the mass sources data.

After executing the analysis of different models stated in the above part we evaluated the deformation at different levels. Figure 4 shows the comparison between the frames provided with or without dampers. Four different cases are considered for different models of building frames. Also description of building model with specifications is given in Table 3.

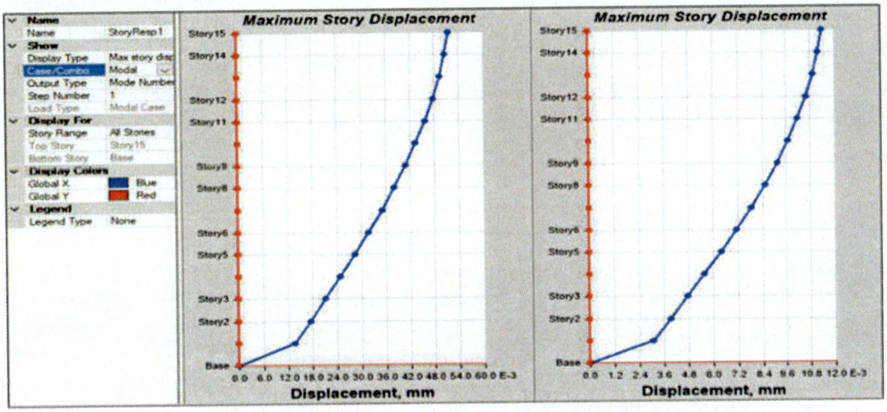

(a) Without Damper (b) With Damper

Fig. 4 Story Displacement for a building frame with equal height and equal floor height

Table 3 Description of Buildings with specifications

Description of building	Building with equal height and equal floor height
Plan dimension	10 × 8 m
Height of 1st building	60 m
Number of floors for 1st building	15
Height of 2nd building	60 m
Number of 2nd building	15
Beam sizes	300 × 300 mm
Column sizes EQ data	230 × 400 mm Zone II

4 Results and Discussions

The seismic analysis is performed considering different cases of building frames undergoing pounding effect. Further Table 4 shows comparison between displacement at various levels for frames subjected to pounding effect with and without dampers. Whereas Table 5 shows Comparison of building frames with different height and different floor height with and without damper for case

To mitigate the effect of seismic pounding; use of dampers is found to be effective for displacement response. The displacement in building 1 at top floor without damper is found to be 530.858mm. And the displacement at top floor of building 1 with damper is found to be 32.125 mm. Thus, the percentage difference in displacement of building 1 with and without damper at top floor is found to be 93.95. As both frames have same height and same storey levels, the results are same as first frame.

Table 4 Comparison of building frames with equal height and equal floor height with and without damper for case 1

With Damper				Without Damper				% Diff (1st Building)	% Diff (2nd Building)
1st Building		2nd Building		1st Building		2nd Building			
Floor	Disp.	Floor	Disp.	Floor	Disp.	Floor	Disp.		
0	0	0	0	0	0	0	0	0.00	0.00
1	9.916	1	9.916	1	164.454	1	164.454	93.97	93.97
2	12.754	2	12.754	2	211.45	2	211.45	93.97	93.97
3	15.563	3	15.563	3	257.997	3	257.997	93.97	93.97
4	18.361	4	18.361	4	304.269	4	304.269	93.97	93.97
5	21.085	5	21.085	5	349.416	5	349.416	93.97	93.97
6	23.662	6	23.662	6	391.906	6	391.906	93.96	93.96
7	26.005	7	26.005	7	430.732	7	430.732	93.96	93.96
8	28.016	8	28.016	8	463.635	8	463.635	93.96	93.96
9	29.585	9	29.585	9	489.586	9	489.586	93.96	93.96
10	30.59	10	30.59	10	505.518	10	505.518	93.95	93.95
11	30.923	11	30.923	11	510.957	11	510.957	93.95	93.95
12	31.28	12	31.28	12	516.025	12	516.025	93.94	93.94
13	31.531	13	31.531	13	521.012	13	521.012	93.95	93.95
14	31.829	14	31.829	14	525.949	14	525.949	93.95	93.95
15	32.125	15	32.125	15	530.858	15	530.858	93.95	93.95

Table 5 Comparison of building frames with different height and different floor height with and without damper

With Damper				Without Damper				% Diff (1st Building)	% Diff (2nd Building)
1st Building		2nd Building		1st Building		2nd Building			
Floor	Disp.	Floor	Disp.	Floor	Disp.	Floor	Disp.		
0	0	0	0	0	0	0	0	0.00	0.00
1	4.898	1	79.278	1	75.952	1	79.278	93.55	0.00
2	8.093	2	237.433	2	125.374	2	237.433	93.54	0.00
3	11.319	3	389.216	3	175.26	3	389.216	93.54	0.00
4	14.533	4	517.207	4	225.045	4	517.207	93.54	0.00
5	17.66	5	413.273	5	273.444	5	602.642	93.54	31.42
6	20.625	6	609.696	6	319.09	6	609.696	93.54	0.00
7	23.318	7	614.674	7	361.09	7	614.674	93.54	0.00
8	25.628	8	-	8	397.256	8	-	93.55	
9	27.429	9	-	9	425.156	9	-	93.55	-
10	28.583	10	-	10	443.753	10	-	93.56	-
11	28.961	11	-	11	449.683	11	-	93.56	-
12	29.31	12	-	12	455.1	12	-	93.56	-
13	29.653	13	-	13	460.412	13	-	93.56	-
14	29.992	14	-	14	465.659	14	-	93.56	-
15	30.329	15	-	15	470.875	15	-	93.56	-

5 Conclusions

The present study is performed to evaluate the seismic behavior of building frames subjected to pounding effect with and without dampers. Hence, displacement response is observed for different models under damping with close spacing between the frames. Thus, present study explains the pounding effect on closely spaced buildings with different cases of frames having equal and unequal height. To mitigate the seismic pounding effect we employed dampers to these models. The results reveal that the dampers drastically mitigate the storey displacement of frames at various storey levels. According to seismic analysis the percentage of reduction is 31.42% for displacement by the use of damper at intermediate level for frames with different height and varying the storey level also. The minimization in structural response is; for the axial load of columns is 50.75%; for shear load is 30.75% and for the bending moment is 27.24%.

References

1. Ormondroyd, J., & Den Hartog, J. P. (1928). The Theory of Dynamic Vibration Absorber. *ASME Journal of Applied Mechanics*, 50(7), 9–22. PEER Ground Motion Database, Berkeley, CA (2010). http://peer.berkeley.edu/Ramllo
2. Randall, S. E., Halsted, D. M., & Taylor, D. L. (1981). Optimum vibration absorbers for linear damped systems. *ASME Journal of Mechanical Design, 103*, 908–913.
3. Kelly, J. M., & Beucke, K. E. (1983). A friction damped base isolation system with fail-safe characteristics. *Earthquake Engineering and Structural Dynamics, 11*(1), 33–56.
4. Mokha, B. A., Constantinou, M., & Reinhorn, A. (1990). Teflon bearings in base isolation. I: testing. *Journal of Structural Engineering., 116*, 438-454. 121.
5. Anagnostopoulos, S. A., & Spilliopoulos, K. V. (1992). An investigation of earthquake induced pounding between adjacent buildings. *Earthquake Engineering & Structural Dynamics., 21*(4), 289–302.
6. Tsai, H., & Lin, G. (1993). Optimum tuned-mass dampers for minimizing steady-state response of support-excited and damped systems. *Earthquake Engineering and Structural Dynamics, 22*, 957–973.
7. Sladek, J. R., & Klingner, R. E. (1993). Effect of tuned-mass dampers on seismic response. *Journal of Structural Engineering, 109*(8), 2004–2009.
8. Leibovich, E., Rutenberg, A., & Yankelevsky, D. Z. (1996). On Eccentric seismic pounding of symmetric Buildings. *Earthquake Eng. Struct. Dyn., 25*(3), 219–233.
9. Joshi, A. S., & Jangid, R. S. (1997). Optimum parameters of multiple tuned mass dampers for base-excited damped systems. *Journal of Sound and Vibration, 202*(5), 657–667. https://doi.org/10.1006/jsvi.1996.0859
10. Jeng, V., & Tzeng, W. L. (1998). Assessment of Seismic Pounding Hazard for Taipei City. *Engineering Structure, 22*(5), 459–474.
11. Whittaker, A., Constantnou, M., & Tsopelas, P. (1998). Displacement estimates for performance based seismic design. *Journal of Struct Division ASCE., 124*(8), 905–912.
12. Ghobarah, A., AbouElfath, H., & Biddah, A. (1999). Response-based damage assessment of structures. *Earthquake Engineering and Engineering Seismology, 28*(4), 79–104.
13. Chopra, A. K., & Chintanapakdee, C. (2001). Comparing response of SDF Systems to near-fault and far-fault earthquake motions in the context of spectral regions. *Earthquake Engineering and Structural Dynamics, 30*, 1769–1789.
14. Xue, Q. (2001). A direct displacement based seismic design procedure of inelastic structures. *Engineering Structures, 23*(11), 1453–1460.
15. Lin, J. H., & Weng, C. C. (2001). Spectral analysis on pounding probability of adjacent buildings. *Engineering Structures, 23*(7), 768–778. https://doi.org/10.1016/S0141-0296(00)00098-5
16. Elenas, A., & Meskouris, K. (2001). Correlation study between seismic acceleration parameters and damage indices of structures. *Engineering Structures, 23*(6), 698–704. https://doi.org/10.1016/S0141-0296(00)00074-2
17. Hong, H. P., Wang, S. S., & Hong, P. (2003). 'Critical building separation distance in reducing pounding risk under earthquake excitation.' *Structural Safety, 25*(3), 287–303.
18. Lin, J.-H., & Weng, C.-C. (2001). 'Probability analysis of seismic pounding of adjacent buildings.' *Earthquake Engineering and Structural Dynamics, 30*, 1539–1557.
19. Bhaskararao, A. V., & Jangid, R. S. (2006). Seismic analysis of structures connected with friction dampers. *Engineering Structures, 28*(5), 690–703. https://doi.org/10.1016/j.engstruct.2005.09.020
20. Li, C., & Zhu, B. (2006). Estimating double tuned mass dampers for structures under ground acceleration using a novel optimum criterion. *Journal of Sound and Vibration, 298*(1–2), 280–297. https://doi.org/10.1016/j.jsv.2006.05.018
21. Almazán, J. L., De la Llera, J. C., Inaudi, J. A., López-García, D., & Izquierdo, L. E. (2007). A bidirectional and homogeneous tuned mass damper: a new device for passive control

of vibrations. *Engineering Structures, 29*(7), 1548–1560. https://doi.org/10.1016/j.engstruct.2006.09.005
22. Kravchuk, N., Colquhoun, R., & Porbaha, A. (2008). Development of a Friction Pendulum Bearing Base Isolation System for Earthquake Engineering Education. In *Proceedings of the 2008 American Society for Engineering Education Pacific Southwest Annual Conference*, pp. 1–16
23. Matta, E., & De Stefano, A. (2009). Robust design of mass-uncertain rolling-pendulum TMDs for the seismic protection of buildings. *Mechanical Systems and Signal Processing, 23*(1), 127–147. https://doi.org/10.1016/j.ymssp.2007.08.012
24. Gewei, Z., & Basu, B. (2010). A study on friction-tuned mass damper: harmonic solution and statistical linearization. *Journal of Vibration and Control, 17*(5), 721–731. https://doi.org/10.1177/1077546309354967
25. Moon, K. S. U. N. (2010). Vertically distributed multiple tuned mass dampers in tall buildings: performance analysis and preliminary design. *The Structural Design of Tall and Special Buildings, 19*, 347–366.
26. Sayani, P. J., Erduran, E., & Ryan, K. L. (2011). Comparative response assessment of minimally compliant low-rise base isolated and conventional steel moment resisting frame buildings. *Journal of Structural Engineering. ASCE, 137*(10), 1118–1131.
27. Gray, B. M., Christopoulos, C., Ph, D., Eng, P., Packer, J., & Gray, M. (2012). A New Brace Option for Ductile Braced Frames. *Modern Steel Construction.* 40–43
28. Mohebbi, M., Shakeri, K., Ghanbarpour, Y., & Majzoub, H. (2012). Designing optimal multiple tuned mass dampers using genetic algorithms (GAs) for mitigating the seismic response of structures. *Journal of Vibration and Control, 19*, 1–21. https://doi.org/10.1177/1077546311434520
29. Wankhade, R. L. & Landage, A. B. (2014). Static analysis for fixed base and base isolated building frame. In *Proceedings of National Conference on Advances in Civil and Structural Engineering (NCACSE-2014)*
30. Wankhade, R.L. (2014). Performance based design and estimation of forces for building frames with earthquake loading. In *International Conference on Recent Trends and Challenges in Civil Engineering December 12–14, 2014, MNNIT Allahabad, India*
31. Wankhade, R. L., & Landage, A. B. (2016). Performance based analysis and design of building frames with earthquake loading. *International Journal of Engineering Research., 5*(1), 106–110.
32. Wankhade, R. L. (2017) Performance analysis of RC moment resisting frames using different rubber bearing base isolation techniques. In *International Conference on Innovations in Concrete for Infrastructure Challenges Nagpur, India October 6–7*
33. Wankhade, R. L., & Bajoria, K. M. (2017). Numerical optimization of piezolaminated beams under static and dynamic excitations. *Journal of Science: Advanced Materials and Devices,* 2(2), 255–262.
34. Bendine, K., & Wankhade, R. L. (2017). Optimal shape control of piezolaminated beams with different boundary condition and loading using genetic algorithm. *International Journal of Advanced Structural Engineering, 9*(4), 375–384.
35. Wankhade, R. L., & Bajoria, K. M. (2019). Vibration analysis of piezolaminated plates for sensing and actuating applications under dynamic excitation. *International Journal of Structural Stability and Dynamics, 19*(10), 1950121. https://doi.org/10.1142/S0219455419501219
36. Wankhade, R. L., & Niyogi, S. B. (2020). Buckling analysis of symmetric laminated composite plates for various thickness ratios and modes. *Innov. Infrastruct. Solut., 5*, 65. https://doi.org/10.1007/s41062-020-00317-8
37. Wankhade, R. L., & Bajoria, K. M. (2021). Vibration attenuation and dynamic control of piezolaminated plates for sensing and actuating applications. *Archive of Applied Mechanics, 91*(1), 411–426.
38. Wankhade, R. L., Niyogi, S., & Gajbhiye, P. (2021). Buckling analysis of laminated composites considering the effect of orthotropic material, IOP Science. *Journal of Physics: Conference Series, 1706*(1), 012188.

39. Bendine, K., & Wankhade, R. L. (2021). Piezoelectric energy harvesting from a curved plate subjected to time-dependent loads using finite elements, IOP Science. *Journal of Physics: Conference Series, 1706*(1), 012008.
40. Jia, H.-Y., Lan, X.-L., Luo, N., Yang, J., Zheng, S.-X., & Zhang, C. (2019). Nonlinear pounding analysis of multispan and simply supported beam bridges subjected to strong ground motions. *Shock and Vibration, 2019*, 1–11. https://doi.org/10.1155/2019/8759428
41. Nayak, C. B., et al. (2021). Experimental, analytical and numerical performance of RC beams with V-shaped reinforcement. *Innovative Infrastructure Solutions, 6*(1), 2. https://doi.org/10.1007/s41062-020-00363-2
42. Wankhade, R. L., Sawarkar, A., Chandwani, A., Chavan, S., Malkar, P., & Sawarkar, G. (2022). Seismic fragility of buildings subjected to pounding effects with soil–structure interaction. In A. K. Gupta, S. K. Shukla, & H. Azamathulla (Eds.), *Advances in Construction Materials and Sustainable Environment: Select Proceedings of ICCME 2020* (pp. 425–434). Singapore: Springer. https://doi.org/10.1007/978-981-16-6557-8_34

An Experimental Investigation on the Partial Substitution of Cement with Sawdust Ash at Higher Temperatures

M. Sweety Poornima Rau and Y. M. Manjunath

Abstract Partial replacements are being investigated to minimize the negative effect on the environment by utilizing cement in concrete. The aim of this study is to evaluate the fresh and hardened characteristics of M30 grade concrete subjected to higher temperatures with sawdust ash (SDA) as partial substitute to cement. In this regard, different concrete mixtures were designed maintaining the same w/c (water-cement) ratio of 0.45. The saw dust ash content is varied with cement at different proportions of 0%, 5%, 10%, 15% and 20%. The workability, compressive, tensile and flexural strength of SDA concrete were decreasing with the increment in the proportions of replacement of SDA in cement. The consistency, initial and final setting time of concrete varied with an increased SDA substitution is within the permissible limits of Indian standards. After 28 days of curing, the SDA concrete with 5% replacement achieved higher compressive, tensile, and flexural strengths of 37.44 N/mm^2, 2.83 N/mm^2 and 5.6 N/mm^2 respectively. The mechanical behavior of conventional and optimal concrete subjecting to higher temperatures varying from 100 °C to 600 °C at 100 °C interval for 2 h duration of exposure were studied. The compressive, split tensile and flexural strengths for optimal replacement concrete were increased as 3.9%, 11.4% and 15.3% respectively after subjecting to higher temperatures. It supported the environmental advantages of partial SDA substitution by 2.2% less of original material usage.

Keywords Cement · Compressive strength · Economical · Eco-friendly · Elevated temperature · Flexural strength · Sawdust Ash · Split-tensile strength

M. S. P. Rau (✉) · Y. M. Manjunath
The National Institute of Engineering, Mysuru, Karnataka 570008, India
e-mail: poornima.merugu71@gmail.com

Y. M. Manjunath
e-mail: ymm@nie.ac.in

A. Vilventhan et al. (eds.), *Advances in Construction Materials and Management*, Lecture Notes in Civil Engineering 346,
https://doi.org/10.1007/978-981-99-2552-0_33

1 Introduction

Concrete is being used rapidly in construction industries world-wide because of its excellent mechanical properties. Due to the enormous demand for concrete, cement, fine and coarse aggregates are becoming very costlier to get. The development of these original materials also has an adverse effect on the environment. According to Flower and Sanjayan [14], 0.8 ton of CO_2 was emitted, for every 1 ton of cement produced. Additionally, excessive use of common aggregates such as gravel, granite etc. contribute to the natural resource depletion, the extraction of these materials may give the negative environmental effects according to Blankendaal et al. [7]. Utilizing industrial waste materials in concrete has generated interest in research as a solution to these problems. This led to an investigation in the exploration of substitute materials for concrete and considered agricultural residues as the replacement for aggregates and substituent materials in mortar and concrete. Agricultural wastes are easily accessible, inexpensive and have good reactivity in concrete according to Ataie and Riding [3]; hence, researchers are paying more attention to the constituents like this in concrete. The effective usage of these ingredients to concrete results in a green manner of removing huge amounts of waste, which would have caused ecological problems. A few agricultural waste products which have been investigated in the usage of mortar and concrete are rice husk ash according to Raheem and Sargın [30, 35], corn cob ash according to Adensya and Ganesan [4, 5, 15], wood ash according to Abdullahi and Raheem [6, 29, 31], saw dust ash according to Raheem [28, 32], and sugarcane bagasse according to Chowdhury and Imran [11, 16]. Each of these agricultural wastes will perform much better in concrete if the appropriate treatment techniques and waste are selected for a particular application according to Luhar [21]. Based on the saying, "a hungry man is an angry man," agriculture is a primary source. Agriculture supports the growth and maintenance of the world's economic, environmental, and social systems based on the triple-line model of sustainability according to Blandford [8]. Agricultural residues will always be available on a global scale because agriculture produces both food and raw materials for industry. However, the type of residue that is accessible varies by country depending on the viability of its agricultural production. To avoid environmental damage, it is necessary to regularly update the knowledge on potential uses of agricultural wastes.

Recently, agricultural waste has been incorporated into concrete in a variety of ways, such as replacement of cement according to Adesina, Chindaprasirt, Luhar, Mosaberpanah, Olutoge, Rattanachu and Shakouri [2, 9, 21, 23, 26, 27, 34], replacement of aggregates according to Kumar V. P and Memon [20, 24], geo-polymer according to Liang and Nuaklong [22, 25] and reinforcement according to Akinyemi and Kumar G. R [1, 19]. This study provides an overview of investigations that have been done so far on specific agricultural wastes as partial substitute to cement in mortar and concrete. Mostly, researches were concentrated on the replacement with waste materials at normal temperatures, and very less findings were there regarding the concrete behavior at higher temperatures.

From all the agricultural waste mentioned above, sawdust is selected for the present experimental investigation and discussed below. Sawdust is a byproduct produced at sawmills when the timber is cut into logs. It is available in extensive forested areas with full of trees. As output increases, the accumulation of sawdust at sawmills increases, creating an environmental problem. Burning sawdust at 650 °C temperature for 8 h approximately results in the production of sawdust ash (SDA). The elemental composition of SDA was extensively investigated by various researchers as shown in Table 1. Silicon dioxide (SiO_2) is the most important elemental oxide with elemental composition varying from 65.3% to 78.92% as shown in Table 1. All the researchers fulfilled the requirement of ASTM C 618 i.e. the combination of $SiO_2 + Al_2O_3 + Fe_2O_3$ should be greater than 70%. As a result, SDA is a good pozzolonic material. It can be observed that, to improve the strength and durability of concrete through interface reinforcement, the pozzolona is added to OPC in concrete. To produce cementitious materials, the chemical constituents of pozzolona react with the chemical constituents of cement (constituents released during the hydration of cement by the process of chemical reaction). Calcium hydroxide ($Ca(OH)_2$) is removed from the mixture as a result of the chemical reaction between the cement and pozzolona, which enhances the concrete's resistance to harsh environments. Moreover, this chemical reaction impacts other properties of concrete, such as durability and thermal resistance, and strengthens the concrete that is formed.

In the present study, the SDA is used as the partial substitution for cement. So, the literature based on SDA is discussed below in detail. SDA has the following physical properties like bulk density (ρ), specific gravity (g), mean size (η) etc. according to Chowdhury [10]: 'g' is 2.16, 'η' is 170 μm and 'ρ' is 720 kg/m3. According to Raheem [32], concrete stiffens when the percentage of SDA rises. Hence, the quantity of water should be increased in order to make workable concrete. The results of this study showed that after 28 days, a larger percentage improvement in strength was observed. The continual hydration of cement and the pozzolonic reaction of SDA were responsible for this enormous increase in strength.

Raheem and Ige [28] looked into the flexure strength of SDA mortar. According to him, the flexure strength of mortar was increased with curing period and declined with rise in percentage of SDA, with 15% SDA as the optimal replacement. Chowdhury

Table 1 Chemical composition of SDA

References	Elemental Oxide Composition (%)							
	SiO_2	Al_2O_3	Fe_2O_3	CaO	MgO	K_2O	Na_2O	LOI
Chowdhury et al. [10]	65.30	4.25	2.24	9.98	5.32	1.90	2.60	4.67
Elinwa et al. [12]	67.20	4.10	2.30	10.00	5.80	0.10	0.10	–
Elinwa and Mahmood [13]	67.20	4.09	2.26	9.98	5.80	–	0.08	4.67
Raheem and Ige [28]	69.29	5.22	2.03	10.07	5.48	0.86	9.68	3.64
Raheem et al. [32]	65.75	5.23	2.09	9.62	4.09	2.43	0.06	4.30
Udoeyo and Dashibil [36]	78.92	0.89	0.85	0.58	0.96	–	0.43	8.40

[10] investigated the XRD analysis and splitting tensile of SDA blended concrete. The splitting tensile was observed to be decreased with increasing SDA percentage. In comparison to the result of compressive strength, lowering the effect of rise in percentage of SDA was smaller. The SDA particle filler activity in the concrete may have contributed to the reduction by Udoeyo F. F [36].

Elinwa et al. [12] analyzed the SDA sample's XRD pattern, which revealed a hump including sawdust ash. The findings suggested that SDA might be used as a powdered material along with cement and super plasticizers to create pumpable concrete. In addition to that, Raheem and Sulaiman [33] produced hollow sand-Crete blocks using SDA instead of cement in different cases. The results have shown that both the compressive strength and density improved with the age of curing and declined with increasing SDA content. Sand-Crete hollow blocks for non-load-bearing walls should ideally contain 10% SDA as per recommendation. It can be concluded that SDA is a good pozzolona for substitution of cement and for better concrete and mortar performance, in which the replacement up to 15% SDA is acceptable.

According to the literature mentioned above, sawdust can be burned into ash and used in place of cement in concrete. Utilizing SDA pozzolona derived from agro-waste natural resources has many advantages. The processing of saw dust into SDA pozzolona is beneficial to the environment, i.e. by preserving natural resources and preventing the disposal of forest (timber) byproducts in landfills. As a result, it is necessary to enhance the performance of concrete at high temperatures by combining cement with saw dust ash (SDA) pozzolona.

Concrete may deform without any constraint under higher temperatures. That deformation may remain same even after the concrete is cooled down to the normal temperature. In general, the type of aggregate used has an impact on how fire affects concrete. 1) Carbonate aggregate such as limestone i.e. light weight aggregate retain most of their compressive strength up to 650 °C. 2) Concrete with siliceous aggregate, such as granite, quartzite, silica materials retain only about 55% of their strength at 650 °C. Advantages of heat resistant concrete are shrinkage, green strength and fired strength, toughness. The effects of concrete when subjected to higher temperatures are reduction in strength, micro-cracks within the microstructure of concrete, change in color, increase in porosity and spalling from certain temperature. The applications of the concrete subjected to higher temperatures are runway pavements, thermal power plants etc.

The sawdust ash used in the present study is sieved through 90μ sieve, partially replaced with cement by weight at various proportions in concrete. The mechanical behavior of concrete (conventional and replaced) subjected to higher temperatures for 2 h duration of exposure was studied. Holistic characteristics of saw dust ash were shown. Also, its effect on the mechanical behavior (workability, compressive, split tensile and flexural strengths) of concrete is discussed in detail.

Table 2 Physical property of the materials

Materials	Physical properties		
	Specific gravity	Water absorption (%)	Fineness modulus (%)
Cement	3.1	-	-
Sawdust ash	2.2	1	-
Coarse aggregate	2.65	0.5	5.3%
M-sand	2.59	2	3.9
CONPLAST SP430	1.202	2.5	-

2 Materials and Methodology

2.1 *Materials*

2.1.1 Physical Properties

Materials used in the experimentation and their physical property were shown in the Table 2.

2.2 *Methodology*

Conventional concrete (CC) and Replacement concrete (RC) of M30 grade concrete were designed as per IS specifications [17]. Various concrete specimens were prepared and casted into cubes, cylinders and beams. The specimens were tested under compressive, splitting tensile and flexural strength tests at the age of 7 and 28 days of curing according to ASTME119. The optimal concrete is exposed to various exposure situations for 2 h at high temperatures of 100 °C to 600 °C at 100 °C interval and compared with room temperature. The detailed process followed during the experimental process is as shown in Fig. 1.

2.2.1 Mix Proportioning [17]

According to IS 10262:2019, the M30 grade concrete mix for conventional concrete mix was designed and quantities are estimated for 1m^3 volume of concrete in the Table 3.

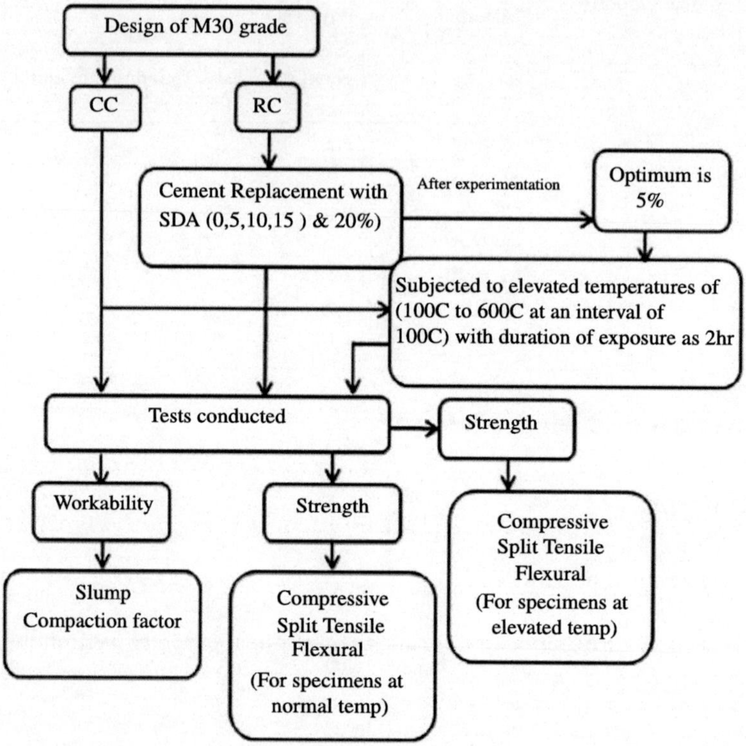

Fig. 1 Schematic representation of experimental investigation

Table 3 Quantity of material per m^3 of concrete for w/c = 0.4

Mix design data	Quantity(kg /m^3)
Cement	373
Fine aggregate (M-sand)	7 69
Coarse aggregate	1075
Water	168
Super Plasticizer	1.5

Note: The quantity of cement in the replacement concrete mixture will change when different percentages of the cement are replaced with the saw dust ash

2.3 *Description of Specimen in Detail*

The detailed description of specimens in the experimentation and its count is mentioned in the Table 4.

Table 4 Description of specimen in detail

S.No	Specimen	Size (mm)
1	Cube	150 × 150 × 150
2	Cylinder	150 × 300
3	Beam	100 × 100 × 500

3 Results and Discussions

In accordance with IS 1199:1959, the tests of slump and compaction factor were performed to evaluate the workability of fresh concrete for conventional concrete. When it comes to hardened concrete, controlled concrete is cast, after 28 days of curing specimens were tested for compressive, splitting tensile and flexural strengths.

Table 5 illustrates how slump and compaction factor changes as replacement percentage increases.

It is observed from Table 5 that the medium and high degree of workability was maintained throughout the experimentation and satisfied the workability of Ready-Mix Concrete i.e., the concrete can be used in construction purposes also.

Also, the concrete mix is cohesive, further it involved in the process of segregation and bleeding. Initial slumps were of 125 mm and 110 mm respectively. Slumps of 95 mm and 90 mm were observed for 15% and 20% replacements. The Compaction factors were observed as 0.95, 0.93 and 0.92 for different replacements of 5%, 10%, 15% & 20%.

3.1 *Compressive, Tensile and Flexural Strengths at Normal Temperatures*

The results of the compressive split tensile and flexural strength at normal temperatures for both conventional and replacement concretes (using SDA as a partial substitute of cement at various percentages) at an age of 7 & 28-days curing was demonstrated in Table 6 [18].

Table 5 Workability test

Percentage of replacement (%)	slump (mm)	compaction factor
0	125	0.95
5	110	0.95
10	95	0.93
15	90	0.92
20	85	0.92

Table 6 Compressive, tensile and flexural strength at normal temperatures

% of replacement of SDA	Compressive strength (MPa)		Tensile strength(MPa)		Flexural strength(MPa)	
	7 days	**28 days**	**7 days**	**28 days**	**7 days**	**28 days**
0	29.08	35.08	2.26	2.64	4.37	5.02
5	30.92	37.44	2.47	2.83	5.10	5.60
10	28.30	35.77	2.38	2.50	4.85	5.41
15	26.85	33.41	2.10	2.31	4.76	5.37
20	25.70	32.53	2.08	2.22	4.64	4.98

From the results of Table 6, the optimum replacement at normal temperatures can be concluded as 5% SDA replacement with cement.

3.2 *Observations Made with Respect to Elevated Temperatures and Its Comparisons*

The higher temperatures of 100 °C to 600°Cat 100 °C are considered to investigate the relation among compressive, tensile and flexural strengths of concrete for different replacements of saw dust ash. Different numbers of specimens were kept in furnace for duration of exposure of 2 h at each selected temperature. After reaching the particular exposure time, the specimens are left to cool down to ambient temperature for 24 h. The weight loss was then taken for all the specimens and non-destructive tests like rebound hammer and ultrasonic pulse velocity test and destructive tests like compressive, tensile and flexural strength tests were conducted.

3.2.1 Residual Compressive Strength at Elevated Temperatures

The change in compressive strength of conventional concrete (0%) and optimal concrete (5% SDA) with varying temperature from 100 °C to 600 °C at 2 h duration of exposure is as shown in Fig. 2.

The SDA pozzolona that is present in the pores of the concrete may have undergone a pozzolonic reaction that increased strength up to 200 °C.

3.2.2 Residual Tensile Strength at Elevated Temperatures

The change in tensile strength of conventional concrete (0%) and optimal concrete (5% SDA) when temperature rises from 100 °C to 600 °C at 2 h duration of exposure is as shown in Fig. 3.

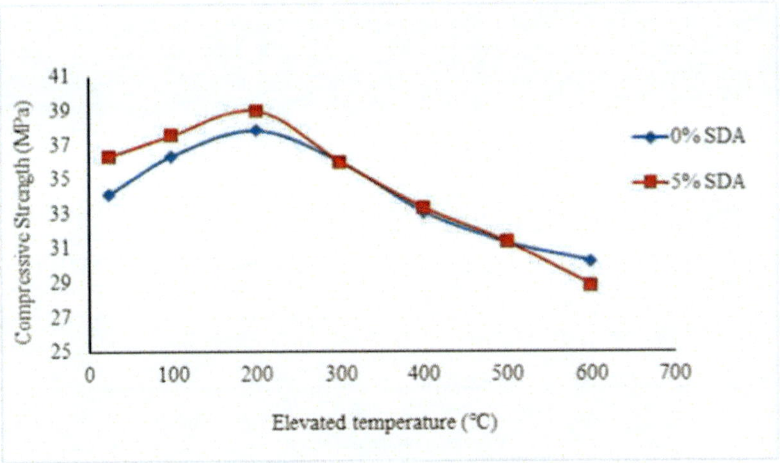

Fig. 2 Compressive strength at normal & elevated temperatures

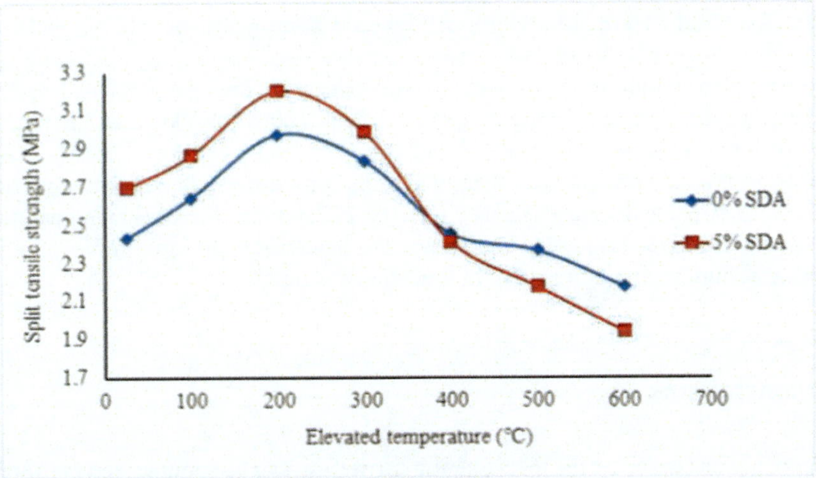

Fig. 3 Tensile strength at normal & elevated temperatures

The residual tensile strength of concrete increases up to certain temperature, then declined with an increase in temperature. This is due to an increment in percentage weight loss and porosity with temperature. The strength of conventional concrete is lower compared to blended mix; this can be attributed to pozzolonic activity of sawdust ash.

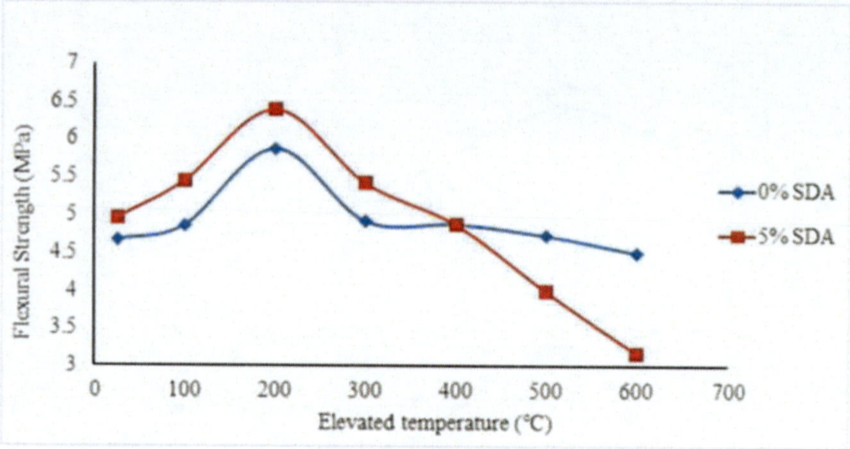

Fig. 4 Flexural strength at normal & elevated temperatures

3.2.3 Residual Flexural Strength at Elevated Temperatures

The change in compressive strength of conventional concrete (0%) and optimal concrete (5% SDA) with varying temperature from 100 °C to 600 °C at 2 h duration of exposure is as shown in Fig. 4.

It is observed from Fig. 4 that, there was a slight increase in flexural strength from 100 °C to 200 °C but decreases further, up to 600 °C, with an increase in temperature. The decrease in strength may be due to flexural behavior is more sensitive to micro-cracking at high temperatures than in low temperatures.

4 Conclusions

The conclusions based on the investigation carried out can be summarized as follows:

1. The optimum percentage of replacement of cement with saw dust ash is concluded as 5% SDA. As the replacement percentage of SDA increases, the overall water absorption and workability of concrete were decreased when compared to the normal concrete.
2. The compressive strength has increased from 35.08 MPa for 0% to 37.44 MPa for 5% concrete with a gain of 6.73%. Split tensile strength has increased from 2.64 MPa for 0% to 2.83 MPa for 5% with a gain of 7.2%. Similarly, a gain of 11.6% is observed from 5.02 MPa to 5.60 MPa in flexural strength for 0% and 5% replacement, respectively.
3. The concrete with 5% SDA has shown increase in strengths at 200 °C. Further increase in temperature results in the decrement of strengths.

4. Compared to conventional concrete, concrete made with 5% SDA maintained more of its strength when subjected to higher temperatures.
5. Based on the test results, it can be concluded that the performance of concrete at higher temperatures was enhanced by replacing OPC with about 5% SDA. In order to use OPC as a fire-resistant bonding material in concrete, blend it with 5% SDA.

References

1. Akinyemi, B. A., & Dai, C. (2020). Development of banana fibers and wood bottom ash modified cement mortars. *Construction and Building Materials, 241*, 118041.
2. Adesina, P. A., & Olutoge, F. A. (2019). Structural properties of sustainable concrete developed using rice husk ash and hydrated lime. *Journal of Building Engineering, 25*, 100804.
3. Ataie, F. F., & Riding, K. A. (2016). Influence of agricultural residue ash on early cement hydration and chemical admixtures adsorption. *Construction and Building Materials, 106*, 274–281.
4. Adesanya, D. A., & Raheem, A. A. (2009). A study of the workability and compressive strength characteristics of corn cob ash blended cement concrete. *Construction and Building materials, 23*(1), 311–317.
5. Adesanya, D. A., & Raheem, A. A. (2009). Development of corn cob ash blended cement. *Construction and Building Materials, 23*(1), 347–352.
6. Abdullahi, M. (2006). Characteristics of wood ash/OPC concrete. *Leonardo Electronic Journal of Practices and Technologies, 8*, 9–16.
7. Blankendaal, T., Schuur, P., & Voordijk, H. (2014). Reducing the environmental impact of concrete and asphalt: a scenario approach. *Journal of cleaner production, 66*, 27–36.
8. Blandford, D. The contribution of agriculture to green growth. *Report to the OECD*, pp. 1–36. (2011).
9. Chindaprasirt, P., Sujumnongtokul, P., & Posi, P. (2019). Durability and mechanical properties of pavement concrete containing bagasse ash. *Materials Today: Proceedings, 17*, 1612–1626.
10. Chowdhury, S., Maniar, A., & Suganya, O. M. (2015). Strength development in concrete with wood ash blended cement and use of soft computing models to predict strength parameters. *Journal of advanced research, 6*(6), 907–913.
11. Danso, H. (2017). Properties of coconut, oil palm and bagasse fibres: as potential building materials. *Procedia Engineering, 200*, 1–9.
12. Elinwa, A. U., Ejeh, S. P., & Mamuda, A. M. (2008). Assessing of the fresh concrete properties of self-compacting concrete containing sawdust ash. *Construction and Building Materials, 22*(6), 1178–1182.
13. Elinwa, A. U., & Mahmood, Y. A. (2002). Ash from timber waste as cement replacement material. *Cement and Concrete Composites, 24*(2), 219–222.
14. Flower, D. J., & Sanjayan, J. G. (2007). Greenhouse gas emissions due to concrete manufacture. *The International Journal of Life Cycle Assessment, 12*(5), 282–288.
15. Ganesan, K., Rajagopal, K., & Thangavel, K. (2008). Rice husk ash blended cement: assessment of optimal level of replacement for strength and permeability properties of concrete. *Construction and Building Materials, 22*(8), 1675–1683.
16. Imran, M., & Khan, A. A. (2018). Characterization of agricultural waste sugarcane bagasse ash at 1100 C with various hours. *Materials Today: Proceedings, 5*(2), 3346–3352.
17. IS 10262–2019: Concrete Mix Proportioning - Guidelines, Bureau of Indian Standards, New Delhi, India
18. IS 516–1959: Methods of Tests for Strength of concrete, Bureau of Indian Standards, New Delhi, India

19. Kumar, G. R., & Kesavan, V. (2020). Study of structural properties evaluation on coconut fiber ash mixed concrete. *Materials Today: Proceedings, 22*, 811–816.
20. Kumar, V. P., Gunasekaran, K., & Shyamala, T. (2019). Characterization study on coconut shell concrete with partial replacement of cement by GGBS. *Journal of Building Engineering, 26*, 100830.
21. Luhar, S., Cheng, T. W., & Luhar, I. (2019). Incorporation of natural waste from agricultural and aquacultural farming as supplementary materials with green concrete: a review. *Composites Part B: Engineering, 175*, 107076.
22. Liang, G., Zhu, H., Zhang, Z., & Wu, Q. (2019). Effect of rice husk ash addition on the compressive strength and thermal stability of metakaolin based geopolymer. *Construction and Building Materials, 222*, 872–881.
23. Mosaberpanah, M. A., & Umar, S. A. (2020). Utilizing rice husk ash as supplement to cementitious materials on performance of ultra high performance concrete:–a review. *Materials Today Sustainability, 7*, 100030.
24. Memon, S. A., Javed, U., & Khushnood, R. A. (2019). Eco-friendly utilization of corncob ash as partial replacement of sand in concrete. *Construction and Building Materials, 195*, 165–177.
25. Nuaklong, P., Jongvivatsakul, P., Pothisiri, T., Sata, V., & Chindaprasirt, P. (2020). Influence of rice husk ash on mechanical properties and fire resistance of recycled aggregate high-calcium fly ash geopolymer concrete. *Journal of Cleaner Production, 252*, 119797.
26. Olutoge, F. A., & Adesina, P. A. (2019). Effects of rice husk ash prepared from charcoal-powered incinerator on the strength and durability properties of concrete. *Construction and Building Materials, 196*, 386–394.
27. Rattanachu, P., Toolkasikorn, P., Tangchirapat, W., Chindaprasirt, P., & Jaturapitakkul, C. (2020). Performance of recycled aggregate concrete with rice husk ash as cement binder. *Cement and Concrete Composites, 108*, 103533.
28. Raheem, A. A., & Ige, A. I. (2019). Chemical composition and physicomechanical characteristics of sawdust ash blended cement. *Journal of Building Engineering, 21*, 404–408.
29. Raheem, A. A., & Orogbade, B. O. (2018). Characteristics of blended cement produced from selected hardwood ashes. *FUOYE Journal of Engineering and Technology, 3*(1), 61–66.
30. Raheem, A. A., & Kareem, M. A. (2017). Chemical composition and physical characteristics of rice husk ash blended cement. *International Journal of Engineering Research in Africa, 32*, 25–35. https://doi.org/10.4028/www.scientific.net/JERA.32.25
31. Raheem, A. A., & Adenuga, O. A (2013) Wood ash from bread bakery as partial replacement for cement in concrete
32. Raheem, A. A., Olasunkanmi, B. S., & Folorunso, C. S. (2012). Saw dust ash as partial replacement for cement in concrete. *Organization, Technology & Management in Construction: An International Journal, 4*(2), 474–480.
33. Raheem, A. A., & Sulaiman, O. K. (2013). Saw dust ash as partial replacement for cement in the production of sandcrete hollow blocks. *International Journal of Engineering Research and Applications, 3*(4), 713–721.
34. Shakouri, M., Exstrom, C. L., Ramanathan, S., & Suraneni, P. (2020). Hydration, strength, and durability of cementitious materials incorporating untreated corn cob ash. *Construction and Building Materials, 243*, 118171.
35. Sargın, Ş, Saltan, M., Morova, N., Serin, S., & Terzi, S. (2013). Evaluation of rice husk ash as filler in hot mix asphalt concrete. *Construction and Building Materials, 48*, 390–397.
36. Udoeyo, F. F., & Dashibil, P. U. (2002). Sawdust ash as concrete material. *Journal of Materials in Civil Engineering, 14*(2), 173–176.

Paste Studies on Indigenous Materials – Effects of Calcined Clays and Limestone

K. Gowri and A. Abdul Rahim

Abstract Now-a-days, lessening the clinker content in blending cement binders or the cement content in concrete, augmented calcined clays binders are the only choice of material that can satisfy the needs of eco-efficient cement-based products in sufficient quantity. Current developments in the potential use of Limestone Calcined Clay Cement (LC^3) as a common cement is gaining popularity around the world. Over the years, several studies indicated that various parameters of the LC^3 binder and concrete were evaluated and examined across the universe. But this study examines the development of indigenous material to make sure the replacement level of clays and limestone. In the neighbourhood, obtained clays were heated up to 700 °C, and then they were grinded to achieve a specific surface area to improve the pozzolanic activity. The DoE-Taguchi method was used to optimize the material. The results of this study demonstrated that even the combination of calcined clay with a low kaolinite concentration showed considerable pozzolanic properties and synergy with the limestone filler in a ternary binder. This study compares the strength behaviour of Limestone - Calcined Clay Cement binder with regular Portland cement. The characterization and microstructural studies were also conducted to analyse both the raw samples and their calcined derivatives to identify the significant changes that occurred in the samples that have been observed and discussed. The optimised sample has resulted in promising outcomes as the early strength is more than the control binder and 30% clay,15% limestone powder and 2% gypsum substitution have achieved similar strength with OPC for 28 days.

Keywords Pozzolanic · Optimize · Synergy · Strength · and Gypsum

K. Gowri (✉) · A. A. Rahim
Vellore Institute of Technology, Vellore, Tamil Nadu, India
e-mail: gowrik.2019@vitstudent.ac.in

A. A. Rahim
e-mail: abdulrahim.a@vit.ac.in

A. Vilventhan et al. (eds.), *Advances in Construction Materials and Management*,
Lecture Notes in Civil Engineering 346,
https://doi.org/10.1007/978-981-99-2552-0_34

1 Introduction

Cement is a long chronicle and is used as a basic building material in the world for making a variety of building establishments and infra projects. the process of making cement emits greenhouse gas emissions and global warming will be impacted the environment. According to the Environmental Protection Agency (EPA) data, every tonne of cement production is accountable for the release of around one tonne of carbon dioxide(CO_2) into the atmosphere [1]. The manufacturing of ordinary Portland cement (OPC) provides for almost 8 percent of total global carbon dioxide emissions [2] and is reasonable for global warming. Researchers and working groups keep on putting their efforts to develop a variety of cement/concrete more sustainable and eco-friendly by switching to consuming additional cementitious materials resulting from industries that can be minimised or have no Portland cement and instead depend on other materials [3]. Nowadays, mostly natural and industrial by-products are commonly used as supplementary cementitious materials (SCMs) for a variety of binder applications. Principally, substituting SCMs in cement clinker fractions significantly improves the quality of binders and concrete, also helpful to cut down CO_2 emissions, both immediately and over the long term [4]. Still, the usage of SCMs in cement/concrete manufacturing helps us to enhance the durability of cementing medium and obtain additional benefits from environmental concerns. To meet the latest requirements/ demands of sustainable cement production and application, it is necessary to identify and develop a novel binder that can provide good mechanical performance at early and later ages, which may be an equivalent or alternative to the Ordinary Portland Cement (OPC), and remain stable in varied service settings. The precise application of numerous dissimilar SCMs can increase the effectiveness of each one by holding a wide range with differing degrees of prominence and reaction potential [5]. Some of the SCMs currently in use are natural pozzolanic materials like calcined clays and limestone and industrial by-products like fly ash, GGBS, sugarcane bagasse, silica fume, etc. Natural pozzolana is commonly used as a replacement for Portland cement in concrete. Among its many positive effects on the environment and economy is a lowering of permeability [].

In the last two centuries, clays used in the cement industry for the manufacturing of Portland cement clinker [6]. The most widely utilized natural SCMs are calcined clays and limestone, both of which are readily available all over the world. Limestone calcined clay cement (LC^3) has received a lot of attention in recent years. By substituting clinker with SCMs such as calcined clays, limestone, and gypsum, an LC^3 binder can be made [7]. In most countries around the world variety of clays are widely available. Generally, clay, once calcined, can be utilized as a pozzolanic material in cement/concrete to enhance its properties. At temperatures between 500 °C and 900 °C, kaolinite-containing clays can be employed as SCMs [8]. Clays with variable proportions of minerals like kaolinite, montmorillonite, and illite display different properties such as strength, porosity, and shrinkage. When compared to other calcined clays, kaolinite was found to have superior SCM properties due to its higher pozzolanic activity [9]. During the calcination, the breakdown is happen in the kaolinitic clay as the release of water into the atmosphere instead of CO2, which is what happened when CO2 was removed from limestone during the clinker process [10].

kaolinite is the most reactive clay mineral after the calcination of the clay. Many studies have investigated the impact of calcination factors of clay like time and temperature on the physical properties of the blends [11]. Clays with kaolinite have a wide variety of potential uses due to their favourable physicochemical properties. Mainly because it contains significant amounts of both silica and alumina. Clay minerals like kaolinite, which are abundant in kaolin, have been applied to a wide range of applications [12]. It is interesting to note that, when grinding the kaolinite materials for finer fractions may increase the surface area and enhance the reactivity. These findings can be explained by the fact that grinding meta-kaolinite clay resulted in a smoother surface [13].

Limestone is a common raw material that has always been used to make OPC. it is available in its natural state easily and lot, because, it is used for replacement in the production plant itself. The direct addition of uncalcined crushed limestone in combination with calcined clay can have a significant impact on phase compositions and early age characteristics [14]. According to the study from Dave et al. [1], when a less reactive pozzolan is used in ternary mixtures with a more reactive one, (materials that are rich in silica content) such as calcined clays and silica fume, there is a synergy between these pozzolans, resulting in a result that is higher than that obtained in the respective binary mixtures; this result is known as the synergic effect. This study hopes that the results will lead to a new way for countries like India to get rid of their trash and also help reduce the amount of carbon dioxide (CO2) that concrete uses.

The present study investigates the potential use of pozzolanic material especially low- grade kaolinite type by clay calcining around 700 °C temperature of locally obtained clays with the combination of uncalcined limestone powder [8, 16] and residence time was maintained as one hour. The work includes a study of the mechanical properties of different types of low- grade calcined clay and limestone mixed with cement and an evaluation of the characterization studies on calcined and uncalcined materials and microstructural properties of binders were discussed.

1.1 Design of Experiments

In the 1920s, Ronald. A. Fisher developed a way of experimentation called Design of Experiments (DoE) to analyse the interaction of several variables involved in an investigation to identify the plant's requirements like precipitation, irrigation, fertilizer and sunlight to equip the best harvest yield in an ideal condition. The DoE needs careful preparation, strategic experiment design, and professional evaluation of data. Dr Genechi Taguchi, Fisher's follower has improved and defined the methodology for all the above DoE application processes based on decades of research and applications. Taguchi's DoE technique for problem-solving and product/process design optimization initiatives can meet those needs at a reasonable cost. Engineers can benefit from understanding and using this method. the application of DoE by the way of Taguchi, the method has grown into a highly desirable resource for working scientists and engineers [16]. So that in this study DoE- the Taguchi method was widely used to optimize the trial mixes of the project and to reduce the cost. This article reports information about the experiments that were selected by DoE to optimize the materials and mixes that were analysed.

2 Materials and Methods

2.1 Materials

The materials used in this experimental study is OPC, with the combination of locally available materials such as calcined clay, limestone powder, gypsum, and LC^2. Which are discussed below.

2.1.1 Ordinary Portland Cement

Ordinary Portland cement (OPC) – 53 grade conforming to IS 12269 – 2009 was used as a primary binder, Sagar Cement of India supplied the cement of OPC 53 grade that was utilized.

2.1.2 Calcined Clays (CC)

Kaolinitic clays typically occur in soils formed by chemical weathering in hot and humid climates, such as those found in the tropics and subtropics. It's interesting to note that kaolinitic clays are found in regions where cement demand is expected to rise the most in the coming decades. The unique structure of kaolinite, which consists of only one octahedral aluminate sheet and one tetrahedral silicate sheet, sets it apart from other clays like illite and smectite [17]. The reactivity of clays in cement is zero. To be reactive, calcination is required. At temperatures between 450 °C to 800 °C clays undergo de-hydroxylation during the calcination process.

In the present study, the low-grade kaolinite clay was collected from the lakes from different sources C1- Lake clay which is collected in the Tiruvallur district, Tamilnadu. C2- Clay collected from fields of the village near Tiruvallur district. C3- clay collected from VIT lake clay, here all the used clays were Calcined initially then it was used as supplementary cementitious material as per IS 1344 (1981) for partial replacement of cement.

2.1.3 Limestone Powder (LSP)

Limestone is a sedimentary rock; limestone is mostly broken pieces of thin organisms. Unhydrated Limestone powder was procured from Pallaniappa minerals, Salem. The powdered form of limestone is a milky white colour, and its particles are uniform in size and shape.

2.1.4 LC^2

A commercially available LC^2 ready mix was collected from (JK Lakshmi Cements Ltd.). which is equipped with OPC and is used as a (50% LC^2 + 50% OPC) second binder.

2.1.5 Gypsum (GY)

Cement setting time can be controlled by adding gypsum, which gives it the common name of "retarding agent of cement." Cement clinker can rapidly condense without gypsum by combining with water and releasing heat. The "setting of cement" can be managed with the use of gypsum. The gypsum (calcium sulphate) was brought from the local market in Vellore, Tamilnadu.

2.1.6 Superplasticizer (SP)

The water-reducing agent poly-carboxylic ether-based superplasticizer (CONPLAST-SP430) was added to improve the mixture's workability. The percentage of superplasticizers was optimized by using DoE.

2.1.7 Water

The water source available in the concrete lab at Vellore institute of technology was used to prepare the paste mixes (pH value is 6.5). The amount of water to binder (W/B) was kept according to the DoE Taguchi method for all mixtures.

2.2 Methods

2.2.1 Calcined Clays with Calcination and Grinding Technology

Clays typically take place in soils by chemical weathering due to high temperate and their effects in the tropical and subtropical regions. Clays are chiefly composed

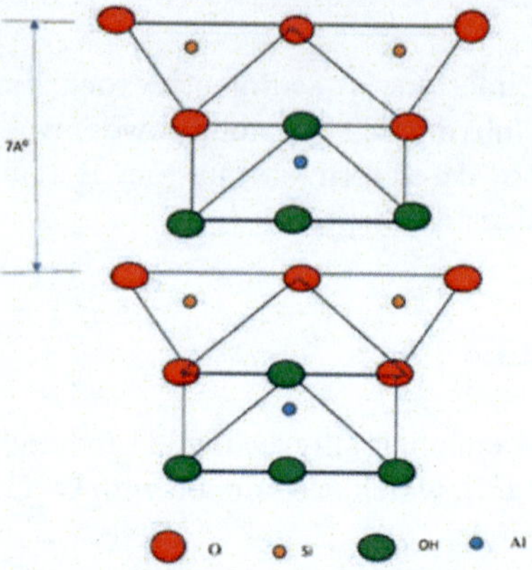

Fig. 1 Kaolinite structure of clays

of aluminium mineral particles with a particle size of less than 45μm. They are typically categorized as 1:1 (kaolinite & halloysite) or 2:1 (illite, montmorillonite, pyrophyllite) clay minerals depending on their internal structure. Kaolinite is an example of a 1:1 clay mineral, which consists of tetrahedral-octahedral sheets of silica and alumina that form a layer through shared oxygen atoms and are held together through hydrogen bonding. Kaolinite is a primary component is a hydrous aluminium silicate, or (Al2O3.2SiO2.2H2O). Clays have a plate-like structure, especially that of calcined kaolinite clays, and with its negative surface charge, calcined clay has a great potential to flocculate [1]. The structural arrangement of the kaolinite is shown in Fig. 6. Because of its strong pozzolanic activity, kaolinitic clay (1:1) is one of the most widely popular types of pozzolanic material utilized around the world. With an appropriate heat treatment, the kaolinite loses its crystalline structure and produces amorphous (pozzolanic reactivity) nature. Among calcined clays, calcined kaolinitic clay is said to have the highest pozzolanic reactivity.

In India, there are many kinds of clays with available different minerals composition. Especially, the low-grade kaolinite clays are easily availed. In this study, clays were collected from various places, like three low-grade kaolinite clay quarries, with different sources and compositions named C1, C2, and C3, which are available locally in Tamil Nadu, India. The clays were first allowed to sun dry for 24 hours before being pulverized with a pulveriser, Then, the grinded material was calcined at 700 °C, and the temperature was raised to 5 °C/min after reaching the 700 °C it was kept for 1 h at a constant rate based on earlier studies [18]. The calcination of the clay was done by flash calcination. Flash calcination happens in special machines called "flash calciners". A high-temperature muffle furnace (1700 °C) was used for the calcination of clay material. After the calcination, the materials were allowed to reach room temperature as shown in Fig. 2 before being ground in a rotary mortar so that they reached a fineness comparable to OPC. During the calcination period, the iron oxide in the samples oxidized and caused to change in the C1 sample to colour from brown to red. In the case of C2 and C3 were changed to a dark red colour indicating low concentrations of iron oxide in the clays. The oxidation state of the colours

Fig. 2 **a** Left: uncalcined clay **b** Right: calcined clay

in clays, like Fe, Ti, and Mn, can vary during the calcination and cooling processes, resulting in a change in the final colour. Up to 80% of the calcined clay mass was processed in a rotary mill (Pulveriser machine) until it went through a 45μm sieve, which is a common goal during the grinding process for blended cement.

3 Characterization of Raw Materials

3.1 Thermogravimetric and Differential Thermal Analyses (TG/DTA)

The Thermogravimetric and Differential thermal studies were conducted to identify the physical and chemical changes occurring in the clay samples. Figure 3 showed that the mass loss and the dynamic temperature range during the thermal response that occurred varied for the different clays. Initially, the first endothermic reaction was observed in C1, C2, and C3 clays curves are 84 °C, 179.59 °C, and 166.73 °C. During the reaction, the water was lost, which showed adsorption and interlayer dehydration (nH2O). The second small endothermic reaction between 430 °C–500 °C was observed in C1 it may be clarified by the simultaneous de- hydroxylation of trace amounts of kaolinite and other clay minerals present in the clay and the transition from α- Quartz SiO2)to β-Quartz (SiO2) [12]. TG and DTA curves exhibited a very little amount of weight loss 0.08 mg (C1), 0.05 mg(C2), and 0.04mg(C3) for temperatures between 100 °C and 200 °C. Water loss in C1, C2, and C3 clay samples are relatively low, indicating the presence of a layer of hydrated 1:1 clay material. The mass loss of the C1, C2, and C3 clays from 200°C to 600 °C was recorded as 6.87%, 4.44%, and 3.34%, respectively. The substantial mass loss indicated that highly reactive kaolinite is present in the clays because of the de-hydroxylation at the temperature. C1 had the highest mass loss out of all the clays, indicating that it has undergone the most de-hydroxylation.

In the case of C2 and C3 clays, a low-intensity endothermic peak with a small amount of mass loss was exposed between the temperatures 450–500 °C this shows that de- hydroxylation has happened and comparatively this was slightly more in C1. After these two endothermic reactions around 570 °C, there is no mass change happened this indicates the presence of trace amounts of kaolinite and other clay minerals presented in the clays.

The kaolinite content was determined by TGA by using the below Eq. (1). From the TGA the mass loss between 200 °C to 600 °C was taken as (*wt%kao*h) the fully

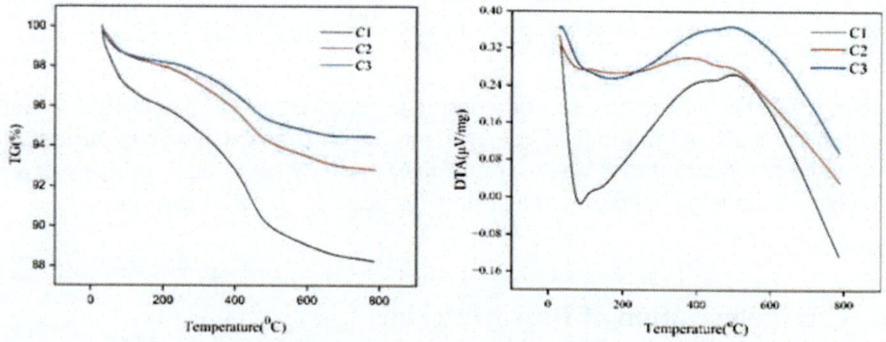

Fig. 3 TG /DTA curves for raw clays

Fig. 4 Morphology of clay samples Left: uncalcined(a,c) Right: calcined(b,d) oven dried for 48 h and then gold coating was done to examine at various magnifications

endorsed to the de-hydroxylation of clays, from the below equation, the weight of percentage of kaolinite content can be calculated as shown below,

$$wt\%kaolinite = wt\%kaoh{-}OH\frac{MKaolinite}{2Mwater} \quad (1)$$

where

M kaolinite is the molecular mass of kaolinite (i.e. 258),

M water is the molecular mass of water (i.e. 18).

using the above Eq. (1), the kaolinite content of three clays was determined as C1–49%, C2–31.8%, and C3–23.9%.

3.2 Composition of Materials

Particle size distribution and specific powder surface area of cementing binders are highly influential in the fresh and hardened properties of cement-based binders. In this research, a Malvern Microron particle sizer with laser diffraction was used to calculate the median particle size in the distributed particles. In the same way, to determine the chemical composition, X- ray fluorescence (XRF) spectroscopy was used. The collected chemical and mineralogical properties are revealed in the Table 1.

Table 1. Chemical Composition of Materials

Oxides (%)	OPC	C1	C2	C3	LS
SiO2	18.6	62.90	68.20	71.80	38.10
Al2O3	4.3	20.70	16.50	13.50	8.82
Fe2O3	2.2	8.22	5.51	5.17	3.41
CaO	62.63	1.62	1.85	2.18	45.00
Na2O	0.18	1.45	2.19	1.96	1.40
K2O	0.85	2.05	3.27	2.98	1.74
TiO2	0.28	0.77	0.61	0.43	0.27
MgO	2.12	1.34	0.86	0.96	0.77
SO3	3.00	-	-	-	-
P2O5	0.27	0.05	0.10	0.12	ND
MnO	-	0.08	0.06	0.03	0.02
LOI	1.36	2.2	2.8	1.8	0.6
Physical Properties					
d10 μm	6.08	2.24	2.5	2.36	2.80
d50 μm	12.94	6.74	8.6	8.00	6.40
d90 μm	24.78	17.84	20.3	23.5	13.94
Specific surface (m^2/g)	0.57	1.22	2.3	1.90	1.14

3.3 Scanning Electron Microscope (SEM)

The morphology of raw materials was examined using electron microscopy with a QUANTA FEG 250 microscope from the FEI company. before the investigation, the clay samples were

However, the morphological characteristics of the clay are shown in Fig. 4. From the image there is a difference has been observed between a raw and a calcined sample, a Kaolin crystal has a typical structure, consisting of false hexagonal plates in the shape of tiny flakes that are uneven at the edges and poorly aggregated. After the application of thermal treatment, it leads to damage to the kaolinite plates, particle aggregation, and uneven, rounded forms are formed.

3.4 X-Ray Powder Diffraction (XRD)

X-ray powder diffraction test was performed to ensure a quantitative breakdown of the minerals present in the raw clays and calcined clays. The identified minerals, the crystalline and amorphous content of each SCM were quantified using X-ray diffraction (XRD). PANanlytical X'pert^3 Powder equipment was utilized for the X-ray diffraction (XRD) investigations. The scanning procedure involved the diffraction angle being tested from 5°– 90° at a rate of 0.2° 2θ per second and a step time is 20 s. using an X-ray tube (CuK-target). X'pert^3 software was used to do the phase quantification analysis with an external standard. To confirm the consistent

promise conditions and a uniform scanning process of XRD analyse, all the samples were tested on the same day. The observed changes from a crystalline to an amorphous form of clay samples are depicted in the XRD graphs of raw and calcined clays (Fig. 5). In addition, X'pert high score software is used for minerals analysis. According to the XRD and TGA results data, the clay (C1) indicated that it completely changed from crystalline to the amorphous state after the calcination process, which is represented as a pozzolanic material, and it has a kaolinite content of about 49%. But in the case of the Other two clays, it has shown some crystalline peaks after the calcination, hence, in this study, the whole experimental work was carried with clay (C1). Even from the above results, the conformation of the suitability of clay has been done by compression test for all the 3 clays by using DoE.

3.5 Design of Mixes- Taguchi Method

A technique design of experiments (DoE) was engaged to construct an experimental setup to fix the optimal binder proportion for producing a desirable mix. Taguchi's statistical experimental design starts with the robust decision of selection parameters/

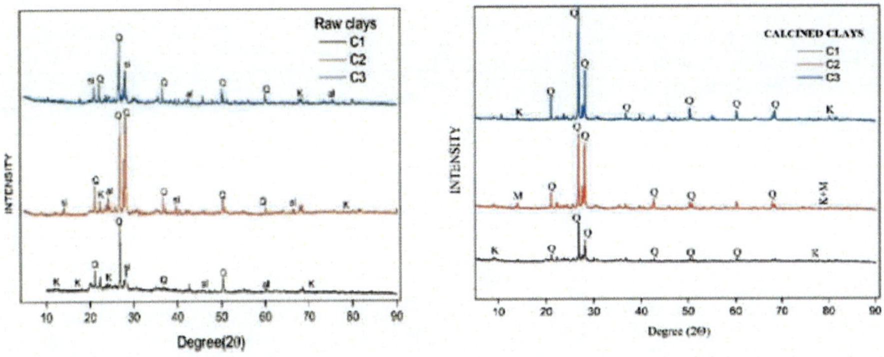

Fig. 5 XRD graphs for Materials

variables as well as their levels. In the research work, a partial factorial design experiment was designed with Taguchi's Orthogonal Array (TOA). An L9(3^3) orthogonal array, which is a combination of 9 trials with three parameters with a maximum of three levels was set up for this experimental investigation. The chosen parameters and appropriate levels were determined by previous studies. In this research, the three design factors: MIX proportion content (%), W/B (%), and superplasticizer (SP%), were considered with three levels, which makes a maximum of nine trials. Table 3 showed the experimental design setup with factors and their variation levels.

The design of the experiment was carried out on all three clays to confirm the pozzolanic behaviour of clay material to the extent remaining investigation. Variable levels of mixed parameters were selected based on the existing literature and preliminary laboratory studies. The optimization was done for a different source of clays C1, C2, and C3 were considered 3 levels and 3 different variables by using the DoE - Taguchi method as shown in Table 2. Based on the previous works, the laboratory trails were fixed as calcined clay 30%, limestone 15%, and gypsum 2% were replaced with OPC. As can be seen in Table 3, the experimental trial runs are constructed based on a standard L9 (3^3) orthogonal array (OA) with three factors and three levels.

The above 9 trail experiments were built and cast. The cured test samples were tested to compressive strength after the age of 7 and 28 days to get an optimized clay mix binder. The preparation of the binder and casting of the cubes were explained below.

Table 2 Mix parameters and their levels

Levels	MIX	W/B	SP
1	C130 LS 15 G2	0.25	0.2
2	C230 LS 15 G2	0.3	0.3
3	C330 LS 15 G2	0.35	0.4

Table 3 Design matrix of L9 (3^3) orthogonal array with parameters and coding.

S.no	coding	W/B	SP
1	C1a	0.25	0.1
2	C1b	0.30	0.2
3	C1c	0.35	0.3
4	C2a	0.25	0.2
5	C2b	0.3	0.3
6	C2c	0.35	0.1
7	C3a	0.25	0.3
8	C3b	0.30	0.1
9	C3c	0.35	0.2

4 Indigenous Lab Scale Production and Preparation of Ternary Blended LC^3 Cement Binders

The preparation of LC^3 by using locally available material is shown in Fig. 6. In cementitious systems, to prepare 50 mm cubes, the paste mixture was prepared by replacing the 2:1 ratio of calcined clay, limestone powder and gypsum as 2% can replace up to 53 wt% of the cement as well as the water-to-binder ratio is kept according to DoE mixes. For homogeneity, the ingredients were dry-mixed in the mixer for one minute. After that, water containing PC- Portland Cement, CC- Calcined clay, LSP- Limestone powder, GY-Gypsum

superplasticizer was added to the mixer, at last, a laboratory bench mixer was used to thoroughly mix the constituents per ASTM C305-06. And the content was blended again for a further three minutes. Thereafter, the uniformly blended paste was poured into steel moulds of size 50 × 50 ×50 mm, wrapped with plastic wrap, and allowed to cure for one day at a temperature of 20 ± 2. The cast samples were demoulded the very next day, after that the samples were put in the water for an age 7, and 28 days. After the attainment of the specified age, the samples were employed

PC- Portland Cement, CC- Calcined clay, LSP- Limestone powder, GY-Gypsum

Fig. 6 Lab Preparation of LC^3, Casting and Compression testing

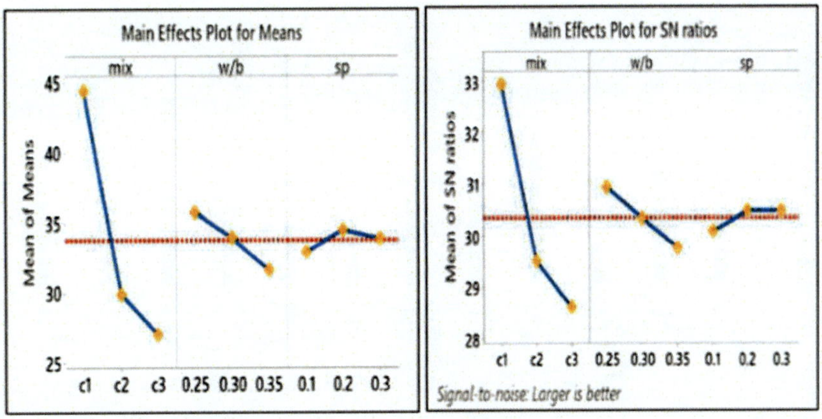

Fig. 7 Signal-to-Noise ratio plots

for compressive strength test a room temperature as per ASTM C31/ C31M -19 and compared with controlled binder and LC^2+OPC binder samples.

4.1 Taguchi Analysis by Signal to Noise Ratio (S/N)

Following the experimentation expertise, Taguchi method, The effectiveness of parameters can be evaluated using three different performance characteristics: larger is better, smaller is better, and nominal is better. The study's goal was to determine the best binder mixing ratios for maximum residual compressive strength. The higher the compressive strength, the more likely it is to have been chosen. Using the programme depicted in Fig. 7, the S/N ratio was computed, and the resulting optimal mix through the main effects of Signal to Noice (S/N) ratio was revealed to be: The best clay to use is C1 level, which has used a 0.25 water to binder (W/B) ratio and requires a dosage of only 0.2 superplasticizers (SP) content. Hence it is confirmed that C1 is suitable for the research work which has been carried out for further studies.

5 Results and Discussions

5.1 Compressive Strength

The samples prepared based on the requirement of Taguchi's $L9(3^3)$ OA was tested at the age of 7 and 28 days. The results of compressive strength for all trials with different proportions as shown in Fig. 8. The 7 days outcomes were between the range of 16 to 36.9 MPa and the 28 days strength values were between 25 to 49.6

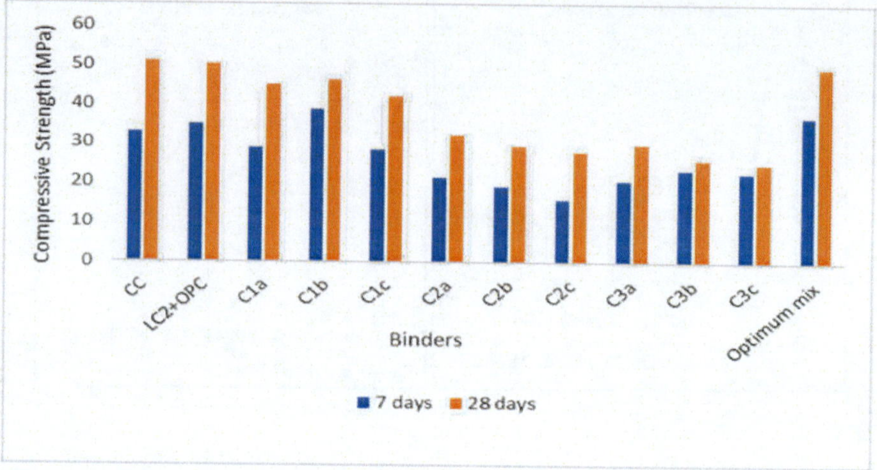

Fig. 8 Compressive strength of binders

MPa respectively. The tested sample's result of compressive strength confirms that the indigenously prepared mixes presented well a result near about comparable to OPC (51.2 MPa). Since, the fineness and calcination of different types of calcined clays, as well as the amount and mineralogical composition of the reactive silica, are the primary factors of the clay activity. On the other hand, the kaolinite group C1b shows identical strength at 28 days, but slightly lower strength than OPC, whereas ($LC^2$50+OPC50) reaches a compressive strength equivalent to that comparable to OPC binder (51.2 MPa) after 28 days. Generally, the nine mixes show different strength gain

depending upon the type of clay, different W/B and SP as per Table 3. The kaolinite clays have less responsive at an earlier age and later strength development at 28 days is phenomenal. But the optimum confirmation mix (C1 clay type, 0.25% of W/B ratio and 0.2% SP), has shown early strength development as compared to trial mix 1 also it was 70% of the 28 days strength of controlled OPC binder.

5.2 *Microstructure*

These investigations showed that the principally hydrated products present in Limestone Calcined Clay (LC3) systems are C-A-S-H gels, ettringite, CH, C-S-H gel pores, Hemi- carbo- aluminate, and mono-carbo-aluminate, all of which contribute to the formation of a dense and porous microstructure of LC system. Figure 8 displays SEM images of the samples at ages 7 and 28 Days of the Optimised Mix.

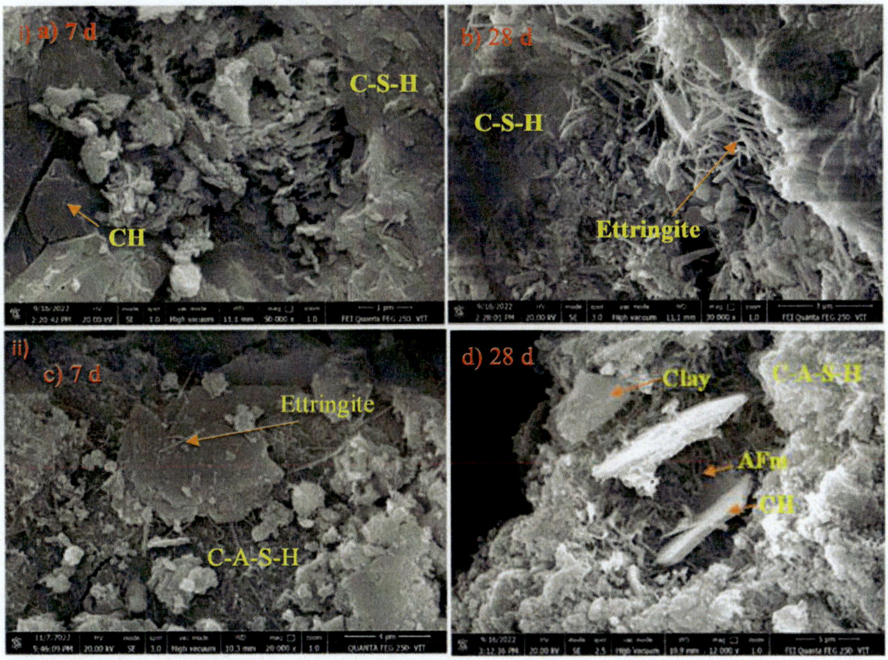

Fig. 9 SEM images i) OPC- a, b ii) LC system - c, d

The optimized mix (C1, W/B- 0.25, SP- 0.2) mix with 30% clay, 15% limestone powder, 2% gypsum systems having Carbo-aluminate phases (Fig. 9d) and C-A-S-H gel spots (Fig. 9c) was numerous in LC systems than in OPC. Furthermore, ettringite, AFt, and AFm were found in the LC system While both calcium hydroxide and ettringite were easily observable in OPC cement paste, only a trace amount of calcium hydroxide was found in LC systems due to their better reactive mineralogical composition of the mixture with pozzolanic materials.

5.3 *XRD Analysis*

X-ray diffraction (XRD) is a flexible technique to recognise the fundamental phase changes happening in crystalline materials and other structural parameters. C-A-S-H, Calcium hydroxide, C-S-H, calcite, ettringite, AFt phases, and all other forms at various times, as shown by the XRD patterns in (Fig. 10). Hemi-carbo-aluminate (Hc) and mono-carbo-aluminate (Mc) are produced when limestone combines with the active alumina in C3A and the calcined clay. Due to the high concentration of active alumina and silica in the calcined clay, it is capable of reacting with the Ca(OH)2 as a hydration product to produce the C-S-H/C-A-S-H gel [18].

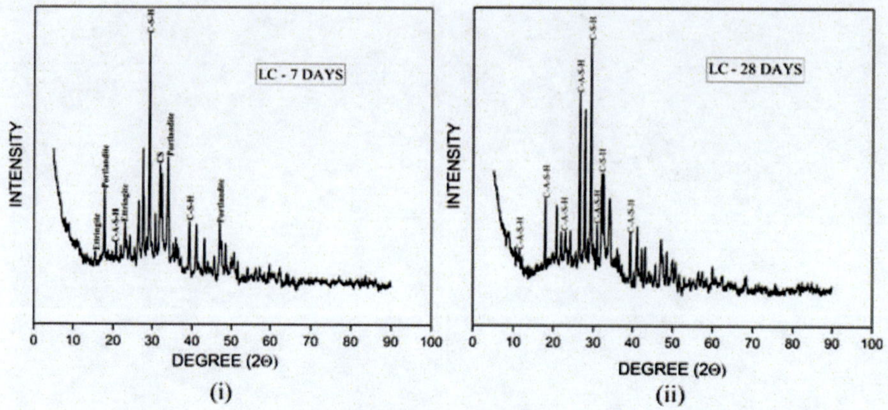

Fig. 10 XRD images i) LC – 7 days ii) LC system- 28 days

Because these high carbo-aluminates can enhance the mechanical properties of LCC paste cubes. Therefore, in the LC systems, after 7 days, not enough calcium hydroxide was present to react with the calcined clay. At early ages, a lesser amount of calcined clay is involved in the pozzolanic reaction and the delayed growth of strength at later ages [19].

Besides, the alumina and silica in the calcined clay can react with the cement Ca(OH)2 to generate the C–A–S–H gel, which brings about the development of strength in the cubes.

6 Conclusion

The effect of different calcined clays and limestone used in place of cement was studied in detail by an organised experimental design method, to improve the better understanding of indigenous materials affect the mechanical properties and microstructure of cementitious materials. Some inferences that can be made are as follows:

1) Research has been conducted to describe the properties of low-grade kaolinite clays derived from a wide variety of local sources. A systematic experimental technique- Taguchi method was carried out to ascertain the type of clay to enhance the better strength properties. Subsequently, at the end of the test, the C1 binder composition has identified as a suitable material for future study.
2) At an early age of the nine mixes, the compressive strengths of the low-grade clay systems are shown more in C1 and lower strength in C2 and C3 than in OPC cement paste. However, the strengths grow at a faster rate from 7 to 28 days. In the case of a commercial product of LC^2 combination with OPC 50%, the

strength attained is (50.3 MPa). In the LC^3 system with indigenous material the strength obtained was 49.6 MPa i.e.comparable to in OPC and LC^2 systems.

3) The hydration studies were analysed as the calcined clay and limestone powder work together to provide a synergistic effect that causes the carboaluminate content to gradually increase while the calcium hydroxide concentration lowers. The filler effect impact of calcined clay and limestone powder causes the LC system "Induction period" to become less than that of cement paste binder at W/B of 0.25
4) The characterization studies were done on the hydrated products of 7 and 28 days, the SEM images give information on the formation of Hemi-carboaluminate and mono- carboaluminate occupied more space in 28 days, and very low CH was presented after 7 days because the reaction between aluminosilicates in calcined clay and Ca(OH)2 in cement has slowed.
5) Considering all the aspects, the maximum percentage of substitution level of low-grade materials was considered as 47% even though the compressive strength is lower at an early age, the later strength is similar to that of OPC

Acknowledgement The authors are grateful to the School of Civil Engineering, VIT University, Vellore and School of advanced Sciences, VIT University for extending the facilities to carry out this investigation

Credit authorship contribution statement **K. Gowri:** Methodology, Visualization, Writing – original draft.

A. Abdul Rahim: Conceptualization, Supervision.

Declaration of Competing Interest The authors declare that they have no known competing financial interests or personal relationships that could have appeared to influence the work reported in this paper.

References

1. Dave, N., Kumar, A., Srivastava, A., & Kumar, S. (2016). Experimental analysis of strength and durability properties of quaternary cement binder and mortar. *Construction and Building Materials, 107*, 117–124. https://doi.org/10.1016/j.conbuildmat.2015.12.195
2. Akhlaghi, O., et al. (2017). Modified poly(carboxylate ether)-based superplasticizer for enhanced flowability of calcined clay-limestone-gypsum blended Portland cement. *Cement Concr Res, 101*(February), 114–122. https://doi.org/10.1016/j.cemconres.2017.08.028
3. Akin, O. O., Ocholi, A., Abejide, O. S., & Obari, J. A. (2020). Prediction of the compressive strength of concrete admixed with metakaolin using gene expression programming. *Adv Civil Eng, 2020*, 1–7. https://doi.org/10.1155/2020/8883412
4. Dhandapani, Y., Sakthivel, T., Santhanam, M., Gettu, R., & Pillai, R. G. (2018). Mechanical properties and durability performance of concretes with limestone calcined clay cement (LC3). *Cement Concr Res, 107*(February), 136–151. https://doi.org/10.1016/j.cemconres.2018.02.005
5. Dhandapani Y., & Santhanam, M. (2020) The role of pore network parameters to link physical structure and transport properties in ternary binders involving aluminosilicate- limestone combination The role of pore network parameters to link physical structure and transport properties in ternary , April 2021

6. Hanein, T., et al. (2021). Clay calcination technology: state-of-the-art review by the RILEM TC 282-CCL. *Materials and Structures, 55*(1), 3. https://doi.org/10.1617/s11527-021-01807-6
7. Dhandapani, Y., Vignesh, K., & Santhanam, M. (2020) Anomalous Early Increase in Concrete Resistivity with Calcined Clay Binders, April 2020. https://doi.org/10.1007/978-981-15-2806-4
8. Dixit, A., Du, H., & Pang, S. D. (2021). Performance of mortar incorporating calcined marine clays with varying kaolinite content. *Journal of Cleaner Production, 282*, 124513. https://doi.org/10.1016/j.jclepro.2020.124513
9. Du, H., & Pang, S. D. (2020). High-performance concrete incorporating calcined kaolin clay and limestone as cement substitute. *Construction and Building Materials, 264*, 120152. https://doi.org/10.1016/j.conbuildmat.2020.120152
10. Tironi, A., Castellano, C. C., Bonavetti, V. L., Trezza, M. A., Scian, A. N., & Irassar, E. F. (2014). Kaolinitic calcined clays - Portland cement system: Hydration and properties. *Construction and Building Materials, 64*, 215–221. https://doi.org/10.1016/j.conbuildmat.2014.04.065
11. Fernandez, R., Martirena, F., & Scrivener, K. L. (2011). The origin of the pozzolanic activity of calcined clay minerals: a comparison between kaolinite, illite and montmorillonite. *Cement Concr Res, 41*(1), 113–122. https://doi.org/10.1016/j.cemconres.2010.09.013
12. Adeniyi, F. I., Ogundiran, M. B., Hemalatha, T., & Hanumantrai, B. B. (2020). Characterization of raw and thermally treated Nigerian kaolinite-containing clays using instrumental techniques. *SN Appl. Sci., 2*(5), 1–14. https://doi.org/10.1007/s42452-020-2610-x
13. Gamelas, J. A. F., Ferraz, E., & Rocha, F. (2014). An insight into the surface properties of calcined kaolinitic clays: The grinding effect. *Colloids Surfaces Physicochem Eng Asp, 455*(1), 49–57. https://doi.org/10.1016/j.colsurfa.2014.04.038
14. Dhandapani, Y., Santhanam, M., Kaladharan, G., & Ramanathan, S. (2021). Towards ternary binders involving limestone additions — a review. *Cement Concr Res, 143*, 106396. https://doi.org/10.1016/j.cemconres.2021.106396
15. Rahim, A., Sharma, U. K., Murugesan, K., Sharma, A., & Arora, P. (2013). Multi-response optimization of post-fire residual compressive strength of high performance concrete. *Construction and Building Materials, 38*, 265–273. https://doi.org/10.1016/j.conbuildmat.2012.08.048
16. Tironi, A., Castellano, C. C., Bonavetti, V., Trezza, M. A., Scian, A. N., & Irassar, E. F. (2015). Blended cements elaborated with kaolinitic calcined clays. *Procedia Mater Sci, 8*, 211–217. https://doi.org/10.1016/j.mspro.2015.04.066
17. Adekitan, O., Polytechnic, M.A., Popoola, M. & Polytechnic, M.A. (2020) Potentials of calcined clay as a pozzolan. January, pp. 2–4. https://doi.org/10.14382/epitoanyag-jsbcm.2020.11
18. Li, Q. (2019). The microstructure and mechanical properties of cementitious materials comprised of limestone, calcined clay and clinker. *Ceramics-Silikaty, 63*(4), 356–364. https://doi.org/10.13168/cs.2019.0031
19. Ruan, Y., Jamil, T., Hu, C., Gautam, B. P., & Yu, J. (2022). Microstructure and mechanical properties of sustainable cementitious materials with ultra-high substitution level of calcined clay and limestone powder. *Construction and Building Materials, 314*, 125416. https://doi.org/10.1016/j.conbuildmat.2021.125416

Assessment of Micro-Strength Properties and Strength Enhancement of the Biomass Aggregate Concrete

Akula Vishal and N. Chandana

Abstract Lightweight concrete materials have rapidly evolved in recent years as they can act as disaster-resistant materials. However, very few studies have been undertaken on micro-strength properties and enhancement. This study has conducted experimental investigations on the micro–macro-strength properties of biomass aggregate concrete developed by replacing mineral aggregates with coconut shells (5, 10, 20%). In addition, strength and mechanical properties enhancement is tried by using fiber reinforcement. Results show that as the replacement percentage of coconut shells increases in conventional concrete, the unit weight, compressive strength, and tensile strength decrease. In addition, in higher coconut shells replaced concrete, as the curing time increases, unit weight increases. It is mainly because of the water absorption properties of coconut shells. The addition of PP fibers, steel fibers, and the combination has reduced compressive strength by 10, 10, and 22%, respectively.

In contrast, tensile strength was increased by PP fibers by 14 and increased by 43 and 22% by steel and combination fibers. The concrete mix's microstructure shows that the coconut shell found a better bonding because of high moisture observation. The mineral composition of biomass aggregate concrete showed the percentage of coconut shells directly proportional to the percentage increase in micro-element increase compared to conventional concrete.

A. Vishal
Department of Civil Engineering, Gandhi Institute of Technology and Management, School of Technology, Hyderabad Campus, Rudraram, Hyderabad, Telangana 502329, India
e-mail: vakula3@gitam.in

N. Chandana (✉)
Center for Emerging Technologies for Sustainable Development, Indian Institute of Technology Jodhpur, NH 62, Karwar, Jodhpur, Rajasthan 342030, India
e-mail: chandana2n@gmail.com; chandana@iitj.ac.in

A. Vilventhan et al. (eds.), *Advances in Construction Materials and Management*, Lecture Notes in Civil Engineering 346,
https://doi.org/10.1007/978-981-99-2552-0_35

1 Introduction

Concrete has become one of the basic needs of modern society. Extreme demand on natural aggregates and Portland cement has led to increased research on the replacement of these materials [1]. From a few decades, a lot of research has been going on usage of different materials to reduce the natural materials and keep ecological balance [2, 3]. Mineral aggregates which form 70% of concrete dead weight has given more importance in recent decades [2, 3]. Studies worldwide are being conducted to replace mineral aggregates in concrete with different industrial and agricultural wastes [5, 6]. Ecologically friendly aggregates like foundry sand, coconut shell, clay sand, recycled aggregates, crumb rubber etc., are used for sustainable development [7].

On the other hand, biowaste usage has been considered a good path for sustainable development and a circular economy [9–11]. Previous researchers have proposed coconut shells to replace mineral aggregates and reduce agricultural waste [12]. In addition, coconut shell concrete has less bulk density (i.e., 1800 - 2000 kg/m^3) than normal conventional concrete bulk density (i.e., 2400 kg/m^3) and has proven to be a potential lightweight concrete (LWC) material [10].

Structural LWC offers design flexibility, improved seismic structural response, and lower foundation costs [5]. Application-wise, LWC was found to be good material for resistance to seismic damage to a structure and in disaster management. However, LWC has shown less compressive strength and previous studies focused only on strength properties [7, 8]. It was found that the strength of LWC is categorized into three different ranges i.e., low-density concrete with a range of 0.8 - 2.0 N/mm2, moderate strength 7 - 15 N/mm2 and structural concrete with 16 - 60 N/mm2 [4]. Though LWC will make a potential material for disaster relief and low-cost housing and can reduce agricultural waste, LWC strength enhancement is the need of hour.

Previous studies [12–19] have shown the addition of fibers (like steel, and polypropylene) can contribute to energy absorption mechanism, bridging action, delaying the crack formation, flexural strength, toughness, plastic shrinkages, concrete life etc. Using fibers/ fibers reinforced biomass aggregate concrete will be a potential solution for disaster resistance structures, reduce solid waste and create a circular economy. The current study tried to evaluate the micro–macro-strength properties of the biomass aggregate lightweight concrete (BLWC) prepared from coconut shells (varying the replacement percentage) compared with conventional concrete. In addition, strength enhancement has been tried with the addition of steel and polypropylene fibers. This study can help government agencies, non-profit organizations, and disaster management agencies use LWC as a sustainable material to replace conventional concrete.

2 Methodology

2.1 Materials

Commercially available ordinary portland cement of 43 grade manufactured as per IS:269–2015 was used in this study. The Physical properties of OPC are shown in the Table 1 below.

Crushed Coconut Shells of size between 4.75–12.5 mm and the lengths were restricted to a maximum of 12 mm by crushing the coconut shells as shown in Fig. 1 and properties are presented in Table 2.

Commercially available steel and polypropylene fibers are used for strength enhancement, and the properties of fibers have been provided in Table 3, and the preparation of fibers is shown in Fig. 1.

(a) Coconut shell crusher (b) Coconut Shell aggregates (c) Fibre-reinforced coconut shell

Fig. 1 Coconut shell and Fibre reinforced coconut shell lightweight concrete preparation

Table 1 Properties of cement

Details	Normal Consistency	Fineness of cement	Specific Gravity	Setting time (min)	
				Initial	Final
OPC 43 grade	30%	2.5%	2.94	40	600

Table 2 Physical properties of crushed coconut shell in comparison with natural aggregates

	Abrasion %	Bulk Density (kg/m3)	Water Absorption (%)	Aggregate Crushing Value (%)	Shell Thickness (mm)	Specific gravity	fineness modulus
Coconut shell	1.65%	605	23%	1.53%	2–6 mm	1.2	6.15
Natural aggregates	27%	1660	0.6%	23.68%		2.85	6.925

Table 3 Properties of the fibers used in the study

Fiber Type	Density Kg/ m^3	Elastic modulus GPA	Tensile Strength MPA	Length (mm)	Fracture elongation (%)
Polypropylene	900	4.35	550–700	12	18.1
Steel	7850	210	250	50	

2.2 Mix Proportions

The mix proportions for M40 grade concrete with a mix proportion of 1:1.6:3.3 by weight was used for the design according to IS: 10,262–2009 as shown in Table 4 and Table 5.

Table 4 Mix Proportions of concrete per cubic meter

Mix No	Concrete	Concrete (5% coconut shell)	Concrete (5% coconut shell)	Concrete (20% coconut shell)
Cement (kg/m3)	380	380	380	380
River Sand (kg/m3)	620	620	620	620
Coarse Aggregate (kg/m3)	1250	1125	1000	1187.5
Coconut Shell	0	125	250	62.5
w/c ratio	0.45	0.45	0.45	0.45

Table 5 Mix Proportions of fiber-reinforced concrete per cubic meter

Mix No	*Concrete (10% coconut shell and pp & Steel fibers)*	*Concrete (10% coconut shell and PP fibers)*	*Concrete (10% coconut shell and steel fibers)*
Cement (kg/m3)	380	380	380
River Sand (kg/m3)	620	620	620
Coarse Aggregate (kg/m3)	1125	1125	1125
Coconut Shell	125	125	125
w/c ratio	0.45	0.45	0.45
PP fibers	0.2	0.2	0
Steel fibers	1	0	1

Note: Steel Fibers 1% to the volume of concrete and Polypropylene fibers – 0.2% to the volume of concrete

2.3 Analytical Methods

A compressive strength test was conducted according to IS 516:1959. The cube specimen is (150 × 150 × 150) mm. Concrete's tensile strength test was conducted on cylinders of size 100 mm diameter and 200 mm height as per IS 5816:1999. The setting time of cement was conducted as per IS 4301-part 5.

2.4 Statistical Analysis

The results are reported as mean values or ± standard error of duplicates and are analyzed using IBM SPSS 21.0 software. The strength values of lightweight coconut shell concrete for different time zones and macro structures are analyzed using the non-parametric Friedman test. Only 95% confidence level with a $p = 0.05$ values were considered for analyzing the results.

3 Results

3.1 Weight properties of biomass aggregate lightweight concrete in comparison to conventional concrete

Concrete has been divided into 3 categories based on its density, i.e., (i) heavy weight concrete (i.e., 3200 to 4000 kg/m3) used in nuclear power plants, (ii) normal weight concrete (i.e., 2400 to 2600 kg/m3), used for general constructions and (iii) lightweight concrete (less than 2400 kg/m3). Unit weight of biomass aggregate lightweight concrete (BALWC) formed from coconut shell in comparison with conventional concrete is shown in Fig. 2. In addition, only 10% coconut shells replacement on conventional concrete showed an optimum decrease in unit weight as curing time increased, whereas 20% showed a unit weight increase, which may be because of the water absorption capacity of coconut shells. It can be inferred that as the replacement percentage of coconut shells increases, the unit weight also decreases. In addition, higher coconut shells replaced concrete; as the curing time increases, unit weight increases, mainly because of coconut shells water absorption properties. Does more research need to be conducted on the classification inside lightweight concrete, as its properties differ based on type of LWCs, which can be used for different applications like floors in high rise buildings, frames for multistorey buildings, improve thermal performance, long-span bridges and shore structures.

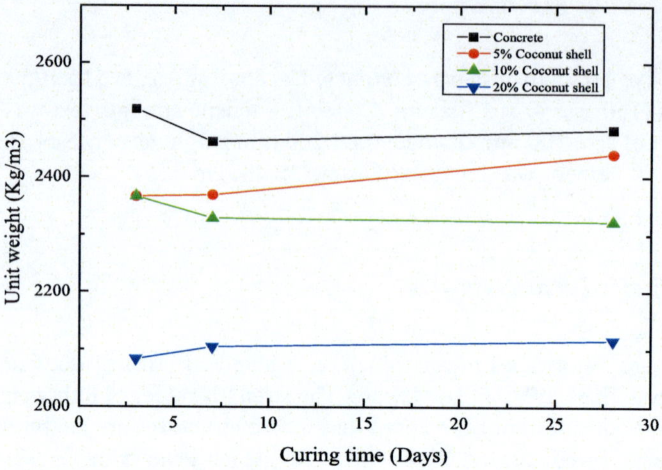

Fig. 2 Unit weight of biomass aggregate lightweight concrete in comparison to conventional concrete

3.2 *Microstructure Properties of Biomass Aggregate Lightweight Concrete in Comparison to Conventional Concrete*

Figure 3 provides the microstructure of the biomass aggregate lightweight concrete prepared from coconut shells in comparison with conventional concrete. It is very clear from Fig. 3 that as mineral aggregates replaced with coconut shell provides higher water absorption leading to higher hydration of concrete leading to good bonding. The BALWC can be a good applicant for floors in high rise buildings, frames for multistorey buildings, improve thermal performance, long-span bridges, and shore structures.

3.3 *Mineral composition of biomass aggregate lightweight concrete in comparison to conventional concrete.*

The mineral composition of biomass aggregate light weight concrete has shown direct proportion increases of elements like Aluminum, Silicon, Sodium, Carbon, Potassium and Fluorine as shown in Fig. 4.

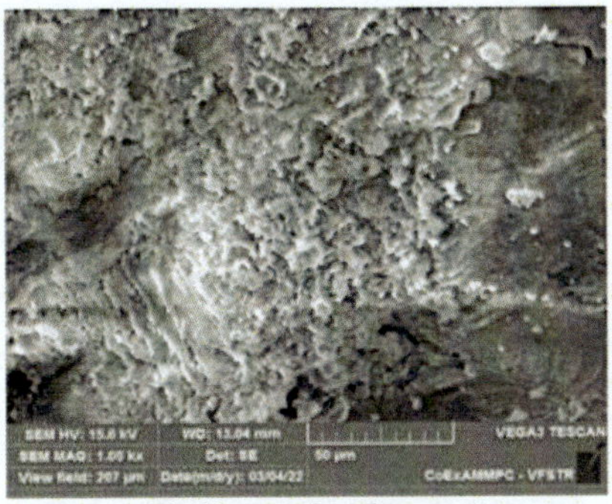

(a) Conventional concrete

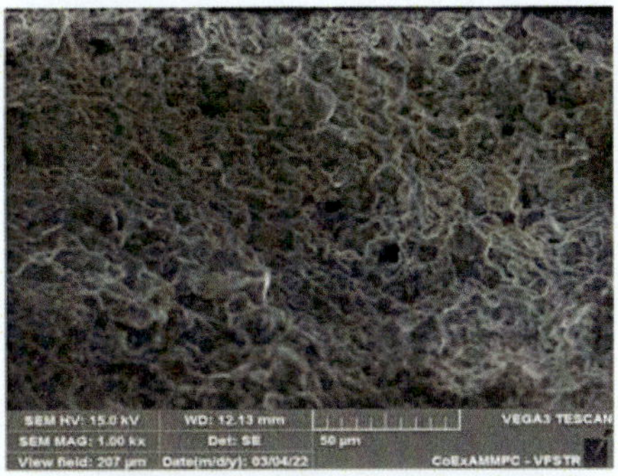

(b) Concrete with 10 % coconut shell replacement

Fig. 3 Microscope structure of biomass aggregate lightweight concrete in comparison to conventional concrete

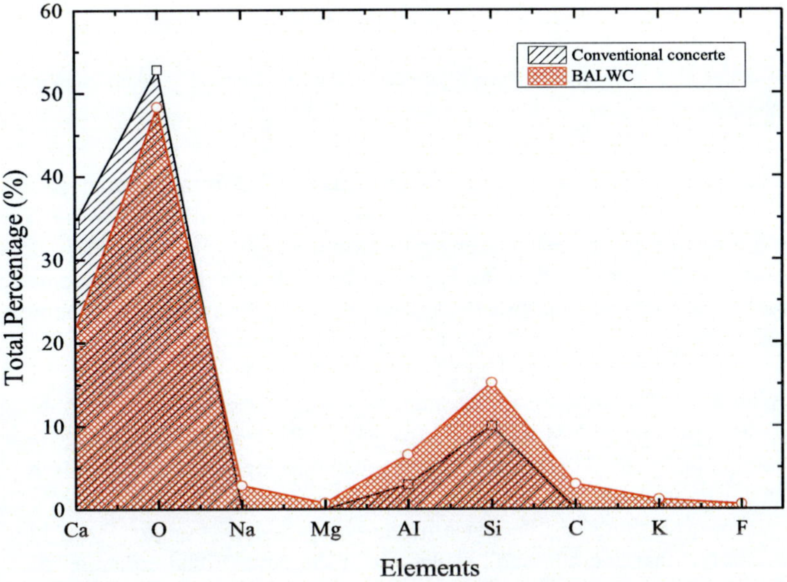

Fig. 4 Elemental composition of of biomass aggregate lightweight concrete in comparison to conventional concrete

3.4 *Strength Properties of Biomass Aggregate Lightweight Concrete in Comparison to Conventional Concrete*

Compressive strength plays an important role in deciding the performance of lightweight concrete. The previous researcher showed that low-density concrete

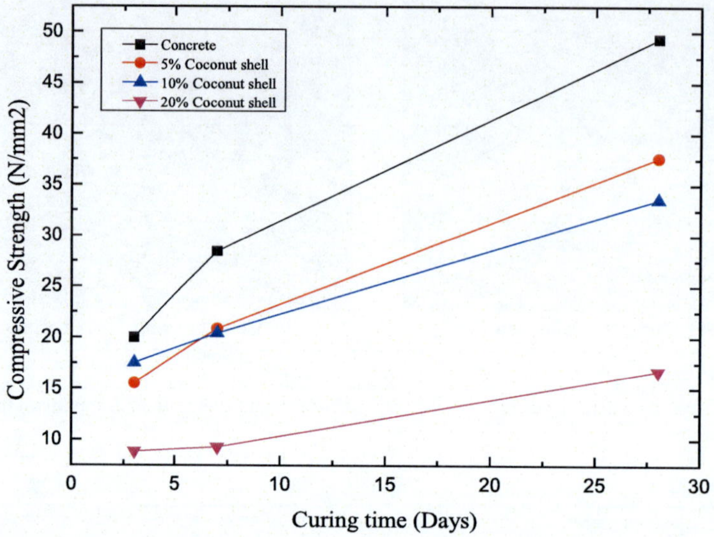

Fig. 5 Compressive strength of biomass aggregate lightweight concrete in comparison to conventional concrete

would have a compressive strength range of 0.7–2.0 N/mm2, moderate-strength concrete will have 7–14 N/mm2 and structural concrete will have 17–63 N/mm2. Figure 5 gives the compressive strength of biomass aggregate lightweight concrete compared to conventional concrete. Figure 5 shows that the BALWC prepared from coconut shells gives compressive strength between moderate-strength concrete and structural concrete.

Conventional aggregate concrete compressive strength for 28 days curing is 49.33 Mpa. With 5%,10%, and 20% replacement of coconut shells to conventional aggregate the strength was 37.6 Mpa, 33.56 Mpa and 16.78 MPa. The percentage decrease was found to be 23.78, 31.97 and 65.99, respectively for 5%,10% and 20% replacement of coconut shells. So here it is observed that with the increase of coconut shells percentage, there is a decrease in the compression strength. From Fig. 5, it can be inferred that coconut shell replacement in conventional concrete, by 5, 10, and 20%, has led to reduction of 32, 48, 79%, strength, respectively, compared to conventional concrete. Figure 6 shows the bond between concrete, coconut shell and mineral aggregates, it can be observed in the cubes that the coconut shell still has moisture absorption.

Like compressive strength, tensile strength plays an important role in deciding to employ concrete in the construction industry. Figure 7 provides the Tensile strength of biomass aggregate lightweight concrete in comparison to conventional concrete. It can be inferred from Fig. 7 that, 5% replacement of coconut shells in conventional concrete has not reduced any tensile strength. 10% replacement of coconut shell in conventional concrete has no reduction of tensile strength in early curing times, whereas curing time increases tensile strength reduces. In contrast, 20% replacement

Fig. 6 Coconut shell bonds with the cement and aggregates

has led to a reduction of tensile strength by 56% by the end of 28 days of curing time.

It can be inferred from all these properties that, 10% replacement is the optimum replacement of coconut shells in place of miner aggregates. Strength enhancement by steel fibers and PP fibers are tried for 10% coconut shells replaced in conventional concrete. From the literature it was found that, the optimum % of the addition was selected based on the literature, which is 0.2% PP and 1% steel fibers.

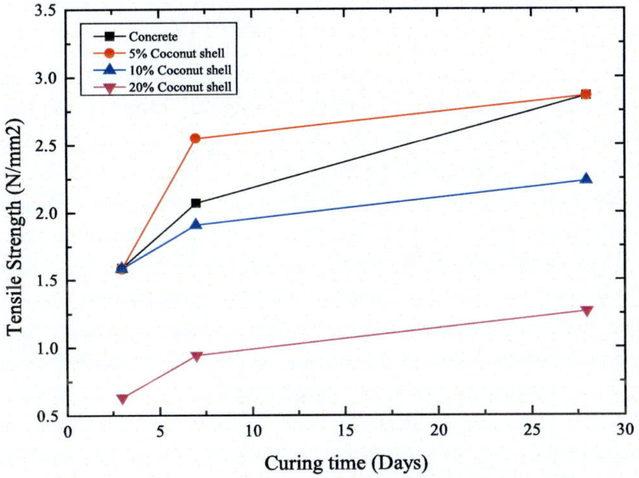

Fig. 7 Tensile strength of biomass aggregate lightweight concrete in comparison to conventional concrete

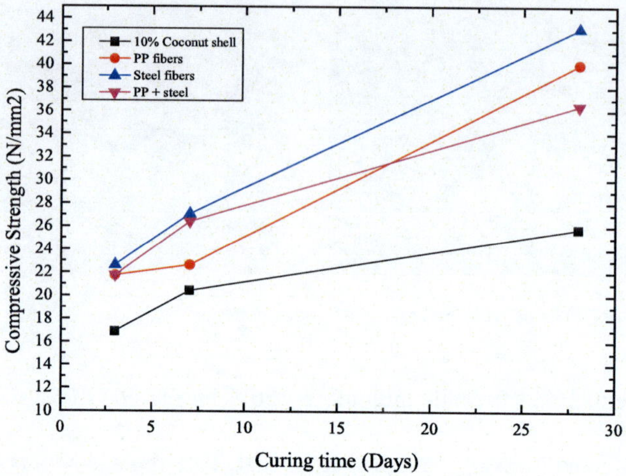

Fig. 8 Compressive strength enhancement by the addition of fibersfibers for coconut biomass aggregate concrete

3.5 Strength Enhancement of Coconut Aggregate Lightweight Concrete by using PP and Steel Fibers

The experimental analysis shows that 10% replacement is the optimum percentage of replacement of coconut shells. However, the overall strength decreased with CS replacement when compared to conventional concrete, in a similar manner, the split tensile strength also decreases with an increase in CS replacement. Steel fibers have been used to recover the tensile strength of plain cement concrete with coconut shells. The addition of PP fibers, steel fibers, and the combination has reduced compressive strength by 10, 10, and 22%, respectively as shown in Fig. 8. As the curing time increases, the compressive strength of the concrete has increased.

In Fig. 8, the compressive strength of PP fibers and steel fibers are high at 28 days curing. But the combination was increased in 7 days and lowered in 28 days. It may be because of both types of fibers were added may be associated to the consistency differences that the addition of fibers caused. Moreover, the random arrangement of fibers can cause the dissolution of research results. In contrast, tensile strength was decreased by PP fibersfibers by 14% and increased by 43 and 22 by steel and combination fibers as shown in Fig. 9. The Split tensile strength of PP fibers are high and steel fibers are high at 28 days curing. But the combination was increased in 7 days and lowered in 28 days. In concrete modified with steel and polypropylene fibers, the increase in the content of air-voids in the concrete mixes may be related to the decrease in the compressive/split strength of the concrete. For the polypropylene fibers, the addition of too many fibers to the mix causes a decrease in the split tensile strength of the concrete. As tensile strength depends on the shape, length and type

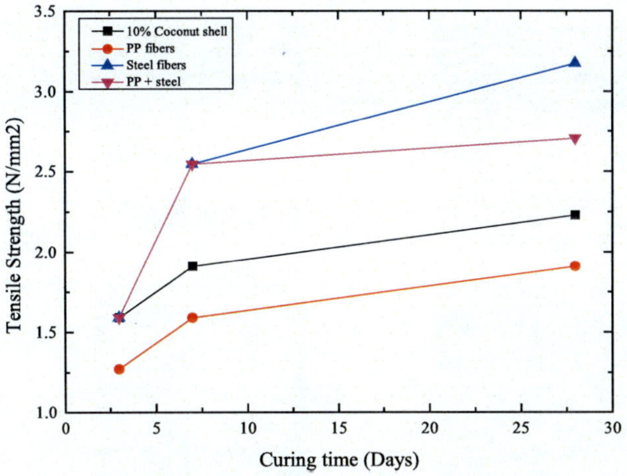

Fig. 9 Tensile strength enhancement by the addition of fibers for coconut biomass aggregate concrete

of fibers, and also on their interaction with the coarse aggregate and coconut shell aggregate matrix in the concrete.

Use of steel fibers have increased the toughness of the specimen there by subsiding the crack propagation from a micro to macro level as shown in Fig. 10. Use of steel fibers have increased the toughness of the specimen there by subsiding the crack propagation from a micro to macro level. Adding 1% steel fibers to the concrete has increased the impact resistance of the concrete specimen, making it more suitable for industrial flooring with heavy loads. In rural areas and other locations where coconuts are abundant, as well as in locations where the cost of using conventional aggregates is an issue, coconut shell concrete may be used. Light weight aggregate concrete can help in the development of a society with affordable homes and rural roads. A large quantity of coconut shell replacement can be evaluated from a single cube computation, which lowers the overall building cost.

Cube failure with 10% Coconut shell

Cube failure with 10% Coconut shell and 0.2% PP fiber

Cube failure with 10% Coconut shell and 1% Steel fiber

Cube failure with only 10% Coconut shell, 1% Steel fiber and 0.2% PP fiber

Fig. 10 Failure patterns observed in cube specimens with different combinations of PP fibers and steel fibers showing the changes in crack propagation

4 Conclusion

This study has conducted experimental investigations on the micro–macro-strength properties of biomass aggregate concrete developed by replacing mineral aggregates with coconut shells (5, 10, 20%). In addition, strength, and mechanical properties (cracking, impact, fatigue bending, and lightweight) enhancement is tried by using fibre reinforcement. The popular polypropylene (PP) fibers and steel fibers in an optimum ratio of 0.2% and 1%, respectively, are tested in this study. As the replacement percentage of coconut shell increases the unit weight also decreases. In addition, higher coconut shells replaced concrete, as the curing time increases unit weight increase, it is mainly because of water absorption properties of coconut shells. Does more research need to be conducted on the classification inside light weight concrete, as its properties differ based on type of LWCs, which can be used for different applications like floors in high rise buildings, frames for multistorey buildings, improve thermal performance, long-span bridges and shore structures. The BALWC prepared from coconut shells gives compressive forms between moderate-strength concrete and structural concrete. The 5% replacement of coconut shell in conventional concrete has not reduced any tensile strength. 10% replacement of coconut shell in conventional concrete has no reduction of tensile strength in early curing times, where as curing time increases, tensile strength reduces. Whereas 20% replacement has led to reduction of tensile strength by 56% by the end of 28 days curing time. It can be inferred from all these properties that, 10% replacement is optimum replacement of coconut shells in place of miner aggregates. The addition of PP fibers, steel fibers, and the combination has reduced compressive strength by 10, 10, and 22%, respectively. As the curing time increases, the compressive strength of the concrete has increases. In rural areas and other locations where coconuts are abundant and where the cost of using conventional aggregates is an issue, coconut shell concrete may be used. Light weight aggregate concrete can help in the development of a society with affordable homes and rural roads. Coconut shell biomass aggregate concrete reinforced with steel fibers can be a good solution for the strength reduction issues in lightweight concretes. However, more research needs to be conducted on the classification in lightweight concrete, as its properties differ based on type of LWCs, which can be used for different applications like floors in high rise buildings, frames for multistorey buildings, improve thermal performance, long-span bridges and shore structures.

References

1. Gerritse, A. (1981). Design consideration for reinforced lightweight concrete. *Cem. Comp. Lightweight Conc., 3*(1), 57–69.
2. T.P. Dolen, Performance of fly ash in roller compacted concrete at upper stillwater dam, *Proceedings: Eighth International Ash Utilization Symposium* (1987)
3. Olanipekun, E. A., Olusola, K. O., & Ata, O. (2006). A comparative study of concrete properties using coconut shell and palm kernel shell as coarse aggregates. *Build. Environ., 41*(3), 297–301. https://doi.org/10.1016/j.buildenv.2005.01.029
4. Duzgun, O. A., Gul, R., & Aydin, A. C. (2005). Effect of steel fibers on the mechanical properties of natural lightweight aggregate concrete. *Mater. Lett., 59*, 3357–3363.
5. Singh, N. K., & Rai, B. (2018). A review of fiber synergy in hybrid fiber reinforced concrete. *J. Appl. Eng. Sci., 8*(2), 41–50. https://doi.org/10.2478/jaes-2018-0017
6. Z. Hongbo, Z. Haiyun, G. Hongxiang Characteristics of ductility enhancement of concrete by a macro polypropylene fiber Results Mater. 100087 (2020). https://doi.org/10.1016/j.rinma.2020.100087
7. BIS 383: 1970 Coarse and fine aggregate from natural sources for concrete 9th Revision, BIS, New Delhi (1970)
8. Li, J. J., Niu, J. G., Wan, C. J., Jin, B., & Yin, Y. L. (2016). Investigation on mechanical properties and microstructure of high performance polypropylene fiber reinforced lightweight aggregate concrete. *Construct. Build. Mater., 118*, 27–35.
9. Filipponi, P., Polettini, A., Pomi, R., & Sirini, P. (2003). Physical and mechanical properties of cement based products containing incineration bottom ash. *Waste Manage., 23*(2), 145–156.
10. Dhir, R. K., Paine, K. A., Dyer, T. D., & Tang, M. C. (2004). Value-added recycling of domestic, industrial and construction arisings as concrete aggregate. *Concr. Eng. Int., 8*(1), 43–48.
11. Poon, C. S., Shui, Z. H., Lam, L., Fok, H., & Kou, S. C. (2004). Influence of moisture states of natural and recycled aggregates on the slump and compressive strength of concrete. *Cem. Concr. Res., 34*(1), 31–36.
12. Khatib, Z. M. (2005). Properties of concrete incorporating fine recycled aggregate. *Cem. Concr. Res., 35*(4), 763–769.
13. Andrade, L. B., Rocha, J. C., & Cheriaf, M. (2007). Evaluation of concrete incorporating bottom ash as a natural aggregates replacement. *Waste Manage., 27*(9), 1190–1199.
14. Jones, R., Zheng, L., Yerramala, A., & Rao, K. S. (2012). Use of recycled and secondary aggregates in foamed concretes. *Mag. Concr. Res., 64*(6), 513–525. https://doi.org/10.1680/macr.11.00026

15. Aslani, F., & Nejadi, S. (2013). Self-compacting concrete incorporating steel and polypropylene fibers: compressive and tensile strengths, moduli of elasticity and rupture, compressive stress-strain curve, and energy dissipated under compressiondi. *Compos. Part B Eng., 53*, 121–133.
16. Shafigh, P., Mahmud, H., & Jumaat, M. Z. (2011). Effect of steel fiber on the mechanical properties of oil palm shell lightweight concrete. *Mater. Des., 32*, 3926–3932.
17. Gao, J., Suqa, W., & Morino, K. (1997). Mechanical properties of steel fiber-reinforced, high-strength, lightweight concrete. *Cem. Concr. Compos., 19*, 307–313.
18. Chandana, N., & Rao, B. (2021). Status of sustainable sanitation chain in rural, semi-urban, and urban regions: a case study of maharashtra, India. *J. Water Sanit. Hyg. Dev., 11*, 112–125. https://doi.org/10.2166/washdev.2020.020
19. Chandana, N., Goli, V. S. N. S., Mohammad, A., & Singh, D. N. (2021). Characterization and utilization of landfill-mined-soil-like-fractions (LFMSF) for sustainable development: a critical appraisal. *Waste Biomass Valoriz., 12*, 641–662. https://doi.org/10.1007/s12649-020-01052-y

Compatibility Study on Combinations of Redmud and Flyash Based Geopolymer Bricks

T. Sowmyashree, S. Muralidhara, Ahamed Sharif, and Vali Mohammed

Abstract The construction sector now has a new way to utilise industrial wastes and byproducts thanks to geopolymerization technology. In addition to using leftover mineral ingredients, the method emits less greenhouse gas than OPC by a smaller percentage. Thus, "green cement" is another name for the cementitious byproduct of geopolymerization. An estimated 290 million tonnes of this inorganic industrial waste are produced in India each year. A significant global concern is how these non-hazardous industrial wastes are treated, handled, transported, and disposed of. A severe environmental concern is posed by their rising.

In current study the properties of water absorption and compressive strength of Geo-polymer bricks with Red mud and fly ash as ingredients in different proportions is studied. These raw materials were activated with chemical activators with NaOH of 8Molarity and NaOH to Na_2Sio_3 ratio of 2.5. The results suggest that maximum CS of 12 N/mm^2 is achieved with 70% fly ash and 30% Red mud and found to decrease with in percentage of Red mud. Water absorption showed decreasing trend with decrease in percentage of Red mud. Water absorption of 9.78% is recorded for 80% fly ash and 20% Red mud and the result satisfies the requirement of less than 20% of water absorption as given in IS 3495–1992.

Red mud, Fly ash, GGBS and Silica fume is used as a binder material. Chemical activator was tried for lower molarity of 2. With red mud content is from 25 to 90% and alkaline activator content, compressive strength obtains in range of 0 to 8.6 N/mm^2, water absorption in the range 5.3% to 18%. It can be concluded that geo polymerization of red mud, fly ash, GGBS and silica fume can be used as sustainable alternative material for conventional bricks. Selected combinations of mixes were later used to prepare bricks of size 225 × 100 x 70mmThe studies have

T. Sowmyashree (✉) · A. Sharif · V. Mohammed
Presidency University, Bengaluru 560064, India
e-mail: sowmyashreet@bmsce.ac.in

A. Sharif
e-mail: ahamedsharif@presidencyuniversity.in

S. Muralidhara
BMS College of Engineering, Bengaluru 560019, India
e-mail: murali.civ@bmsce.ac.in

A. Vilventhan et al. (eds.), *Advances in Construction Materials and Management*, Lecture Notes in Civil Engineering 346,
https://doi.org/10.1007/978-981-99-2552-0_36

shown that 9.8Mpa compressive strength bricks and cubes made of geopolymer are feasible. Geo polymerization of red mud, GGBS, fly ash and Silica fume can be used as sustainable substitute material for conventional bricks. Although, the strength gained is significant and with acceptable water absorption percentage, in rural locations, bricks made from red mud and fly ash may be utilised as sustainable building materials. Hence the problem of redmud waste disposal for aluminium industries also can be resolved without the soil and water contamination which is responsible for environmental pollution. As these raw material are industrial by-products and the waste materials, the cost of production of bricks worked out to be lesser compared to conventional materials.

1 Introduction

As a result of the overpopulation, there is a demand for basic necessities, which leads to construction activity and the production of industrial waste. The use of materials for a housing facility is part of the construction process [1]. To meet the demand of Construction materials there is need to find alternative building materials. Consequently, in this study, an effort is undertaken to test the different industrial wastes materials to produce masonry bricks mainly redmud, flyash. That reduces the use of high energy material usage and also can effectively handle the global waste disposal issues being an efficient method to use waste by-products and to be an eco-friendly idea.

Burnt clay bricks are used extensively as construction material. The use of industrial waste products can be the alternative for construction. The unrestricted use of clay is hazardous to society because all of India's brick kilns rely on good grade clay [3]. Making bricks is both environmentally and economically beneficial, as it saves valuable top agricultural soil while also meeting the social goal of disposing of industrial waste. The ever-increasing amount of fly ash produced around the world has not been nearly matched by its use. However, it is unjustifiable for the brick business to adopt a similarly cautious stance [6].

In some places of the world, where coal is the primary source of energy and bricks are the primary building material, environmental concerns have been raised. As a result of these concerns, legislation has been enacted requiring the brick industry to integrate at least 25% by weight of fly ash and/or bottom or pond ash in the brick producing mixture if the industry is located within 50 kms of a coal power station. Fly ash is the by-product of powdered coal burning. There have been some successful initiatives where fly ash was included into the mixture [9]. Even when the law was followed, there is scant indication that the amount of fly ash in the brick mixture exceeded 30 percent by volume.

The study has got the main problem context towards sustainability to use industrial wastes without any recycling process to prevent hazardous impacts on environment. That indirectly saves time and reduces, or eliminates pollution.

2 Objectives

The study is anticipated to establish a solid foundation for the use of industrial wastes, notably red mud, in the production of geopolymer building products and to open the door to a broad range of prospective uses for red mud geopolymer composites in the construction market. The main objectives of this study are listed below.

- To explore possibility of using red mud with significant of mineral admixtures as ingredient in geo-polymer composites.
- To check compatibility of combination of red mud, flyash, silica fume, GGBS and M-sand in the production of geo-polymer masonry bricks.
- To produce geo-polymer masonry bricks with different combinations of raw material with addition of chemical activator.
- To find optimum percentage of combinations of ingredients to obtain quality geo-polymer masonry bricks on the bases of their mechanical property such of compressive strength with water absorption.

3 Methodology

To achieve the desired qualities, namely strength, binders are produced in a precise percentage. To combine the right binders and obtain the best value of characteristics, experiments must be carefully planned. The same experiment is conducted in such a way that a small number of experiments can get the desired results. The following steps are followed to accomplish the study for the preparation of Geo-polymer Mix.

1. Procuring raw materials such as Red mud, GGBS, NaOH, Na_2SiO_3 and Water.
2. Characterization of raw materials mainly redmud, flyash.
3. Determining the mix quantity for all combinations of Red mud and GGBS to produce geo-polymer mix for bricks production.
4. Producing geo-polymer cubes using different proportion of Red mud, M sand, Metakaolin and GGBS using NaOH and Na_2SiO_3 as alkaline activators.
5. Determination of strength properties of produced bricks.

3.1 Materials Used

Material characterisation of red mud revealed that it had silica mostly in the crystalline form, and it possessed negligible pozzolanic behaviour as indicated by lime reactivity test.

Nevertheless, its alkalinity and fineness were favorable for geopolymerisation. The role of the predominant phase, iron oxide had not been clearly understood. An attempt was made in the subsequent sections to explore the possibility of using red mud as base material for geopolymerisation. Fly ash, microsilica and GGBS

Table 1 Geotechnical properties of Red mud

Sl No	Particular of Test	Values
1	Specific Gravity	2.80
2	OMC	15.10%
3	Maximum Dry Density	2.50 g/cm^3

Table 2 Geotechnical properties of FA

Sl No	Particular of Test	Values
1	Specific Gravity	2.2
2	Colour	Light Grey
3	Liquid Limit	31.10%
4	Density	1.50 g/cm^3

have been blended along with red mud to explore the possibility of enhanced polymerization.

There are different types of materials used in preparation of geo-polymer brick as follows.

- Red Mud
- Fly Ash
- Silica Fume
- GGBS
- M Sand, Water
- Chemical Activators
- Sodium Hydroxide
- Sodium Silicate.

3.2 *Material Characterisation*

REDMUD: The bauxite residue by-product of the Bayer process. It is collected from Hindalco Industries Pvt Limited, Belagavi: fine red mud which is passing through 425 micron (Table 1).

FLYASH: It is obtained from the Raichur Thermal Power Plant in India's Karnataka state. The material property is as follows (Table 2).

3.3 *Alkaline Solution*

The sodium hydroxide (1) to sodium silicate (2.5) ratio is used. Sodium hydroxide solution with a concentration of 8 molar is used to study the workability of mix and to produce the bricks.

Fig. 1 Geopolymer brick specimens

Fig. 2 Produced geopolymer brick specimens kept for drying

4 Production of Bricks

Distilled water and Sodium Hydroxide is mixed first one day prior. Then sodium silicate solution was added before adding it to constituents produce mix. The solid constituents are mixed dry for 5 min. Chemical activator is then added, and the mixture was stirred for another four minutes. Mixed composites are directly poured into a 190 × 90 × 90 mm brick mould by blowing 25 times in three layers. These samples are kept for drying for one day and remolded after that (Fig. 1).

The moulded after 24 h and cured for seven, fourteen, and twenty-eight days. Samples were placed in water for 24 h after the cure phase., Water absorption and wet compressive strength were tested on the specimens (Fig. 2).

5 Results and Discussion

This section clearly presents all the obtained results of different test conducted and redmud silt geopolymer bricks for the physical and mechanical properties checked after ambient curing, then submerging the specimens in water for 24 h to find the properties for its wet condition of saturation. As per following standard procedure, the tests were conducted:

- Tested according to IS: 3495 (Part 2)-1992 for water absorption.
- Compressive strength was evaluated in accordance with IS: 3495 (Part 1)-1992.

Experiments are done on many combinations composition prior to which the flyash alone was combined to partially replace the redmud content. Figure 3 and Fig. 4 shows the test results of redmud and flyash combination. Further Table 3 presents the results of all combined admixtures with redmud.

All results so obtained were greater than 3.5 MPa which satisfied code provisions IS 3495:1992 Part 2 (Figs. 5 and 6).

Compatibility of the combination checked by mechanical properties of geopolymer mix.

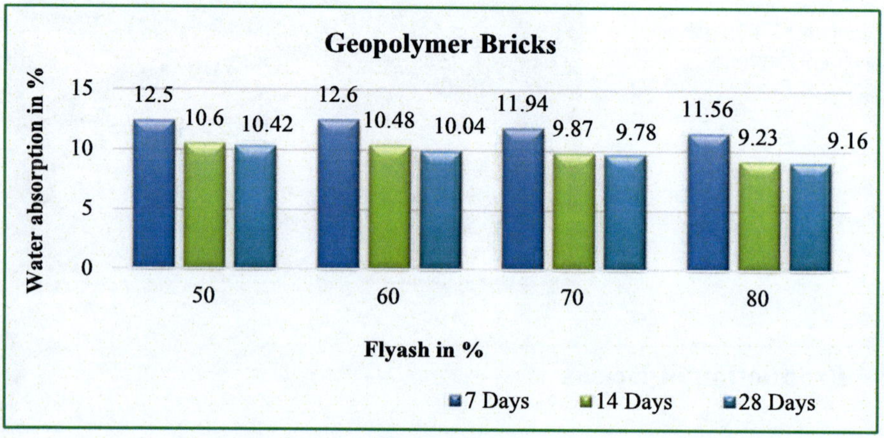

Fig. 3 Water absorption test results of produced bricks

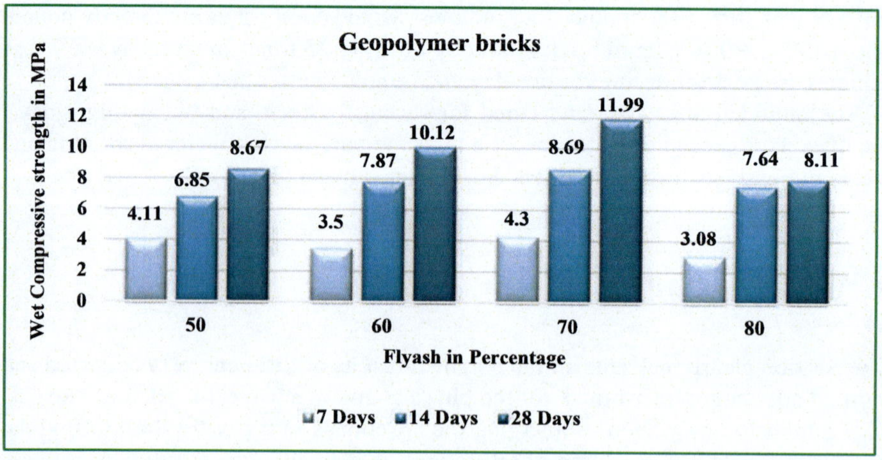

Fig. 4 Wet compressive strength test results of produced bricks

Table 3 Test results for different material composites to produce geopolymer bricks

Sl No	Composition of Geopolymer cubes	Material proportion (%)					Alkaline solution in %	Water absorption in %(7 days)	Compressive strength in Mpa (7 days)
		Red mud	**Fly ash**	**Silica fume**	**GGBS**	**M sand**			
1	Mix1	66	30	4	-	-	32	15.4	5.2
2	Mix 2	78	20	2	-	-	33	16.4	4.3
3	Mix 3	45	20	35			30	3.1	0
4	Mix 4	25	25	-	-	50	26	12.9	3.5
5	Mix 5	30	20	-	-	50	24	12.5	4.1
6	Mix 6	68	24	8	-	-	34	16.8	2.8
7	Mix 7	75	18	7	-	-	36	18.0	1.4
8	Mix 8	90	5	5	0	-	34	17.1	3.2
9	Mix 9	85	5	5	5	-	35	14.6	4.1
10	Mix 10	80	5	5	10	-	34	11.2	4.5
11	Mix 11	75	5	5	15	-	33	8.4	5.8
12	Mix 12	70	5	5	20	-	32	5.3	8.6

(a)

(b)

Fig. 5 **a** Flow test for red mud paste to find required alkaline solution. **b** Flow test for red mud paste to find required alkaline solution

Fig. 6 Produced brick specimen with brick failure pattern under compressive load

To prove the highest redmud replacement in masonry units production Mix 12 was selected as optimum with higher strength and geopolymer bricks were produced and tested for physical and mechanical properties (Tables 4 and 5).

Table 4 Mechanical properties for Mix12

Trial Mix	Average Initial weight (g)	Average Final weight (g)	Water absorption in %	Average compressive strength in Mpa (7 days)
Mix 12	2966.66	3093.33	4.3	9.8

Table 5 Physical Test results of produced brick specimens

Test on bricks	Result
Hardness test	Hard brick
Shape and size	Qualified
Colour test	Bright and uniform colour
Soundness test	Passed
Structure of brick	Free from lumps and homogeneous
Efflorescence test	No efflorescence
Impact test	Impact value is high

6 Discussion

All specimens with red mud percentage exhibited excellent strength properties as compared to the one having only fly ash and GGBS. The paste incorporating 30% red mud possessed maximum strength of 35.8 MPa. Moreover, all variations of red mud demonstrated better strength than GRM0, where only fly ash was used with GGBS. Presence of GGBS aided in the setting of red mud, which was otherwise difficult to set at ambient temperature.

It is important to note that the presence of red mud impacted the strength for all variations. This was rather unprecedented because it was the fly ash that contained much greater percentage of reactive silica than red mud. Hence, Given that fly ash had a higher alumina percentage than red mud, it was possible to draw the conclusion that iron in red mud was what made the difference. The investigation of red silt geopolymeric determines its potential application in the building sector. A crucial component of building constructions are bricks. The experiment to make bricks by red mud and examine their mechanical qualities was motivated by the short supply for masonry units.

Brick strength is influenced by the raw ingredients and manufacturing process. It is not necessary for bricks to have extremely high compressive strength. According to IS 1077 (1992), the specified minimum strength is 3.5 N/mm2.

7 Concluding Remarks

The following is a list of results reached as a result of the research on geopolymer composites for civil engineering applications. From the experimental results, the conclusion reached on the basis of evidence and reasoning is as follows,

- Its alkalinity and fineness, however, were advantageous for geopolymerization. Iron oxide, the dominating phase, had not been given a clear understanding of its function. In the sections that followed, an effort was made to investigate if red mud might be used as a basic material for geopolymerization. Red mud has been combined with fly ash, microsilica, and GGBS to investigate the possibility of improved polymerization.
- Geo polymerization of red mud and FLYASH can be used as sustainable alternative material for conventional bricks.
- The use of red mud and Flyash based on geo polymer technology to play a major role in the mechanical qualities of brick, with red mud content ranging from 20 to 60% and alkaline activator, compressive strength is achieved in the range of 3.08 to 11.99 N/mm^2, water absorption gets 9.16 to 12.5%.
- The greatest compressive strength was achieved in this study was at 70% Fly ash and 30% Red mud i.e. 11.99 N/mm2. Percentage of water absorbed is lower for mix with 70% Fly ash and 30% Red mud i.e. 9.01%. Use of Flyash and Red mud in bricks can solve the waste disposal problem.
- The analysis showed that it was feasible to manufacture a high-quality, first-class red mud-based brick that complied with the Indian standard code's standards. For 2molarity, with 70% redmud the highest strength was 9.8 MPa. When another industrial waste, GGBS, was used to upto 20% of the sand, the strength raised.
- Although, the strength gained is significant and with acceptable water absorption percentage, in rural locations, bricks made from red mud and fly ash may be utilised as sustainable building materials. Hence the problem of redmud waste disposal for aluminium industries also can be resolved without the soil and water contamination which is responsible for environmental pollution. As these raw materials are industrial by-products and the waste materials, the cost of production of bricks worked out to be lesser compared to conventional materials.

References

1. Davidovits, J.: (1999), Chemistry of Geopolymeric Systems, Terminology, Geopolymers 99 International Conference, France
2. Palomo, A., Grutzeck, M. W., & Blanco, M. T. (1999). Alkali-activated fly ashes. *Cement and Concrete Research, 29*(8), 1323–1329. https://doi.org/10.1016/S0008-8846(98)00243-9
3. FernándezJiménez, A., & Palomo, A. (2003). Characterisation of fly ashes. Potential reactivity as alkaline cements. *Fuel, 82*(18), 2259–2265. https://doi.org/10.1016/S0016-2361(03)00194-7

4. Hardjito, & Rangan B.V, (2005) Development and properties of Low Calcium Fly ash based Geopolymer Concrete, Reseach Report GCI, Faculty of Engineering, Curtin University of Technology, Perth
5. Radhakrishna, S., & Udayashankar, B. C. (2008). Phenomenological model to re-proportion geopolymer compressed blocks. *Concrete Technology Today, 7*(3), 46–48.
6. Rangan, V. (2008). Low-calcium, fly-ash-based geopolymer concrete. In E. Nawy (Ed.), *Concrete Construction Engineering Handbook.* CRC Press. https://doi.org/10.1201/9781420007657.ch26
7. Akhtar, J. N., Alam, J., & Akhtar, M. N. (2010). An experimental study on fibre reinforced fly ash based lime bricks. *International Journal of thePhysical Sciences, 5*(11), 1688–1695.
8. Ananya, A., Avarebeel, P. B., & Sambasivam, M. C. (2015). Experimental study on red mud based geo-polymer bricks. *IJAREST., 2*(8), 76–83.
9. Banupriya, C., SharonJohn, R., Suresh, E., & Divya, D. V. (2016). Experimental Investigations on Geopolymer Bricks/Paver Blocks. *Indian Journal of Science and Technology, 9*(16), 1–5. https://doi.org/10.17485/ijst/2016/v9i16/92209
10. Subramani, T. & Sakthivel, P. (2016) experimental investigations on fly ash based geopolymer bricks. *IJAIEM*
11. Smitasingh, R. B., & Aswath, M. U. (2016) Experimental study on Red mud based geo-polymer concrete with fly ash & GGBS in ambient temperature curing. *Indian Journal of advances in Mechanical and Civil Engineering*
12. Sowmyashree, T., Arpitha, D. J., Manjunath, C. S., & SidduKarthik, C. S. (2016). Study on mechanical properties of red mud as a partial replacement of cement with hydrated lime for M40 grade concrete with superplasticizer. *International Journal of Research in Engineering and Technology, 5*(4), 2319–1163.
13. Sowmyashree, T., Manjunath, C. S., Arpitha, D. J., & Panditharadhya, B. J. (2016). Comparison on durability properties of red mud as a partial replacement of cement with hydrated lime for different grades of concrete with and without superplasticiser. *Int J Eng Res Technol, 5*(5), 572–577.
14. Waghmareonkar, D., & Hake, S. L. (2018). Evaluate properties of Fly ash based Geopolymer Bricks. *IRJET., 5*(6), 1–9.
15. Pavithra, N., et al. (2018). Experimental study on the behaviour of geopolymer bricks. *International Journal of Civil Engineering and Technology (IJCIET)., 9*(2), 87–95.
16. IS:2720–1983: Methods of tests for Soils-Specification, Bureau of Indian Standards, New Delhi
17. IS: 3495 (Part 1 and 2)-1992: Methods of tests of Burnt Clay Building BricksSpecification, Bureau of Indian Standards, New Delhi

High Performance FRC Tunnel Lining with Alkali Activator and Cementitious Materials

R. B. Shivali, Rajan L. Wankhade, and Dulal Goldar

Abstract Nowadays major tunneling projects focuses on low-carbon concrete as a waste by-product and co-product materials from Industry. Also, mineral resources called cementitious materials like Fly Ash, slag, micro silica etc. are useful in partial replacement of cement. Alkali activator can be used to fully replace the cement. For the tunneling projects, the biggest leap forward would be to use high-performance FRC tunnel lining with alkali activator and cementitious materials. As concrete dominates the carbon footprint of segmentally lined/sprayed concrete tunnels replacing standard concrete with an alkali activator would be of great use. Such FRC concrete is capable of having better strength properties. The variety of cementitious waste products available is Fly ash (FA), Micro silica/Silica Fume (SF), Ground Granulated Blast Furnace Slag (GGBS) and Rice Husk Ash (RHA). These are high in alumino-silicates and therefore must be activated by the addition of an alkali activator. In this work efforts are made to consider cementitious materials to be mixed to achieve high-performance FRC for tunnel lining projects.

1 Introduction

Many industrial waste products results in emission of greenhouse gases which includes carbon dioxide (CO_2), into the atmosphere. Out of these hazardous gases, CO_2 resembles for 65% of global warming. For some amount of CO_2 emission the cement industry is also responsible. It can be stated that the manufacture of one ton of Portland cement is accountable for the emission of nearly one ton of CO_2 to the atmosphere.

As concrete is the main material used for construction, the requirement of Portland cement is increasing enormously. It was assessed that the requirement of cement is going to increase day by day due to meet heavy demand. Hence, several efforts are being made to replace the cement by cementitious materials and alkali activator.

R. B. Shivali (✉) · R. L. Wankhade · D. Goldar
Civil Engineering Department, Netaji Subhas University of Technology, New Delhi, India
e-mail: rbshivali1@gmail.com

A. Vilventhan et al. (eds.), *Advances in Construction Materials and Management*, Lecture Notes in Civil Engineering 346,
https://doi.org/10.1007/978-981-99-2552-0_37

Removing and replacing cement will reduce approximately 8% of the world's carbon emissions. Stress and strain are improved with the use of steel FRC under compression [1]. Steel fiber can be used to achieve high shear strength in such FRC without stirrups [2]. High fiber volume fractions are required to be considered for achieving such high performance concrete [3]. Polypropylene fiber plays vital role to resists spalling of concrete, shrinkage and also provides lateral confinement [4–6]. Fatigue behavior of polypropylene fiber can be seen as modified than that of convensional concrete [7]. For this creep and shrinkages of concrete with and without fibers are need to be checked properly [8]. Fly and silica mixed with polypropylene fiber forms a composite for better durability [9]. Non-destructive tests are also required to be performed to verify the strength characteristics [10]. Mechanical and durability characteristics of concrete are shown to be enhanced by use of steel and polypropylene fibers to form a HPFRC [11–13]. Experimental investigation is done to verify the effect of addition of modified polymer and steel fibers on properties of concrete [14]. Nano and micro-silica affects the bond strength of steel and polypropylene fibers mixed in high volume fraction [15]. Beam specimens are experimentally tested to examine the strength characteristics of RCC with polypropylene fibers [16]. Impact and blast loads are also given to show mechanical and structural characteristics of HPFRC [17].

Ternary blended concrete can mitigate thermal cracking in massive concrete structures [18]. Fiber volume and hybridization impacts the strength characteristics of high performance concrete [19]. For a low reinforcement ratio also steel fiber mixed in concrete considerably improves the flexural strength [20]. Hybrid concrete with steel fibers, SBR and polypropylene fibers shows improvement in strength of concrete [21]. Hence in the present work, cementitios materials are used to predict the strength of high performance concrete used for tunnel lining projects.

2 Benefits of High-Performance FRC for Tunnel Lining

The study aims on the investigation of the effects of high performance fiber reinforced concrete used for tunnel lining. Following are the benefits of use of high performance FRC.

a) Optimal reinforcement (multidirectional & protects against spalling and impact forces)
b) Safe-(safer) and reliable over entire service life & traditional reinforcement are avoided.
c) Easy storage and reduced carbon footprint-(packaging, less space, store and transport).
d) Easy, time-saving, cost-effective-(Eliminating the time and labour needed to

transport, install & check, storing, handling and mounting of traditional steel rebar reinforcement.

e) Eco-friendly

This represents an effective solution where the following advantages can be obtained with the use of high performance concrete:

- Fibers (steel & polypropylene) reduces the steel bar congestion which makes it easy to place the concrete.
- It reduces the time of construction as because of absence of steel reinforcements
- Safer working environment is provided for the operatives and hence operational risks are diminished.
- Shear strength is also increased in longitudinal cross-section
- Material ductility is achieved more in tension as well as in compression.
- Control in crack propagation and water tightness is achieved

The use of high strength concrete can effectively increases durability and steel corrosion and damages in concrete are greatly reduced.

3 Objectives of Using High-Performance FRC for Tunnel Lining

The main objective of this study is the replacement of cement wan ith alkali activator and cementitious materials (indigenously available) like fly ash, GGBS, rice husk ash, silica fume/microsilica and fibers ((steel & polypropylene) reinforcement on an experimental basis with admixtures/superplasticizer for the use High-Performance FRC Tunnel Lining with alkali activator and cementitious materials in field of constructions of Highway/Expressway/Metro rail/Railway tunnels structures projects for Green Industrial revolution (low carbon technology).

This study focuses on the use of combined steel fibers and polypropylene fibers with alkali activator {sodium hydroxide (NaOH) solution and sodium silicate} and cementitious materials like fly ash and GGBS with as zero cement content for tunnel lining for study the flexural behaviour of high-performance concrete beams. Different percentages of these fibers (steel & polypropylene) reinforcement with different combinations and proportions of cementitious materials are considered to examine experimentally the strength of concrete to be used for tunnel lining. The main objective is at first to verify whether the addition of fibers conveys the excellence in compression and flexure strength with cementitious materials (Fly ash, GGBS, Silicafume & RHA) with alkali activator. The second objective is to take the suitable percentage of cementitious materials GGBS, Fly ash, Slicafume/RHA with alkali activator. The last objective of the present study aims at the mix of steel fiberss (60%) with synthetic polypropylene fibers (40%) in different percentages. The Forth objective to take cementitious materials (Fly ash, GGBS, Silica fume & RHA) with

Alkali Activator in the combination of steel fibers and Polypropylen fibers. Preparation of laboratory specimens is done for different percentages and proportions. The experimentation procedure comprises cube, cylinder,r and beam specimens to be fabricated with GGBS (75%), Fly ash (25%), Silicafume (5–10%), Steel fibers (60%) & Polypropylene fibers (40%). Further with Steel fiber in varying percentages and 0.15% to 0.25% polypropylene is added to get a desired grade of mix for better results under flexure.

Hence in the present study we considered different variation of these ingredients while concentrating the workability. To achieve the above aim, the main objectives of the present study are given as follows:

(1) To examine the influence of alkali activator with by-product cementitious materials along with Slica fume/RHA and Steel-poly-propylene fibers, on workability characteristics, compressive strength characteristics, flexural strength behaviour of concrete mixes.
(2) To examine the effect of alkali activator with by-product cementitious materials along with Slicafume/RHA and Steel-polypropylene fibers on the ability toughness characteristics to absorb energy after cracking under the stress–strain relationship. Toughness also increases with the combination of fibers.
(3) To assess the effect of alkali activator with by-product cementitious materials along with Slica fume/RHA and Steel-polypropylene fibers, on concrete cubes and beams using nondestructive techniques (NDT).
(4) To assess the effect of alkali activators with cementitious materials along with Slica fume/RHA and steel-polypropylene fibers, on the durability characteristics of concrete mixes.

4 Results and Discussions

Various components show changes in their constituent's properties for chemical reactions. Hence the comparison of OPC, Fly-ash, GG, BS, RHA, and micro silica is shown in Table 1. Table 2 shows the physical as well as chemical properties of Cement, Fly ash, GBFS, RHA and Micro silica.

When cement is mixed with lime and other binder materials, it shows pazzolonic reaction. Figure 1 shows pozzolanic reaction between different ingredients. Figure 2 explains mix design for conventional concrete as well as for high performance concrete (M50 Grade).

Conventionally, tunnels are constructed with initially sprayed concrete with a temporary sprayed concrete lining and further a permanent concrete lining is installed for durability and long term. Now days, it is focused on the use of low-carbon concrete. Alkali Activator as an alternative is used to make zero-cement concrete. Alkali activator activates inactive waste products like fly ash, GGBS, rice husk ash, micro silica etc. which are high in alumino-silicates to form an alumino-silicate gel.

Table 1. Comparison of OPC, Fly ash, GGBS, RHA and Micro silica

	OPC	Fly ash	GGBS	RHA	Micro silica
SiO_2	20	50	38	22	89
Fe_2O_3	3.5	10.4	0.3		0.8
Al_2o_3	5	28	11	5	0.3
CaO	65	3	40	0.3	0.4
MgO	0.1	2	7.5		0.3
Na_2O	0.1	0.7	0.4		0.2
K_2O	0.7	2.5	0.8		0.9
Fineness CM^2/Kg test method	300–400 Blaine	400–700 Blaine	350–600 Blaine	15,000–20,000 Blaine	15,000–25,000 BET

Table 2. The physical and chemical properties of Cement, Fly ash, GBFS, RHA, and Micro silica

Description of Test	Ordinary Portland Cement 43 Gr.	Fly-ash	GBFS	RHA	Micro silica
Physical Test:					
Fineness					
- Color	Grey	Grey	Grayish white	Grey	Greyish white
- Standard consistency, %	29	-	-		-
- Fineness, m^2/kg	350.0	-	-	-	-
- Soundness, mm	2.3	-	-	-	-
- Specific Gravity	3.15	2.27	2.08	-	2.21
- Setting time, minutes				2.3	
a. Initial	136	125	-		53
b. Final	215	296	-		101
- 45 m□(passing),%	88.1	-	86.3		97.0
- Surface area by Blaine's, m^2/kg	355	358	451		-
- Surface area by Nitrogen adsorption, m^2/kg	-	-	446		20000
Compressive Strength					
3 days	25.6 Mpa	-	-	-	-
7 days	35.2 Mpa	-	-	-	-
28 days	48.6 Mpa	-	-	-	-
Chemical properties					
- Insoluble residue	0.48 %	-	-	-	-
- Silicon dioxide	21.30 %	54.0	38.5 %	91 %	89.0 %
- Aluminum oxide	5.8 %	23.2	7.0 %		0.06 %
- Ferric oxide	4.5 %	4.0	0.9 %		0.03 %
- Calcium oxide	60.5 %	2.1	35.8 %		0.5 %
- Magnesium oxide	2.4 %	0.81	11.8 %		1.1 %
- Sulphur trioxide	2.0 %	0.31	3.2 %		0.5 %
- Sodium oxide	0.11 %	0.60	0.02 %		0.02 %
- Potassium oxide	0.72 %	-	0.02 %		0.20 %
- Loss of ignition	1.9 %	10.6	0.16 %	Max 4.0 %	2.5 %

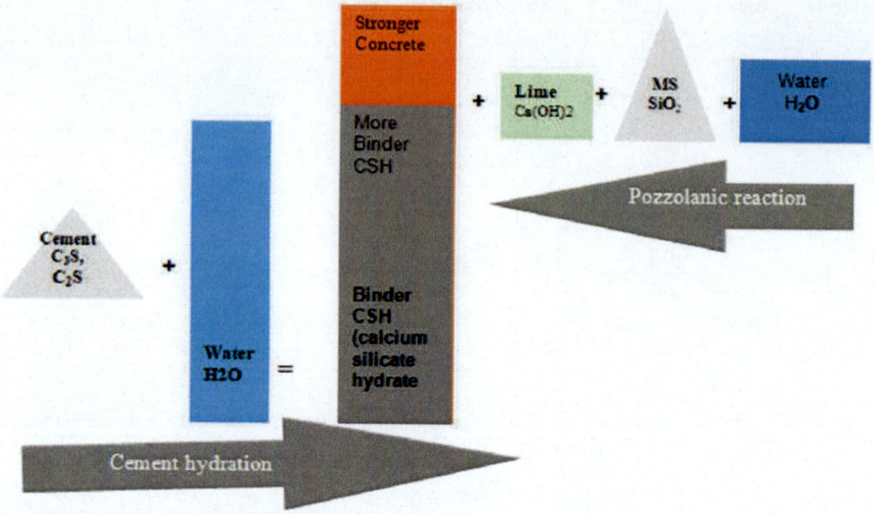

Fig. 1 Pozzolanic Reaction between different ingredients

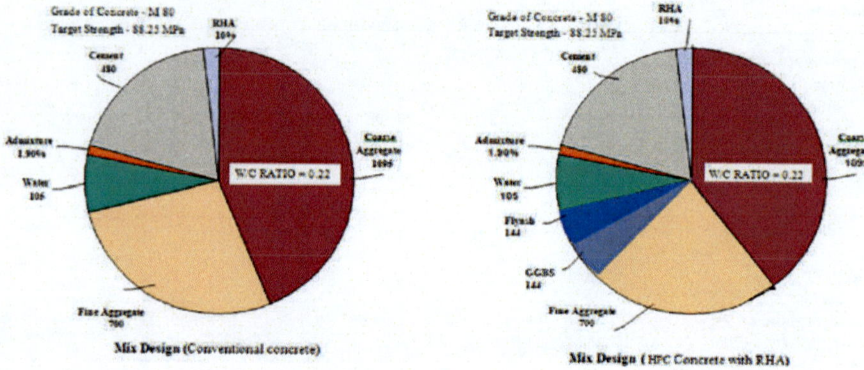

Fig. 2 Mix design for conventional concrete and high performance concrete for M50 Grade

5 Conclusions

In the present work it is tried to formulate mixes for high performance FRC concrete. When slag and Fly ash added with RHA/Micro silica, the hydration process is subsequently affected reflecting the modified chemical composition, reactivity of the components, their particle sizes, size distribution, shape and other factors which affect the final paste structures and its performance as the matrix of concrete. The utilization of SF & Slag and FA & SF therefore presents challenges and opportunities by adding to the concrete mixture affect both the hydration process and the characteristic of the hardened concrete. The compositions are average and slag & Fly ash particle also exist as a range of composition within a single ash sample.

References

1. Nataraja, M. C., Dhang, N., & Gupta, A. P. (1999). Stress strain curves for steel fiber reinforced concrete under compression. *Cement Concrete Composites, 21*, 383–390.
2. Kwak, Y.-K., Eberhard, M. O., Kim, W.-S., & Kim, J. (2002). Shear strength of steel fiber-reinforced concrete beams without stirrups. *ACI Structural Journal, 99*, 530–538.
3. Balaguru, P., & Najm, H. (2004). High-performance fiber-reinforced concrete mixture proportions with high fiber volume fractions. *ACI Materials Journal*, July–August, pp. 281–286
4. Han, C.-G., Hwang, Y.-S., Yang, S.-H., & Gowripalan, N. (2005). Performance of spalling resistance of high performance concrete with polypropylene fiber contents and lateral confinement. *Cement and Concrete Research, 35*(9), 1747–1753.
5. Kang, S. T., Lee, Y. B., Kim, J. K., & Kim, Y. Y. (2011). The effect of fiber distribution characteristics on the flexural strength of steel fiber-reinforced ultra high strength concrete. *Construction and Building Materials, 25*, 2450–2457.
6. Saje, D., Bandelj, B., Šušteršič, J., Lopatič, J., & Saje, F. (2011). Shrinkage of polypropylene fiber-reinforced high-performance concrete. *Journal of Materials in Civil Engineering, 23*(7), 941–952. https://doi.org/10.1061/(ASCE)MT.1943-5533.0000258
7. Mohamadi, M., Mohandesi, J. A., & Homayonifar, M. (2013). Fatigue behavior of polypropylene fiber reinforced concrete under constant and variable amplitude loading. *Journal of Composite Materials, 47*(26), 3331–3342.
8. Aslani, F., & Nejadi, S. (2013). Creep and shrinkage of self-compacting concrete with and without fibers. *Journal of Advanced Concrete Technology, 11*(10), 251–265.
9. Zhang, P., & Li, Q.-F. (2013). Effect of polypropylene fiber on durability of concrete composite containing fly ash and silica fume. *Composites Part B: Engineering, 45*(1), 1587–1594.
10. Wankhade, R. L., & Landage, A. B. (2013). Non-destructive testing of concrete structures in Karad region. *Procedia Engineerin, 51*, 8–18.
11. Sounthararajan, V. M., Sivakumar, A. (2013). Durability evaluation in concrete using cracked permeability and chloride permeability tests. *Advances in Materials Science and Engineering, 2013*, Article ID 754027, 8 p (2013)
12. Afroughsabet, V., & Ozbakkaloglu, T. (2015). Mechanical and durability properties of high-strength concrete containing steel and polypropylene fibers. *Construction and Building Materials, 94*, 73–82.
13. Lavanya, G., Jegan, J. (2015). Durability study on high calcium fly ash based geopolymer concrete. *Advances in Materials Science and Engineering, 2015*, Article ID 731056, 7 p (2015)
14. Wankhade, R. L., Landage, A. B., & Konnur, B. A. (2016). Experimental investigation on combined effect of SBR and steel fiber on properties of concrete. *International Journal of Engineering Research, 5*(1), 257–261.
15. Shaikh, F. U. A., Shafaei, Y., & Sarker, P. K. (2016). Effect of nano and micro-silica on bond behaviour of steel and polypropylene fibers in high volume fly ash mortar. *Construction and Building Materials, 115*, 690–698.
16. Muntean, R., Cazacu, C., Mizgan, P., Galatanu, T., & Tamas, F. (2017). Practical applications of dispersely reinforced concrete with polypropylene fibers: beams. *Procedia Engineering, 181*, 285–292.
17. Yoo, D.-Y., & Banthia, N. (2017). Mechanical and structural behavior of ultra-high-performance fiber-reinforced concrete subjected to impact and blast. *Construction and Building Materials, 149*, 416–431.
18. Jung, S. H., Choi, Y. C., & Choi, S. (2017). Use of ternary blended concrete to mitigate thermal cracking in massive concrete structures-a field feasibility and monitoring case study. *Construction and Building Materials, 137*, 208–215.

19. Smarzewski, P., & Barnat-hunek, D. (2017). Effect of fiber hybridization on durability related properties of ultra-high performance concrete. *International Journal of Concrete Structures and Materials, 11*(2), 315–325.
20. Yoo, D.-Y., & Moon, D.-Y. (2018). Effect of steel fibers on the flexural behavior of RC beams with very low reinforcement ratios. *Construction and Building Materials, 188*, 237–254.
21. Pathan, M. G., Wankhade, R. L., Shende, A. M., Swaroop, A., Mashaan, N. (2022). Experimental analysis for performance of concrete with addition of steel fibres, SBR and polypropylene fibres. *Jurnal Kejuruteraan 34*(3), 429–445, https://doi.org/10.17576/jkukm-2022-34(3)-10

Performance Assessment of Recycled Aggregate Concrete Blended with Supplementary Cementitious Materials and Steel Fibers: An Approach Towards Developing Green and Sustainable Concrete

S. R. R. Teja Prathipati, Yeswanth Paluri, Hanuma Kasagani, and Kunamineni Vijay

Abstract Though the term "sustainability" was heard frequently, few people understand its significance, which is to ensure human safety on the planet. Sustainability has many short- and long-term benefits and ignoring it will result in depletion of natural resources in the long run. In this context utilisation of by-products as well as waste materials into the concrete and thereby decreasing carbon dioxide emissions and virgin aggregate extraction will be a sustainable way of developing a green concrete. In this manuscript, an attempt has been made to develop green and sustainable concrete by including Recycled aggregate (RA), Fly Ash (FA) & Sugar Cane Bagasse Ash (SCBA) as supplementary cementitious materials (SCMs). Also, an attempt is made to access the effect of steel fibers (SF) on the concrete thus made. The variables of the study include RA content (0, 100%), FA content (0 to 40%) and SCBA content (0 to 15%) and SF dosage (0, 1%). The properties examined in the study include compressive strength, split tensile strength, flexural strength, water absorption (WA), and acid attack. The results of the experimental programme show that the use of SCMs and SF can significantly improve the mechanical strength and durability of Recycled Aggregate Concrete (RAC). The use of SCMs improved the durability of the mix by densification of the matrix whereas incorporating steel fibers improved the tensile characteristics of the mix significantly, which means that use

S. R. R. T. Prathipati (✉) · H. Kasagani
Department of Civil Engineering, VR Siddhartha Engineering College, Vijayawada, India
e-mail: teja@vrsiddhartha.ac.in

H. Kasagani
e-mail: hanumakasagani@vrsiddhartha.ac.in

Y. Paluri · K. Vijay
Department of Civil Engineering, Bapatla Engineering College, Bapatla, India
e-mail: vijay.kunamineni@becbapatla.ac.in

A. Vilventhan et al. (eds.), *Advances in Construction Materials and Management*, Lecture Notes in Civil Engineering 346,
https://doi.org/10.1007/978-981-99-2552-0_38

of RA in conjunction with SCMs and SF improves the overall performance of the concrete and might be a green and sustainable alternative to plain concrete.

Keywords Recycled aggregates · Recycled Aggregate Concrete · Sugar Cane Bagasse Ash · Fly Ash · Steel Fiber · Green Concrete · Sustainability

1 Introduction

With the societies becoming modern and urban, a huge junk of Construction & Demolition Waste (C&DW) has been generated due to the need of transformation and renovation of the already existing infrastructure. Till today, in many countries still utilization of C&DW was limited which will be hazardous and adversely effects the environment [1]. Hence there was a need to address the problem of the generated C&DW by re-using it by which the negative environmental impact can be mitigated along with the effective utilization of the materials. For a sustainable development in the construction field, one must slower the rate at which the natural resources were depleting and thereby which the preservation of environment was need of the hour. In a step towards addressing the before said issue, the utilization of C&DW along with industrial wastes into the construction activities was needed [2].

In terms of statistics, the global annual production of concrete is around 5.3 billion m^3 [3], every year on an average of over 1 billion tonnes of C&DW was being produced worldwide and use of virgin aggregate in concrete has resulted in the generation of 0.0046 million tons of carbon. Roughly an 14% of total energy was consumed and about 8% of CO_2 emissions was released by the cement industry [4]. Hence as a measure towards the reduction of carbon footprint and total energy used, a viable option will be utilization of Recycled Aggregate (RA) from C&DW into the concrete in place of virgin aggregate and effective utilization of wastes as a partial replacement to cement, thereby which developing a green and sustainable concrete will be possible. Considering these things, in this study the utilization of RA and blending an industrial waste along with agricultural waste into the concrete was experimented.

Among the different industrial wastes generated around the worldwide, Fly Ash (FA) contributes a major amount and without proper disposal it will harm the environment and hence it was being widely used as Supplementary Cementitious Material (SCM) in the concrete [5, 6]. India being the second largest producer of sugar cane in the world and Sugar Cane Bagasse Ash (SCBA) a byproduct of sugar cane industry if not properly disposed may cause a major negative impact on the environment. Partial substitution of cement with SCBA and FA has proven to be beneficial in view of the cost cuttings and reduction of greenhouse gases [7]. Hence in this study the concrete formed by replacing the Natural Aggregate with RA and partly replacing cement with blended FA and SCBA was investigated. To further boost the performance of the concrete thus formed, there was a need to reinforce it by incorporating fiber which improves the overall performance of the concrete [8–12]. Reinforcing discontinuous steel fiber to the concrete has improved the compressive and tensile properties was already established in the literature [13, 14]. In this study, steel fiber was used as a reinforcement to the green and sustainable concrete formed by utilizing RA, FA, and SCBA. The composite behavior, both at fresh and hardened stage will be evaluated and reported in the manuscript in the further sections.

2 Experimental Methodology

2.1 Materials

OPC 53 grade with a specific gravity of 3.12, fine aggregate (Zone-II River sand) with a specific gravity of 2.68, and coarse aggregate (crushed granite, 20 mm nominal size) with a specific gravity of 2.78 were used. Recycled aggregates (RA) made by crushing up old concrete in a jaw crusher, screening it, and then using the resulting aggregates to fill in the necessary grading gaps. RA was found to have a specific gravity of 2.72 and a water absorption-rate of 5.70 percent. In this research, two different SCMs were used: class F fly ash (FA) with a specific gravity of 2.38 from a thermal power plant in Visakhapatnam, and sugar cane bagasse ash (SCBA) with a specific gravity of 2.21 from a sugar factory in Samalkota. The investigation makes use of crimped steel fibers with the following characteristics: a diameter of 0.5 mm, a length of 36 mm, a tensile strength of 1145 MPa, and an elastic modulus of 205 GPa. Fosroc Complast SP430, a commercially available chemical admixture was used as Superplasticizer (SP).

2.2 Mix Proportions

The control mix of M50 grade concrete was made in accordance with Indian Standards (IS 10262, 2019).

Water-binder ratio of 0.34 and mix proportion of 1:1.58:3.06 were determined from the trail mixes and used in this investigation. Steel fiber dosage (1%) and FA & SCBA content (from 0 to 15%) are also included as independent variables. All the 12 mixes were used in the study, and their respective mix denotations are listed in Table 1.

2.3 Experimental Program

Various tests were performed to understand the fresh, hardened and durability characteristics of the concrete mixes considered. Testing was done on triplicate specimens

Table 1 Mix denotations

Mixes	Natural aggregate (%)	Recycled aggregate (%)	Cement (%)	Fly ash (%)	SCBA (%)	Fiber (%)	SP (%)
NAC-OPC	100	–	100	–	–	–	1
RAC-OPC	–	100	100	–	–	–	1
RAC-F20	–	100	80	20	–	–	1
RAC-F30	–	100	70	30	–	–	1
RAC-F40	–	100	60	40	–	–	1
RAC-S5	–	100	95	–	5	–	1
RAC-S10	–	100	90	–	10	–	1
RAC-S15	–	100	85	–	15	–	1
RAC-F30S5	–	100	70	30	5	–	1
RAC-F30S10	–	100	70	30	10	–	1
FRAC-F30S5	–	100	70	30	5	1	1
FRAC-F30S10	–	100	70	30	10	1	1

Here, NAC denotes natural aggregate concrete, RAC denotes recycled aggregate concrete, FRAC denotes fiber reinforced recycled aggregate concrete, and F30S10 denotes concrete with 30% FA and 10% SCBA

for all the mixes, and their average was reported. The slump of fresh concrete, which is a used to quantify the workability of the concrete was accessed in accordance with Indian standards (IS 1199: Part 2, 2018). Compressive and flexural strengths of hardened concrete mixtures were measured in accordance with Indian specifications (IS 516: Part 1: Sec 1, 2021). Water absorption, porosity, acid mass loss factor (AMLF), and acid strength loss factor (ASLF) were measured to better understand the long-lasting qualities of the considered mixes. The porosity and water absorption were measured in accordance with ASTM criteria (ASTM C642, 2021). Access to hostile environments with resistance per ASTM standards (ASTM C267, 2020). The strength loss and mass loss of concrete exposed to an aggressive environment were measured using AMLF and ASLF, respectively. For 56 days, a concrete specimen was immersed in 5% HCl and 5% H2SO4 to simulate the effects of an aggressive environment on the material's properties. To determine AMLF and ASLF, the following formulas (1) & (2) were used.

$$\text{AMLF} = \frac{\text{Change in Mass}}{\text{Initial Mass}} \times 100 \tag{1}$$

$$\text{ASLF} = \text{Relative strength loss (\%) at N days} \times \frac{\text{N}}{56} \tag{2}$$

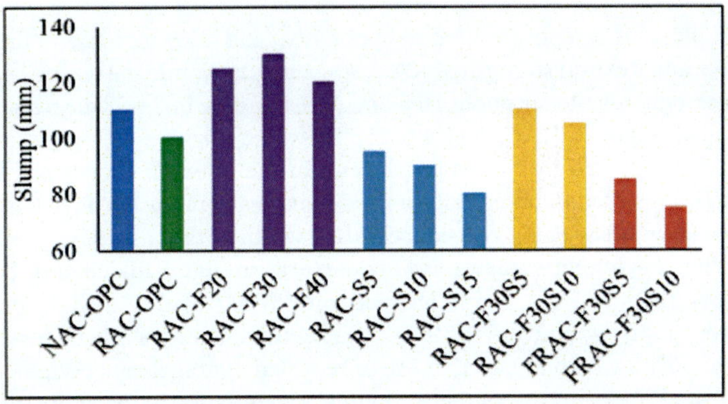

Fig. 1 Slump of NAC, RAC and FRAC mixes

3 Results and Discussion

Testing on triplicate samples specimens for all the mixes was done as per the methodology discussed in the previous section and their variation will be discussed in the subsequent sections.

3.1 Workability

The variation of slump of the considered mixes is as shown in Fig. 1. It can be observed that RAC had lower slump compared to NAC, which might be due to greater porosity and rougher surface texture of RA when compared to NA. The use of FA as SCM improved the workability of RCA mixes whereas the use of SCBA as SCM reduced the workability of RCA mixes. This might be due to spherical shape of FA particles that provide lubricating effect to the mix, while the SCBA particles being hygroscopic in nature, reduces the workability of the mix. Blending FA and SCBA resulted in mixes with slump comparable to that of NAC. It is also worth noting that FRAC specimens had lower workability compared to their RAC counterparts.

3.2 Strength Properties

The variation of compressive and flexural strength of considered mixes after a curing period of 7 and 28 days is shown in Figs. 2 and 3 respectively. The use of RA as a replacement to NA lead to a lower strength both at 7- and 28-days curing, which might be due to presence of old mortar attached to RA and high porosity of RCA. The use of SCMs had a positive effect on the strength properties of RAC. The use of

30% FA in place of cement resulted in about 12% and 20% increase in the 28 days compressive and flexural strength of RAC, respectively. While the use of 10% SCBA in place of cement resulted in about 16% and 30% increase in the 28 days compressive and flexural strength of RAC, respectively. However, the use of higher dosages of FA and SCBA resulted in a downward trend in the strength properties. Blending FA and SCBA resulted in about 30% and 35% increase in the 28 days compressive and flexural strength of RAC, respectively. It is worth noting that RAC-F30S10 mix had 10% and 13% higher compressive and flexural strengths than that of NAC mix. Adding steel fibers to RCA mixes blended with SCMs lead to a slight increase in compressive strength (about 8.5 to 10%) and a significant increase in flexural strength (about 34 to 36%), which means that use of recycled aggregates in conjunction with SCMs and steel fibers improves the strength characteristics of the concrete.

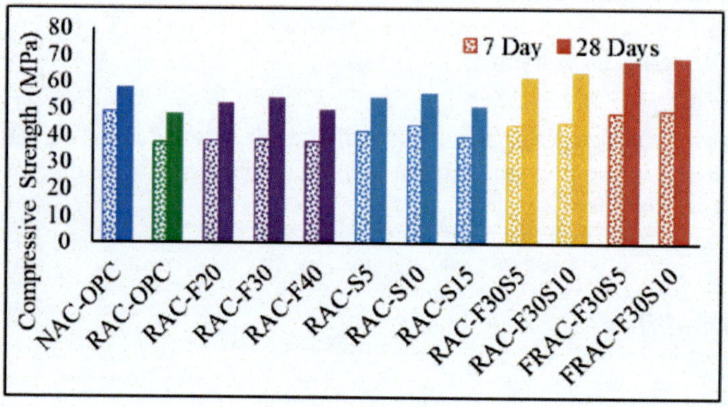

Fig. 2 Compressive Strength of NAC, RAC and FRAC mixes

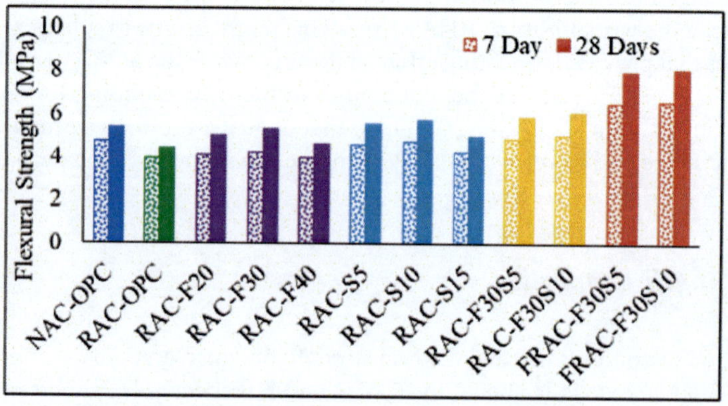

Fig. 3 Flexural Strength of NAC, RAC and FRAC mixes

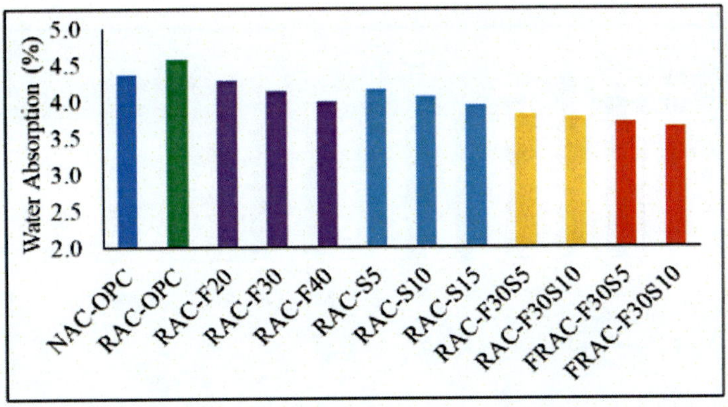

Fig. 4 Water absorption in NAC, RAC and FRAC mixes

3.3 *Durability Properties*

3.3.1 Water Absorption and Porosity

Concrete's water absorption and porosity are two of the most important durability properties since they predict the quality of the material's void network. Figure 4 and Fig. 5 depicts the water absorption and porosity of NAC, RAC and FRAC mixes. The water absorption and porosity of the concrete increased with the use of RA, which may be due to higher porosity of RA and weaker bond between RA and cement mortar. However, the use of SCMs (FA, SBCA or FA+SCBA) decreased the water absorption and porosity of the concrete, this may be due to the fact that the use of SCMs lead to densification of the matrix, thereby decreasing the pore size and hence a smaller void network. It can also be noted that addition of fibers in concrete containing SCMs further decreased the porosity and water absorption due to lower interconnected voids as a result of discrete fiber distribution in concrete.

3.3.2 Acid Attack

To investigate how concrete samples would fare in a hostile environment, authors have immersed them in a mixture of 5% HCl and 5% H2SO4 for 56 days. The strength loss and mass loss of concrete exposed to an aggressive environment were measured using AMLF and ASLF, respectively. Figures 6, 7, 8, and 9 depict the AMLF and ASLF of NAC, RAC, and FRAC mixtures exposed to 5% HCl Solution and 5% H_2SO_4 for varying amount of time. It was discovered that H_2SO_4 is more aggressive toward concrete than HCl, as the strength loss and mass loss of concrete specimens exposed to H_2SO_4 were greater than those of specimens exposed to HCl. AMLF and ASLF variation mirrors that of water absorption variation. Compared to NAC, RAC exhibited greater mass loss and strength; this may be because RAC has a more porous

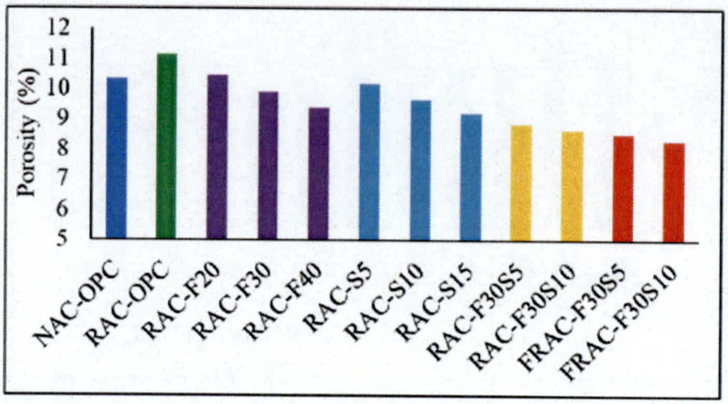

Fig. 5 Porosity in NAC, RAC and FRAC mixes

and brittle matrix. However, the strength loss and mass loss of the concrete were reduced when SCMs were used (either FA, SBCA, or FA+SCBA), possibly because the densification of the matrix afforded greater resistance to acid attack. Adding fibers, which distribute more randomly throughout the concrete, helped mitigate the loss of strength and mass in the specimen.

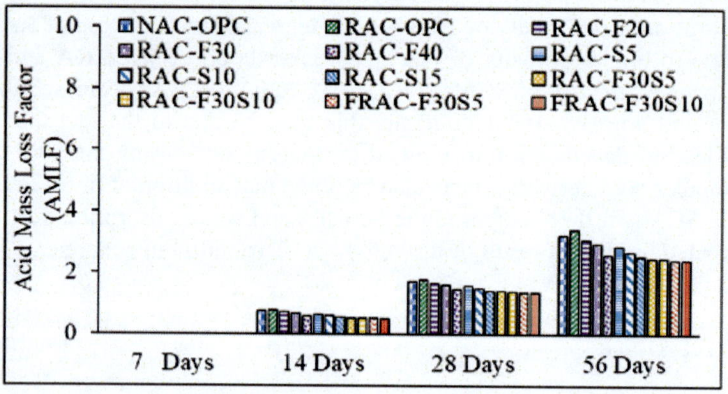

Fig. 6 AMLF in NAC, RAC and FRAC mixes submersed in 5% HCl solution

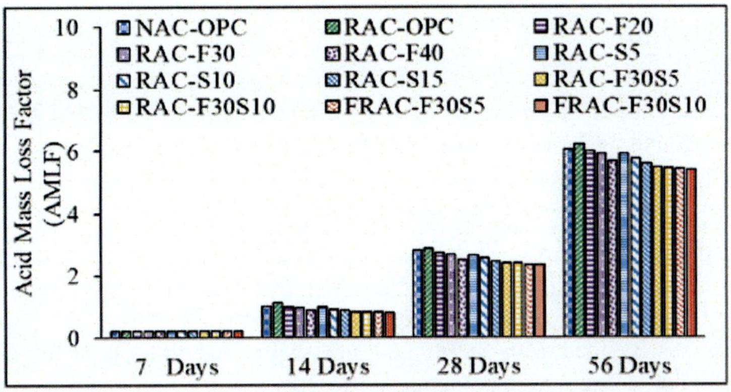

Fig. 7 AMLF in NAC, RAC and FRAC mixes submersed in 5% H_2SO_4 solution

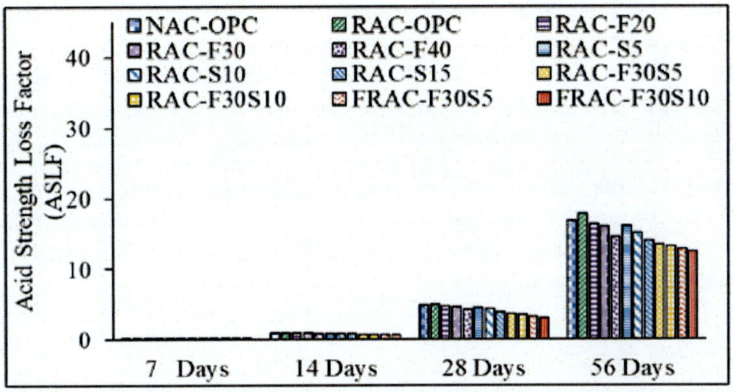

Fig. 8 ASLF in NAC, RAC and FRAC mixes submersed in 5% HCl solution

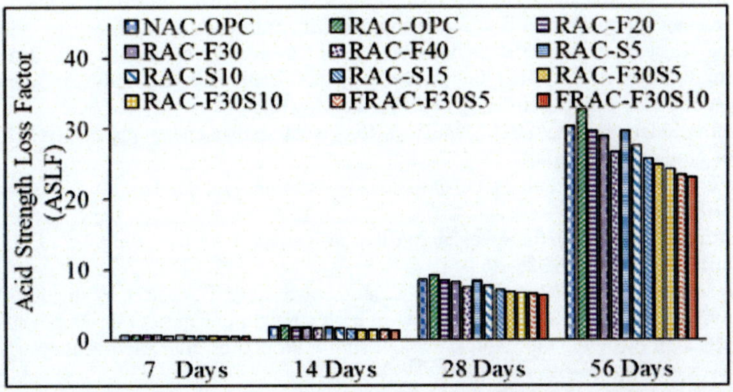

Fig. 9 ASLF in NAC, RAC and FRAC mixes submersed in 5% H_2SO_4 solution

4 Conclusions

The expected output of this study was the effective utilisation of RA along with the industrial and agricultural waste into the concrete reinforced with steel fiber to form a green and sustainable concrete. From the outcomes of this investigation the following conclusions can be drawn:

- Replacing NA with RA into the concrete resulted in the reduction of workability which further reduced by incorporating fiber. Incorporating SCMs has not much affected the fresh properties of the concrete.
- RCA mixes blended with SCMs and steel fiber lead to a slight increment in compressive strength values (about 8.5 to 10%) and a significant increment in flexural strength values (about 34 to 36%), from which it can be inferred that use of recycled aggregates in conjunction with SCMs and steel fibers improves the strength characteristics of the concrete.
- The water absorption and porosity of the concrete increased, and higher mass loss and strength loss was observed with the use of RA and this was compensated by the effective utilization of SCMs in combination with fiber and thereby improving the overall durability performance of the composite.

Overall, from this study it can be understood that the effective use of recycled aggregates in conjunction with SCMs and steel fiber improves the overall performance of the concrete and might be a green and sustainable alternative to the plain concrete in the construction sector.

References

1. Kabirifar, K., Mojtahedi, M., Wang, C., & Tam, V. W. Y. (2020). Construction and demolition waste management contributing factors coupled with reduce, reuse, and recycle strategies for effective waste management: A review. *Journal of Cleaner Production, 263*, 121265.
2. de Brito, J. & Saikia, N. (2013). Recycled aggregate in concrete: use of industrial, construction and demolition waste. *Green Energy and Technology, 54.*
3. Cement production global 2021 | Statista. https://www.statista.com/statistics/1087115/global-cement-production-volume/.
4. Andrew, R. M. (2018). Global CO2 emissions from cement production. *Earth System Science Data, 10*, 195–217.
5. Ahmaruzzaman, M. (2010). A review on the utilization of fly ash. *Progress in Energy and Combustion Science, 36*, 327–363.
6. Prathipati, S. R. R. T., Paluri, Y., Vijay, K., & Chowdary, V. B. (2022). Evaluating the feasibility of blending fly ash and quarry dust in high-strength concrete to develop a sustainable concrete: a Study on the Mechanical and Durability Properties. *IOP Conference Series: Earth and Environmental Science, 1086*, 012060.
7. Fantilli, A. P. & Jóźwiak-Niedźwiedzka, D. (2021). Special issue: supplementary cementitious materials in concrete, Part I. *Mate, **14***, 2291.
8. Teja Prathipati, S. R. R., Rao, C. B. K. & Dakshina Murthy, N. R. (2020). Mechanical behavior of hybrid fiber reinforced high strength concrete with graded fibers. *International Journal of Engineering, **33***, 1465–1471. (2020)

9. Koniki, S., Kasagani, H., Prathipati, S. R. R. T. & Paluri, Y. (2021). Mechanical behavior of triple-blended hybrid fiber-reinforced concrete: an experimental and numerical study. *Innovative Infrastructure Solutions, 6*.
10. Kasagani, H., Teja, P. S. R. R. & Rao, C. B. K. (2021). Influence of graded glass fibres on strain hardening and strain softening behaviour of CGGF under uniaxial stress *73*, 674–700. https://doi.org/10.1680/jmacr.19.00303
11. Paluri, Y., Mogili, S., Mudavath, H., & Pancharathi, R. K. (2020). A study on the influence of steel fibers on the performance of Fine Reclaimed Asphalt Pavement (FRAP) in pavement quality concrete. *Materials Today Proceedings, 32*, 657–662.
12. Paluri, Y., Mogili, S., Mudavath, H., & Noolu, V. (2021). Effect of fibres on the strength and toughness characteristics of recycled aggregate concrete. *Materials Today Proceedings, 38*, 2537–2540.
13. Teja Prathipati, S. R. R. & Rao, C. B. K. (2021). A study on the uniaxial compressive behaviour of graded fiber reinforced concrete using glass fiber/steel fiber. *Innovative Infrastructure Solutions, **6***
14. Paluri, Y., Asce, S. M., Noolu, V., Heeralal, M., & Pancharathi, R. K. (2020). Flexural fatigue behavior of steel fiber-reinforced reclaimed asphalt pavement–based concrete: an experimental study. *Practice Periodical on Structural Design and Construction, 26*, 04020053.

A Numerical Study on the Shear Strength of RC Beams Provided with Welded Wire Mesh as Core Zone Reinforcement

Ch. Manjula, D. Rama Seshu, and T. D. Gunneswara Rao

Abstract In general, the conventional way of resisting shear in RC beams using stirrups leaves the core zone of RC beam section unreinforced. This makes the shear cracks to propagate rapidly through the core zone of the section once initiated. To address this problem a novel method of reinforcing the core zone of RC beam section using welded wire mesh (WWM) has been developed. The efficiency of the proposed method was investigated numerically by using the ABAQUS software. For comparison of the numerically obtained shear strength values, the RC beams provided with WWM as shear reinforcement were cast and tested. The parameter of the study included two different orientations of the WWM (Longitudinal and Transverse). The use of stirrups was completely eliminated in the RC beams provided with WWM as transverse shear reinforcement. In RC beams with longitudinal WWM in the core zone, the stirrups were used at larger spacing. The shear span to effective depth ratio equal to 3 was used in the study. The results of the investigation indicated the effectiveness of WWM over the conventional stirrups in providing the shear resistance. The percent difference between the numerically obtained shear strengths and that observed in experiments ranged from 0.1 to 16.

Keywords RC beams · Shear strength · Welded wire mesh · Core Zone · ABAQUS modelling

---

Ch. Manjula (✉) · D. Rama Seshu · T. D. Gunneswara Rao
Department of Civil Engineering, NIT Warangal, Warangal, India
e-mail: manjulach@student.nitw.ac.in

D. Rama Seshu
e-mail: drseshu@nitw.ac.in

T. D. Gunneswara Rao
e-mail: tdg@nitw.ac.in

A. Vilventhan et al. (eds.), *Advances in Construction Materials and Management*, Lecture Notes in Civil Engineering 346,
https://doi.org/10.1007/978-981-99-2552-0_39

1 Introduction

As safety is the primary concern in the design of reinforced concrete (RC) beams, abrupt failures should be avoided. Since shear failure occurs frequently and suddenly, with little to no other notice, the design for shear must make sure that each member's shear strength exceeds its flexural strength. The geometry, member properties, types of loading and cross-sectional dimensions are the factors which affect the shear failure mechanism significantly [1, 2].

It is known that shear forces will be somewhat resisted by any type of firmly anchored reinforcement that intersects these diagonal cracks [3–5]. The center of the cross section, where there is a significant transverse stress, is left unreinforced by the transverse reinforcement in the manner of conventional stirrups placed close to the outerface of the beam cross section (Fig. 2). Additionally, the stirrups offer only discrete resistance carried by the shear against diagonal tension. This causes cracks to develop and spread suddenly, which results brittle failures. Therefore, it is proposed that by reinforcing the core zone as well, the performance of the RC elements under the influence of transverse loads will be enhanced. The way of reinforcing core zone will be effectively possible with welded wire mesh. The way of reinforcing core zone will be effectively possible with welded wire mesh. Prefabricated reinforcement, such WWM, has also been employed in thin webbed concrete beams as shear reinforcement and in Reinforced Concrete slabs as Flexural reinforcement [6, 7]. According to experimental research on the use of welded wire mesh for main and transverse reinforcement placed close to the cross section's outerface, traditional stirrups and weld mesh together offer modestly greater strength and crack resistance [8, 9]. The WWM was only employed at the outermost portion of the cross section in all of the above investigations. There is relatively less published information on the implementation of such core zone reinforcement where shear stresses are large.

A 3D non-linear analysis called ABAQUS Finite Element Modeling use numerical techniques to evaluate and identify structural element failures [10]. As experimental research involves Complex, expensive and time-consuming for full-scale specimens, FEM Software's plays prominent role in research. The impact of the different parameters on the strength and behaviour of such beams may be better understood using finite element analysis (FEA). Numerous number of commercial Finite element software, including ABAQUS, ANSYS, ATENA, and ADINA of nonlinear analysis can be used, but none of them are custom applications that can run with less information, Instead, they all require accurate input values and a solid knowledge of FE [11]. Beams are analyzed using ABAQUS's concrete damaged plasticity model. Both crushing due to compression and tensile cracking of concrete are potential failure mechanisms taken into account by the concrete damaged plasticity (CDP) model. [12].

In this study, numerical analysis utilising ABAQUS for six beams provided a unique technique for strengthening reinforced concrete elements by resisting shear, employing welded wire mesh as a prefabricated mesh (WWM). The weld mesh is beams are loaded in transverse direction, resulting internal shear stresses and non-uniform used as longitudinal core reinforcement in addition to traditional rectangular stirrups or ties (Fig. 4) and also utilized as shear reinforcement completely replacing the traditional stirrups (Fig. 3).

2 Investigation Significance

In general, when RC bending. As a result, the cross-section of a reinforced concrete beam receives both normal and flexural shear forces. Flexural shear stresses is not evenly distributed over the rectangular cross sections. In homogeneous rectangular cross sections, the flexural shear stress intensity decreases to negligable at the margins and increases near the neutral axis as shown in Fig. 1.

The weld mesh acts as a barrier against diagonal tension due to shear, whether it is positioned longitudinally or transversely in the core zone. Regular Practice of stirrup making involves tedious process and this will not be in the case of weld mesh that can be used as longitudinal or transverse direction in core reinforcement.

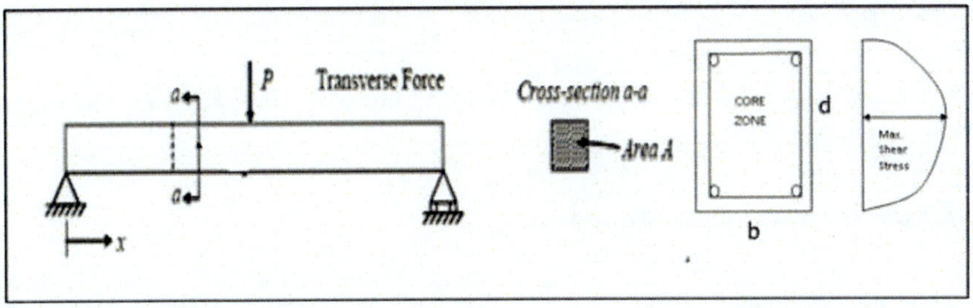

Fig. 1 Rectangular Beam shear stress distribution

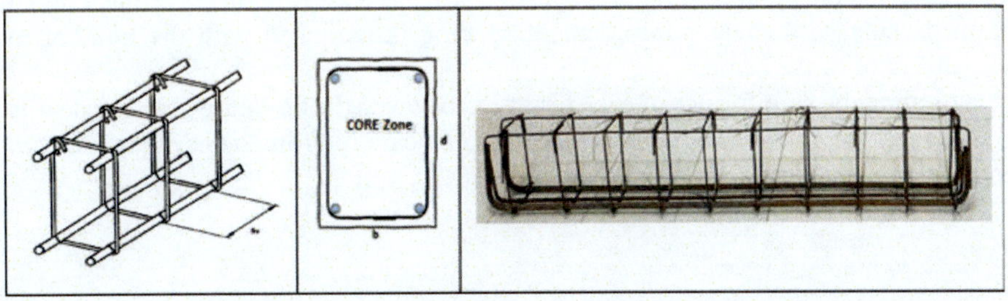

Fig. 2 Conventional Stirrups reinforcement cage and cross section

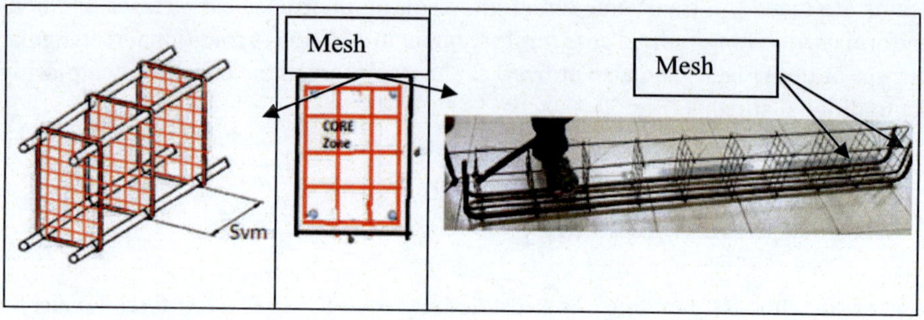

Fig. 3 Mesh as transverse reinforcement replacing the traditional stirrups reinforcement cage and also its cross section

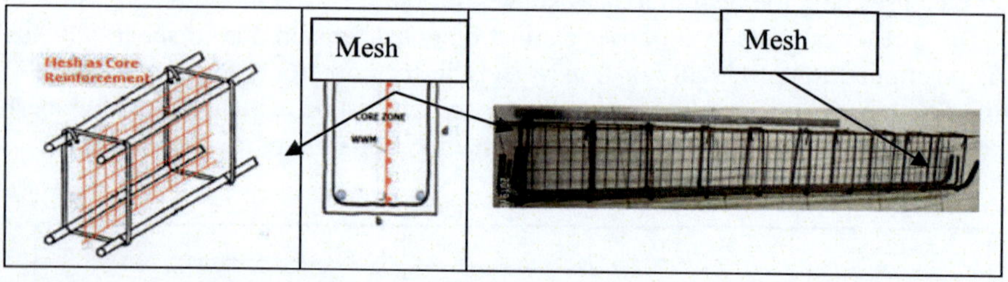

Fig. 4 Mesh as longitudinal reinforcement along with stirrups reinforcement cage and also its cross section

3 Experimental Investigation

The experimental investigation consists of Six RC beams with shear span to effective depth is 3. These beams are named as M160, L160, R160 L0, L180, M100 in which 'L' indicates beams with longitudinal mesh along with stirrup, 'M' indicates beams having transverse mesh completely replacing stirrups and 'R' indicates conventional beam. Beams with shear reinforcement spacing is indicated with the number in nomenclature of the beam.

Experimental significance of each beam can be taken the literature [14] The details of analyzed specimens are given in Table 1. The beam reinforcement cages used are presented in Fig. 5(a)–(d).

Table 1 Data of RC beam specimens

SNo	Beam Nomenclature	Dimensions (mm) (b × d × L)	Longitudinal Reinforcement		Reinforcement in core zone	Shear Reinforcement
			Tension	Comp	WWM	Stirrups
1	R160	140 × 300 × 1650	2–12Φ & 1-16Φ	2–6Φ	–	2 Legged 6Φ @160 c/c
2	M160				WWM (transverse) @160 c/c	–
3	L160				WWM (longitudinal)	2 Legged 6Φ @160 c/c
4	L0				WWM (longitudinal)	
5	L180				WWM (longitudinal)	2 Legged 6Φ @180 c/c
6	M100				WWM (transverse) @100 c/c	-

Table 2 Details of Experimental Results

S.No	Beam	At Service		At Ultimate		% increase in ultimate load	Nature of failure At Ultimate
		Fs(kN)	ds(mm)	Fu(kN)	du(mm)		
1	R160	117.40	3.78	176.04	7.10	–	Shear
2	M160	122.30	3.80	181.55	6.43	3.12	Shear
3	L160	181.50	4.02	272.32	9.86	54.70	Flexure
4	L0	115.98	2.38	173.98	5.45	-	Shear
5	L180	113.7	5.10	245.85	8.25	39.6	Flexure
6	M100	121.03	5.46	233.54	12.7		Flexure

Notation: Fs = Service Load, Fu = Ultimate Load, ds = Service deflection, du = Ultimate deflection

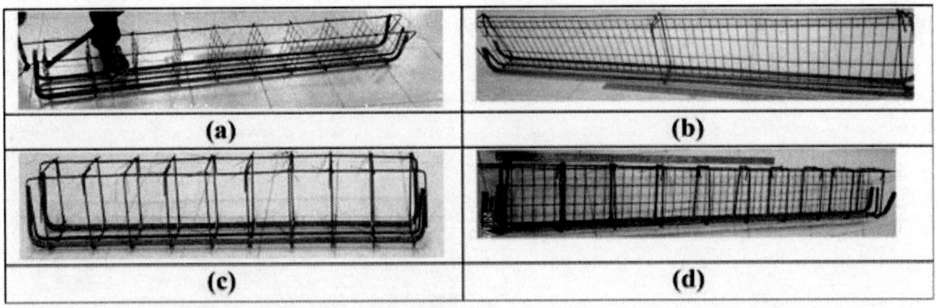

Fig. 5 **a** The Reinforcement cage with Welded wire meshes as transverse steel (M160). **b** The Reinforcement cage with longitudinal mesh only in the core zone (L0). **c** The Reinforcement cage with regular Stirrups (R160). **d** The Reinforcement cage with longitudinal mesh in the core zone along with stirrups (L160)

4 Finite Element Model

In finite element modeling, the modelling developed in ABACUS/standard explicit and the module part helps to create individual parts of the required beam which involves creation of concrete beam, reinforcement bars of 16mm,10mm and 6mm. Development of mesh creation in longitudinal direction and in transverse direction plays vital difference than regular beams and to be properly assembled to attain our beam modelling and also generating mesh deliberately provides accuracy of results and it is presented in Fig. 6(A), 6(B) and Fig. 7. The step module makes it simple to record changes in the model for loading and boundary parameters and also lets you alter the data output, the analysis method, and numerous controls [13]. In load module, two point loading is used and it is presented in Fig. 8 for assigning the boundary conditions.

Concrete Damaged Plasticity (CDP) model, based on numerous data from other researchers, was utilised in ABAQUS to examine the behaviour of complex concrete. Based on the CDP model's variables are dilation angle (ψ), flow potential eccentricity (e), fb0/fc0, the ratio of the stress invariant on the tensile meridian to that on the compressive meridian (K), the viscosity parameter (μ), and the elastic modulus of concrete (Ec). Damage is described separately tension and compression may be

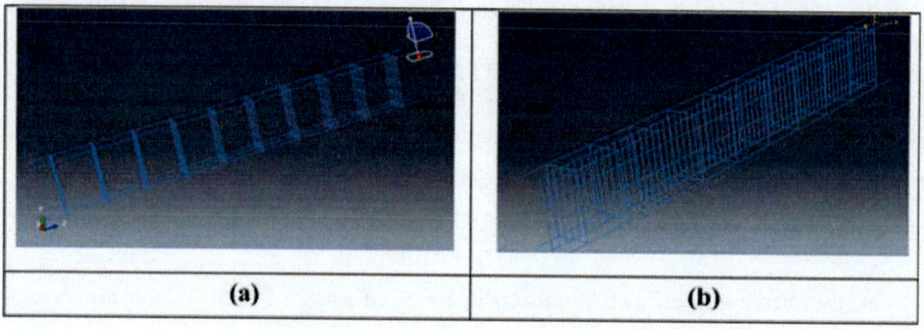

(a) (b)

Fig. 6 **a** Assembled Cage of Transverse mesh. **b** Assembled Cage of longitudinal mesh

Fig. 7 Meshing of RC beams

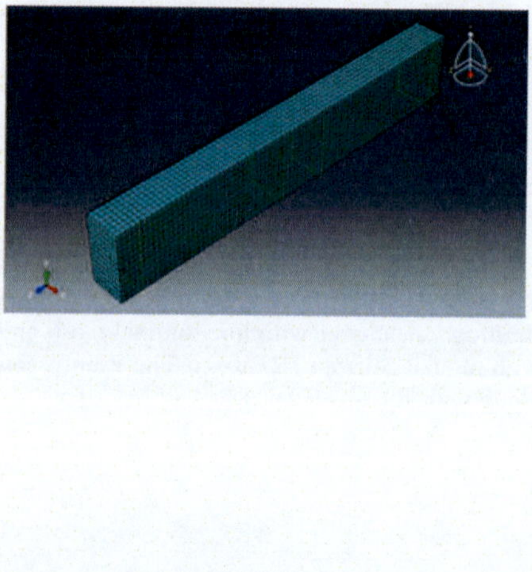

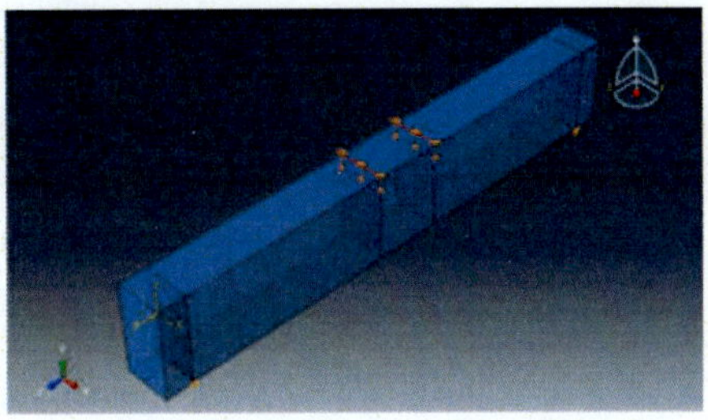

Fig. 8 Applying the boundary conditions

incorporated into the model. The Visualization Module presents numerical and experimental failure patterns, fracture patterns, and findings from finite element models visually are presented in Fig. 9.

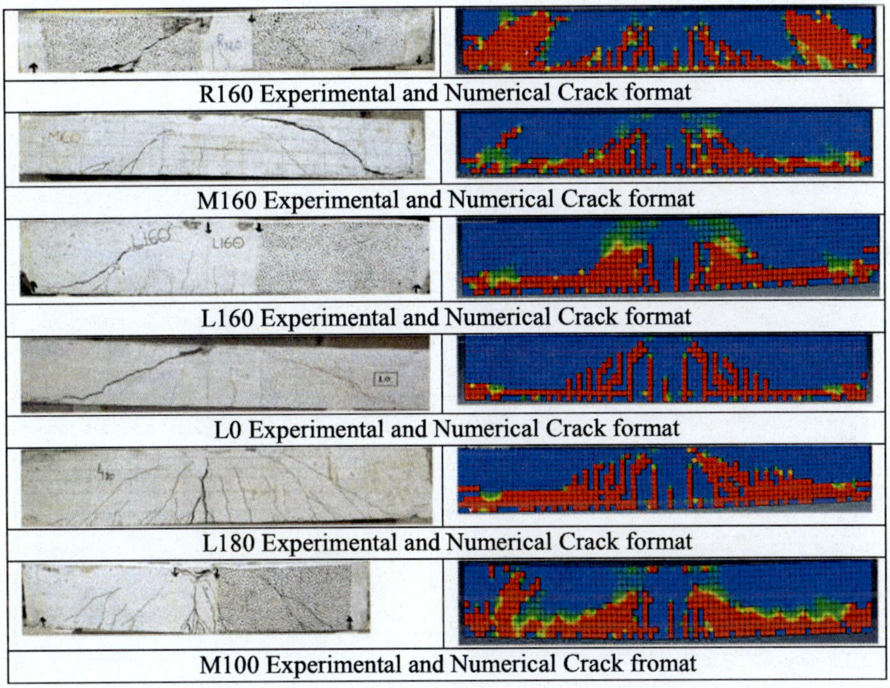

Fig. 9 Comparison of Experimental and numerical crack format

5 Results and Observations

Comparing experimental to Numerical crack and failure patterns seems to be in good correlation. The behavior of beams with longitudinal mesh shows high load carrying capacity and changes its failure pattern to flexure though it is designed to fail under shear and this is due to the arrangement of mesh in continuous manner throughout the beams. On the other hand, transverse mesh beams though failed in shear but able to show high load carrying capacity than conventional beams both in experimental and numerical results and the tabulated results for load and deflection are presented in Table 3. As Beams are modelled using concrete damaged plasticity model leads to present the behavior of beams separately in compression and tension.

From the test results and numerical results load deflection curves are developed presented in Fig. 10 clearly reflects its correlation and the variation of load and deflection values from numerical to experimental ranges from 0.1 to 16%. In longitudinal mesh variation is higher compared to transverse mesh. From the tabulated values and graphs, beams with welded wire mesh shows superior performance than regular beam.

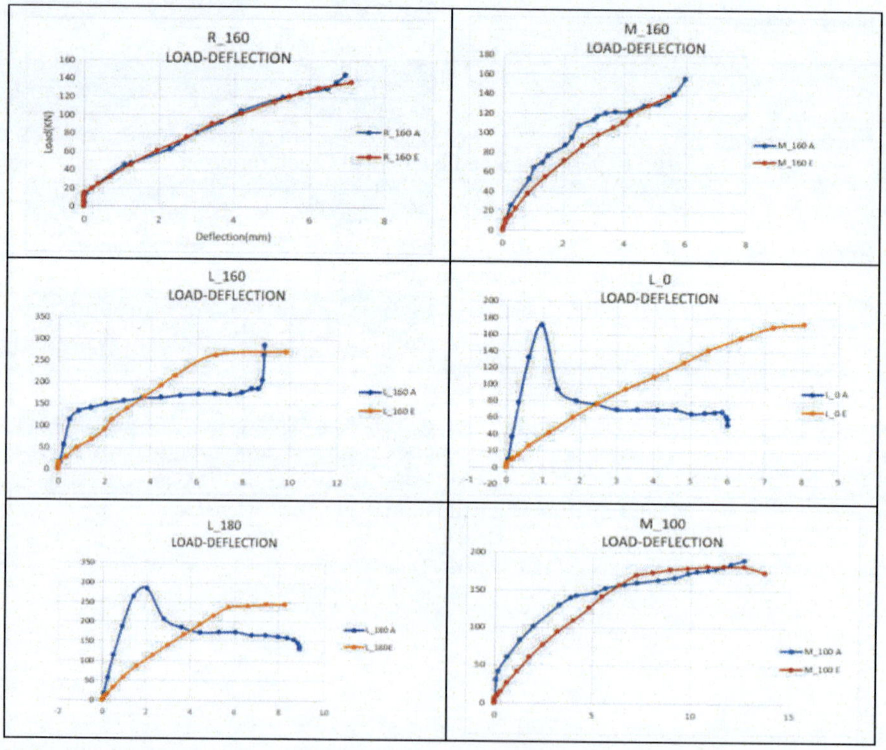

Fig. 10 Load–Deflection Curves

Table 3 Numerical Loads and Deflections

Model	R 160	L160	M160	L0	M100	L180
Numerical Ultimate Load (kN)	186.34	287.38	200.62	175.86	244.46	285.25
Numerical Deflection (mm)	6.939	8.93	5.98	5.99	12.71	8.90
% Variation of Ultimate Loads	5.8%	5.53%	10.5%	1.08%	4.67%	16%
%Variation of Deflection	2.26%	9.43%	7.0%	9.90%	0.1%	7.87%

6 Conclusion

By investigating the use of welded wire mesh (WWM) both as a transverse reinforcement to replace the traditional stirrups or as a longitudinal core reinforcement is analysed in ABACUS with good correlation to experimental results. With the traditional stirrups replaced with weld mesh as transverse reinforcement for the same spacing of steel in the form of mesh/stirrups, the nature of failure/behavior of the RC beam remained more or less the same. Unlike the discontinuous resistance offered by standard stirrups, shear resistance is continuously produced by the application of longitudinal core reinforcement. Crack propagation and type of failures can be predicted easily using concrete compression damage and concrete tension damage and they are almost same as experimental. The percentage variation in Numerical and experimental ultimate loads range from 0.1% to 16%. The CDP model adopted for modelling RC beams has yielded ultimate loads and deflections almost close to the experimental results. Deviation in the value of ultimate load from experimental and analytical analysis ranges from 1.08% to 16%, deviation in the value of defection at from experimental and analytical analysis ranges from 0.1% to 9.90%. The performance of the RC beams under test clearly shown that the weld mesh, used as longitudinal and as transverse reinforcement in core zone reinforcement, outperformed the traditional stirrups.

References

1. Słowik, M. (2014). Shear failure mechanism in concrete beams. *Procedia Materials Science, 3*, 1977–1982. 20th European Conference on Fracture (ECF20).
2. ACI-318 (2011). Building Code Requirements for Structural Concrete (ACI 318-08) and Commentary, ACI, Farmington Hills, MI, USA, p. 510.
3. Al-Nasra, M.M. (2013). Shear failure investigation of reinforced concrete beams with swimmer bars. *Journal of Civil Engineering and Construction Technology, 4*(2), 56–74. http://www.academicjournals.org/JCECT. https://doi.org/10.5897/JCECT2013.0251. ISSN 2141-2634.
4. Zakaria, M., Ueda, T., Wu, Z., & Meng, L. (2009). Experimental investigation on shear cracking behaviour in RC beams with shear reinforcement. *Journal of Advanced Concrete Technology, 7*(1), 79–96.

5. Shuraim, A. B. (2014). A novel approach for evaluating the concrete shear strength in reinforced concrete beams. *Latin American Journal of Solids and Structures, 1*, 093–112.
6. Pincheira, J. A., Rizkalla, S. H., & Attiogbe, E. K. (1989). Performance of welded wire fabric as shear reinforcement under cyclic loading. *ACI Structural Journal, 86*(6), 728–735.
7. Ibrahim, H. M. (2011). Experimental investigation of ultimate capacity of wired mesh-reinforced cementitious slabs. *Construction and Building Materials, 25*, 251–259.
8. Gayathri, S., & Kirthiga, T. (2018). Experimental and analytical investigation of RC beam with welded wire mesh as shear reinforcement. *International Research Journal of Engineering and Technology (IRJET), 5*(5), 0072–2395.
9. Alexander, D., & Ramakrishnan, S. (2016). Design of RC Beam with and without welded mesh as shear reinforcement in flexural and shear behaviour. *International Journal of Advanced Engineering Research and Technology (IJAERT), 4*(6), 2348–8190.
10. Giner, E., Sukumar, N., Tarancón, J. E., & Fuenmayor, F. J. (2009). An Abaqus implementation of the extended finite element method. *Engineering Fracture Mechanics, 76*(3), 347–368.
11. Lee, S. H., Abolmaali, A., Shin, K. J., & Lee, H. D. (2020). ABAQUS modeling for post-tensioned reinforced concrete beams. *Journal of Building Engineering, 30*, 101273.
12. Lee, J., & Fenves, G. L. (1998). Plastic-damage model for cyclic loading of concrete structures. *Journal of Engineering Mechanics, 124*(8), 892–900.
13. Kartheek, T., & Das, T. V. (2020). 3D modelling and analysis of encased steel-concrete composite column using ABAQUS. *Materials Today: Proceedings, 27*, 1545–1554.
14. Rama Seshu, D., Manjula, C., Rao, T. D. G., Rao, C. B. K., Anand, A., & Kumar, K. (2020). A novel means of improving the performance of reinforced concrete beams using the welded wire mesh as core zone reinforcement. *Asian Journal of Civil Engineering, 21*(6), 959–965.

Experimental Analysis of Bubble Deck Slab and Its Application for Sustainability in the Ecology Dimension

C. Akin, V. Preethi, and K. Vignesh

Abstract Concrete is the most widely used construction material in all categories of building construction projects. In Buildings, concrete is used for constructing foundations, columns, beams, and slabs. Slabs consume more concrete for construction. And Concrete consumes more energy for manufacturing its constituents and emits more CO_2. In this paper, the Bubble deck slab is introduced in a different dimension to reduce the consumption of concrete in slabs. The Physical properties of the slab are analyzed, and the sustainable concrete slab is implemented with good compressive strength. Finally, this leads to reducing CO_2 and acts as an environmentally friendly product.

1 Introduction

Concrete is one of the highly demanded and widely used building construction products. It is indeed a must to research alternative ways to reduce the consumption of concrete. Every year concrete is emitting 80% of CO_2 for the manufacturing of cement, Quarry dust, M sand, and aggregates with huge machinery and fuel cost in the construction industry [1]. The ecological dimension is needed to be considered in the material usage and the consumption of concrete in slabs [2, 3]. One such method to consider ecological dimensions for building construction is using a bubble deck slab.

The bubble deck slab was invented in Denmark, and it is a biaxial hollow core slab. The slab has HDPE plastic bubble balls sandwiched in between the steel reinforcement cage reducing the weight of the slab [4]. The middle part of the floor slab will function structurally eliminating concrete in its layer [5]. Therefore, a bubble deck slab is introduced in this concrete structure and the experimental analysis is carried out in this paper by comparing it with a normal floor slab. Finally, the bubble deck

C. Akin (✉) · V. Preethi · K. Vignesh
KCG College of Technology, Chennai, India
e-mail: akin.akin766@gmail.com

A. Vilventhan et al. (eds.), *Advances in Construction Materials and Management*, Lecture Notes in Civil Engineering 346,
https://doi.org/10.1007/978-981-99-2552-0_40

slab has less weight giving good compressive strength and reducing the consumption of concrete and energy which maintains sustainability.

2 Bubble Deck Slab

Bubble deck slab is one of the patented techniques used to reduce the consumption of concrete and introduce HDPE plastic bubble balls in between the reinforcement [4]. Figure 1 shows a schematic diagram of a Bubble deck slab.

The void in the flat slab eliminates 30% of the self-weight of the slab. The system of alignment of the HDPE plastic bubble ball allows a longer span and is more rapid and it is less expensive for an erection. It can reduce 3% of the total cost of the project. The Bubble deck slabs can be used in all floor slabs.

2.1 *Flexural Strength*

The flexural strength of the bubble deck slab is principally un-stressed in flexure and the volume of concrete is consumed less when compared to the solid slab. The depth of compressed concrete in slabs is similarly proportional to the depth of the bubble deck slab [6]. Therefore, there is no major sensitive difference between the behavior of solid slab and bubble deck slab [7]. The outer shell of the concrete on the compression side and steel reinforcement on the tension side acts similarly in the working principle [8]. The shear resistance of the bubble deck slab is 0.65 times the shear resistance of the solid slab with the same thickness property [9].

2.2 *Durable Property*

The durability of the bubble deck slab is also like the solid slab [10]. These slabs do not allow a direct route to air from the surface of the reinforcement rebar, and

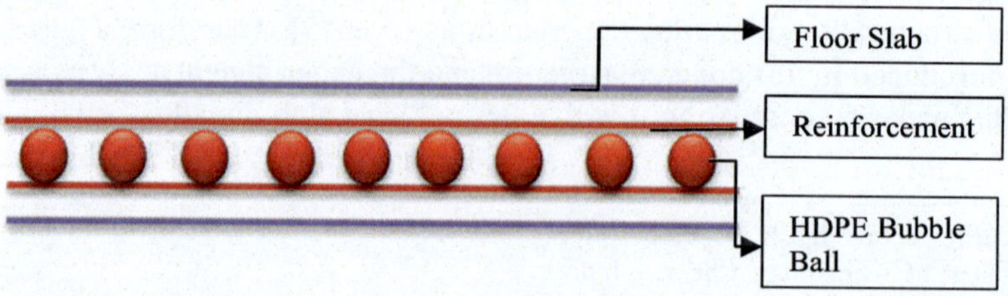

Fig. 1 Bubble Deck Slab

Fig. 2 Methodology of the Work

it is also fire-resistant [11–13]. The cracking does not occur like a solid slab [14] due to more friction and bonding happening between the HDPE bubble ball and concrete and reinforcement rebars [15, 16]. This slab possesses continuous mesh at the top, bottom, and throughout the slab and has high shrinkage restraint in which it maintains minimum cracking with either intrinsic or extrinsic [17].

3 Methodology

In this experimental analysis of the bubble deck slab, the load, and flexural behavior were analyzed and compared with the solid slab. We followed the same design of solid slab for the bubble deck slab and then started with material procurement, mix design of concrete [18], formwork making, Reinforcement, casting of the slab, and then testing and result of the bubble deck slab compared with the solid slab [19, 20]. Figure 2 shows the methodology of the work.

4 Testing

In the testing procedure of the bubble deck slab, four slab specimens were cast and out of which two specimens is made for the conventional slab and two specimens were made for the bubble deck slab. The dimension of the formwork was 70 cm in length and 70 cm in breadth. The thickness of the slab was 12 cm. The cement used for the experiment was ordinary Portland cement and a 10 mm diameter reinforcement rebar of Fe500 was used [21]. The c/c of the bar is 150 mm. As per IS456:2000, M30 grade concrete was used for slabs [22]. The hollow HDPE bubble ball is 70 mm in diameter. The size of the fine aggregate was 4.55 mm to 150 microns. River sand was used for fine aggregate. The size of the coarse aggregate was 10 mm.

Figure 3 shows the formwork and reinforcement rebars kept for testing. The formwork was made up of plywood. And Fig. 4 shows the hollow HDPE Bubble balls kept in the rebars for testing [23]. After placing the balls in the reinforcement, it is placed inside the formwork with a cover block of 25 mm. Figure 5 shows the slab after curing for 28 days. The slab specimen was kept under a UTM of 2000 kN for testing.

Table 1 shows the mix ratio of M30-grade concrete used for slabs. Whereas materials consumed are, cement 200 Kg, water 93 liters, FA 337 Kg, and CA 64.5 Kg for Solid slabs. And materials consumed for the Bubble deck slab are cement 160

Fig. 3 Formwork & Rebars of the slab

Fig. 4 HDPE Bubble Ball kept in Rebar

Fig. 5 Slabs after curing

Table 1 Mix Ratio of M30 grade for slab

Mix	Cement (Kg)	Water (Lit)	FA (Kg)	CA (Kg)
M30 Solid slab	200.25	93	337	64.5
M30 Bubble deck slab	160	74.4	269.6	51.6

Kg, Water 74.4 liters, FA 269.6 Kg, CA 51.6 Kg. Approximately 20% of materials are saved from bubble deck slab.

5 Results

The four slab specimens were tested for weight, load, and deflection. The solid slab-1 gives 481 kN load and the deflection was 13.12 mm, and the weight of the slab was 145 Kg. The solid slab-2 gives 465.23 kN load and the deflection was 13.87 mm, and the weight of the slab was 138 Kg. The Bubble deck slab-1 gives 425.01 kN load and the deflection was 13.97 mm, and the weight of the slab was 126 Kg. The Bubble deck slab-2 gives 422.21 kN load and the deflection was 14.22 mm, and the weight of the slab was 121 Kg.

Fig. 6 Flexural Behavior of Bubble deck slab

Table 2 Comparison table for slab specimens

Slabs	Weight (Kg)	Load (kN)	Deflection (mm)
Solid Slab-1	145	481	13.12
Solid Slab-2	138	465.23	13.87
Bubble deck slab 1	126	425.01	13.97
Bubble deck slab 2	121	422.21	14.22

Figure 6 shows the flexural behavior of Bubble deck slab-2 where the load was 417.15 kN, the deflection was 14 mm, and the weight of the slab was 126 Kg.

Table 2 shows the comparison between solid slabs and bubble deck slabs. Where the bubble deck slab is less in weight and reduces the material consumption by 20%. Bubble deck slab 2 was less in weight when compared to other slabs and the load was 422 kN with a deflection of 14.22 mm. The load and deflection various with different slabs and the weight of bubble deck slab is less when compared to solid slabs. Hence when compared with the result of all slabs bubble deck slab is also safer.

6 Conclusion

The Bubble deck slab is one of the innovative structural elements in the construction industry. It is proven that these slabs are safer for use and consume less quantity of material. The deflection was more in the bubble deck slab when compared to the solid slab since the stiffness in the middle portion of the slab was reduced due to hollow HDPE Bubble balls. Around 20% of materials are consumed in this slab. The total weight of the slab was reduced when compared to the solid slab. The cracks occurred in the slabs while testing results were longitudinal for both slabs. And finally, the cost was also reduced to 10% of the total cost of the slab. Therefore, the bubble deck slab paves way for an ecological dimension and maintains sustainability.

References

1. Akin, C., Devadoss, R.S. (2019). Planning and formulation of environmental management system for Indian infrastructure projects. In *MATEC web of conferences,* vol. 277, p. 02017
2. Akin, C. (2014). Energy Efficient Materials for Sustainable Building. (MDPI AG 2014). https://doi.org/10.3390/ecm-1-b014.
3. Akin, C., Samuel Devadoss, R. (2021). Mitigation of energy consumption impact by planning and formulation of environmental management system for Indian infrastructure projects, pp. 133–139. https://doi.org/10.1007/978-981-15-5101-7_13.
4. Konuri, S., & Vara lakshmi, T. V. S. (2022). Experimental studies for comparing conventional concrete with bubble deck concrete. *Materials Today: Proceedings.* https://doi.org/10.1016/j.matpr.2022.10.198.
5. Li, Z., et al. (2022). A review on substitution of natural sand with granite fines in sustainable concrete. *Construction and Building Materials, 346*, 128417.
6. Liu, Z., Alsomiri, M., Li, M., Chen, X., & Meng, J. (2022). Experimental investigation on the flexural behavior of coarse aggregate reactive powder concrete (CA-RPC) bridge deck. *Engineering Structures, 271*, 114951.
7. Kumar, V. K., Rajan, N. K., Devi, V. V., & Kaushik, G. (2022). Relationship between Gaestel Index and rheology of blended bitumens (p. 020006). https://doi.org/10.1063/5.0118092.
8. Anusha, M., Surendra, H. J., Vinod, B. R., & Bhavya, S. (2022). Modelling using finite element analysis in the structural behavior of bubble deck slab. *Materials Today: Proceedings, 66*, 2397–2404.
9. Habeeb, M., Al-Azzawi, A. A., & Al-Zwainy, F. M. S. (2021). Punching shear behavior of LWA bubble deck slab with different types of shear reinforcement. *Journal of King Saud University - Engineering Sciences, 33*, 15–22.
10. Oukaili, N. K., & Merie, H. D. (2021). CFRP strengthening efficiency on enhancement punching shear resistance of RC bubbled slabs with openings. *Case Studies in Construction Materials, 15*, e00641.
11. Zheng, Y., Zhou, L., Xia, L., Luo, Y., & Taylor, S. E. (2018). Investigation of the behaviour of SCC bridge deck slabs reinforced with BFRP bars under concentrated loads. *Engineering Structures, 171*, 500–515.
12. Wang, H., et al. (2022). Wave impacts on a solid deck in transient wave groups. *Journal of Fluids and Structures, 114*, 103755.
13. Seiffert, B. R., Ertekin, R. C., & Robertson, I. N. (2015). Wave loads on a coastal bridge deck and the role of entrapped air. *Applied Ocean Research, 53*, 91–106.
14. Xiao, J.-L., et al. (2022). Full-scale experimental study on continuous UHPC slabs of steel-UHPC composite deck system. *Structures, 43*, 1716–1731.
15. Kueres, D., Ricker, M., Classen, M., & Hegger, J. (2018). Fracture kinematics of reinforced concrete slabs failing in punching. *Engineering Structures, 171*, 269–279.
16. Wang, J.-J., Tao, M.-X., Zhou, M., & Nie, X. (2018). Force transfer mechanism in RC beams strengthened in shear by means of steel plated concrete. *Engineering Structures, 171*, 56–71.
17. Loaiza, N., Graciano, C., & Casanova, E. (2018). Design recommendations for patch loading resistance of longitudinally stiffened I-girders. *Engineering Structures, 171*, 747–758.
18. Rekha, Y., Suriya, S., & Hamedul Irshad, H. M. (2022). Comparative study on oven curing of Geo-polymer concrete over conventional concrete. *Materials Today: Proceedings, 55*, 462–469.
19. Manju, K., Kalyana Chakravarthy, P. R., & Prasanna Balaji, B. (2021). Investigation of slab for analytical method. *Materials Today: Proceedings, 37*, 3125–3129.
20. Andrews, A. A. E., et al. (2021). Experimental exploration of sodium hydrate – hypermangan treated G/SB/SB/SB/G hybrid composite. *Materials Today: Proceedings, 37*, 59–62.
21. Niyirora, R., et al. (2022). Intelligent damage diagnosis in bridges using vibration-based monitoring approaches and machine learning: a systematic review. *Results in Engineering, 16*, 100761.

22. Alrousan, R. Z., & Alnemrawi, B. R. (2022). The behavior of alkali-silica reaction-damaged full-scale concrete bridge deck slabs reinforced with CFRP bars. *Results in Engineering, 16*, 100651.
23. Al-Rousan, R. Z. (2022). Impact of elevated temperature on the behavior of full-scale concrete bridge deck slabs reinforced with GFRP bars. *Structures, 43*, 621–634.

Evaluation of Self Compacting Concrete with Fiber and Bagasse Ash

G. D. Kumara, V. Sai Kumar, P. V. Sivapullaiah, and A. Sreenivasa Murthy

Abstract The most versatile civil engineering construction material, concrete, because of its good compressive strength and flexibility, in handling as well as ability to shape it to required form, suffers from a few disadvantages because of its low tensile strength and durability. Another issue is that the production of important component of concrete, cement, is very energy consuming and impacts environment in all stages of production. To reduce these issues and also reduce the cost involved in cement, many supplementary cementitious materials are considered to substitute cement to the extent possible without impairing its mechanical properties. The tensile strength of concrete depends only with reinforcement, which is mainly steel bars. In this study industrial fiber i.e. Recron 3S is proposed to use as reinforcement and abundantly available Bagasse ash a pozzolanic materials as a replacing cement to the extent possible. Another problem in using concrete is that it requires compaction after placing, which is not always possible. This issue is circumvented by using self-compacting concrete. A study has been undertaken to produce SCC with different amounts of bagasse ash replacing the cement and with industrial fiber as reinforcement. To reduce high requirement of water for SCC required amount of superplasticizers used. The water content is determined based on slump tests, L box tests and V funnel test to obtain satisfactory passing ability and flowability for M50 concrete, respectively. Mechanical properties (compressive strength, tensile

G. D. Kumara (✉) · V. Sai Kumar
Department of Civil Engineering, GITAM School of Technology, Visakhapatnam, Visakhapatnam, India
e-mail: gdk.gbc@gmail.com

G. D. Kumara
Public Works Department, Government of Karnataka, Bangalore, India

P. V. Sivapullaiah
Indian Institute of Science, Bengaluru, Bengaluru 560012, India

GITAM University, Visakhapatnam, India

Bangalore Institute of Technology, Bengaluru, India

A. Sreenivasa Murthy
Department of Civil Engineering, BMS College of Engineering, Bengaluru, Bengaluru, India

A. Vilventhan et al. (eds.), *Advances in Construction Materials and Management*,
Lecture Notes in Civil Engineering 346,
https://doi.org/10.1007/978-981-99-2552-0_41

strength, and flexural strength) of this concrete with different amounts of bagasse ash are determined after curing for periods up to 90 days. Results showed that 10% replacement of cement with bagasse ash can maintain good mechanical properties.

Keywords Bagasse ash · Self-compacting concrete · Slump test · Compressive strength · Tensile strength · Flexural strength

1 Introduction

Concrete is the most popular and widely used construction material because of its versatility in terms of easy production and flexibility for handling as well as its ability to shape it to the required form. Concrete develops good compressive strength with curing and in fact, has become synonymous with solid. However, the properties of concrete vary widely depending on the components used in concrete production. In its classical form, the important ingredients of concrete are cement, water, and coarse and fine aggregate. To suit the use of concrete in different applications, these proportions are varied as well as the incorporation of many supplementary materials including industrial and agricultural waste as well as some natural materials [1, 2].

Water content plays a very important role in mobilization of strength through pozzolanic reactions as well as to maintain flowability [4, 5]. Many attempts are made to reduce the use of cement which is cost-intensive and environmentally not very desirable cement. Many supplementary cementitious materials are used to substitute cement. Attempts made in the past show that the extent of substitution meets success to different extents on the type, nature and physical and chemical properties [4, 6–9]. Concrete which develops good compressive strength with curing lacks tensile strength which is essential for many structural applications [10, 11]. To overcome this problem, steel is used as reinforcement to attribute tensile strength to concrete. Compaction in the form of vibration of concrete is required to create a hard mass for binding of aggregates by cementitious reaction products generated by hydration of cement phases with time and also the entrapped air is expelled [12, 13] Vibrating concrete during the placing helps to develop the required properties, formwork is fully filled to planned level and entrapped air is expelled, all reinforcement, anchorages, ducts, tendons and embedment are encapsulated, specified finish to formed surfaces of member is provided. Meanwhile, monolithic mass is created between ends of member, planned joints or both. If at certain locations vibration is restricted or not possible the concrete properties cannot have the designed value [14, 15] Therefore, there is a need to use concrete that compact by itself. For this, a class of concrete referred as self-compacting or self-consolidating concrete (SCC) and is developed which can consolidate itself without compacting. Nevertheless, the concrete possesses certain limitations, SCC is used because of the advantages which override these limitations. Important limitations include lack of universally acceptable specification for SCC, slightly higher cost, material selection and monitoring are more stringent than conventional RCC [16, 17]

To achieve workability without segregation, the components of SCC preferred to have high volume paste, high volume fines, higher dosage of super plasticiser, viscosity modifier, less coarse aggregate, and use of admixtures [11, 11]. Therefore, SCC constitutes Portland cement of 43 or 53 grade, aggregates less than 20 mm, mineral admixtures such as fly ash, ground-granulated blast furnace slag (GGBFS), silica fume (SF), stone powder, metakaolin (MK), rice husk ash (RHA) and super plasticiser etc. [18] The specification and guidelines for self-compacting concrete are followed as per Hampshire, U.K. [2002] European Federation of National Associations Representing for Concrete (EFNARC). For this flowability of concrete is very important which is normally achieved through the addition of superplasticizer. Methods are established to determine the flowability for this concrete [19–22].

The studies on SCC with BA are limited, though some data are available on the properties of fresh SC concrete. In this paper an attempt is to develop SCC using agricultural waste bagasse ash to reduce the cement content. Also, instead of conventionally used steel which is used to attribute tensile strength, synthetic fiber is proposed to be used. The amount of bagasse ash that can be used to replace cement needs to be used is arrived after determining the initial flow test values etc., and mechanical properties. The required water content along with superplasticizer is determined based on slump tests, V funnel, and L box tests to evaluate its flowability. However, apart from maintaining flowability of fresh concrete, its mechanical properties viz., compressive, tensile strength and flexural strength has to be determined. For this purpose, normal concrete designed for M30 is tested without supplementary cementitious additives and with different amounts of bagasse ash substituting for cement. Samples with different amounts of replacement have been studied after different curing periods, to find the amount of bagasse ash to maintain flowability and can replace cement without affecting the mechanical properties.

2 Materials and Methods

2.1 Sugarcane Bagasse Ash (SCBA)

The bagasse was procured from the sugar factory located in Mudhol, India. The bagasse ash was obtained during controlled burning of sugarcane after the extraction of juice and the pulp so obtained will be burnt completely into ash at a temperature of 600 °C. Then ashes were placed in an oven at a temperature of 200 °C so that remaining organic matter would be burnt. The remaining ashes were sieved. Table 1 presents physical properties of bagasse ash used for preparation of self-compacting bagasse ash concrete.

Table 1 Physical properties of bagasse ash

Property	Value
Colour	Black
Specific gravity	2.10
Particle size	<90 μ

Table 2 Physical properties of cement

Property	Value
Color	Grey
Type of cement	OPC 53 grade
Particle size	<90 μ
Standard consistency	32%
Initial setting time	33 min
Final setting time	10 h
Soundness of cement	<1 mm
Fineness of cement (by 90μsieve)	0.5%
Specific gravity	3.12

2.2 Cement

The cement used for carrying out the experiments was OPC 53 grade manufactured by ACC company, India. Table 2 presents the physical properties of OPC 53 grade cement.

2.3 Aggregates

The coarse aggregate used in this concrete which was used for carrying out the experimental work was granite from the local Bangalore region. The aggregates have a maximum size of 12.5 mm and specific gravity of 2.74. the fine aggregates chosen for this investigation are of Zone I and were of manufactured sand with a fineness modulus of 2.75. Tests on a coarse and fine aggregates were carried out as per the set standard procedure in BIS codes. The physical properties of coarse aggregate and fine aggregates are given in Tables 3 and 4, respectively.

2.4 Fibers

Engineered fibers are short-cut micro fibers that are scientifically engineered with advanced technology to provide cost-effective solutions to the mixed concrete

Table 3 Physical properties of coarse aggregate

Property	Vale
Shape	Angular
Nominal size	12.5 mm
Specific gravity	2.74
Bulk density	1651 kg/m^3
Impact value	18.6%

Table 4 Physical properties of fine aggregate

Property	Value
Specific gravity	2.69
Bulk density	1557 kg/m^3
Fineness modulus	2.75
Zone of aggregate	Zone - 1

Table 5 Physical properties of fiber

Property	Value
Product code	Recron-3S
Diameter	0.5 mm
Cut length	12.5 mm
Specific gravity	0.92
pH	7.3 $\pm$ 0.5
Elongation	20–60%
Melting point	210–250 °C
Youngs modulus	6000
Tensile strength	4–6 Mpa

industry. It complements structural steel reinforcements as a secondary reinforcement material and provides adequate tensile strength to substantially arrest shrinkage cracks in plastic and drying states of concrete. The physical properties of the fiber are given in Table 5.

2.5 *Mix Proportions of Specimens*

The properties of cement, coarse aggregate, fine aggregate, bagasse ash, super plasticiser, and fiber are summarised in Tables 1, 2, 3, 4 and 5, respectively. Bagasse ash content is varied from 0 to 30% by the weight of the cement. Initial tests for filling ability, flowing ability, passing ability and segregation have been determined

Table 6 Classification of SCC mixes according to EFNARC guidelines

Mix designation	Flowability tests						Passing ability test	
	Slump flow		T_{500} mm (sec)		V-funnel time (sec)		L-box ratio (h_2/h_1)	
	Dia. (mm)	Classification	Avg. Flow time, sec	Classification	Avg. Flow time, sec	Classification	Ratio	Classification
SCCBA0	765	SF3	2.49	VS2	6.72	VF1	0.96	PA2
SCCBA5	760	SF3	2.63	VS2	6.98	VF1	0.93	PA2
SCCBA10	745	SF2	2.76	VS2	7.35	VF1	0.91	PA2
SCCBA15	700	SF2	3.05	VS2	7.87	VF2	0.865	PA2
SCCBA20	690	SF2	3.15	VS2	9.23	VF2	0.85	PA2
SCCBA25	630	SF1	3.56	VS2	12.45	VF2	0.825	PA2
SCCBA30	610	SF1	4.02	VS2	13.26	VF2	0.75	PA2

Note: Slump acceptance criteria of SCC as per EFNARC 2002 650-800 mm

for the SCCBA with varying amounts viz., 5, 10, 15, 20, 25, and 30% of bagasse ash substituting cement and the values are presented in Table 6.

3 Results and Discussion

3.1 *Fresh Properties of M50 Grade Self-compacting Concrete*

The filling ability by slump test, flowing ability by T_{500}, passing ability by L – box, segregation, and V funnel test are determined for fresh concrete.

3.1.1 Slump Test

The slump test is conducted to check the workability of the concrete. For self-compacting concrete, the value of the slump will be more than 650 mm as per the EFNARC guidelines. The horizontal slump will be measured for the self-compacting concrete. The slump values of different mixes are tabulated in Table 6. It is observed from the table that, the slump value is within the satisfactory limit of 650 mm for the designated mix-up to SCCBA20.

3.1.2 Flowability Test

Flowability is the ability of concrete to flow into small spaces within the formwork of the component of the structure to be cast. The T_{500} mm test will be conducted to assess the free flowability in the absence of obstructions. It can be seen from the results the flow time increases as the bagasse ash percentage is increasing. This indicates the flowability gets decreased. But the values are within the allowable limits of up to 30% bagasse replaced concrete as the T_{500} mm values are good within a range of 2 to 5 s for housing structures applications.

3.1.3 V Funnel Test

The V funnel test explicates the ease of flow of concrete. Meanwhile, greater flow ability specifies shorter flow time. A flow value of 10 s is designated as good for self-compacting concrete. As per the results obtained from the experiments and tabulated in Table 6, up to 20% bagasse replacement the flow time is falling within 10 s.

3.1.4 Passing Ability Test

Passing ability indicates capacity of concrete by virtue of which will pass through any space within the formwork by its self-weight. The EFNARC guidelines specify the minimum value between T20 and T40 is 0.8 and it can be accepted. The values of the L-Box test in Table 6 are obtained for SCCBA. The values of SCCBA25 are within the permissible limits. The values are listed in Table 6 and compared with values as per EFNARC guidelines.

3.2 Hardened Properties of Self-compacting Concrete

Based on the results, the concrete mix is designed for strength of M50. Tests are conducted on different forms of the strength of concrete. A detailed testing program is illustrated in Table 7.

3.3 Mix Design

The mixture was intended to acquire slump of 650 mm as per EFNARC recommendations (600–800 mm). The slump flow at the initial stages was deemed to be satisfactory.

3.4 Mix Proportions

The coarse and fine aggregate content was 757 kg/m^3 and 897 kg/m^{3}and binder content was 604 kg/m^3. Mechanical properties of reinforced SCCBA after curing for 7, 14, 28 and 90 days. cement with bagasse ash content was 0 to 30% in

Table 7 Mix proportions of concrete with different BA contents and test details

Mix designation	OPC, 0%	% Bagasse ash	Curing Period (days)			
			Initial tests	Compressive Strength	Tensile strength	Flexural strength
SCCBA0	100	0	nil	7, 14, 28, 90	7, 14, 28, 90	7, 14, 28, 90
SCCBA5	95	5	nil	7, 14, 28, 90	7, 14, 28, 90	7, 14, 28, 90
SCCBA10	90	10	nil	7, 14, 28, 90	7, 14, 28, 90	7, 14, 28, 90
SCCBA15	85	15	nil	7, 14, 28, 90	7, 14, 28, 90	7, 14, 28, 90
SCCBA20	80	20	nil	7, 14, 28, 90	7, 14, 28, 90	7, 14, 28, 90
SCCBA25	75	25	nil	7, 14, 28, 90	7, 14, 28, 90	7, 14, 28, 90
SCCBA30	70	30	nil	7, 14, 28, 90	7, 14, 28, 90	7, 14, 28, 90

varying 5% as per the design test. The amount of fiber for reinforcement is kept at 0.5% of the weight of the concrete. SCCBA sample specimens were prepared and cured for different periods and tested as per standard methods to determine compressive strength, tensile strength and flexural strength and results are summarised in Table 8. The variations in the compressive strength, tensile and flexural strength as determined for different periods are explained.

Table 8 Mechanical properties of M50 grade self-compacting concrete

Mix designation	Compressive strength (N/mm^2) with a curing period of days				Split tensile strength (N/mm^2) with a curing period of days				Flexural strength (N/mm^2) with a curing period of days			
	7	14	28	90	7	14	28	90	7	14	28	90
SCCBA0	32.81	43.92	62.12	66.48	2.75	3.65	6.07	6.52	2.41	3.55	5.75	6.23
SCCBA5	33.15	44.25	63.17	67.19	2.81	3.81	6.15	6.59	2.52	3.72	5.88	6.48
SCCBA10	33.81	44.51	64.23	67.84	2.99	3.92	6.35	6.72	2.69	3.85	5.97	6.59
SCCBA15	32.57	43.28	63.23	65.97	2.78	3.71	6.19	6.43	2.32	3.46	5.72	6.28
SCCBA20	32.12	41.97	61.58	64.81	2.47	3.65	6.02	6.39	2.25	3.05	5.61	6.19
SCCBA25	30.57	39.54	57.61	61.74	2.42	3.54	5.58	6.19	2.09	2.87	5.19	5.77
SCCBA30	28.62	37.17	57.54	58.83	2.39	3.15	5.46	5.96	2.07	2.62	4.89	5.65

3.4.1 Compressive Strength

Previous studies [23–25] with a few investigators with bagasse ash for normal SCM concrete have shown that the amount of bagasse ash that replaces cement in concrete successfully varied depending on the characteristics of bagasse ash and other components of concrete.

Bagasse ash in spite of its crystallinity and carbon content is pozzolanic. This is because a good amount of its silica which is in amorphous form reacts with lime. The availability of reactive silica for reactions increases with an increment in fineness and grinding of bagasse ash. The ash produced by burning at incinerating temperatures converts silica to an amorphous form. The reactivity of BA depends on the fineness of the reactive silica. Chemical activation is another way to increase the reactivity apart from the prolonged grinding and burning at enhanced temperatures [26]. The presence of hydrated lime can help enhance the pozzolanic reactivity of BA by dissolving silica due to its higher alkalinity. However, the presence of a significant amorphous fraction is not essential for reactivity, as crystalline phases (quartz, corundum, and hematite) can involve in pozzolanic reactions though slowly. In this study BA without grinding is used to save energy (cost) and time. This in turn may slow down the mobilization of strength but may be advantageous to consolidate the concrete in case of SSC. The higher water content which dilutes the alkalinity also may slow down the development of strength. It is seen from Table 7 that the compressive strength of SCC with different amounts of substitution develops strength almost similar to that without bagasse ash (control specimen). The strength of SCC specimens with 5, 10, 15% of replacement of cement with bagasse ash were less than SCC without bagasse ash concrete. However, peak strength of concrete was obtained with 10% bagasse ash. The strength increases with an increase in the curing period. But the strength gain is negligible beyond 28 days of curing for all the samples.

3.4.2 Tensile Strength

The variations in the tensile strength of concrete with different amounts of replacement with bagasse ash after curing for 7, 14, 28 and 90 days are presented in Table 8 and the relative variations after curing for different periods. It is seen that the tensile strength of SCC with different levels of replacement of cement with bagasse ash after curing for different periods are slightly higher than SCC without bagasse ash with 5% bagasse ash. However, the highest tensile strength is obtained with 10% replacement of cement with bagasse ash as in the case of compressive strength. Subsequently, tensile strength improves generally with curing period.

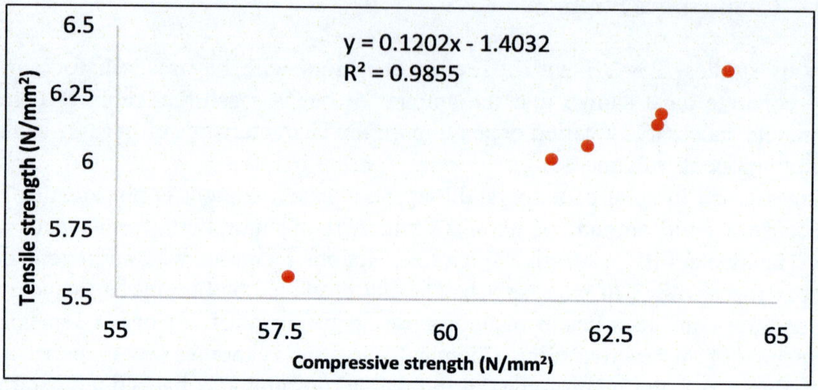

Fig. 1 The relationship between compressive strength and tensile strength

3.4.3 Comparison of Compressive Strength and Tensile Strength (28 days) of Concrete

From the compressive strength of SCC with different amounts of bagasse, one would have expected slightly higher values of SCC with bagasse ash higher than SCC without bagasse ash. Therefore, it is seen that the variations in tensile strength are generally similar to variations in compressive strength. Accordingly, an attempt is made to understand the relationship between compressive strength and tensile strength. It is seen from Fig. 1 that a very good relationship exists between these forms of strength of SCC concrete with bagasse ash after curing for 28 days. The relationship as obtained is $y = 0.1202x - 1.4032$ with a correlation coefficient of 0.98.

3.4.4 Flexural Strength

As seen from Table 8, the variations in flexural strength are similar to variations in tensile strength. Thus, it is interesting to compare the variations between tensile strength and flexural strength. Meanwhile, this relationship is shown for 28 days of cured samples in Fig. 2.

3.4.5 Comparison of Tensile Strength Versus Flexural Strength (28 Days)

Figure 2 shows the comparison of tensile strength and flexural strength of M50 SCCBA. It is seen from this figure that a very good relationship exists between the tensile strength and flexural strength of SCC with different amounts of bagasse ash. This leads to the conclusion that there should be a good relationship

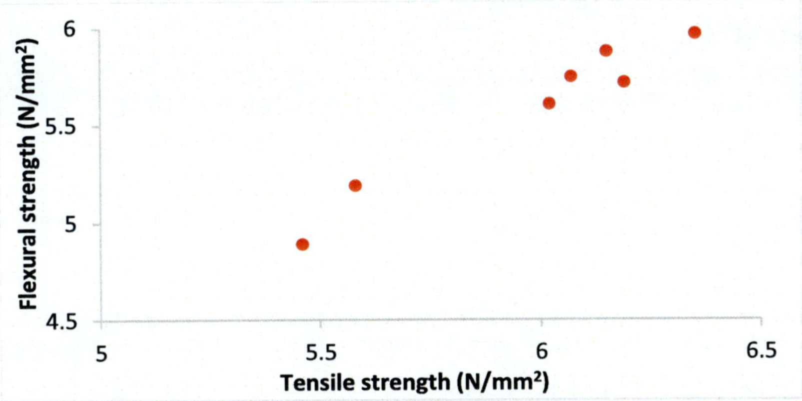

Fig. 2 Comparison of tensile strength and flexural strength of M50 SCCBA

between compressive strength and flexural strength similar to that observed between compressive strength and tensile strength.

4 Conclusion

In this study, performance evaluation of self-compacting concrete containing fiber and bagasse ash were investigated. From, results it was found that, better strength obtained with the self-compacting concrete made with BA replacing up to 30% cement and 0.5% weight of cement as fibers with a desired quantity of superplasticizer. The desired degree of initial workability was achieved with a greater slump value with respect to the other mixes. All the binary mixtures satisfied the limits of the U-box test, and V-funnel test while all fulfilled the requirements of the EFNARC recommendations. The self-compacting concrete produced with BA and fiber develop good strength of M50 in all its form with 30% replacement of cement and had shown reasonable flowability, passing ability, and segregation resistance. The test results showed optimal compressive, tensile and flexural strength of 64.23, 6.35, 5.97 N/mm^2 respectively at 28 days obtained with 10% BA replacement. Meanwhile, a good correlation between flexural and tensile strength for SCC with different amounts of replacement of cement content up to 30% with bagasse ash and after curing for different periods was observed. In general, a relation between compressive strength and flexural/tensile strength of self-compacting bagasse ash concrete was observed in this research.

It was noticed that, self-compacting concrete of M50 mix with SCBA (up to 10% replacement) and fibre (0.5%) satisfying the requirements of a concrete used for urban roads as per Indian Road Congress specifications.

References

1. Abdul-Wahab, S. A., Al-Dhamri, H., Ram, G., & Chatterjee, V. P. (2021). An overview of alternative raw materials used in cement and clinker manufacturing. *International Journal of Sustainable Engineering, 14*, 743–760. https://doi.org/10.1080/19397038.2020.1822949
2. Tabatabaeian, M., Khaloo, A., Joshaghani, A., & Hajibandeh, E. (2017). Experimental investigation on effects of hybrid fibers on rheological, mechanical, and durability properties of high-strength SCC. *Construction and Building Materials, 147*, 497–509.
3. Azmee, N. M., & Shafiq, N. (2018). Ultra-high performance concrete: From fundamental to applications. *Case Studies in Construction Materials, 9*, e00197.
4. Sobolev, K., Kozhukhova, M., Sideris, K., Menéndez, E., & Santhanam, M. (2018). Alternative supplementary cementitious materials. In *RILEM State-of-the-Art Reports* (Vol. 25, pp. 233–282).
5. Luhar, S., Cheng, T. W., & Luhar, I. (2019). Incorporation of natural waste from agricultural and aquacultural farming as supplementary materials with green concrete: A review. *Composites Part B: Engineering, 175*, 107076. https://doi.org/10.1016/j.compositesb.2019.107076
6. Zahedi, M., Ramezanianpour, A. A., & Ramezanianpour, A. M. (2015). Evaluation of the mechanical properties and durability of cement mortars containing nanosilica and rice husk ash under chloride ion penetration. *Construction and Building Materials, 78*, 354–361.
7. Cordeiro, G. C., Toledo Filho, R. D., Tavares, L. M., & Fairbairn, E. M. R. (2012). Experimental characterization of binary and ternary blended-cement concretes containing ultrafine residual rice husk and sugar cane bagasse ashes. *Construction and Building Materials, 29*, 641–646.
8. Shah, V., Scrivener, K., Bhattacharjee, B., & Bishnoi, S. (2018). Changes in microstructure characteristics of cement paste on carbonation. *Cement and Concrete Research, 109*, 184–197.
9. Chusilp, N., Jaturapitakkul, C., & Kiattikomol, K. (2009). Utilization of bagasse ash as a pozzolanic material in concrete. *Construction and Building Materials, 23*, 3352–3358.
10. Karim, M. R., et al. (2014). On the utilization of pozzolanic wastes as an alternative resource of cement. *Materials, 7*, 7809–7827.
11. de Soares, M. M. N. S., Garcia, D. C. S., Figueiredo, R. B., Aguilar, M. T. P., & Cetlin, P. R. (2016). Comparing the pozzolanic behavior of sugar cane bagasse ash to amorphous and crystalline SiO2. *Cement and Concrete Composites, 71*, 20–25.
12. Arenas-Piedrahita, J. C., et al. (2016). Mechanical and durability properties of mortars prepared with untreated sugarcane bagasse ash and untreated fly ash. *Construction and Building Materials, 105*, 69–81.
13. Brogan C. (2021) Best ways to cut carbon emissions from the cement industry explored.
14. Yang, N., & Wang, R. (2015). Sustainable technologies for the reclamation of greenhouse gas CO2. *Journal of Cleaner Production, 103*, 784–792.
15. Sakir, S., Raman, S. N., Safiuddin, M., Kaish, A. B. M. A., & Mutalib, A. A. (2020). Utilization of by-products and wastes as supplementary cementitious materials in structural mortar for sustainable construction. *Sustainability, 12*, 3888.
16. Schuman, L., & Tucker, J. (1943). Tensile and other properties of concretes made with various types of cements. *Journal of Research of the National Bureau of Stands, 31*, 107–124.
17. Abbas, S., Sharif, A., Ahmed, A., Abbass, W., & Shaukat, S. (2020). Prospective of sugarcane bagasse ash for controlling the alkali-silica reaction in concrete incorporating reactive aggregates. *Structural Concrete, 21*, 781–793.
18. Pillai, R. G., Gettu, R., & Santhanam, M. (2020). Use of supplementary cementitious materials (SCMs) in reinforced concrete systems – benefits and limitations. *Revista ALCONPAT, 10*, 147–164.
19. Figueiredo, R. L., & Pavía, S. (2020). A study of the parameters that determine the reactivity of sugarcane bagasse ashes (SCBA) for use as a binder in construction. *SN Applied Sciences, 2*, 1–15.
20. Subedi, S., et al. (2021). Properties of engineered cementitious composites with raw sugarcane bagasse ash used as sand replacement. *Journal of Materials in Civil Engineering, 33*, 04021231.

21. Pacewska, B., & Wilińska, I. (2020) Usage of supplementary cementitious materials: advantages and limitations: Part I. C–S–H, C–A–S–H and other products formed in different binding mixtures. *Journal of Thermal Analysis and Calorimetry, 142*, 371–393.
22. Abdel Rahim Mohammed Abdel Rahim Ahmed, B. & Mahjoub Osman Mahjoub, S. (2001). The Effect of Admixtures on Concrete Properties.
23. Patel, N., Dave, R., Modi, S., Joshi, C., Vora, S., & Solanki, M. (2016). Effect of binary and quaternary blends on compressive strength. *International Journal of Civil Engineering and Technology, 7*, 242–246.
24. Akinpelu, M. A., Odeyemi, S. O., Olafusi, O. S., & Muhammed, F. Z. (2019). Evaluation of splitting tensile and compressive strength relationship of self-compacting concrete. *Journal of King Saud University - Engineering Sciences, 31*, 19–25.
25. Kalyani, V. N., Krishna, K. H., Gunnepalli, C., Pradesh, A., & Sai, A. N. A Comparitative Study of Compressive Strength and Split Tensile Strength on Effect of Size of Coarse Aggregate in Hybrid Fiber Reinforced Concrete with Different Grades. www.ijert.org.
26. Rithuparna, R., Jittin, V., & Bahurudeen, A. (2021). Influence of different processing methods on the recycling potential of agro-waste ashes for sustainable cement production: A review. *Journal of Cleaner Production, 316*, 128242.

Fly Ash and GGBS Based Geopolymer as Alternate Binder for Treating Soft Soils

Aravind Goud Gaddam, Ramana Murthy Varudu, and Sudheer Kumar Yamsani

Abstract Soft clay deposits suffer from inadequate strength with high compressibility due to its high natural water content close to liquid limit. Soft clays are conventionally stabilised with ordinary Portland cement (OPC) or lime additions in many ground improvement works. The manufacturing processes of these traditional stabilisers are highly energy intensive and emit significant amounts of CO_2. Geopolymer is a promising alternative to these stabilisers, as it gives high strength, consumes low energy, and emits low CO_2 during synthesis and application. In this study, geopolymer synthetized from mixture of fly ash, Ground Granulated Blast Furnace Slag (GGBS), and alkaline sodium hydroxide (NaOH) solution was used to stabilize soft clay. The performance of geopolymer is compared with cement mixed with clay, by evaluating their unconfined compressive strength (UCS) properties. The study is carried out with different contents of binder (prepared by varying percentage of fly ash and GGBS) and 8 M NaOH alkaline activator solution, with activator to binder ratio of 1. From this study, it is found that up to 20% binder content, UCS values of stabilised soil mixes have increased significantly, and then the rate of strength gain is slightly reduced. Furthermore, for a given binder content, UCS values increased as the proportion of GGBS to fly ash increased. This study further depicted that geopolymer synthesised from optimum mixture of fly ash and GGBS are superior to cement stabilised soils and hence can be a promising alternative to traditional stabilisers.

Keywords Cement · Fly Ash · GGBS · NaOH · Geopolymer · UCS

A. G. Gaddam (✉) · R. M. Varudu · S. K. Yamsani
Department of Civil Engineering, National Institute of Technology, Warangal 506004, India
e-mail: garavind@student.nitw.ac.in

R. M. Varudu
e-mail: vramana@nitw.ac.in

S. K. Yamsani
e-mail: skyamsani@nitw.ac.in

A. Vilventhan et al. (eds.), *Advances in Construction Materials and Management*, Lecture Notes in Civil Engineering 346,
https://doi.org/10.1007/978-981-99-2552-0_42

1 Introduction

The introduction includes the status of current research and need to identify Flyash and GGBS based geopolymer for improving the performance of Soft soils.

Soft clays are often considered as one of the major challenges for building infrastructure, due to its low shear strength and large deformations under loads. These soils are primarily found beneath layered deposits in coastal regions. The foundation design is complicated because of the poor bearing capacity and the high compressibility. Soft clays have a number of issues, ranging from field uncertainities to their modelling behaviour. Given the significant economic activity that occurs along coastlines around the world it is necessary to improve these soft clay beds in order to construct high-end infrastructure facilities. Numerous researchers have worked in this field to create appropriate ground improvement/foundation approaches. Preloading with vertical drains, electro-osmotic dewatering, pile foundations, stone columns, and deep lime/cement mixing are among the most popular foundation techniques. Admixture stabilisation is generally preferred to stabilise soils to shallow depths, which includes subgrade stabilisation for roadways, airfields, and other similar structures. A novel in situ soil stabilisation approach called deep mixing methodology (DMM) involves delivering cement or lime additions in the form of powder or slurry into the ground, where they are mixed with the local soil using blades to create a hardened soil column [1].

Though lime and cement have been widely used for deep mixing, it is realized that these conventional admixtures are energy intensive and emit large quantity of CO_2 [15]. For the production of one tonne of cement, about one tonne of CO_2 is released into atmosphere [14]. As a result of these material disadvantages, efforts were made to develop alternative binding materials that are equally effective but less harmful to the environment. In this direction, the geopolymer technique was developed, which combines an alkaline activator with aluminosilicate industrial waste products viz., fly ash, ground granulated blast furnace slag (GGBS), rice husk ash, etc., to make a paste that can bond soil particles.

Davidovits introduced the idea of geopolymers. Geopolymers are three-dimensional aluminosilicate binder materials synthesised by mixing variety of raw materials containing reactive silica and alumina with different alkali activators [3, 5]. Geopolymer mechanism involves the polycondensation of alumina and silica and alkali polysiliates, resulting in polymeric Si-O-Al bonds [8]. The general structure of geopolymer is Mn [$-$(Si-O_2) z$-$Al-O] n. wH_2O where M is the alkaline cation, z is 1,2, or 3 and n is the degree of polymerization [11]. Singhi et al., [13] studied the behaviour of soil-geopolymer system using fly ash, GGBS, and a combination of fly ash and GGBS as raw materials and found that when GGBS and fly ash were blended, the most important element affecting UCS was the slag content. Lee and Lee [10] showed in their study that the reaction product formed and their silicate structure were found to be predominantly determined by the amount of additional slag. As the quantity of slag added increased, so did the amount of C-S-H gel, while the amount of aluminosilicate gel decreased and a similar study conducted by

[12] concluded that low activator dosages were required for fly ash/GGBS mixes to achieve high compressive strength, which is advantageous in view of cost and environment. Recently Lal and Patel [9] investigated BCS stabilisation with geopolymer based on fly ash and rice husk ash (RHA) and concluded that BCS stabilised with blended ash geopolymer has the potential to be a more environmentally friendly alternative to conventional stabilisers. However according to research conducted by [4], the activator modulus that is appropriate varies with the GGBS to fly ash ratio. Greater GGBS to fly ash ratios tend to favour higher activator moduli, but either an extremely higher or lower activator modulus has a negative effect on strength.

Although lot of research has been done on geopolymer synthesised from blending of various raw materials, but a thorough investigation of factors influencing the UCS of clayey soil-geopolymer is required. The objective of the current study is to examine the relative effects of geopolymer and cement stabilisation on the behaviour of soft clay. In this study, the proposed geopolymer was created using GGBS and fly ash, which contain significant amounts of reactive silica and alumina. The feasibility of soft clay stabilization with GGBS and fly ash based geopolymer is explored and compared the effect of Geopolymer mix with cement on the strength characteristics of soft clay.

2 Materials and Methodology

The soft soil used in the present study is prepared using the locally available Black cotton soil (BCS) by mixing it with the water content equal to liquid limit of the soil. The properties of soil are determined according to relevant IS codes and the test outcomes are summarized in Table 1. In the present study, fly ash (F) and GGBS (S) were procured from the local suppliers in Warangal. The chemical composition of the fly ash and slag are summarized in Table 2. Alkali activator used for the present study is Sodium Hydroxide of 8 Molar concentration. Sodium hydroxide solution was prepared by dissolving required amount of NaOH pellets of 98% purity in water to obtain desired concentration of 8 M. Alkali activator solution was prepared 24 h prior to use. The study is carried out with different contents of binder 10%, 20%, and 30% by dry mass of soil. The binder is prepared by varying quantities of F and S in proportions of 0:100; 25:75; 50:50; 75:25; 100:0. Then Alkali activator (A) i.e., Sodium hydroxide solution was mixed with the binder(B) (i.e., blend of F and S) at a mass ratio of 1.0 i.e., A/B = 1.0.

Alkali activator, binder, and various concentrations of fly ash and GGBS were combined and thoroughly agitated using a mechanical mixer and then this prepared paste was added to soft clay to create the geopolymer stabilised specimens. These specimens are denoted as F-100, S-0; F-75, S-25; F-50, S-50; F-25, S-75; F-0, S-100. The cement was mixed with soft clay to prepare the cement stabilised samples. UCS specimens were prepared using cylindrical PVC pipes, 100 mm in height and 50 mm in diameter with the bottom closed by thin circular plate of diameter 50 mm and sealed by non-adsorptive tapes. Mixtures in three layers were poured into the

Table 1 Physical and Index properties of BC soil

Properties	Values
Colour	Greyish black
Specific gravity	2.68
Liquid limit (%)	51
Plastic limit (%)	22
Plasticity Index (%)	29
Shrinkage limit (%)	13
Fines (%)	76%
IS Classification	CH
Maximum Dry Density (g/cc)	1.76
Optimum Moisture Content (%)	21

Table 2 Major chemical compositions of BCS, Fly ash and GGBS

Oxide	SiO_2(%)	Al_2O_3(%)	CaO(%)	Fe_2O_3(%)	MgO(%)	Na_2O(%)
BCS	58.2	12.4	4.7	14.2	3.5	0.4
Fly ash	55.3	20.8	4.8	8.6	1.7	0.8
GGBS	30.6	12.4	46.8	0.6	5.8	1.2

pipe, and the pipes were tapped after each layer, to eliminate any remaining air. The density of all mixtures was examined to verify consistency. Immediately after, the samples were encased in plastic film, kept in a humid environment at a constant temperature of 23 ± 1 °C. The next day, the hardened mixtures were removed from PVC pipes and re-wrapped in plastic film. This study was carried out for a curing period of 7 and 28 days. Three specimens of each mixture have been prepared to provide representative results.

3 Results and Discussion

This section consists of all results of the experiments conducted in this study and the possible reasons behind the observed results are discussed accordingly by refereeing published articles in literature.

3.1 Influence of Different Proportions of Fly Ash and Slag on Strength Characteristics

From Fig. 1, 2 and 3, it is observed that strength increases as the percentage of GGBS increases. The development of calcium in the matrix improved the strength with increase in GGBS. The Activator induced the amorphous silica, alumina, and calcium present in fly ash and GGBS to dissolve, as a result of which geopolymeric network is formed and developed all through the soil structure thus stabilizing the soft clay and in turn increasing its compressive strength.

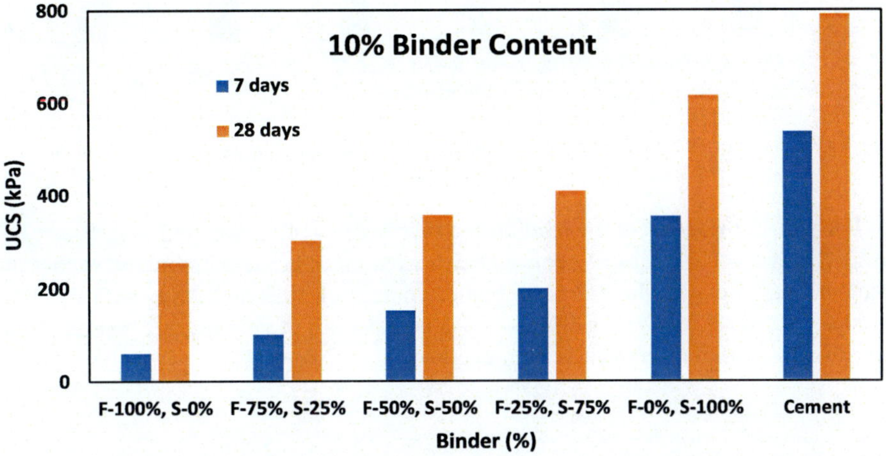

Fig. 1 Response of UCS with varying percentages of Fly ash and Slag @10% binder content

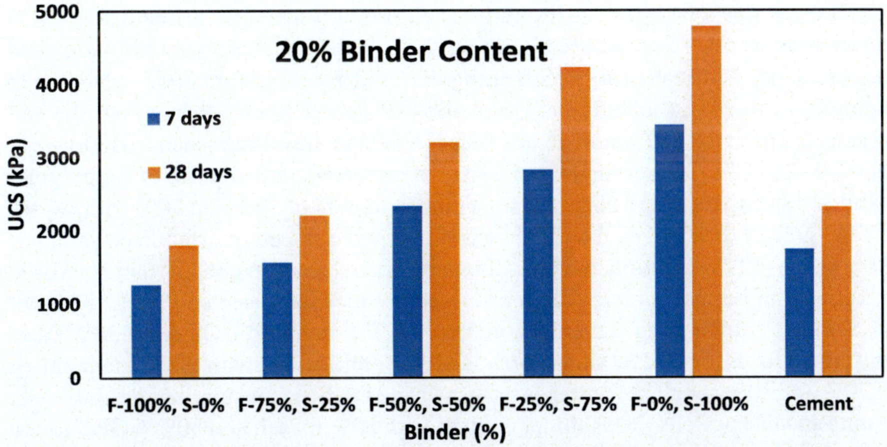

Fig. 2 Response of UCS with varying percentages of Fly ash and Slag @20% binder content

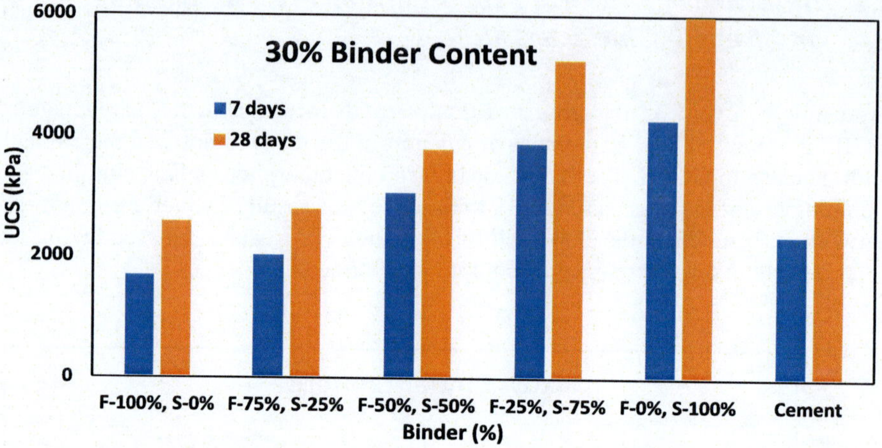

Fig. 3 Response of UCS with varying percentages of Fly ash and Slag @30% binder content

For a binder content of 10%, at all the percentages of fly ash and slag, it can be observed that UCS values of geopolymer treated clays are less compared to that of cement treated clays whereas for a binder content of 20% and 30%, with F-50%, S-50%, F-25%, S-75% and F-0%, S-100% UCS values of geopolymer treated clays are superior to that of cement treated clays.

3.2 *Effect of Binder Content on UCS of Soft Clay*

From Fig. 4, it is noted that a significant increase in strength up to 20% binder, after which there was a less gain in strength for all proportions of fly ash and slag. UCS values were increased up to 20% binder content which indicates more dissolution and maximizing gel formation by accelerating the condensation of monomers leading to a stronger cementing material and formation of denser matrix throughout the soil structure. The existence of unreacted binder content in the mixture leads to less rate of increase in UCS at 30% compared to 20%. An improved performance of geopolymer stabilised specimens were observed for a binder content of 20% and 30% with GGBS dosage greater than or equal to 50%, when compared to cement stabilised samples. As both the polymerization and pozzolanic reactions occur in geopolymer stabilised mixes and higher pozzolanic reactions in geopolymer stabilised soil leads to higher UCS values compared to cement stabilised soil. The required UCS of 1.034 MPa as prescribed for Deep soil mixing (Puppala et al., 2008) is met with the binder content of 20% and 30% for all percentages of fly ash and GGBS. Considering the above test results and discussions, 20% binder content with 50% fly ash and 50% GGBS found to be the most effective for soft clay stabilization and also best suited for Deep soil mixing applications.

Fig. 4 Response of UCS with the binder content after 28 days of curing

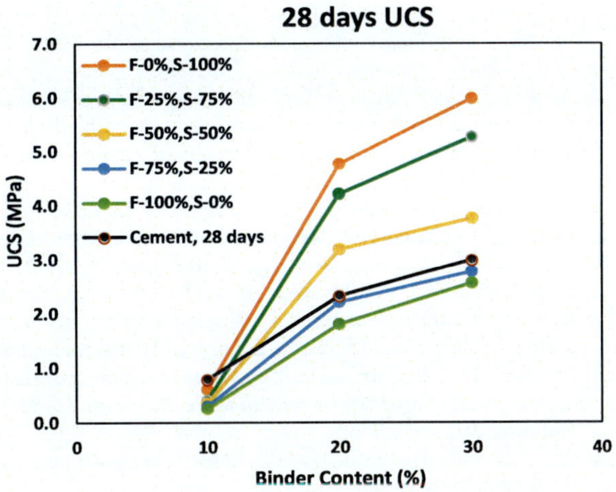

4 Conclusions

In this study, the geopolymer is synthetized using GGBS and fly ash and an alkaline solution (NaOH) is used to stabilize soft clay in order to evaluate its strength. The study is carried out with different contents of binder by varying percentage of fly ash and slag. An 8M activator with the ratio of Alkaline activator content to binder as 1.0 is considered. The performance of geopolymer is compared to the cement. The following are the conclusions drawn out of this study.

1. At a specific binder dosage, UCS values were proportionately increased with GGBS percentage from 0 to 100%.
2. UCS increased with binder content, and a sharp improvement in strength is observed up to 20% binder content, where it is decreased slightly in later stages.
3. The required strength of 1.034 MPa was attained with 20% and 30% binder content.
4. The UCS values of geopolymer stabilized samples with fly ash and GGBS dosage greater than or equal to 50% were superior compared to the samples that were stabilised with cement.

Acknowledgements Authors would like to thank the Ministry of Higher education, India for sponsoring this work.

References

1. Arulrajah, A., Yaghoubi, M., Disfani, M. M., Horpibulsuk, S., Bo, M. W., & Leong, M. (2018). Evaluation of fly ash- and slag-based geopolymers for the improvement of a soft marine clay

by deep soil mixing. *Soils and Foundations, 58*(6), 1358–1370. https://doi.org/10.1016/j.sandf.2018.07.005

2. Bhadriraju, V., Puppala, A.J., Madhyannapu, R.S., & Williammee, R. (2007). Laboratory procedure to obtain well-mixed soil binder samples of medium stiff to stiff expansive clayey soil for deep soil mixing simulation. *Geotechnical Testing Journal, 31*, 225–238. https://www.astm.org/gtj100936.html
3. Duxson, P., Fernández-Jiménez, A., Provis, J. L., Lukey, G. C., Palomo, A., & Van Deventer, J. S. J. (2007). Geopolymer technology: the current state of the art. *Journal of Materials Science, 42*(9), 2917–2933. https://doi.org/10.1007/s10853-006-0637-z
4. Gao, X., Yu, Q. L., & Brouwers, H. J. H. (2015). Reaction kinetics, gel character and strength of ambient temperature cured alkali activated slag – fly ash blends. *Construction and Building Materials, 80*, 105–115. https://doi.org/10.1016/j.conbuildmat.2015.01.065.
5. Heah, C. Y., et al. (2012). Study on solids-to-liquid and alkaline activator ratios on kaolin-based geopolymers. *Construction and Building Materials, 35*, 912–922. https://doi.org/10.1016/j.conbuildmat.2012.04.102
6. IS: 2720 (Part-4) (1985). Methods of tests for soils: Part 4 Grain size analysis. Bureau of Indian Standards, New Delhi.
7. IS: 2720 (Part-5) (1985). Methods of tests for soils: Part 5 Determination of liquid limit and plastic limit. Bureau of Indian Standards, New Delhi.
8. Khale, D., & Chaudhary, R. (2007). Mechanism of geopolymerization and factors influencing its development: a review. *Journal of Materials Science, 42*(3), 729–746. https://doi.org/10.1007/s10853-006-0401-4
9. Lal, A., & Patel, M. A. (2020). Studies on the properties of fly ash – rice husk ash - based geopolymer for use in black cotton soils. *International Journal of Geosynthetics and Ground Engineering, 6*, 1–14. https://doi.org/10.1007/s40891-020-00224-z
10. Lee, N. K., & Lee, H. K. (2015). Reactivity and reaction products of alkali-activated, fly ash/slag paste. *Construction and Building Materials, 81*, 303–312. https://doi.org/10.1016/j.conbuildmat.2015.02.022
11. Li, C., Sun, H., & Li, L. (2010). Cement and concrete research a review: the comparison between alkali-activated slag (Si + Ca) and metakaolin (Si + Al) cements. *Cement and Concrete Research, 40*(9), 1341–1349. https://doi.org/10.1016/j.cemconres.2010.03.020
12. Rafeet, A., Vinai, R., Soutsos, M., & Sha, W. (2019). Effects of slag substitution on physical and mechanical properties of fly ash-based alkali activated binders (AABs). *Cement and Concrete Research, 122*, 118–135. https://doi.org/10.1016/j.cemconres.2019.05.003
13. Singhi, B., Islam, A., & Ali, L. M. (2016). Investigation on Soil – Geopolymer with Slag, Fly Ash and Their Blending Investigation on Soil – Geopolymer with Slag, Fly Ash and Their Blending. https://doi.org/10.1007/s13369-015-1677-y
14. Syed, M., GuhaRay, A., & Kar, A. (2020). Stabilization of expansive clayey soil with alkali activated binders. *Geotechnical and Geological Engineering, 38*(6), 6657–6677. https://doi.org/10.1007/s10706-020-01461-9
15. Zhang, M., Guo, H., El-korchi, T., Zhang, G., & Tao, M. (2013). Experimental feasibility study of geopolymer as the next-generation soil stabilizer. *Construction and Building Materials, 47*, 1468–1478. https://doi.org/10.1016/j.conbuildmat.2013.06.017

Gradation of Aggregates Using Standard Codes and Particle Packing Methods - A Comparative Study

Madhavi Latha Kasulanati and Rathish Kumar Pancharathi

Abstract Concrete being the second most consumed material by mankind', the material selection for its manufacturing process certainly influences the quality and performance. Aggregate form the skeleton of concrete and the proportioning of different sizes of aggregate used in the volume of concrete, affects the packing density of aggregate. This in turn, affects volume of voids in the aggregate fraction, and so, on the amount of paste content to fill these voids in concrete. The standard codes of practice determine the guidelines for the gradation of both coarse and fine aggregate according to the requirement of a certain concrete. In this context, Particle packing methods proved to be effective for gradation of aggregate, where the optimum proportions of aggregate can be tailor made to increase the packing density of aggregate in concrete, thereby, reducing the void ratio. In the present paper, the aggregate gradations of coarse and fine aggregate as stipulated in Indian Standard code (IS 383:2016), American Society for Testing and Materials (ASTM C33/C33M-18) and European Norms (EN 12620:2016) are compared with the gradations determined using particle packing methods like MTM (Modified Toufar Method), J D Dewar method (JDD) and Compressible Packing Model (CPM). The particle packing methods were found to effectively improve the aggregate gradation, minimize the void content and improve the aggregate skeleton, thus contributing to the strength and sustainability of concrete by reducing the paste content required to fill the voids in the volume of concrete.

Keywords Packing Density · Void content · Paste content · Sustainability

M. L. Kasulanati (✉) · R. K. Pancharathi
Department of Civil Engineering, NIT Warangal, Warangal, India
e-mail: madhavi@student.nitw.ac.in

A. Vilventhan et al. (eds.), *Advances in Construction Materials and Management*, Lecture Notes in Civil Engineering 346,
https://doi.org/10.1007/978-981-99-2552-0_43

1 Introduction

Aggregates contribute to 60 to 70% of the volume of concrete, and form the main skeleton of concrete. Aggregates, due to their shape, texture and proportioning influence the fresh state properties like workability, segregation, temperature retention, hardened state properties, durability and porosity. The properties of concrete can therefore be improved by improving the quality of aggregates and gradation of aggregate. The gradation of coarse and fine aggregate according to the different size fractions is determined during the mix proportioning process of concrete according to the standard stipulations of aggregate.

The optimum aggregate gradation not only improves the properties of concrete but also reduce the paste content by reducing the void ratio of aggregate thereby, contributing to the sustainability of concrete. The packing theories form the basics of concreting, where the materials are packed effectively to reduce the voids in the order of coarse aggregate, fine aggregate, cement and water. Many of the prominent packing theories are being developed by the researchers to attain the optimum packing density and gradation, some of them including the method of particle packing by Aim and Goff (1967) [1], Toufar (1976) [2], J D Dewar (1986) [3], De Larrard (1989) [4], Andersen (1996) [5], etc.,

In the present study, three of these particle packing methods i.e., Modified Toufar Method (MTM), J D Dewar method (JDD) and Compressible Packing Model (CPM) are used for the determination of gradation of coarse and fine aggregate and the optimum method of gradation is chosen based on the required type of concrete. The particle packing methods utilize the empirical relations using the physical properties of aggregate to determine the aggregate proportions and the packing density of aggregate in concrete. The proportioning of aggregate according to the standard Codes of IS 10262;2019 [6], considers the volume proportioning of aggregate in the unit volume of concrete from 0.6 to 0.7 as compared to the particle packing methods where, the volume of aggregate is about 0.75 to 0.85 [7]. This means that the paste content is being reduced in the volume of concrete.

2 Methodology

The details of physical properties of natural aggregate are considered according to the size classes of different codes from the previous studies [8–10] for this study purpose. The physical properties according to the size classes of IS 383:2016 [11], ASTM C33-18 [12] and EN 12620:2016 [13] are shown in Table 1.

The aggregate proportions and the packing density of the aggregate are determined from the empirical relations using the MTM, JDD and CPM methods. The gradation curves are then developed for the particle packing methods and are compared with the gradation of aggregate according to the provisions made in standard codes. The optimum method of gradation yielding minimum void ratio and well graded aggregate

Table 1 Physical properties of Aggregate as per size classes (Different Codes)

IS 383:2016			ASTM C33-18			EN 12620:2016		
Sieve size (mm)	γ_b	G	Sieve size (mm)	γ_b	G	Sieve size (mm)	γ_b	G
20–16	1412.00	2.62	37.5–25	1430.00	2.65	38.10–25.40	1434.00	2.70
16–12.5	1395.00	2.61	25–19	1414.00	2.64	25.40–19.05	1420.00	2.68
12.5–10	1392.00	2.60	19–12.5	1376.00	2.62	19.05–12.70	1382.00	2.67
10–4.75	1360.00	2.61	12.5–9.5	1392.00	2.64	12.70–9.53	1402.00	2.68
4.75–2.36	1490.00	2.43	9.5–4.75	1360.00	2.60	9.53–4.76	1385.00	2.65
2.36–1.18	1516.00	2.42	4.75–2.36	1490.00	2.45	4.76–2.80	1502.00	2.48
1.18–0.6	1492.00	2.41	2.36–1.18	1516.00	2.46	2.80–1.40	1531.00	2.46
0.6–0.15	1454.00	2.40	1.18–0.60	1492.00	2.44	1.40–0.71	1512.00	2.45
0.15–0.075	1395.00	2.40	0.60–0.30	1450.00	2.45	0.71–0.30	1470.00	2.45
-	-	-	0.30–0.15	1420.00	2.44	-	-	-

γ_b – Bulk Density, G – Specific gravity

gradation is then determined. The empirical relations to determine the packing density of aggregate using the particle packing methods are as follows.

2.1 *Modified Toufar Method (MTM)*

The method considers the packing density of individual size classes using the bulk density and specific gravity of the aggregate and the binary packing density of aggregate is determined by simultaneously considering the smaller size aggregate size fraction until the final aggregate size fraction is reached.

The binary packing density of aggregate is determined using the formula as shown in Eq. (1).

$$\phi = \frac{1}{\left[\frac{y_1}{\phi_1} + \frac{y_2}{\phi_2} - y_2\left(\frac{1}{\phi_2} - 1\right)K_d K_s\right]} \tag{1}$$

where, y_1 and y_2 are the volumetric proportions of fine and coarse aggregate fractions, ϕ_1 and ϕ_2 are the individual packing densities of fine and coarse aggregate fractions, K_d and K_s are the diameter and statistical factors respectively.

2.2 J D Dewar Model

The physical properties of aggregate are used for the determination of void ratio of binary size fraction of aggregate which can be used for the determination of packing density. The void ratio of finer particle size fraction is calculated as in Eq. 2.

$$U_1'' = \frac{(1 + U_1) * U_0''}{\left(1 + U_0''\right) - (1 + Z)^3} - 1 \tag{2}$$

where, U_0'' is the void ratio of coarser particle size fraction, Z is the notional width factor which depends on the log mean diameter of the aggregate size fraction.

2.3 Compressible Packing Model

This model considers the wall effect, loosening effect and compaction effect in determining the packing density of aggregate and is considered to attain the real packing density as compared to the other methods where, the packing density of aggregate is over estimated. The virtual packing (γ_i) density of aggregate is determined as in Eq. 3.

$$\gamma_i = \frac{\beta_i}{1 - \sum_{j=1}^{i-1}\left[1 - \beta_i + b_{ij}\beta_i\left(1 - \frac{1}{\beta_j}\right)\right]y_j - \sum_{j=i+1}^{n}[1 - a_{ij}\frac{\beta_i}{\beta_j}]y_j} \tag{3}$$

where, β_i is the virtual packing density of individual aggregate size fraction, a_{ij} is the loosening effect, b_{ij} is the wall effect.

The real packing density is then determined after considering the compaction effect for the manufacture of concrete.

The aggregate proportions for the optimum packing density are determined using MTM, JDD, CPM and for the standard codes IS 383:2016, ASTM C33-18 and EN 12,620:2013 for natural aggregate. The gradation curves are then compared for coarse and fine natural aggregate.

3 Results and Discussion

The aggregate proportions for the optimum packing density determined using MTM, JDD and CPM method as compared to codal stipulations are as shown in the Table 2.

From the Table 2, it can be observed that the maximum packing density is attained using MTM method, followed by JDD method in turn followed by CPM method.

Table 2 Evaluation of Packing density based on packing theories and codal provisions

Sieve size (mm)	IS 383:2016 [11]			Sieve size (mm)	ASTM C33-18 [12]			Sieve size (mm)	EN 12620:2016 [13]		
	MTM	JDD	CPM		MTM	JDD	CPM		MTM	JDD	CPM
20–16	0.09	0.28	0.16	37.5–25	0.07	0.32	0.11	38.10–25.40	0.09	0.28	0.01
16–12.5	0.04	0.18	0.08	25–19	0.04	0.21	0.07	25.40–19.05	0.05	0.12	0.06
12.5–10	0.05	0.14	0.1	19–12.5	0.08	0.11	0.08	19.05–12.70	0.06	0.13	0.02
10–4.75	0.11	0.09	0.12	12.5–9.5	0.06	0.06	0.01	12.70–9.53	0.07	0.14	0.12
4.75–2.36	0.15	0.14	0.07	9.5–4.75	0.11	0.02	0.22	9.53–4.76	0.09	0.10	0.30
2.36–1.18	0.20	0.05	0.12	4.75–2.36	0.13	0.01	0.18	4.76–2.80	0.15	0.03	0.07
1.18–0.6	0.16	0.02	0.09	2.36–1.18	0.18	0.08	0.09	2.80–1.40	0.17	0.08	0.11
0.6–0.15	0.13	0.03	0.18	1.18–0.60	0.11	0.05	0.05	1.40–0.71	0.12	0.04	0.05
0.15–0.075	0.07	0.07	0.08	0.60–0.30	0.1	0.07	0.12	0.71–0.30	0.09	0.06	0.10
-	-	-	-	0.30–0.15	0.12	0.06	0.07	0.30–0.15	0.10	0.02	0.16
φ_{max}	0.82	0.84	0.75	φ_{max}	0.86	0.87	0.78	φ_{max}	0.85	0.85	0.76
U_{min}	0.22	0.19	0.33	U	0.16	0.15	0.28	U	0.18	0.18	0.32

φ_{max} - Maximum Packing Density, U_{min} – Minimum void ratio

Individually, the MTM method however yielded greater proportion of finer sized aggregate and JDD method yielded higher proportion of coarser sized aggregate fractions. CPM method is based on trial-and-error procedure and the proportions of the aggregate, can therefore be adjusted according to the availability and requirement by changing the dominant aggregate fraction. Also, CPM method is considered to be more reliable and optimum method as the loosening, wall and compaction effect are considered. While, MTM and JDD methods overestimate the packing density as the practical discrepancies are not considered. Thus, CPM method of gradation is considered to be a well-rounded method. The optimum method is further be justified by the experimental determination of the packing density of aggregate from the previous studies [15].

The variation of the optimum packing density designed based on standard codes using particle packing methods are shown in Fig. 1.

From the plot, it can be observed that the MTM and JDD methods yielded higher packing density compared to CPM method. This can be attributed to over estimation of the packing density using MTM and JDD as the practical variation due to loosening effect, wall effect and compaction effect are not considered like in the case of CPM method.

Similarly, the variation of minimum void ratio of aggregate using the particle packing methods is shown in Fig. 2.

From the figure, it can be seen that the void ratio is maximum when designed using CPM method and the packing density based on numerical relations is minimum. The variation of void ratio using different particle packing methods is shown in Fig. 3.

From the Figure, it can be seen that the void ratio is maximum using the IS 383:2016 method of gradation as some of the size fractions are not considered in the gradation of aggregate in the code. This is the limitation in the code.

The variation of fineness modulus of coarse aggregate determined using the particle packing methods is shown in Fig. 4.

From the figure, it can be observed that the fineness modulus of coarse aggregate is maximum using JDD method as this yields greater proportion of coarser sized fractions compared to finer sized fractions. There is no definite way of proportioning using CPM method, as the method is flexible with proportioning and so different proportions can be considered for the same packing density to yield better results. The method of proportioning can therefore be chosen based on the requirement of workability and aggregate size fractions.

The variation of fineness modulus of fine aggregate using the particle packing methods is shown in Fig. 5.

From the figure, it can be seen that JDD yielded the minimum value of fineness modulus in all the standard codes, as the proportion of finer sized aggregate fraction is less compared to coarser sizes. Similar to the coarse aggregate, the fineness modulus determined using CPM method is flexible and can be altered according to the requirement. MTM method yielded higher fineness modulus, as the finer sized proportion is greater compared to the coarser sized fractions.

The gradation curves of coarse and fine aggregate developed using MTM, JDD and CPM and IS 383:2016 proportions are shown in Fig. 6.

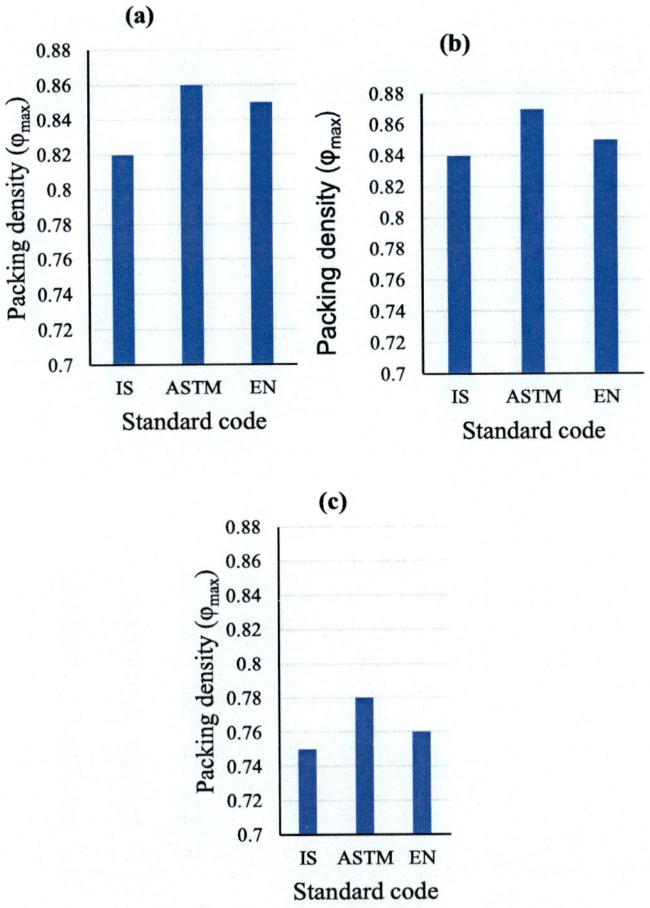

Fig. 1 Packing density of aggregate using **a** MTM, **b** JDD and **c** CPM methods

From Fig. 6(a), it can be observed that the IS 383:2016 [11] stipulations yielded gap graded coarse aggregate as compared to that based on particle packing methods yielding well-graded aggregate curves. Some of the size fractions are not considered in the proportioning of aggregate in IS 383:2016 and this can be modified using the particle packing methods. The fine aggregate curves as seen from Fig. 6(b) showed well graded curves using IS and particle packing methods and CPM method showed nearer values to the IS gradation curve.

Similarly, the gradation curves of coarse and fine aggregate developed using MTM, JDD and CPM and ASTM C33-18 proportions are shown in Fig. 7.

From Fig. 7(a), it can be observed that the gradation curves of coarse aggregate yielded well graded aggregate except in case of JDD method which yielded uniform graded aggregate. Also, the CPM and ASTM gradation curves almost showed identical variation in the proportioning. The fine aggregate gradation curves as seen from

Fig. 2 Void ratio of aggregate using **a** MTM, **b** JDD and **c** CPM methods

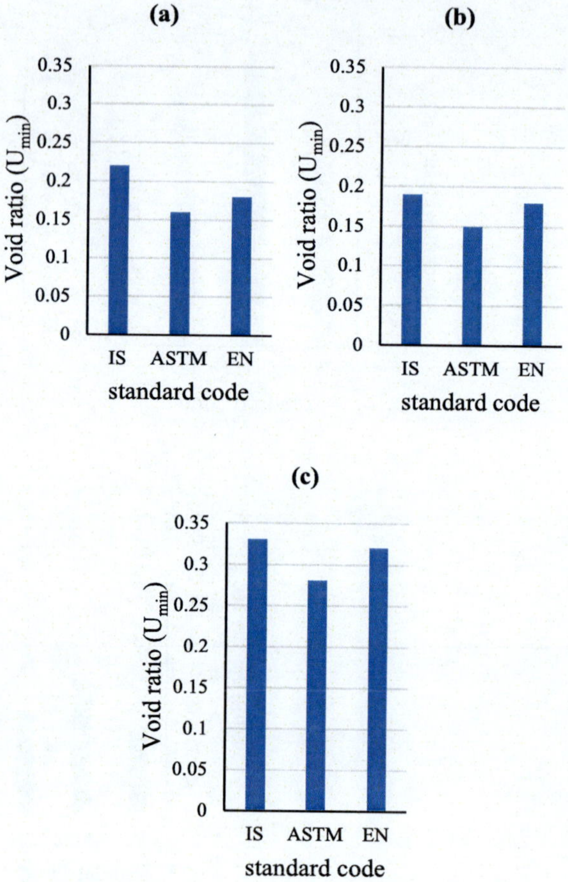

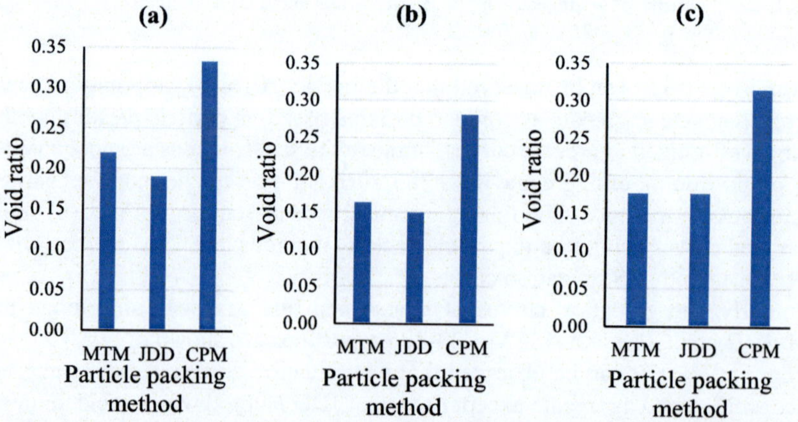

Fig. 3 Void ratio of aggregate using **a** IS 383:2016, **b** ASTM C33 and **c** EN 12620 methods

Fig. 4 Variation of fineness modulus of coarse aggregate using **a** IS 383, **b** ASTM C33 and **c** EN 12620

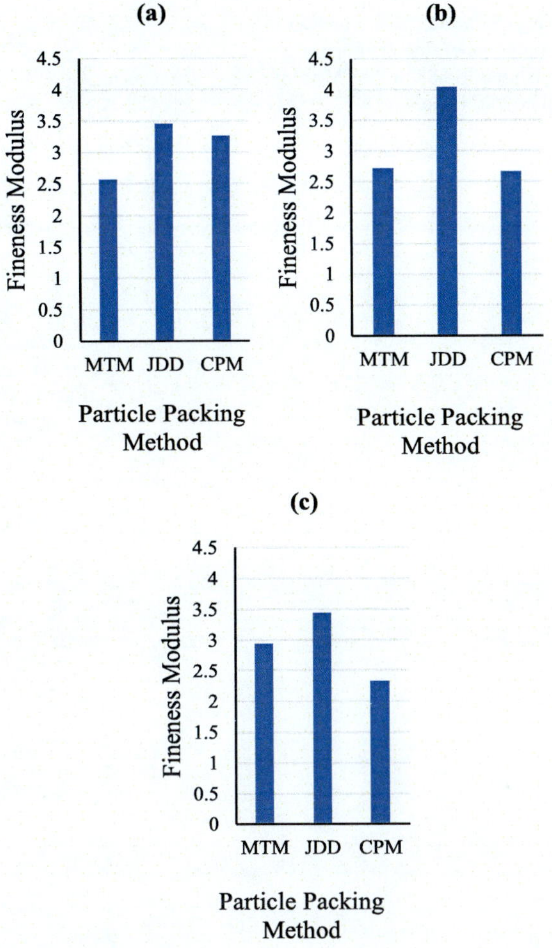

Fig. 7(b), showed well graded curves using ASTM proportions and particle packing methods. The gradation curves according to European Norms and particle packing methods are shown in Fig. 8.

From Fig. 8, it can be observed that the gradation curves showed irregular variation as compared to the other grades. As can be seen form Fig. 8(a), EN gradation curve is uniformly graded curve and some of the sizes are not considered in the proportioning of aggregate which can be overcome using particle packing methods where, all the aggregate sizes can be considered for determining the optimum proportions. From Fig. 8(b), CPM and EN gradation curves showed nearly well graded aggregates compared to MTM and JDD methods.

The particle packing methods can therefore be chosen for the proportioning of aggregates, as per availability of aggregate size fractions, required workability of

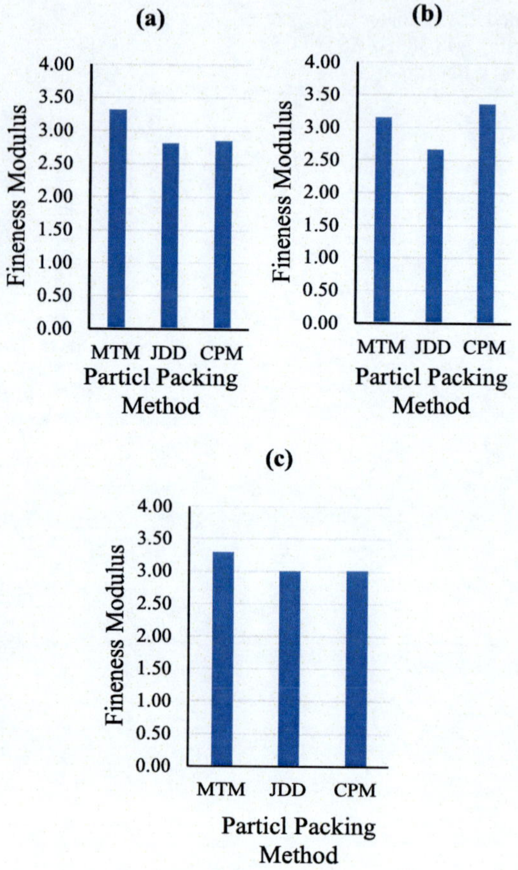

Fig. 5 Variation of fineness modulus of fine aggregate using **a** IS 383, **b** ASTM C33 and **c** EN 12620

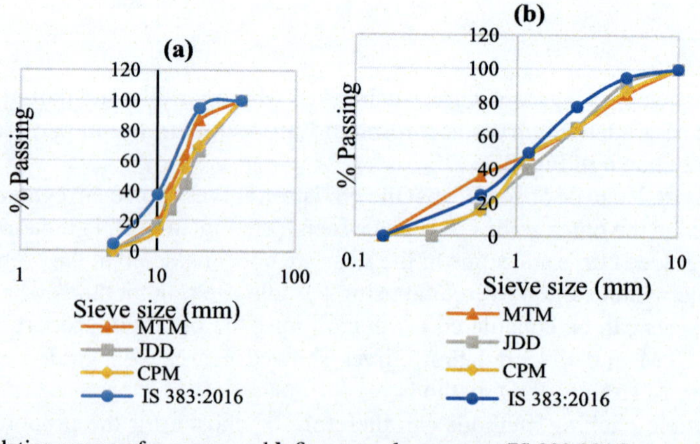

Fig. 6 Gradation curves of **a** coarse and **b** fine natural aggregate (IS 383:2016)

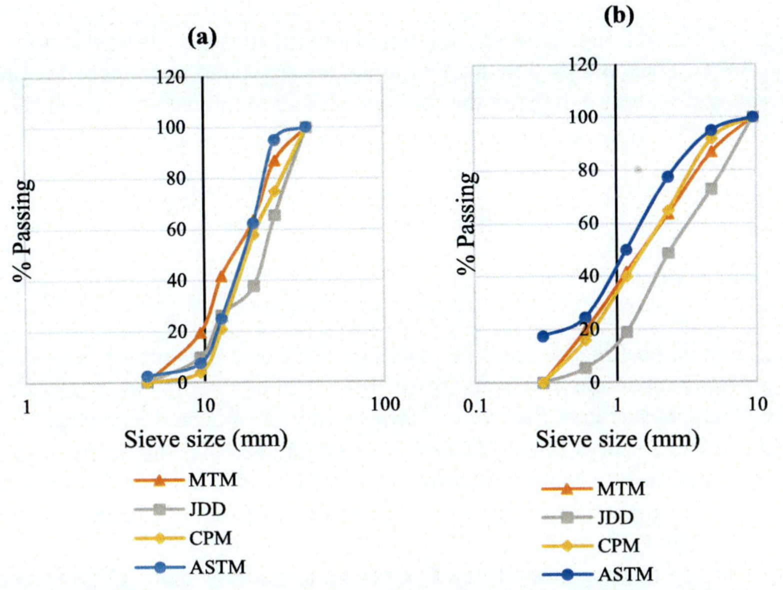

Fig. 7 Gradation curves of **a** coarse and **b** fine natural aggregate (ASTM C33-18)

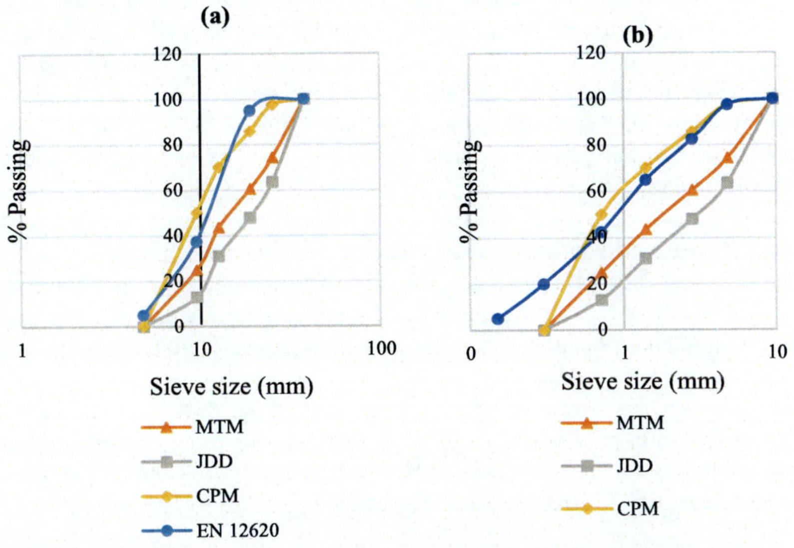

Fig. 8 Gradation curves of **a** coarse and **b** fine natural aggregate (EN 12620:2016)

concrete and the packing density of aggregate making the proportioning a more flexible approach. Thus, the particle packing methods are found to enhance the proportioning of aggregate and increase the packing density of aggregate in the skeleton of concrete, thereby, reducing the void ratio and the amount of paste content required to fill the voids in concrete accounting for sustainability of concrete [14].

4 Conclusions

1) The particle packing methods of gradation increased the volume percentage of aggregate in concrete from usual 60 to 70% to about 85%, thereby strengthening the aggregate skeleton, thus, the strength and performance of concrete.
2) The packing density of the aggregate is maximum and the void ratio is minimum using MTM and JDD methods. The practical losses due to loosening effect, wall effect and compaction effect are not being considered in the determination of packing density, in these methods.
3) The packing density is about 9.5% lower on an average using CPM method as compared to MTM and JDD methods, however the CPM method is considered to be practically reliable method.
4) The fineness modulus of coarse aggregate is about 25% higher based on CPM on an average, while in case of fine aggregate it is about 14% lower using JDD method as compared to MTM method. JDD method yields higher proportion of coarser fraction of aggregate than finer sized fractions.
5) The gradation of coarse natural aggregate determined using IS 383:2016 stipulations is more of gap graded and can be improved by using CPM method of particle packing.
6) ASTM gradation curves showed a gap gradation type of distribution using MTM and JDD methods while it is well graded curve using CPM method.
7) The European Norms showed gap graded coarse and fine aggregate and some of the size fractions are not considered in proportioning of aggregate. The particle packing methods considered the optimum proportions of all the aggregate sizes and yielded a well graded curve.
8) The increased packing density of aggregate to about 85% decreased the void ratio in the volume of concrete, thereby, reducing the paste content required in the volume of concrete to 15 to 20% compared to 30 to 40% as per standard codes accounting for the sustainability of the concrete.

In general, it can be concluded that irrespective of the gradation of coarse and fine aggregates proportioned based on different codes of practice, Particle Packing methods can enhance the particle size distribution, reduce the paste content and thus contribute to overall sustainability.

References

1. Aim, R. B., & Le Goff, P. (1968). Effet de paroi dans les empilements désordonnés de sphères et application à la porosité de mélanges binaires. *Powder Technology, 1*(5), 281–290.
2. Toufar, W., Born, M., & Klose, E. (1976). Contribution of optimisation of components of different density in polydispersed particles systems. In Freiberger Booklet A 558 (VEB Deutscher Verlag fur Grundstoffindustrie, pp. 29–44).
3. Dewar, J. D. (1997). Development of a theory of particle mixtures and its application to aggregates, mortars and concretes (Doctoral dissertation, City University London).
4. De Larrard, F. (1999). *Concrete mixture proportioning: a scientific approach*, 1st edn. CRC Press.
5. Andreasen, A. H. M. (1930). Ueber die Beziehung zwischen Kornabstufung und Zwischenraum in Produkten aus losen Körnern (mit einigen Experimenten). *Kolloid-Zeitschrift, 50*(3), 217–228.
6. Indian Standards (2019). IS: 10262:2019, "Concrete mix proportioning-Guidelines (second revision)", Bureau of Indian Standards, New Delhi, India.
7. Mangulkar, M. N., & Jamkar, S. (2013). Review of particle Packing theories used for concrete mix proportioning. *International journal of Scientific and Engineering Research, 4*(5), 143–148.
8. Salesa, Á., et al. (2017). Physico – mechanical properties of multi – recycled concrete from precast concrete industry. *Journal of Cleaner Production, 141*, 248–255. https://doi.org/10.1016/j.jclepro.2016.09.058
9. Zhu, P., Hao, Y., Liu, H., Wei, D., Liu, S., & Gu, L. (2019). Durability evaluation of three generations of 100% repeatedly recycled coarse aggregate concrete. *Construction and Building Materials, 210*, 442–450.
10. Silva, S., Evangelista, L., & de Brito, J. (2021). Durability and shrinkage performance of concrete made with coarse multi-recycled concrete aggregates. *Construction and Building Materials, 272*, 121645. https://doi.org/10.1016/j.conbuildmat.2020.121645
11. Indian Standards (2016), IS: 383:2016, "Coarse and Fine Aggregate for Concrete Specification", Bureau of Indian Standards, New Delhi, India.
12. ASTM, A., C33/C33M-18. 2018. Standard Specification for Concrete Aggregates.
13. European Committee for Standardization (2016) EN 12620:2016 aggregates for concrete. CEN, Brussels.
14. Fennis, S., & Walraven, J. (2012). Using particle packing technology for sustainable concrete mixture design. *Heron, 57*, 73–101.
15. Kasulanati, M. L., & Pancharathi, R. K. (2022). Particle packing approach for proportioning recycled aggregate from building demolition waste (BDW)-a sustainable solution. *Journal of The Institution of Engineers (India): Series A* 1–16.